Metal Organic Framework (MOF)-Based Micro/Nanoscale Materials

Metal Organic Framework (MOF)-Based Micro/Nanoscale Materials

Guest Editors

Jian Wang
Pengyan Wu
Wenqian Chen

MDPI

Basel • Beijing • Wuhan • Barcelona • Belgrade • Novi Sad • Cluj • Manchester

Guest Editors
Jian Wang
School of Chemistry and Materials Science
Jiangsu Normal University
Xuzhou
China

Pengyan Wu
School of Chemistry and Materials Science
Jiangsu Normal University
Xuzhou
China

Wenqian Chen
School of Environmental and Chemical Engineering
Shanghai University
Shanghai
China

Editorial Office
MDPI AG
Grosspeteranlage 5
4052 Basel, Switzerland

This is a reprint of the Special Issue, published open access by the journal *Nanomaterials* (ISSN 2079-4991), freely accessible at: https://www.mdpi.com/journal/nanomaterials/special_issues/2X14M2ENI9.

For citation purposes, cite each article independently as indicated on the article page online and as indicated below:

Lastname, A.A.; Lastname, B.B. Article Title. *Journal Name* **Year**, *Volume Number*, Page Range.

ISBN 978-3-7258-7516-0 (Hbk)
ISBN 978-3-7258-7517-7 (PDF)
https://doi.org/10.3390/books978-3-7258-7517-7

Contents

About the Editors . . . . **vii**

Preface . . . . **ix**

Jian Wang, Wenqian Chen and Pengyan Wu
Metal–Organic Framework (MOF)-Based Micro/Nanoscale Materials
Reprinted from: *Nanomaterials* **2025**, *15*, 1096, https://doi.org/10.3390/nano15141096 . . . . . . **1**

Marcin Kamiński and Rafał Leszek Ossowski
Shannon Entropy in Uncertainty Quantification for the Physical Effective Parameter Computations of Some Nanofluids
Reprinted from: *Nanomaterials* **2025**, *15*, 250, https://doi.org/10.3390/nano15030250 . . . . **5**

Talha Bin Nadeem, Muhammad Imran and Emad Tandis
Applications of MOF-Based Nanocomposites in Heat Exchangers: Innovations, Challenges, and Future Directions
Reprinted from: *Nanomaterials* **2025**, *15*, 205, https://doi.org/10.3390/nano15030205 . . . . **20**

Andrés González-Banciella, David Martinez-Diaz, Adrián de Hita, María Sánchez and Alejandro Ureña
M-Doped (M = Zn, Mn, Ni) Co-MOF-Derived Transition Metal Oxide Nanosheets on Carbon Fibers for Energy Storage Applications
Reprinted from: *Nanomaterials* **2024**, *14*, 1846, https://doi.org/10.3390/nano14221846 . . . . . . **59**

Marvin Kloß, Lara Schäfers, Zhenyu Zhao, Christian Weinberger, Hans Egold and Michael Tiemann
Water Sorption on Isoreticular CPO-27-Type MOFs: From Discrete Sorption Sites to Water-Bridge-Mediated Pore Condensation
Reprinted from: *Nanomaterials* **2024**, *14*, 1791, https://doi.org/10.3390/nano14221791 . . . . . . **75**

Yafang Guan, Xiaoshan Huang, Fangyi Xu, Wenfei Wang, Huilin Li, Lingtao Gong, et al.
Data-Driven and Machine Learning to Screen Metal–Organic Frameworks for the Efficient Separation of Methane
Reprinted from: *Nanomaterials* **2024**, *14*, 1074, https://doi.org/10.3390/nano14131074 . . . . . . **97**

Simon Lukato, Michał Wójcik, Agnieszka Krogul-Sobczak and Grzegorz Litwinienko
Enhancing the Green Synthesis of Glycerol Carbonate: Carboxylation of Glycerol with CO_2 Catalyzed by Metal Nanoparticles Encapsulated in Cerium Metal–Organic Frameworks
Reprinted from: *Nanomaterials* **2024**, *14*, 650, https://doi.org/10.3390/nano14080650 . . . . **113**

Jun Yeong Kim, Jiwon Kang, Seungheon Cha, Haein Kim, Dopil Kim, Houng Kang, et al.
Stability of Zr-Based UiO-66 Metal–Organic Frameworks in Basic Solutions
Reprinted from: *Nanomaterials* **2024**, *14*, 110, https://doi.org/10.3390/nano14010110 . . . . **132**

Ashraf Ali, Yaser E. Greish, Reem H. Alzard, Lamia A. Siddig, Ahmed Alzamly, Naser Qamhieh and Saleh T. Mahmoud
Bismuth-Based Metal–Organic Framework as a Chemiresistive Sensor for Acetone Gas Detection
Reprinted from: *Nanomaterials* **2023**, *13*, 3041, https://doi.org/10.3390/nano13233041 . . . . . . **145**

Bo Wen, Yunzi Miao, Zhijie Zhang, Na Li, Jiyuan Xiao, Yushuo Li, et al.
Tuning Electromagnetic Parameters Induced by Synergistic Dual-Polarization Enhancement Mechanisms with Amorphous Cobalt Phosphide with Phosphorus Vacancies for Excellent Electromagnetic Wave Dissipation Performance
Reprinted from: *Nanomaterials* **2023**, *13*, 3025, https://doi.org/10.3390/nano13233025 . . . . . . **157**

Chen Ji, Li Pei, Junyi Qin, Pengyan Wu, Nuo Su, Ting Zhang, et al.
Post-Synthetic Modification of an Amino-Functionalized Metal–Organic Framework for Highly In Situ Luminescent Detection of Mercury (II)
Reprinted from: *Nanomaterials* **2023**, *13*, 2784, https://doi.org/10.3390/nano13202784 . . . . . . **170**

Qiuyu Wu, Feiyang Tian, Wenqian Chen, Jianying Wang and Bo Lei
Specific Recognition and Adsorption of Volatile Organic Compounds by Using MIL-125-Based Porous Fluorescence Probe Material
Reprinted from: *Nanomaterials* **2023**, *13*, 2732, https://doi.org/10.3390/nano13192732 . . . . . . **182**

Dasom Jeong, Seong Cheon Kim, Taeseop An, Dongho Lee, Haejin Hwang, Siyoung Q. Choi and Jeasung Park
Synthesis of Aluminum-Based Metal–Organic Framework (MOF)-Derived Carbon Nanomaterials and Their Water Adsorption Isotherm
Reprinted from: *Nanomaterials* **2023**, *13*, 2351, https://doi.org/10.3390/nano13162351 . . . . . . **196**

Shunyu Zhang, Jinxia Li, Liang Yan, Yue You, Feng Zhao, Jixing Cheng, et al.
Zeolitic Imidazolate Framework-8 (ZIF-8) as a Drug Delivery Vehicle for the Transport and Release of Telomerase Inhibitor BIBR 1532
Reprinted from: *Nanomaterials* **2023**, *13*, 1779, https://doi.org/10.3390/nano13111779 . . . . . . **217**

About the Editors

Jian Wang

Jian Wang is an associate professor at the School of Chemistry and Materials Science at Jiangsu Normal University. He is a senior member of the Chinese Chemical Society. His research focuses on the development of functional porous materials and their applications in clean energy and environmental sustainability. His primary research interests include the controllable synthesis of porous frameworks such as MOFs, COFs, and HOFs; CO_2 capture, activation, and catalytic conversion; photocatalytic hydrogen production; and the detection and removal of environmental pollutants. He has received the First Prize for Natural Science from Liaoning Province and the Second Prize for Scientific and Technological Research Achievements from the Jiangsu Provincial Department of Education.

Pengyan Wu

Pengyan Wu is a professor at the School of Chemistry and Materials Science at Jiangsu Normal University. She was a JSPS Postdoctoral Fellow in Japan, working under the supervision of Nobel Laureate Susumu Kitagawa. She is also a recipient of the Outstanding Youth Foundation of Jiangsu Province. Her research focuses on the controllable synthesis of Metal–Organic Frameworks (MOFs) and their applications in fluorescence sensing and biomimetic catalysis. Her primary research interests include the controlled preparation of MOFs, fluorescence sensing for the detection of environmental pollutants and biomolecules, photocatalytic hydrogen production, and CO_2 conversion. She has received the Second Prize for Scientific and Technological Research Achievements from the Jiangsu Provincial Department of Education and the First Prize for Science and Technology from the Jiangsu Society of Materials Science and Technology.

Wenqian Chen

Wenqian Chen is a Research Professor at Shanghai University. She has been selected for the Shanghai Overseas High-Level Talent Program and has received the Second Prize of the Shanghai Science and Technology Progress Award, the Young Elite Award at the International Conference on Advanced Glass, the JSPS Award for Excellent Young Scientists, the Best Presentation Award at the 4th International Conference on MOFs, and the Wiley China Open Science Highly Contributing Author Award. Her research focuses on the deep integration and innovation of nuclear technology and synchrotron radiation techniques, with particular emphasis on the local structural characterization of ordered–disordered materials, particle beam–matter interactions, and their applications.

Preface

The 2025 Nobel Prize in Chemistry being awarded to the pioneers of Metal–Organic Frameworks (MOFs) marks a historic moment, recognizing the transformative impact of these porous materials across science and technology. This Special Issue, "Metal–Organic Frameworks (MOFs)-Based Micro/Nanoscale Materials", was conceived to capture the vibrant interdisciplinary research that has propelled the field to this milestone. Its scope encompasses the design, synthesis, and multifaceted applications of MOF-based micro/nanomaterials, ranging from fundamental investigations into stability and sorption mechanisms to advanced applications in energy storage, catalysis, environmental sensing, and biomedicine. The contributions herein reflect the remarkable structural tunability and functional versatility of MOFs, showcasing innovations such as machine learning-guided material screening, defect-engineered composites for electromagnetic wave dissipation, and nanocarrier-based drug delivery systems. This collection is intended for researchers, scientists, and students engaged in materials chemistry, nanotechnology, and related fields, offering a comprehensive overview of current frontiers and future directions. It serves as both a celebration of past achievements and an inspiration for continued exploration in this dynamic and impactful domain.

Jian Wang, Pengyan Wu, and Wenqian Chen

Guest Editors

nanomaterials

MDPI

Editorial

Metal–Organic Framework (MOF)-Based Micro/Nanoscale Materials

Jian Wang [1,*], **Wenqian Chen** [2,*] **and Pengyan Wu** [1,*]

[1] Jiangsu Key Laboratory of Green Synthetic Chemistry for Functional Materials, School of Chemistry and Materials Science, Jiangsu Normal University, Xuzhou 221116, China

[2] Key Laboratory of Organic Compound Pollution Control Engineering, Ministry of Education, School of Environmental and Chemical Engineering, Shanghai University, Shanghai 200444, China

* Correspondence: wjian@jsnu.edu.cn (J.W.); wenqianchen@shu.edu.cn (W.C.); wpyan@jsnu.edu.cn (P.W.)

This Special Issue showcases cutting-edge advances in the design, synthesis, and multifaceted applications of Metal–Organic Framework (MOF)-based micro/nanoscale materials. The research collected within highlights the remarkable versatility of these porous architectures, spanning fundamental property investigations to groundbreaking functional applications. Contributions delve into critical areas such as enhancing material stability in challenging environments (e.g., basic solutions) [1], elucidating intricate sorption mechanisms (including water-bridge-mediated pore condensation) [2,3], and leveraging computational methods like machine learning [4] and Shannon entropy [5] for property prediction and material screening, and MOF-based nanocomposites in heat exchangers [6]. Furthermore, the Issue explores the transformative potential of MOF derivatives and composites in energy storage (transition metal oxides on carbon fibers) [7], catalysis (glycerol carbonate synthesis) [8], sensitive chemical detection (acetone, VOCs, Hg^{2+}) [9–11], electromagnetic wave dissipation [12], and innovative drug delivery systems [13], setting the stage for significant technological impacts.

We summarize the scientific contributions below.

A systematic study was conducted to assess the stability of Zr-based UiO-66 MOFs under basic conditions, addressing a critical limitation of their practical applications in catalysis and industrial processes. The authors comprehensively evaluated 11 inorganic and organic bases, correlating framework degradation with base strength (pK_b) and concentration. It was revealed that strong bases (e.g., KOH, DBU) rapidly degrade UiO-66, while weaker bases like KOAc and pyridine exhibit higher tolerance. The integration of PXRD, NMR, BET, and ICP-OES data provided robust validation for structural dissolution and porosity loss. The proposed stability guidelines offer practical insights for the design of Zr-MOF applications in basic environments [1].

By studying the water absorption mechanism of isoreticular CPO-27-type Ni-MOFs, a further study effectively correlated pore size and linker hydrophobicity with hydration behavior. The authors adeptly combined water vapor sorption, FT-IR, and defect analysis via ^{1}H NMR to demonstrate a transition from discrete site binding to water-bridge-mediated pore condensation. The findings elucidate how defects influence sorption in Ni_2 (dhtp), while larger pores exhibit bulk-like water behavior. This work contributes to the rational design of MOFs for humidity-dependent applications, though hydrolytic stability remains a notable limitation. The experimental rigor and mechanistic clarity of this study are commendable [2].

A third study reported the synthesis of aluminum-based MOF-derived carbons (C-MDC and A-MDC) via the carbonization/activation of CAU-10-H and Al-fu precursors.

C-MDC exhibits hydrophobic microporosity with rapid water uptake at P/P_0~0.6, while A-MDC shows micro–mesoporous behavior. The water adsorption aligns excellently with the Do-Do model (type 5), highlighting C-MDC's potential for humidity control. This represents a significant and innovative contribution to the design of tunable adsorbents [3].

By integrating computational screening with machine learning, a fourth study establishes a robust framework for optimizing MOFs for the purpose of methane separation. The authors effectively employed LGBM and SHAP to identify pore-limiting diameter (PLD) as the key descriptor for diffusivity and diffusion selectivity across six binary mixtures. The work stands out for its data-driven insights into separation mechanisms and practical design principles, advancing MOF optimization for clean energy applications. The approach is innovative and offers valuable guidance for future research [4].

Presenting a novel probabilistic approach, the fifth study employs Shannon entropy and Monte Carlo simulation to quantify uncertainty in key nanofluid properties (density, heat capacity, thermal conductivity, viscosity) arising from Gaussian-distributed nanoparticle volume fraction. The approach demonstrates entropy as a unified uncertainty measure, offering a significant simplification over traditional multi-moment analyses. While the methodology is robust and implemented effectively in MAPLE, the analysis is constrained to low volume fractions ($\leq$5%) and limited input variation (CoV = 5%). This work valuably advances uncertainty quantification in nanofluidics but requires validation under broader parameter ranges for wider applicability [5].

The sixth submission, a review, rigorously examines advancements in MOF–nanocomposite applications for heat exchangers, specifically addressing thermal conductivity enhancement, fouling mitigation (through the use of hydrophobic MOFs such as ZIF-8), and facilitating scalable synthesis via microwave-assisted methodologies. The review effectively addresses critical challenges, including industrial integration costs and long-term stability under harsh conditions, while highlighting future directions like multifunctional hybrid composites. Case studies, including BASF's energy-efficient dehumidification system, further strengthen its practical implications. Although comprehensive in scope, a deeper discussion on standardized testing protocols would enhance its contribution to the field. Overall, it offers valuable insights for advancing sustainable thermal management and merits publication [6].

The seventh submission to this Special Issue is a well-structured study on M-doped Co-MOF-derived TMO nanosheets on carbon fibers for energy storage. The authors systematically compare the effects of Zn, Mn, and Ni doping, demonstrating Ni-doped samples' superiority in both supercapacitors (13.3 F/g at 50 mA/g) and Li-ion batteries (410.5 mAh/g at 25 mA/g). The work innovatively leverages ion exchange to enhance conductivity and capacity, offering valuable insights for multifunctional structural energy devices. The methodology is robust, and the findings advance the field effectively [7].

In the eighth contributed study, Au nanoparticles encapsulated in Ce-based MOFs provided an innovative catalytic platform for sustainable glycerol carbonate synthesis from glycerol and CO_2. A key strengths of the study is the achievement of the highest reported TOF (78 h^{-1}) under mild conditions (1.5 MPa CO_2), enabled by synergistic Au-CeO interfaces and $MgCO_3$ as an eco-friendly dehydrant. The catalysts, characterized via XRD, TEM, XPS, and physisorption, show uniform NP distribution within MOF cavities. While the yields ($\leq$44%) require optimization, this work demonstrates significant progress in green catalysis by valorizing waste glycerol and CO_2. Extension to Ag/Cu NPs further highlights the system's versatility for industrial adaptation [8].

In a ninth study, breakthrough chemiresistive acetone sensing was realized via a Bi-gallate MOF/chitosan/ionic liquid composite membrane, delivering 10 ppm detection, 15 s/3 s response/recovery, high selectivity, and flexible/biocompatible properties. Its application as a non-invasive breath analyzer for diabetes diagnosis is highly relevant and the mechanistic insight it provides into the hydrogen-bonding network is valuable. While operation at 60 °C is noted, the material shows significant promise for real-time sensing applications [9].

In the tenth contribution, novel composite FGFL-B_1—fabricated by encapsulating thioflavin T dye in MIL-125—serves as a porous fluorescence probe for VOC detection/adsorption. It exhibits exceptional selectivity: 36-fold fluorescence enhancement for THF and yellow-to-yellowish-green shift for CCl_4. The adsorption capacities reached 655.4 mg g^{-1} (THF) and 811.2 mg g^{-1} (CCl_4) with excellent recyclability. While surpassing conventional MOFs' VOC response, MIL-125 dependence (vs. NH_2-MIL-125) and untested generalizability require further study. Test strip applications demonstrated promising on-site monitoring potential [10].

In the eleventh contribution to this Special Issue, the post-synthetic modification of UiO-66-NH_2 with 2-(Methylthio)benzaldehyde yielded sulfur-functionalized UiO-66-NSMe, enabling highly selective aqueous Hg^{2+} detection via luminescent quenching (K_{sv} = 2.5 × 10^4 M^{-1}, LOD = 20 nM). Rigorous characterization confirms 87% modification retention and S-Hg^{2+} coordination. The sensor achieved 96.1–99.5% recovery in real water and enabled rapid in situ imaging on lettuce/test paper. The conversion of Hg^{2+}-loaded material into a Lewis acid catalyst enhanced resource utilization, showcasing a strategically valuable PSM approach for environmental monitoring with circular economy potential [11].

In the twelfth contribution, Co-MOF-derived nanorods were phosphate to fabricate amorphous Co@CoPx@C composites with engineered phosphorus vacancies. This design leverages synergistic dual polarization: vacancies enhance defect polarization while P-O bonds strengthen dipole polarization, coupled with interfacial polarization at Co/C/CoPx heterojunctions and conductive loss. The resultant optimal electromagnetic parameters enable exceptional performance in Co@CoPx@C-5: −55 dB reflection loss at 2.0 mm thickness and 5.5 GHz bandwidth. The vacancy-engineering strategy effectively tunes dielectric loss/impedance matching, offering novel methods for high-efficiency amorphous absorbers, though the effects of vacancy concentration on scalability require further exploration [12].

In the final submission to this Special Issue, ZIF-8 nanocarriers were used to deliver the hydrophobic telomerase inhibitor BIBR 1532, overcoming solubility and cellular delivery barriers. The synthesized BIBR 1532@ZIF-8 nanoparticles (165 nm) enhanced drug stability and enabled pH-responsive lysosomal escape, boosting nuclear accumulation. In vitro studies demonstrated significantly amplified cytotoxicity, hTERT suppression, G0/G1 arrest, and senescence versus free drug, validated through rigorous physicochemical characterization. This innovative MOF-based strategy advances oncology therapeutics, though future work requires in vivo validation and exploration of combinatorial approaches [13].

The research presented in this Special Issue vividly demonstrates the immense potential and rapidly expanding frontiers of MOF-based micro/nanoscale materials. Significant strides have been made in understanding fundamental behaviors (sorption mechanisms, stability), developing sophisticated synthesis and modification strategies (doping, post-synthetic modification, derivation to oxides/carbons), and achieving high performance in diverse applications—from advanced energy storage and efficient catalysis to ultra-sensitive chemical sensing and targeted drug delivery. Looking ahead, key challenges remain in scaling up production, further enhancing long-term stability under operational conditions, and deepening our understanding of structure–property relationships at the nanoscale.

Future directions should prioritize integrating multiple functionalities into single platforms, exploring hybrid or multi-MOF systems, advancing AI-guided design for specific applications (as exemplified in separations and EM wave dissipation), and rigorously assessing the environmental impact and lifecycle of these promising materials, particularly for large-scale implementations like heat exchangers. The journey from tailored nanoscale design to macroscopic functional integration continues to be an exciting and fruitful endeavor.

Author Contributions: J.W., conceptualisation; writing—original draft preparation; writing—review and editing. W.C., conceptualisation; writing—original draft preparation; writing—review and editing. P.W., conceptualisation; writing—original draft preparation; writing—review and editing. All authors have read and agreed to the published version of the manuscript.

Funding: This research received no external funding.

Acknowledgments: I am grateful to all the authors who contributed to this Special Issue. I also acknowledge the referees for reviewing the manuscripts. With their professionalism and expertise, the reviewers helped to improve the quality and impact of all submitted manuscripts. Finally, I sincerely thank all the editorial staff of *Nanomaterials* for their support during the development and publication of this Special Issue.

Conflicts of Interest: The authors declare no conflicts of interest.

References

1. Kim, J.Y.; Kang, J.; Cha, S.; Kim, H.; Kim, D.; Kang, H.; Choi, I.; Kim, M. Stability of Zr-Based UiO-66 Metal–Organic Frameworks in Basic Solutions. *Nanomaterials* **2024**, *14*, 110. [CrossRef] [PubMed]
2. Kloß, M.; Schäfers, L.; Zhao, Z.; Weinberger, C.; Egold, H.; Tiemann, M. Water Sorption on Isoreticular CPO-27-Type MOFs: From Discrete Sorption Sites to Water-Bridge-Mediated Pore Condensation. *Nanomaterials* **2024**, *14*, 1791. [CrossRef] [PubMed]
3. Jeong, D.; Kim, S.C.; An, T.; Lee, D.; Hwang, H.; Choi, S.; Park, Q.J. Synthesis of Aluminum-Based Metal–Organic Framework (MOF)-Derived Carbon Nanomaterials and Their Water Adsorption Isotherm. *Nanomaterials* **2023**, *13*, 2351. [CrossRef] [PubMed]
4. Guan, Y.; Huang, X.; Xu, F.; Wang, W.; Li, H.; Gong, L.; Zhao, Y.; Guo, S.; Liang, H.; Qiao, Z. Data-Driven and Machine Learning to Screen Metal–Organic Frameworks for the Efficient Separation of Methane. *Nanomaterials* **2024**, *14*, 1074. [CrossRef] [PubMed]
5. Kamiński, M.; Ossowski, R.L. Shannon Entropy in Uncertainty Quantification for the Physical Effective Parameter Computations of Some Nanofluids. *Nanomaterials* **2025**, *15*, 250. [CrossRef] [PubMed]
6. Nadeem, T.B.; Imran, M.; Tandis, E. Applications of MOF-Based Nanocomposites in Heat Exchangers: Innovations, Challenges, and Future Directions. *Nanomaterials* **2025**, *15*, 205. [CrossRef] [PubMed]
7. González-Banciella, A.; Martinez-Diaz, D.; de Hita, A.; Sánchez, M.; Ureña, A. M-Doped (M = Zn, Mn, Ni) Co-MOF-Derived Transition Metal Oxide Nanosheets on Carbon Fibers for Energy Storage Applications. *Nanomaterials* **2024**, *14*, 1846. [CrossRef] [PubMed]
8. Lukato, S.; Wójcik, M.; Krogul-Sobczak, A.; Litwinienko, G. Enhancing the Green Synthesis of Glycerol Carbonate: Carboxylation of Glycerol with CO_2 Catalyzed by Metal Nanoparticles Encapsulated in Cerium Metal–Organic Frameworks. *Nanomaterials* **2024**, *14*, 650. [CrossRef] [PubMed]
9. Ali, A.; Greish, Y.E.; Alzard, R.H.; Siddig, L.A.; Alzamly, A.; Qamhieh, N.; Mahmoud, S.T. Bismuth-Based Metal–Organic Framework as a Chemiresistive Sensor for Acetone Gas Detection. *Nanomaterials* **2023**, *13*, 3041. [CrossRef] [PubMed]
10. Ji, C.; Pei, L.; Qin, J.; Wu, P.; Su, N.; Zhang, T.; Zhang, Y.; Wang, J. Post-Synthetic Modification of an Amino-Functionalized Metal–Organic Framework for Highly In Situ Luminescent Detection of Mercury (II). *Nanomaterials* **2023**, *13*, 2784. [CrossRef] [PubMed]
11. Wu, Q.; Tian, F.; Chen, W.; Wang, J.; Lei, B. Specific Recognition and Adsorption of Volatile Organic Compounds by Using MIL-125-Based Porous Fluorescence Probe Material. *Nanomaterials* **2023**, *13*, 2732. [CrossRef] [PubMed]
12. Wen, B.; Miao, Y.; Zhang, Z.; Li, N.; Xiao, J.; Li, Y.; Feng, J.; Ding, S.; Yang, G. Tuning Electromagnetic Parameters Induced by Synergistic Dual-Polarization Enhancement Mechanisms with Amorphous Cobalt Phosphide with Phosphorus Vacancies for Excellent Electromagnetic Wave Dissipation Performance. *Nanomaterials* **2023**, *13*, 3025. [CrossRef] [PubMed]
13. Zhang, S.; Li, J.; Yan, L.; You, Y.; Zhao, F.; Cheng, J.; Yang, L.; Sun, Y.; Chang, Q.; Liu, R.; et al. Zeolitic Imidazolate Framework-8 (ZIF-8) as a Drug Delivery Vehicle for the Transport and Release of Telomerase Inhibitor BIBR 1532. *Nanomaterials* **2023**, *13*, 1779. [CrossRef] [PubMed]

nanomaterials

Article

Shannon Entropy in Uncertainty Quantification for the Physical Effective Parameter Computations of Some Nanofluids

Marcin Kamiński * and Rafał Leszek Ossowski

Faculty of Civil Engineering, Architecture and Environmental Engineering, Lodz University of Technology, 93-590 Łódź, Poland; rafal.ossowski@gmail.com
* Correspondence: marcin.kaminski@p.lodz.pl; Tel.: +48-42-631-35-64

Abstract: The main aim of this study is probabilistic computer simulation of the effective physical parameters of fluids containing nanoparticles. A deterministic model following the rule of mixtures and some semi-empirical formulas are employed to calculate effective density, heat conductivity, heat capacity, as well as viscosity for the given nanofluid. This models is randomized here using the Monte-Carlo simulation apparatus for estimation of the Shannon entropy of all these physical parameters, which is the crucial novelty of this study. The volume fraction of the nanoparticles is assumed for this purpose as the Gaussian uncertainty source with the given first two moments. The basic probabilistic characteristics of the nanofluids' homogenized parameters have also been determined here for some validation of Shannon entropy variations in addition to the statistical disorder of the nanoparticle fraction. These research findings contribute to advancing nanofluidic and microfluidic research, offering robust tools for uncertainty analysis and enhancing the reliability of physical parameter predictions in applications requiring high numerical and/or experimental precision.

Keywords: nanofluids; Shannon entropy; effective properties; Monte-Carlo simulation; homogenization

1. Introduction

Determination of the effective parameters of composites composed of coupled solid components, as well as multiphase mixtures, is a fundamental topic within homogenization theory [1]. This approach has a well-established history and encompasses a range of mathematical, computational, and semi-experimental methods, including applications in nanofluidics [2]. As known from the literature, all these homogenization models could be divided into a direct algebraic approximation of the effective parameters like a rule of mixtures and their improvements, upper and lower bound estimates [3] as well as a variety of numerical implicit methods generally based upon some physical equivalence and machine learning apparatus [4]. This equivalence most frequently includes equity of deformation energy of the real and homogenized medium and may be provided using different boundary conditions imposed on the Representative Volume Element (RVE). The concept of nanofluids represents a significant advancement in the targeted modification, enhancement, and design of the physical properties of fluid-based mixtures [5]. These mixtures are engineered to exhibit optimized density, thermal conductivity [6,7], viscosity, and heat capacity by incorporating nanoparticles into the base fluid. Such precise control over these parameters has far-reaching implications, particularly in the development of advanced cooling systems, energy-efficient thermal management solutions, and enhanced material processing techniques. Nanofluids provide a platform for tailoring the thermal and rheological behavior of fluids to meet specific industrial and technological demands, thereby broadening their applications in sectors such as machining [8], electronics, transportation, and renewable energy systems [9]. A nanofluid is a liquid substance composed of a liquid base with an admixture of solids in the form of nanoparticles; the size of the molecules is smaller than 100 nm, where the content usually does not exceed 4% of the total volume. The base of such a nanofluid may be any organic liquid or water [10,11], while the most commonly used molecules are metals (Au, Cu, C), oxides (CuO, Al_2O_3 [12], SiO_2, TiO_2 [13], ZrO_2 [14]), and carbon nanotubes. The process of enriching fluids with solid particles to alter their properties dates back over a century and has evolved significantly with advancements in materials science. Early studies focused on suspensions of micron-sized particles to enhance the density and viscosity of fluids for specific industrial applications, such as drilling muds and concrete slurries. For example, the use of suspended solids to modify fluid flow and improve performance in hydraulic systems was documented as early as the early twentieth century. Modern advancements have transitioned from using micron-sized particles to nanoscale particles, which offer superior control over fluid properties due to their high surface area-to-volume ratio and unique thermal and mechanical behaviors [15]. The inventor of this idea was most probably J.C. Maxwell, but only the development of electron and tunneling microscopy in the second half of the previous century allowed for full experimental verification of theoretical predictions [16,17]. Calculation of the effective material and physical characteristics of the nanofluids still remains important and a challenging research problem due to the fact that such characteristics avoid multiscale and multiphase computational modeling of such media and enables relatively faster solutions of the flow problems described using the Navier–Stokes equations [18,19]. The main shortcoming of many of the existing models is that statistical scattering of nanofluids' properties is usually postponed even if some experimentation is presented; the same concerns an impact of this scattering on or within their effective characteristics (Figure 1).

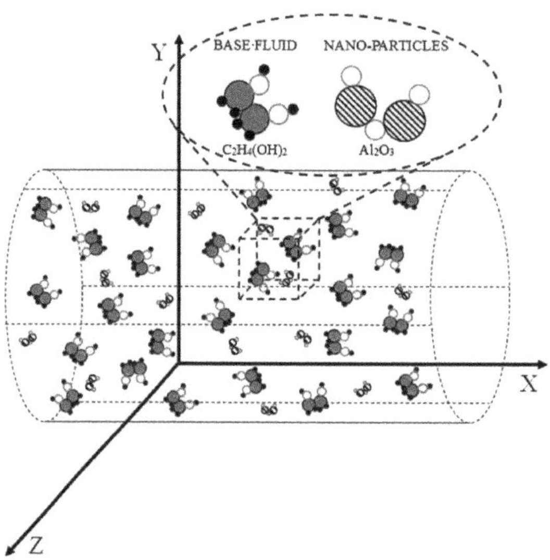

Figure 1. The general structure of the nanofluid.

Taking the above into account, the deterministic philosophy presented in [2] is extended towards a probabilistic model, where the basic random parameter is the volumetric ratio of the nanoparticles. This is proposed to make the model consistent with the needs of uncertainty quantification, propagation, and also with the stochastic reliability models; this need is also driven by some engineering applications of this type of fluids [8,20]. Further, it is almost impossible to insert the very specific and precisely measured, deterministic amount of the nanoparticles and to mix all of them with a large amount of the liquid base, which leads to relatively large variations (frequently uncertain) in the effective properties of nanofluids in addition to small changes in input parameters. The observed nonlinear entropy fluctuations indicate critical thresholds at which minor variations in input uncertainty result in disproportionately large changes in the effective properties. These thresholds may be associated with physical phenomena such as percolation transitions, nanoparticle clustering, or modifications in heat transfer mechanisms.

The homogenization approach for calculation of effective physical properties of the nanofluid is presented first [21], and it is extended next using statistical Monte-Carlo methodology [22] towards statistical estimation of the first two probabilistic moments of the effective parameters of the nanofluids. Additionally, Shannon entropy is computed using numerical recovery and specific further processing of the resulting histograms of effective parameters separately. Using additional partition of such a histogram, it was possible to implement the Shannon entropy [23] scheme in the system MAPLE 2024 [24]. This numerical tool was engaged in the calculation of these basic statistics for effective physical parameter uncertainties of the glycol ethylene fluid base enriched with the alumina particles. A final comparison of the expectations and coefficients of variation with the Shannon entropy fluctuations obtained for different mean values of the volumetric ratio of these nanoparticles shows uncertainty propagation in nanofluids' properties. It also enables a validation of the Shannon entropy apparatus in such statistical analysis, which has been applied before in some nonlinear problems of solid mechanics [25]. Let us note that the uncertainty analysis was based here upon the assumption that the input statistical scattering is in nanoparticle volume fraction having Gaussian distribution, while Weibull probability distribution was also employed [26]. This choice is driven by the Maximum Entropy Principle, where Gaussian uncertainty causes largest uncertainties in physical systems. A new element of this work is the uncertainty quantification of the effective parameters in nanofluids using a single parameter—Shannon entropy—instead of multiple graphs of probabilistic moments as applied before. It remarkably simplifies any stochastic analyses in the area of homogenization in nanofluidics, whereas applications of any probabilistic entropy in stochastic computational dynamics still remain very scarce.

2. Basic Equations for the Uncertainty Quantification in Nanofluid Parameters

Let us consider fundamental Navier–Stokes flow equations in the given computational domain Ω for the unknown fluid state functions, i.e., velocities $v_i = v_i(x)$, pressure $p = p(x)$, and temperature $\theta = \theta(x)$ in their following classical formulation [19,27]:

$$\rho\left(\frac{\partial v_i}{\partial t} + v_{i,j}v_j\right) = \sigma_{ij,j} + \tilde{f}_i, \tag{1}$$

$$v_{i,i} = 0, \tag{2}$$

$$\sigma_{ij} = -p\delta_{ij} + 2\mu\varepsilon_{ij}, \tag{3}$$

$$\rho c\left(\frac{\partial\theta}{\partial t} + \theta_i v_i\right) = (k\theta_i)_i + \widetilde{q}_i, \tag{4}$$

where strain tensor is described as

$$\varepsilon_{ij} = \frac{1}{2}(v_{i,j} + v_{j,i}) = \frac{1}{2}\left(\frac{\partial v_i}{\partial x_j} + \frac{\partial v_j}{\partial x_i}\right), \quad i = 1,2,3. \tag{5}$$

The tensor σ_{ij} represents here the stress state in the given fluid, whose viscosity is denoted by μ, its heat conductivity by k, heat capacity as c, and mass density by ρ. These equations are relevant to macroscopically homogeneous single-phase fluids but can also be applied to heterogeneous systems as long as they exhibit similar behavior on a macroscopic scale. In the context of nanofluids, such an approach allows for modeling systems where nanoparticles are unevenly distributed, yet their impact on thermal or rheological properties is significant at the macroscopic level. The following general boundary conditions are applicable.

$$\begin{gathered} v_i = \hat{v}_i; \mathbf{x} \in \partial\Omega_v, \sigma_{ij}n_j = \hat{f}_i; \mathbf{x} \in \partial\Omega_\sigma, \theta = \hat{\theta}; \mathbf{x} \in \partial\Omega_\Theta \\ k\frac{\partial\theta}{\partial x} = \hat{q}; \mathbf{x} \in \partial\Omega_q. \end{gathered} \tag{6}$$

Let us assume further that the computational domain for these NS equations is filled with the nanofluid, which means that this is a mixture of the homogeneous liquid, which is a continuous basis (f) and a discontinuous solid in the form of molecules (p), and its flow at the macroscale is similar in a physical sense to a homogeneous fluid having effective parameters calculated from the corresponding characteristics of the original fluid and the nanoparticles. Therefore, instead of Equations (1)–(4), one solves the equivalent NS equations system with the homogenized coefficients (with the subscripts 'nf') as

$$\rho_{nf}\left(\frac{\partial v_i}{\partial t} + v_{i,j}v_j\right) = \sigma_{ij,j} + \widetilde{f}_i, \tag{7}$$

$$v_{i,i} = 0, \tag{8}$$

$$\sigma_{ij} = -p\delta_{ij} + 2\mu_{nf}\varepsilon_{ij}, \tag{9}$$

$$\rho_{nf}c_{nf}\left(\frac{\partial\theta}{\partial t} + \theta_i v_i\right) = \left(k_{nf}\theta_i\right)_i + \widetilde{q}_i, \tag{10}$$

with the same boundary conditions as before except for the heat flux, where the fluid heat conductivity is replaced with the effective one. The effective density, which opens a collection of the effective characteristics, is calculated according to the simple mixtures rule, where φ serves as the volume ratio of the nanoparticles; this holds similarly to heterogeneous solids models [1,2]:

$$\rho_{nf} = \left(\frac{m}{V}\right)_{nf} = \frac{m_f + m_p}{V_f + V_p} = \frac{\rho_f V_f + \rho_p V_p}{V_f + V_p} = (1-\phi)\rho_f + \phi\rho_p \tag{11}$$

For typical nanofluids, where the percentage of fixed molecules does not exceed 1% by volume of the substance, the observed change in density does not usually exceed 5%. Then, the effective heat capacity is calculated in quite a similar way, i.e.,

$$\begin{aligned}(\rho c_p)_{nf} &= \rho_{nf}\left(\frac{Q}{m\Delta T}\right)_{nf} = \rho_{nf}\frac{Q_f+Q_P}{(m_f+m_p)\Delta T} = \rho_{nf}\frac{(mc_P)_f\Delta T+(mc_P)_p\Delta T}{(m_m+m_p)\Delta T} = \\ &= \rho_{nf}\frac{(\rho c_P)_f V_f+(\rho c_P)_p V_p}{\rho_f V_f+\rho_p V_p} = (1-\phi)(\rho c_P)_f + \phi(\rho c_P)_p\end{aligned} \tag{12}$$

so that the following holds true:

$$c_{nf} = \frac{(1-\phi)(\rho c_P)_f + \phi(\rho c_P)_p}{(1-\phi)\rho_f + \phi\rho_p} = \frac{(1-\phi)(\rho c_P)_f + \phi(\rho c_P)_p}{\rho_{nf}} \tag{13}$$

Using Equation (13), one may predict the changes in the specific heat value assuming a uniform dispersion of molecules in a liquid. Effective thermal conductivity has a more complex determination. In general nanofluids, it depends upon eight key parameters:

(a) The volume concentration of molecules, where an increase in the concentration of the molecules usually increases the thermal conductivity values for the nanofluid.
(b) Particle size, where in most cases, an increase in their diameter yields the additional increase in nanofluid conductivity; however, some specific experimental works demonstrate an inverse relation.
(c) The shape of nano molecules, where elongated shapes increase the effective thermal conductivity concerning spherical particles.
(d) Material properties of particles.
(e) Temperature, whose increase yields an additional increase in effective conductivity,
(f) Material properties of the liquid base (an increase in thermal conductivity coefficient of the nanofluid in addition to the based fluid).
(g) Extra components, which are added to the liquid to maintain the nanoparticles in suspension and prevent them from clumping; enrichment of the base fluid increases the coefficient of thermal conductivity for the nanofluids.
(h) Acidity, whose increase is proportional to an increase in the overall conductivity coefficient.

It should be noted that the chemical composition of a given nanofluid may significantly influence the sensitivity of the physical parameters under consideration. While modeling the effective thermal conductivity coefficient, several algebraic approximations can be applied. Among these, the rule of mixtures is often the initial choice; however, it is widely acknowledged as insufficiently accurate for capturing the complex interactions inherent in nanofluid systems. Alternative approaches, incorporating particle–fluid interactions and nonlinear effects, have demonstrated superior performance in accurately predicting thermal behavior across various experimental setups.

$$k_{nf} = (1-\phi)k_f + \phi k_p \tag{14}$$

Alternatively, one may employ the model proposed by Maxwell based on the potential theory, which treats nanoparticles as homogeneous spheres, which does not interact with the other spheres, and all of them are embedded into the homogeneous viscous Newtonian fluid with no gas bubbles [2]. The following holds:

$$\frac{k_{nf}}{k_f} = 1 + \frac{3\left(\frac{k_p}{k_f}-1\right)\phi}{\left(\frac{k_p}{k_f}+2\right)-\left(\frac{k_p}{k_f}-1\right)\phi} \tag{15}$$

for complex models, which may be an example of a model including the Brownian motion

$$\frac{k_{nf}}{k_f} = \left(1 + A\mathrm{Re}^{\mathrm{m}}\mathrm{Pr}^{0.333}\Phi\right)\frac{\left(1+2\frac{k_p}{k_f}\right)+2\Phi\left(1+\frac{k_p}{k_f}\right)}{\left(1+2\frac{k_p}{k_f}\right)+\phi\left(1+\frac{k_p}{k_f}\right)} \tag{16}$$

The effective viscosity determination looks similar to the case of thermal conductivity, the appearance of the nanoparticles in a liquid base should result in viscosity increase in the nanofluids. Change in viscosity depending on the content of the nanoparticles in a liquid base can be predicted by the Einstein equation as

$$\mu_{nf} = (1 + 2.5\phi)\mu_f \text{ gdy } \phi < 0.05 \tag{17}$$

This has some limitations with respect to the relatively small values of the volume fraction φ. Alternatively, one considers the Brinkman model, which is less sensitive to this volumetric ratio:

$$\mu_{nf} = \frac{\mu_f}{(1-\phi)^{2.5}} \tag{18}$$

Further, we consider the volume fraction φ of the nanoparticles as the input uncertainty source, whose probability distribution causing the largest uncertainty in effective parameters according to the Maximum Entropy Principle should be the Gaussian one. Such a computer analysis is carried out in the following steps: (1) generation of a population of the volume fraction according to its statistical parameters (expected value and standard deviation); (2) sequential calculation of the effective parameters; and (3) statistical estimation of the first two probabilistic moments of these parameters [22]. However, it is known that uncertainty analysis based upon probabilistic moments and coefficients has some limitations and may be biased, for instance, by estimation numerical errors; therefore, a concept of Shannon probabilistic entropy h was additionally employed [23,25]. Its basic formula for the given statistical function $f = f(\varphi)$ has been proposed in the literature as follows:

$$h(f(\varphi)) = -\sum_{i=1}^{n} p_i(f(\varphi))\ln(p_i(f(\varphi))) \tag{19}$$

where n stands for the number of possible different states of this system. Since the coefficient of variation (CoV) is dominantly used in stochastic computational mechanics to discuss uncertainty importance and propagation in the given engineering problem, a comparison of Shannon entropy fluctuations with analogous changes in the CoV is investigated here.

The computer simulation of the effective physical characteristics, their probabilistic moments and Shannon entropy in the presence of some input uncertainty was programmed in the computer algebra system MAPLE 2024. This system was selected due to its Statistics library, its simplicity of statistical functions programming, its efficacy and satisfactory speed in probabilistic analyses programming and performance as well as nice and easy visualization of the resulting probabilistic characteristics. The statistical numerical simulation prepared for the needs of this study was based upon the traditional Monte-Carlo scheme and consisted of the following steps:

(i) Initial generation of random population for the given input parameter(s) of the nanofluid;
(ii) Sequential recalculation of a few populations of its effective physical characteristics;
(iii) Usage of statistical estimators for up to the fourth-order probabilistic characteristics of these effective parameters;

(iv) Creation of the histogram of the probability distributions for effective characteristics;
(v) Partition of this histogram into a few subintervals and final calculation of Shannon entropy.

The entire numerical simulation is parametrized with a few inputs and these are as follows: (i) input coefficient of variation in the given uncertainty source (whose impact has been demonstrated in all the resulting graphs); (ii) total number of random trials in Monte-Carlo scheme, whose impact was studied in the literature many times; and (iii) a number of subintervals proposed for the PDF partition at the last phase, which were adopted after [25]. Some additional internal parameters of symbolic computations precision like the number of digits may also have some importance in this analysis.

3. Numerical Results

The main goal of the numerical results was to check whether uncertainty quantification in effective characteristics of the nanofluids can be performed using a single statistical disorder function (Shannon entropy) instead of a widely applicable multiple parametric study (expected values and standard deviations, for instance). The computer simulation enabling such a contrast was performed using the computer algebra system MAPLE 2024. However, analogous numerical calculations can be carried out with other available tools such as MATLAB R2024b, Mathematica 13.3, or MathCAD 10, offering flexibility in computational approaches depending on the specific requirements of the analysis. The computer algebra system MAPLE 2024 is well known for its powerful symbolic computation abilities and is widely employed in solving differential equations, simplifying algebraic expressions, and conducting dimensional analysis [24,25]. In all numerical experiments, the volume fraction, φ, is modeled as a Gaussian random variable characterized by its first two moments. A typical variation range for the volume fractions φ for the fluids with the nanoparticles is between 0.01 and 0.04 so the range 0.005–0.055 was initially preselected. The fluid base was chosen in the ethyl glycol $C_2H_4(OH)_2$ with the following physical parameters: d = 1113 kg/m^3, mi = 0.018396 Pa·s, C = 2382 J/kg·K, k = 0.25 W/m·K. But for the needs of sensitivity verification, it was additionally accepted that C = {1900, ..., 2700} J/kg·K. Nanoparticles' characteristics were adopted with the values typical for the alumina Al_2O_3, which means d = 3900 kg/m^3, k = 29.5 W/m·K and as a reference point for determining the direction and magnitude of entropy changes, copper Cu (d = 8960 kg/m^3, k = 400 W/m·K) was chosen, considering its practical applications [12,28]. The Monte-Carlo simulation as well as statistical estimation contained in the *Statistics* library were engaged in the computer program, where the total number of random trials for all simulations was fixed as equal to 200,000. The standard deviation of the fluid volumetric ratio was predefined as 5% of the corresponding expected value in all simulations. The results in the form of expected values, coefficients of variations as well as Shannon entropy of the effective nanofluid density are presented in Figure 2a,b and Figure 3 as the functions of both the particles' density as well as their volume fraction. There is no doubt that the key parameter here is the volume fraction of the particles, and even a small increase in this leads to a rapid increase in the resulting density statistical parameters. The expected values and coefficients of variation, which were determined based on the model outlined in [22], and changes concerning the particle density are also remarkable and linear, whereas Shannon entropy fluctuations are nonlinear (Figure 3a,b), which is even more pronounced in the case of copper nanoparticles (Figure 3b). Additionally, they converge to some extreme value obtained at the upper end at the horizontal axis but for larger volumetric ratios only, which is also clearly observed for nanoparticles with higher density (Figure 3b). The trends observed for lower volume fractions in both cases continue to exhibit an increasing tendency in this region. This phenomenon may be attributed to the ordering of nanoparticles upon exceeding a certain volume fraction threshold—around 4%—which becomes more pronounced when there is a significant difference in relative density between the base fluid and the nanoparticles (Figure 3b). The significantly higher Shannon entropy observed for Cu nanoparticles compared to Al_2O_3 nanoparticles at the same φ parameter value (Figure 3a,b) can be attributed to Brownian effects. The size of Cu nanoparticles is approximately 45 nm, whereas Al_2O_3 nanoparticles range between 0.5 and 1.5 μm. In the initial stage of homogenization, Cu nanoparticles are more susceptible to thermal motion induced by the base fluid than Al_2O_3 particles. Since the Shannon entropy is proportional to the standard deviation for the given population, it is clear that the extreme uncertainty would be obtained for the extreme value expectation of the particle density and their extreme volumetric ratio expectation.

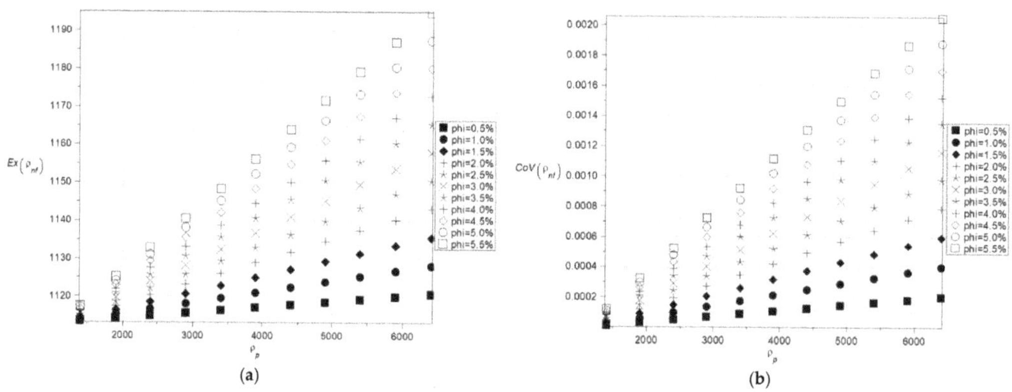

Figure 2. Expected values (**a**) and coefficients of variation (**b**) of the effective density (the density of the particles varies (1400–6400) kg/m^3).

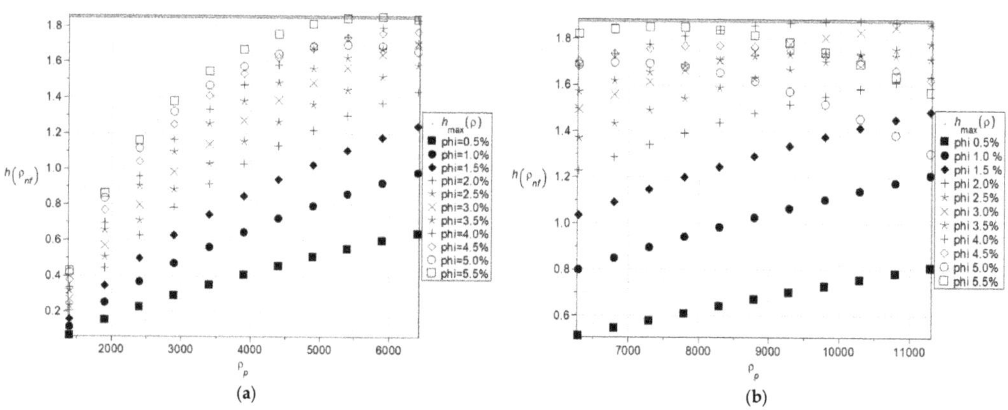

Figure 3. Shannon entropies for the effective density of the nanofluids with Al_2O_3 (**a**) and Cu (**b**) (the density of the particles varies (1400–11,400) kg/m^3).

The next two figures, cf. Figures 4 and 5, show expectations (Figure 4a) and coefficients of variation (Figure 4b) as well as Shannon entropies of the effective specific heat capacity of the nanofluid with Al_2O_3 (Figure 5a) where the specific heat of the particles varies (380–1380) J/kg·K and with Cu (Figure 5b) where the specific heat of the particles varies (160–640) J/kg·K and volumetric ration of the nanoparticles does not exceed 5%. Now, the largest expectations of the effective heat capacity of the nanofluid are obtained with the smallest concentration of the nanoparticles, whereas minimum expectations are obtained for their largest amount. This expectation of effective capacity is almost insensitive to the input particles' heat capacity for their smaller volume fractions in the given fluid base and linearly increases together with the particle's heat capacity in the opposite situation.

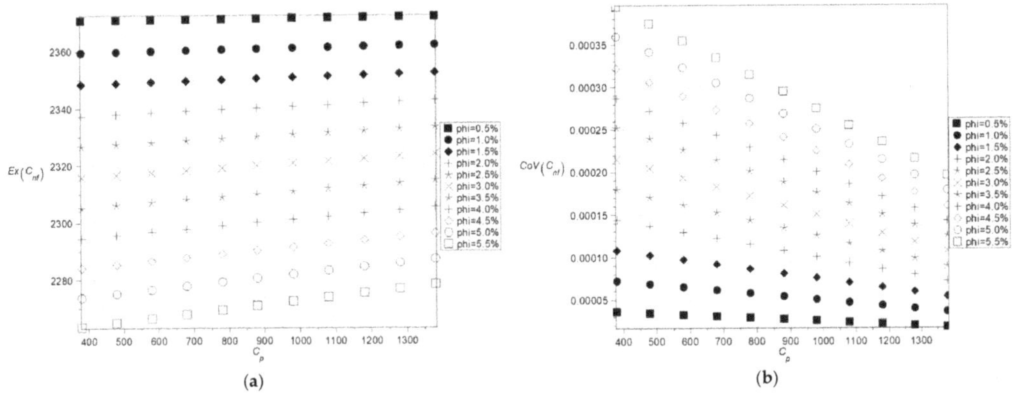

Figure 4. Expected values (**a**) and coefficients of variation (**b**) of the effective specific heat capacity.

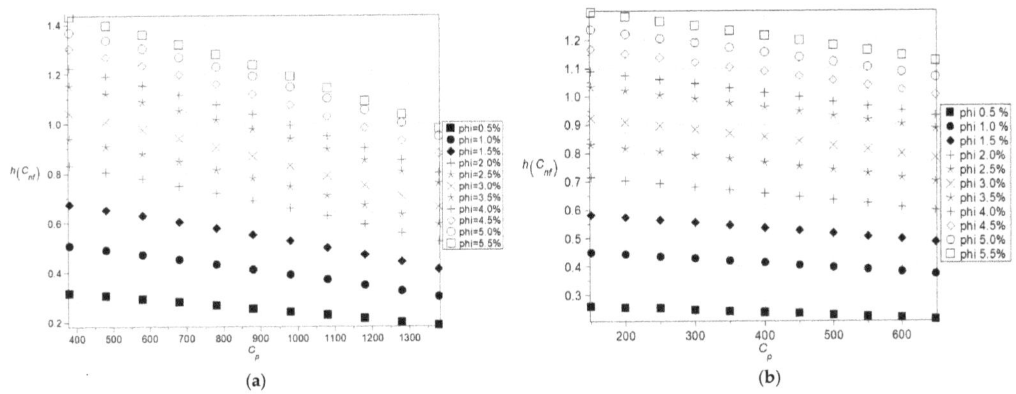

Figure 5. Shannon entropies for the effective specific heat capacity of the nanofluids with Al_2O_3 (**a**) and Cu (**b**) where the specific heat of the particles varies (380–1380) J/kg·K (**a**) and (160–640) J/kg·K (**b**).

Coefficients of variation as well as Shannon entropies decrease while increasing the input mean value of the particle's heat capacity, and this inverse proportionality has an almost linear character. The largest values in both cases are obtained with the largest volumetric ratio of the particles and the smallest mean value of the particle's heat capacity. It can be observed that, within the analyzed range, an increase in the average heat capacity and the volumetric fraction of particles leads to a decrease in entropy, which is clearly noticeable in (Figure 5a). This phenomenon may be related to the emergence of nonlinearities at critical values of the φ coefficient from the perspective of nanofluid theory or to physical processes such as aggregation or percolation.

In the case of poorly dispersed Al_2O_3 nanoparticles, molecular aggregation may occur, leading to a higher Shannon entropy compared to the corresponding situation for Cu nanoparticles (Figure 5a,b). This, in turn, can result in local fluctuations in the effective specific heat capacity of the nanofluid. This result is expected because the particle fraction is the uncertainty source and increasing its heat capacity mean value reduces the impact of the Gaussian volume fraction φ. Nevertheless, the coefficient of variation and Shannon entropy exhibit similar variations with almost constant expectations; probabilistic entropy has slightly smaller values than the effective nanofluid density in Figure 3.

Further, Figure 6a,b and Figure 7a,b show an analogous set of the results, the basic statistical characteristics of the effective heat conductivity. The expected values now exhibit a linear increase, closely resembling the trends observed for the effective density shown in Figure 2a. This behavior is accompanied by the thermal conductivity of the base fluid. However, the range of admissible values for different volumetric fractions is considerably narrower. Notably, the influence of these two parameters appears to be largely independent, underscoring the complexity of their interaction in nanofluid systems. These results are consistent with previous studies carried out using analytical methods [29,30] as well as due to an application of neural networks [31]. Coefficients of variation of this effective conductivity do not change while increasing input mean conductivity of the fluid, but remarkably increase while increasing the parameter φ. As these coefficients are constant and expectations linearly increase, one expects Shannon entropy to proportionally increase together with input fluid conductivity. This is noticeable in Figure 7a,b, where additionally, one notices expectedly that the larger the volumetric ratio of the nanoparticles, the higher the resulting entropy. However, Shannon entropy is remarkably smaller here, so uncertainty in effective heat conductivity is smaller than for effective density and heat capacity. Noteworthy is the entropy behavior observed under a significant disparity between the thermal conductivity of the base fluid and that of the nanoparticles (Figure 6b). When the volume fraction exceeds $\varphi = 5\%$, a slight increase in entropy can be observed. This may indicate the onset of dominant interactions between nanoparticles within the nanofluid. Cu nanoparticles have a significantly smaller active surface area compared to Al_2O_3 nanoparticles, which enhances their thermal conductivity. This effect is reflected in the observed increase in Shannon entropy (Figure 6a,b). However, it should be noted that considering the time evolution of the system could lead to opposite conclusions due to the higher susceptibility of Cu nanoparticles to oxidation or percolation effects, which would contribute to an increase in entropy.

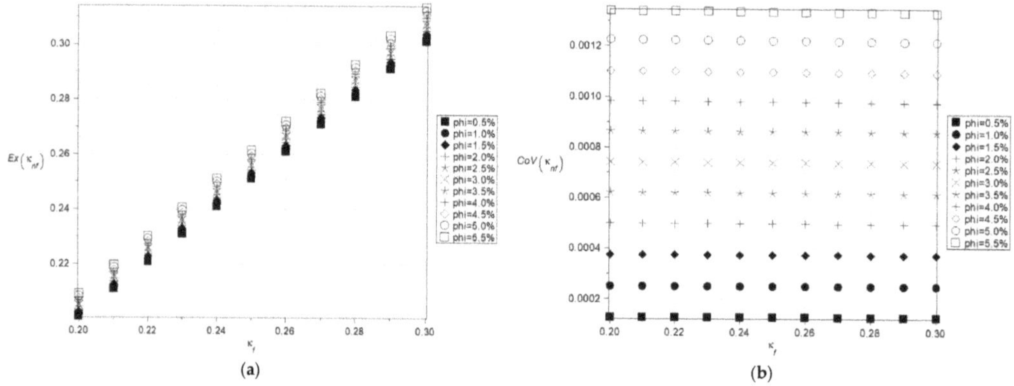

Figure 6. Expected values (**a**) and coefficients of variation (**b**) of the effective thermal conductivity.

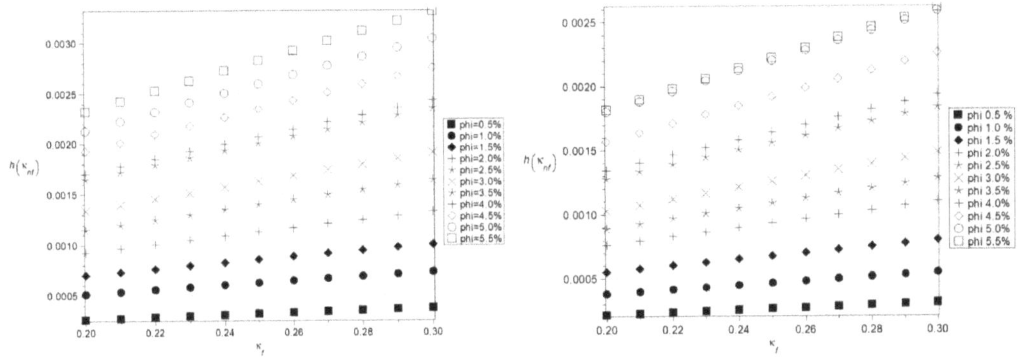

(**a**) (**b**)

Figure 7. Shannon entropies for the effective thermal conductivity of the nanofluid with Al_2O_3 (**a**) and Cu (**b**) nanoparticles (the thermal conductivity of the base fluid varies (0.2–0.3) W/m·K).

The last effective parameter under consideration is the effective viscosity calculated here in the context of both Einstein and Brinkmann models specified in the previous section; the corresponding expectations, coefficients of variations, and Shannon entropies are presented in Figures 8–11. The left series is adjacent to the Einstein theory, whereas the right column corresponds to the Brinkmann idea. The first and most important observation is that, for the given nanofluid, these two models return almost the same statistical parameters. Expected values of the effective viscosity increase linearly together with the fluid base viscosity; it is observed for relatively small variations in the latter. Additionally, the larger the volume fraction of nanoparticles, the larger the resulting effective viscosity. The coefficients of variation (cf. Figure 9) are insensitive to any variations in the fluid base viscosity but still increase in the function of the parameter φ. It is noticeable that by changing this parameter from 0.5% up to 5.5%, one can increase this CoV more than five times. As a result of the expectations and coefficients of variation, one notices that Shannon entropy increases moderately while increasing input fluid viscosity, and also, remarkably, when increasing the nanoparticles' volume ratio. Based on Figure 11, it can be concluded that the Brinkman's model exhibits a higher entropy value than the Einstein's model under the same input conditions. This phenomenon is clearly visible only for large values of the phi coefficient. The larger size of Al_2O_3 nanoparticles may initially result in a higher Shannon entropy of the effective viscosity of the nanofluid compared to Cu nanoparticles, even at relatively low volume fractions (φ) (Figures 10a,b and 11). The greater entropy increase at high concentrations ($\varphi > 1\%$), observed in both models for aluminum oxide, may be associated with its tendency to aggregate, indicating a potential loss of stability in the designed nanofluid parameters.

In summary, the uncertainty in the effective physical parameters of nanofluids is relatively small, typically calculated for nanoparticle volumetric ratios not exceeding 5% and their coefficient of variation (CoV) limited to 5%. However, it is important to consider that practical applications of nanofluids may require larger CoVs and could also involve an uncontrolled increase in uncertainty when the effects of temperature are taken into account. A key finding of this study is that the homogenization procedure evaluated herein does not amplify the uncertainty inherent in the enrichment of base fluids with nanoparticles. Therefore, this procedure can be efficiently applied in multiscale modeling of nanofluid flows using the Navier–Stokes equations, even in the presence of certain physical uncertainties, by common usage with the polynomial chaos approach [32], stochastic collocation method [33], stochastic perturbation method [19], and even neural network-based algorithms [34].

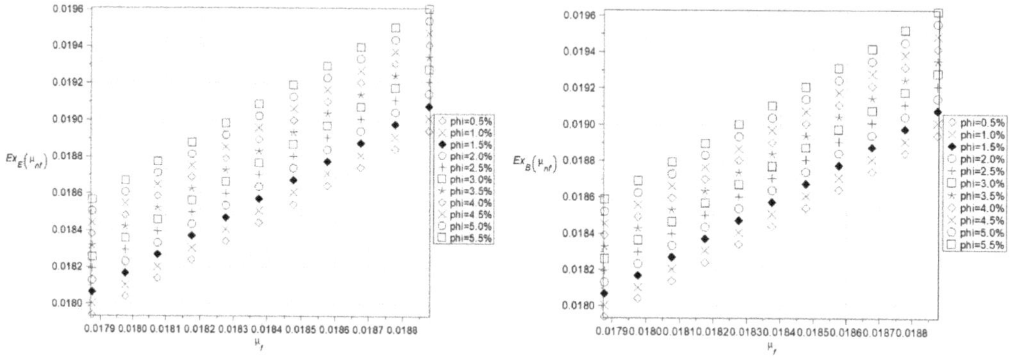

(a) (b)

Figure 8. Expected values of the effective viscosity of the nanofluid in Einstein's model (**a**), and in Brinkman's model (**b**).

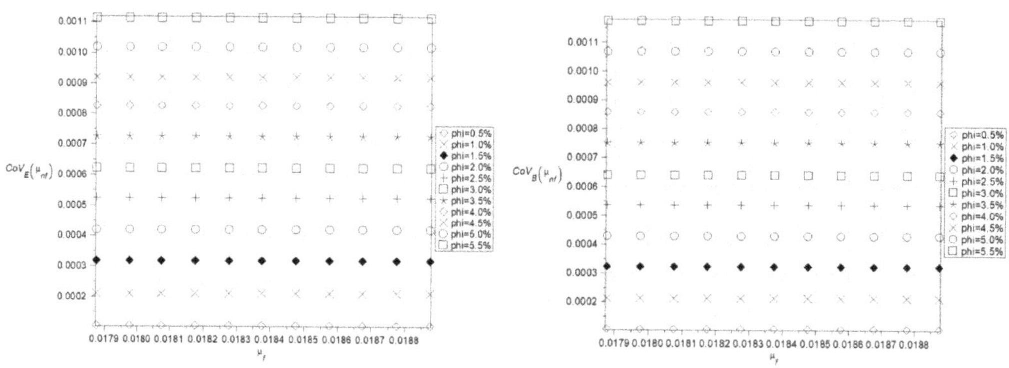

(a) (b)

Figure 9. Coefficients of variation of the effective viscosity of the nanofluid in Einstein's model (**a**), and also in Brinkman's model (**b**).

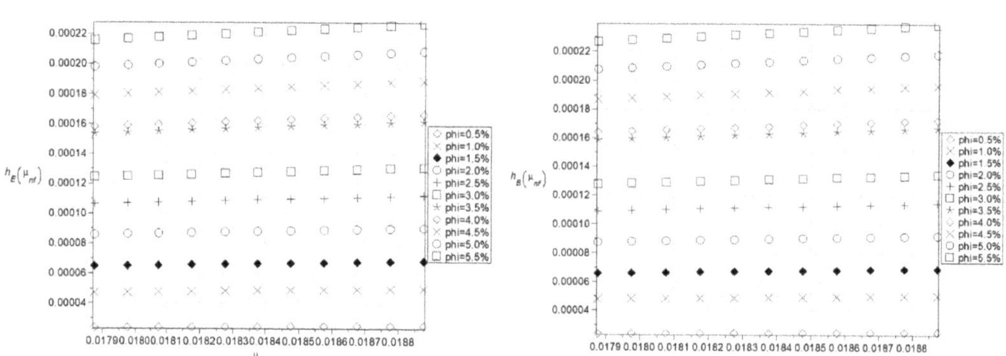

(a) (b)

Figure 10. Shannon entropies for the effective viscosity of the nanofluid in Einstein's model (**a**) and Brinkman's model (**b**) for Al_2O_3.

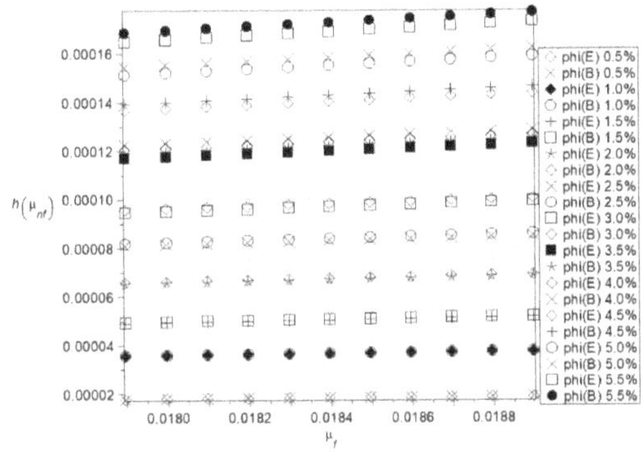

Figure 11. Shannon entropies for the effective viscosity of the nanofluid in Einstein's and Brinkman's model for Cu—comparison.

4. Concluding Remarks

A probabilistic numerical approach to the determination of the effective parameters of the nanofluids exhibiting some uncertainty in the volumetric ratio of the nanoparticles has been presented in this paper. It is based upon temperature-independent analytical formulas following some previous experimental works, Monte-Carlo simulation of Shannon entropy, and also the first two probabilistic characteristics of effective heat capacity and conductivity, mass density as well as viscosity. It has been demonstrated that Shannon entropy may serve as a single and universal uncertainty measure while analyzing statistical scattering of effective parameters contrary to multiple functions of positional statistics. This hypothesis needs to be validated for other physical problems in nanofluids, and also with larger input uncertainty levels. It is also seen that random fluctuations in effective parameters resulting from the considered uncertainty in particles' volume ratio have rather limited values, which enables further application of the stochastic perturbation method while solving NS Equations (17)–(19) in nanofluidics for the coupled heat and mass transfer. Such an approach can be relatively easily implemented in any computer algebra system or any programming language having a statistical library together with symbolic differentiation and integration tools.

The numerical simulation reported in this paper shows that the expected value of the volumetric ratio of the nanoparticles has a noticeable impact on the resulting uncertainty of all effective physical characteristics of the given nanofluid, as detailed in [22]. This impact is many times larger than the importance of the same fluid base physical parameters. Noticeably, further increases in input uncertainty may lead to nonlinear Shannon entropy fluctuations, consistent with the physical model, and they arise when the relationship between input uncertainty and the effective properties of a nanofluid becomes increasingly complex, deviating from a straightforward proportionality. These fluctuations are consistent with the underlying physical models accounting for the nonlinearity in particle–fluid interactions and the dynamic redistribution of nanoparticles. In the context of nanofluids, the thermal and rheological properties depend heavily on nanoparticle concentration φ, size, shape, and distribution. As the Gaussian input uncertainty increases, its influence on the system's effective parameters changes from linear to nonlinear behavior. This is caused due to multiple factors, such as the aggregation of nanoparticles, interparticle forces, and thermal interactions between the nanoparticles and the base fluid. These effects are captured in the variations in Shannon entropy, which encapsulates the probabilistic representation of system uncertainty, as demonstrated in this study through the analysis of effective density, specific heat capacity, thermal conductivity, and viscosity. The nonlinear entropy fluctuations highlight critical thresholds where small changes in input uncertainty lead to disproportionately large variations in effective properties. These thresholds may correspond to physical phenomena such as percolation thresholds (Cu-doped nanofluids), nanoparticle clustering (Al_2O_3-based nanofluids), or modifications in heat transfer pathways when particle–particle interactions become dominant.

Future work should focus on validating these findings across broader parameter ranges, incorporating temperature sensitivity, and comparing results with experimental data to ensure consistency with real nanofluid behavior. It would be interesting with no doubt to see a comparison of the NS equations' solutions having practical importance to contrast realistic fluctuations of the nanofluid coupled behavior with their homogenized counterpart. An important direction would be an extension of the algebraic formulas for effective physical characteristics towards their temperature sensitivity [35,36] as well as applications of the nanofluidics in cellular structure analysis and biomechanics [37].

Author Contributions: Conceptualization, M.K.; Methodology, M.K. and R.L.O.; Software, R.L.O.; Validation, R.L.O.; Formal analysis, M.K.; Investigation, M.K. and R.L.O.; Resources, R.L.O.; Data curation, M.K. and R.L.O.; Writing—original draft, M.K. and R.L.O.; Supervision, M.K.; Project administration, M.K.; Funding acquisition, M.K. All authors have read and agreed to the published version of the manuscript.

Funding: This work contains the results obtained in the framework of the research grant OPUS no. 2021/41/B/ST8/02432 "Probabilistic entropy in engineering computations" sponsored by the National Science Center in Cracow, Poland, in the period of 2022–2025.

Data Availability Statement: Data is contained within the article.

Conflicts of Interest: The authors declare no conflict of interest.

References

1. Christensen, R.M. *Mechanics of Composite Materials*; Wiley: New York, NY, USA, 1979.
2. Vasu, V.; Krishna, K.R.; Kumar, A.C.S. Analytical prediction of thermophysical properties of fluids embedded with nanostructured materials. *Int. J. Nanopart.* **2008**, *1*, 32–49. [CrossRef]
3. Selvakumar, R.D.; Dhinakaran, S. A multi-level homogenization model for thermal conductivity of nanofluids based on particle size distribution (PSD) analysis. *Powder Technol.* **2016**, *301*, 310–317. [CrossRef]
4. Ma, T.; Guo, Z.; Lin, M.; Wang, Q. Recent trends on nanofluid heat transfer machine learning research applied to renewable energy. *Renew. Sustain. Energy Rev.* **2021**, *138*, 110494. [CrossRef]
5. Sheikh, N.A.; Chuan Ching, D.L.; Khan, I. A comprehensive review on theoretical aspects of nanofluids: Exact solutions and analysis. *Symmetry* **2020**, *12*, 725. [CrossRef]
6. Awais, M.; Bhuiyan, A.A.; Salehin, S.; Ehsan, M.M.; Khan, B.; Rahman, M.H. Synthesis, heat transport mechanisms and thermophysical properties of nanofluids: A critical overview. *Int. J. Thermofluids* **2021**, *10*, 100086. [CrossRef]
7. Angayarkanni, S.A.; Philip, J. Review on thermal properties of nanofluids: Recent developments. *Adv. Colloid Interface Sci.* **2015**, *225*, 146–176. [CrossRef]
8. Wang, X.; Song, Y.; Li, C.; Zhang, Y.; Ali, H.M.; Sharma, S.; Li, R.; Yang, M.; Gao, T.; Liu, M.; et al. Nanofluids application in machining: A comprehensive review. *Int. J. Adv. Manuf. Technol.* **2024**, *131*, 3113–3164. [CrossRef]
9. Colangelo, G.; Favale, E.; Miglietta, P.; Milanese, M.; de Risi, A. Thermal conductivity, viscosity and stability of Al_2O_3-diathermic oil nanofluids for solar energy systems. *Energy* **2016**, *95*, 124–136. [CrossRef]
10. Sharifpur, M.; Tshimanga, N.; Meyer, J.P.; Manca, O. Experimental investigation and model development for thermal conductivity of α-Al_2O_3-glycerol nanofluids. *Int. Commun. Heat Mass Transf.* **2017**, *85*, 12–22. [CrossRef]
11. Esfe, M.; Firouzi, M.; Afrand, M. Experimental and theoretical investigation of thermal conductivity of ethylene glycol containing functionalized single-walled carbon nanotubes. *Phys. E Low-Dimens. Syst. Nanostruct.* **2018**, *95*, 71–77. [CrossRef]
12. Barbés, B.; Pàramo, R.; Blanco, E.; Casanova, C.J. Thermal conductivity and specific heat capacity measurements of CuO nanofluids. *Therm. Anal. Calorim.* **2014**, *115*, 1883–1891. [CrossRef]
13. Murshed, S.M.S.; Leong, K.C.; Yang, C. Enhanced thermal conductivity of TiO_2—Water based nanofluids. *Int. J. Therm. Sci.* **2005**, *44*, 367–373. [CrossRef]

14. Iqbal, S.M.; Raj, C.S.; Michael, J.J.; Irfan, A.M. A comparative investigation of Al_2O_3/H_2O, SiO_2/H_2O, and ZrO_2/H_2O nanofluid for heat transfer applications. *Dig. J. Nanomater. Biostruct.* **2017**, *12*, 255–263.
15. Choi, U.S.; Eastman, J. Enhancing thermal conductivity of fluids with nanoparticles. In Proceedings of the ASME International Mechanical Engineering Congress and Exposition, San Francisco, CA, USA, 12–17 November 1995; Volume 66.
16. Pryazhnikov, M.I.; Minakov, A.V.; Rudyak, V.Y.; Guzei, D.V. Thermal conductivity measurements of nanofluids. *Int. J. Heat Mass Transf.* **2017**, *104*, 1275–1282. [CrossRef]
17. Waqas, H.; Farooq, U.; Liu, D.; Abid, M.; Imran, M.; Muhammad, T. Heat transfer analysis of hybrid nanofluid flow with thermal radiation through a stretching sheet: A comparative study. *Int. Commun. Heat Mass Transf.* **2022**, *138*, 106303. [CrossRef]
18. Zienkiewicz, O.C.; Taylor, R.; Nithiarasu, P. *The Finite Element Method for Fluid Dynamics*; Elsevier Butterworth-Heinemann: Amsterdam, The Netherlands, 2005.
19. Kamiński, M.; Carey, G.F. Stochastic perturbation-based finite element approach to fluid flow problems. *Int. J. Numer. Methods Heat Fluid Flow* **2005**, *15*, 671–697. [CrossRef]
20. Srikant, R.R.; Rao, D.N.; Subrahmanyam, M.S.; Krishna, V.P. Applicability of cutting fluids with nanoparticle inclusion as coolants in machining. *Proc. Inst. Mech. Eng. Part J J. Eng. Tribol.* **2009**, *223*, 221–225. [CrossRef]
21. Xie, C.; Li, H. Multiscale simulations of nanofluidics: Recent progress and perspective. *Wiley Interdiscip. Rev. Comput. Mol. Sci.* **2023**, *13*, e1661. [CrossRef]
22. Kamiński, M.; Ossowski, R.L. Prediction of the effective parameters of the nanofluids using the generalized stochastic perturbation method. *Phys. A Stat. Mech. Appl.* **2014**, *393*, 10–22. [CrossRef]
23. Shannon, C.E. A mathematical theory of computation. *Bell Syst. Tech. J.* **1948**, *27*, 379–423, 623–656. [CrossRef]
24. Cornil, J.M.; Testud, P. *An Introduction to MAPLE V.*; Springer: Berlin/Heidelberg, Germany, 2001.
25. Kamiński, M. On shannon entropy computations in selected plasticity problems. *Int. J. Numer. Methods Eng.* **2021**, *122*, 5128–5143. [CrossRef]
26. Zhou, Y.; Sui, S.Y.; Li, J.; Wang, Z.Y.; Cui, W.; Lv, Y.Z.; Li, C.R. Statistical analysis of moisture's effect on AC breakdown strength of TiO_2 nanofluids. *J. Mol. Liquids* **2018**, *249*, 420–428. [CrossRef]
27. Kamiński, M.; Ossowski, R.L. Navier-Stokes problems with random coefficients by the Weighted Least Squares Technique Stochastic Finite Volume Method. *Arch. Civ. Mech. Eng.* **2014**, *14*, 745–756. [CrossRef]
28. Parsian, A.; Akbari, M. New experimental correlation for the thermal conductivity of ethylene glycol containing Al_2O_3–Cu hybrid nanoparticles. *J. Therm. Anal. Calorim.* **2017**, *131*, 1605–1613. [CrossRef]
29. Simpson, S.; Schelfhout, A.; Golden, C.; Vafaei, S. Nanofluid thermal conductivity and effective parameters. *Appl. Sci.* **2019**, *9*, 87. [CrossRef]
30. Khodadadi, H.; Aghakhani, S.; Majd, H.; Kalbasi, R.; Wongwises, S.; Afrand, M. A comprehensive review on rheological behavior of mono and hybrid nanofluids: Effective parameters and predictive correlations. *Int. J. Heat Mass Transf.* **2018**, *127 Pt B*, 997–1012. [CrossRef]
31. Pourpasha, H.; Farshad, P.; Heris, S.Z. Modeling and optimization the effective parameters of nanofluid heat transfer performance using artificial neural network and genetic algorithm method. *Energy Rep.* **2021**, *7*, 8447–8464. [CrossRef]
32. Wan, X.; Karniadakis, G.E. Long-term behavior of polynomial chaos in stochastic flow simulations. *Comput. Methods Appl. Mech. Eng.* **2006**, *195*, 5582–5596. [CrossRef]
33. Feireisl, E.; Lukáova-Medvidová, M. Convergence of a stochastic collocation finite volume method for the compressible navier-stokes system. *Ann. Appl. Probab.* **2023**, *33*, 4936–4963. [CrossRef]
34. Yang, H.; Ou, J.; Wang, Y. Neural-physics multi-fidelity model with active learning and uncertainty quantification for GPU-enabled microfluidic concentration gradient generator design. *Comput. Methods Appl. Mech. Eng.* **2023**, *417*, 116434. [CrossRef]
35. Agarwal, R.; Verma, K.; Agrawal, N.; Singh, R. Sensitivity of thermal conductivity for Al_2O_3 nanofluids. *Exp. Therm. Fluid Sci.* **2017**, *80*, 19–26. [CrossRef]
36. Li, C.H.; Peterson, G.P. Experimental investigation of temperature and volume fraction variations on the effective thermal conductivity of nanoparticle suspensions (nanofluids). *J. Appl. Phys.* **2006**, *99*, 084314. [CrossRef]
37. Basson, N.; Peng, C.-H.S.; Geoghegan, P.; van der Lecq, T.; Steven, D.; Williams, S.; Lim, A.E.; Ho, W.H. A computational fluid dynamics investigation of endothelial cell damage from glaucoma drainage devices. *Sci. Rep.* **2024**, *14*, 3777. [CrossRef] [PubMed]

nanomaterials

MDPI

Review

Applications of MOF-Based Nanocomposites in Heat Exchangers: Innovations, Challenges, and Future Directions

Talha Bin Nadeem [1,2], Muhammad Imran [1,3,*] and Emad Tandis [1]

1 Department of Mechatronics and Biomedical Engineering, College of Engineering and Physical Sciences, Aston University, Birmingham B4 7ET, UK; 250000069@aston.ac.uk or talhanadeem@neduet.edu.pk (T.B.N.); e.tandis@aston.ac.uk (E.T.)

2 Department of Mechanical Engineering, NED University of Engineering and Technology, Karachi 75270, Pakistan

3 Energy Systems Group, Energy and Bioproduct Research Institute, College of Engineering and Physical Sciences, Aston University, Birmingham B4 7ET, UK

* Correspondence: m.imran12@aston.ac.uk

Abstract: Metal–organic frameworks (MOFs) have garnered significant attention in recent years for their potential to revolutionize heat exchanger performance, thanks to their high surface area, tunable porosity, and exceptional adsorption capabilities. This review focuses on the integration of MOFs into heat exchangers to enhance heat transfer efficiency, improve moisture management, and reduce energy consumption in Heating, Ventilation and Air Conditioning (HVAC) and related systems. Recent studies demonstrate that MOF-based coatings can outperform traditional materials like silica gel, achieving superior water adsorption and desorption rates, which is crucial for applications in air conditioning and dehumidification. Innovations in synthesis techniques, such as microwave-assisted and surface functionalization methods, have enabled more cost-effective and scalable production of MOFs, while also enhancing their thermal stability and mechanical strength. However, challenges related to the high costs of MOF synthesis, stability under industrial conditions, and large-scale integration remain significant barriers. Future developments in hybrid nanocomposites and collaborative efforts between academia and industry will be key to advancing the practical adoption of MOFs in heat exchanger technologies. This review aims to provide a comprehensive understanding of current advancements, challenges, and opportunities, with the goal of guiding future research toward more sustainable and efficient thermal management solutions.

Keywords: metal–organic frameworks; MOFs; heat exchanger; nanocomposites; thermal conductivity; energy efficiency; heat transfer enhancement; fouling resistance

1. Introduction

Heat exchangers are essential components across numerous engineering applications, including HVAC systems, automotive designs, and industrial manufacturing. These systems demand materials that can effectively conduct heat at high rates while maintaining stability and minimal thermal expansion across a range of operating temperatures [1]. In applications such as air conditioning, refrigeration, and energy recovery, heat exchangers are critical to optimizing system efficiency, cost, and compactness. Similarly, heat exchangers play a pivotal role in both cooling and heating applications, acting as the core component for energy transfer between fluids in these systems. Similarly, in indirect evaporative cooling, heat exchangers enable the transfer of heat from the warm air inside a building to the cooler exhaust air outside, facilitating temperature reduction without adding moisture to the air [2]. Figure 1a represents the schematic diagram of an indirect evaporative cooling system, depicting its working principle. The efficiency of this process heavily depends on the heat exchanger's design, as it determines the thermal performance and overall energy efficiency of the system. A well-designed heat exchanger maximizes the surface area for heat transfer while minimizing the energy required to drive the process, thereby improving the cooling performance and reducing operational costs. However, the design must be optimized to balance thermal conductivity, airflow, and the resistance to fouling, which can significantly impact system efficiency over time [3]. Consequently, in heat pumps, heat exchangers are crucial for transferring heat between the refrigerant and the environment. In heating, heat exchangers extract heat from the outside air or ground and release it inside [4]. The effectiveness of heat exchangers in heat pumps directly influences the system's coefficient of performance (COP), which is a measure of efficiency. A higher COP translates to better energy utilization, which is particularly important in residential and commercial applications where energy efficiency and cost savings are prioritized [5]. In both applications, selecting the right type of heat exchanger—whether air-to-air, air-to-water, or ground-coupled—is crucial for optimizing the system's performance [6]. Figure 1b represents the schematic diagram of a heat pump, depicting its working principle.

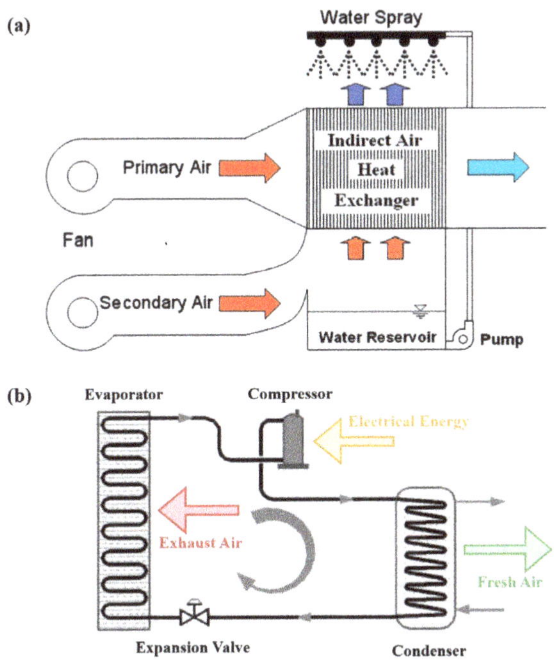

Figure 1. Schematics of (**a**) indirect evaporative cooling system [7] and (**b**) heat pumps [8], illustrating their working principles (reproduced with permission from Wiley and Elsevier).

Traditional designs, like fin-and-tube or plate-fin heat exchangers made from materials such as copper, aluminum, or steel, are widely used. Figure 2 provides illustrations of widely used heat exchangers in HVAC. Numerous studies focus on improving heat transfer in these conventional systems. However, the performance limitations of metal-based heat exchangers in certain applications have driven interest in exploring alternative designs with different materials [9]. Despite their widespread use, several efficiency bottlenecks limit the performance, longevity, and reliability of these devices.

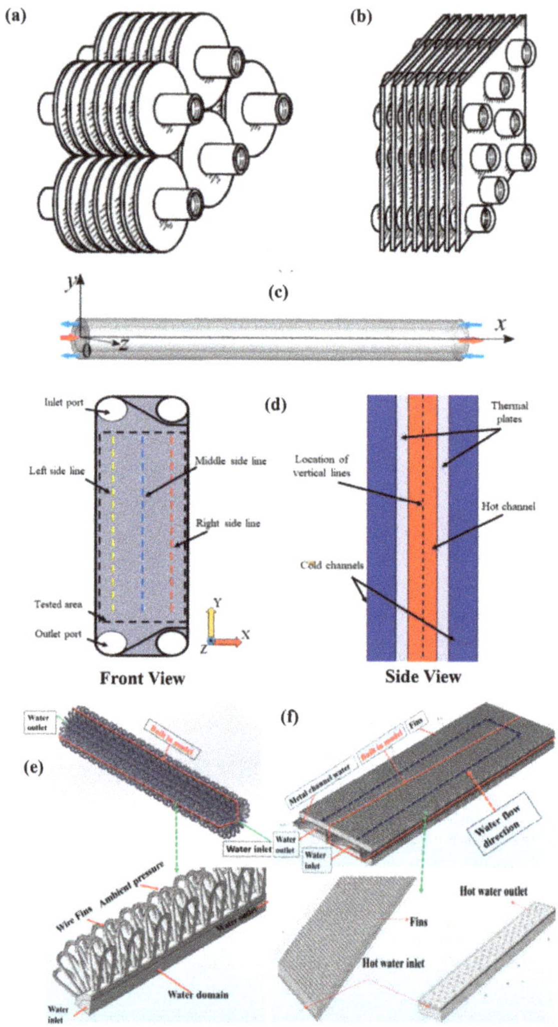

Figure 2. Schematics of (**a**) individual fin [10], (**b**) continuous fins [10] in tube finned, (**c**) double pipe [11], (**d**) plate type [12], (**e**) wire-finned [13], and (**f**) micro-channel heat exchanger [13] commonly used in HVAC applications (reprinted with permission from Wiley and Elsevier).

Key issues include thermal resistance, fouling, and material degradation, each of which affects the heat exchanger's efficiency, operational cost, and maintenance needs [14]. Thermal resistance, a key limitation in heat exchangers, hinders efficient heat transfer between mediums. It is influenced by the heat exchanger's design, material thermal conductivity, and surface thermal boundary layers, directly affecting overall system performance [15]. In metallic heat exchangers, materials like copper and aluminum are used due to their high thermal conductivity. However, limitations arise, as these metals exhibit high thermal expansion and can degrade over time under harsh conditions, leading to increased thermal resistance [16]. Advanced materials, such as MOFs, have shown promise in enhancing thermal conductivity while maintaining structural stability. Researchers have also explored nanocomposites to improve the heat transfer rate, especially in systems where high-temperature stability is critical [17]. Fouling, caused by the buildup of materials like dust, biofilms, or mineral deposits on heat exchanger surfaces, adds thermal resistance and reduces heat transfer efficiency. This can decrease heat transfer rates by 30–50%, significantly increasing energy consumption to maintain desired performance [18,19]. Fouling is often more prevalent in heat exchangers used in environments with high concentrations of particles or biological matter, such as wastewater treatment facilities and cooling towers [20]. The cost of dealing with fouling, through either frequent maintenance or more complex cleaning processes, significantly adds to the operational cost of heat exchanger systems [21]. Antifouling coatings and self-cleaning mechanisms are emerging as potential solutions to mitigate these effects. For example, nanocomposite coatings that combine MOFs with hydrophobic materials can minimize fouling while maintaining high thermal conductivity [22]. Biofouling involves microbial growth forming biofilms on surfaces. Biofouling can obstruct flow channels, decrease membrane permeability, and necessitate frequent cleaning, thereby elevating maintenance costs and downtime [23,24]. Mechanical failure in water treatment systems can result from material degradation, corrosion, or stress-induced damage. Factors such as biofouling exacerbate these issues by creating differential pressures and localized corrosion sites, leading to equipment malfunction or failure [25]. For instance, biofouling-induced corrosion can compromise the structural integrity of pipelines and membranes, resulting in leaks or ruptures [26]. Fouling layers in heat exchangers decrease thermal conductivity, reducing energy efficiency. Hybrid coatings and periodic cleaning methods have been implemented to counteract fouling, ensuring consistent heat transfer performance. Antibacterial treatments also prevent biofilm-induced corrosion, extending equipment life [27]. Biofouling in water-related processes, such as reverse osmosis membranes and pipelines, reduces water flow and increases energy consumption. Antimicrobial coatings, such as the hybrid polyaniline (PANI)-halamine introduced by Weiss et al. [28], effectively prevent biofilm formation, improving system reliability and minimizing downtime.

Material degradation is another major bottleneck in heat exchanger performance, particularly in systems exposed to corrosive or high-temperature environments. Common materials like aluminum, copper, and steel are prone to corrosion, which can reduce their effectiveness and shorten the lifespan of the heat exchanger. Corrosion not only lowers thermal conductivity but also makes the material more susceptible to mechanical failure [29]. In high-temperature applications, such as those found in power generation or chemical processing, degradation can also occur due to thermal cycling, which leads to metal fatigue and cracking [30]. Researchers are exploring MOF-based composites and other advanced materials to combat these issues by improving thermal stability and reducing degradation rates [31]. For instance, integrating MOFs with carbon-based materials like graphene can offer a combination of high thermal conductivity and resistance to thermal and mechanical stress, making them more durable than conventional materials in challenging environments [32,33]. Advancing heat exchanger technology requires addressing challenges like thermal resistance, fouling, and material degradation. MOFs and MOF-based nanocomposites, with their high surface area and customizable properties, show potential in enhancing heat transfer efficiency, resisting corrosion, and reducing fouling for improved performance [34,35]. For example, in one study, a MOF composite integrated with graphene achieved a 25% reduction in thermal resistance and a 40% improvement in fouling resistance compared to conventional metal-based heat exchangers [34]. Such innovations highlight the potential of MOF-based materials to overcome long-standing challenges in heat exchanger design. Figure 3 illustrates the categorization of desiccant materials that can be used as coating material in heat exchangers.

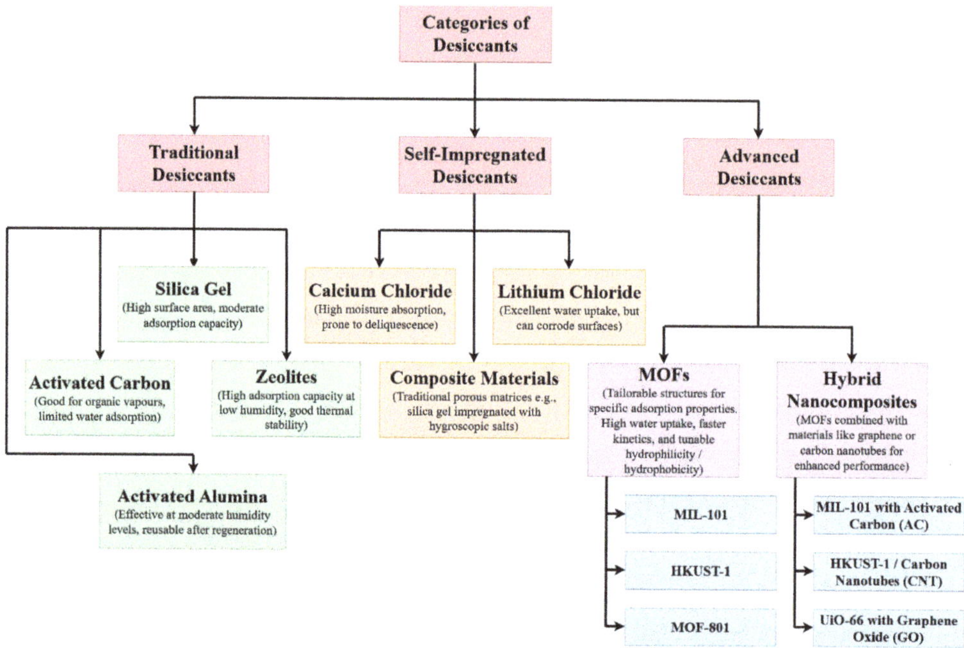

Figure 3. Categorization of desiccant materials that can be used as coating material in heat exchangers.

MOFs are highly porous materials consisting of metal ions coordinated to organic ligands, forming intricate network structures with exceptionally high surface areas and tunable properties. Due to these characteristics, MOFs have gained attention in applications such as gas storage, catalysis, and thermal management. However, by combining MOFs with other materials—such as metals, polymers, or carbon-based substances—researchers have created MOF-based nanocomposites that can further enhance thermal properties, making them especially promising for applications in heat exchangers and other energy-intensive systems [36]. MOFs are known for their vast internal surface area, often exceeding that of traditional porous materials, which enables greater interaction with heat transfer media. This high surface area plays a key role in MOF-based nanocomposites' ability to improve thermal management in heat exchangers [37]. For instance, studies have demonstrated that MOFs with large surface areas can effectively adsorb and release thermal energy, making them suitable for thermal energy storage and transfer. Additionally, the adjustable pore size in MOFs allows them to accommodate various guest molecules and even modify the thermal flow rate by controlling gas or fluid adsorption [38]. Pure MOFs often have lower thermal conductivity than metals, which limits their direct use in heat exchangers. However, by integrating MOFs with conductive fillers such as graphene oxide (GO), carbon nanotubes (CNTs), or metallic particles, researchers can significantly enhance their thermal conductivity, achieving composite materials that efficiently manage heat [39]. Graphene-based MOF composites, for example, leverage the thermal conductivity of graphene while retaining the high porosity and adsorption properties of the MOF framework, creating materials suitable for high-performance thermal applications. This approach addresses one of the primary limitations of traditional MOFs, making them more viable for practical applications in thermal management systems [40]. Figure 4 represents some of the research areas where MOFs have been used in recent years.

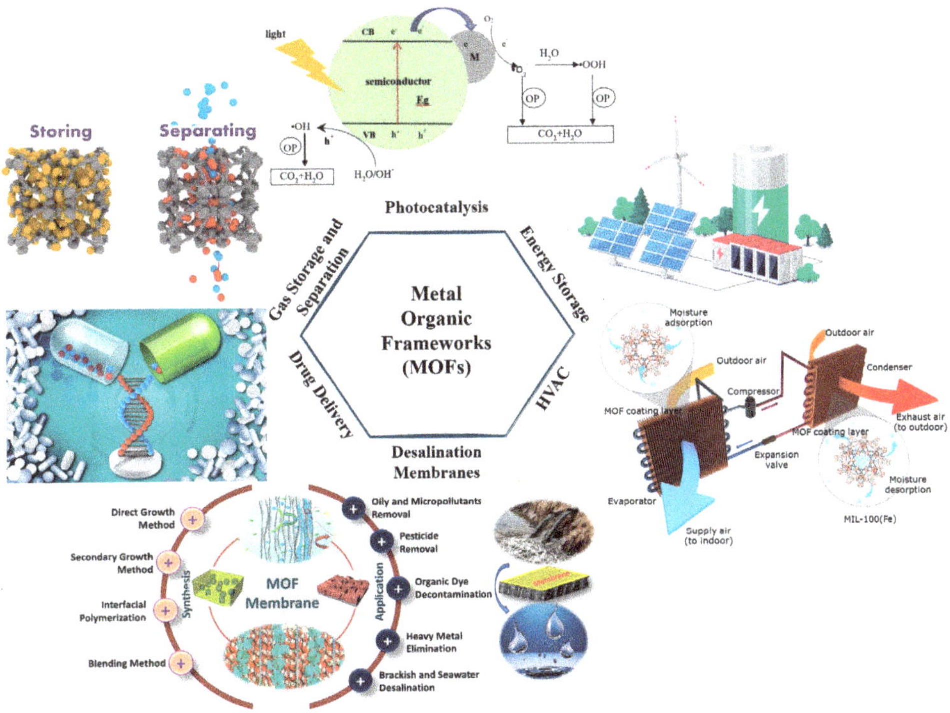

Figure 4. Applications of metal–organic frameworks (MOFs) in different research areas.

MOFs have gained attention for their versatility across various applications beyond heat exchangers such as water treatment and gas separation. In water treatment, MOFs demonstrate exceptional performance in removing heavy metals, organic pollutants, and dyes due to their high surface area and customizable pore structure [41]. For instance, Zr-based MOFs like UiO-66 effectively adsorb arsenic and lead ions from contaminated water [42] due to their chemical stability, high surface area, and abundant functional groups such as hydroxyl (-OH) and carboxyl (-COOH) on their frameworks. These functional groups interact strongly with toxic metal ions through coordination bonds or electrostatic attraction, facilitating their removal from contaminated water. Additionally, UiO-66's tunable pore size allows for efficient diffusion and trapping of these ions, making it an excellent candidate for water purification applications [43]. Similarly, MOFs are highly effective in gas separation, where their tunable pore sizes enable selective adsorption of gases like CO_2, CH_4, and H_2. ZIFs are particularly notable for their efficiency in CO_2 capture and storage [44]. Their imidazolate linkers create a hydrophobic environment, enhancing CO_2 selectivity while minimizing interference from water vapor, making them ideal for carbon capture applications [45]. Furthermore, MOFs are being explored in drug delivery, catalysis, and energy storage due to their structural flexibility and functional diversity. Addressing these broader applications highlights the multifunctionality of MOFs and underscores their potential for innovative solutions across diverse fields.

A major advantage of MOF-based nanocomposites is their chemical and structural tunability, which allows for customized thermal properties. MOFs can be synthesized with a wide range of metal nodes and organic linkers, providing flexibility in terms of pore size, surface chemistry, and thermal response [46]. Furthermore, when MOFs are combined with polymers, the resulting composites gain mechanical stability and can be engineered to resist degradation under thermal cycling. For instance, incorporating polymers such as polydimethylsiloxane (PDMS) with MOFs can produce materials that maintain structural integrity at elevated temperatures, extending the lifespan of heat exchangers in high-temperature environments [47]. Another significant advantage of MOF-based nanocomposites lies in their excellent adsorption capacity, which enhances their thermal management capabilities [48,49]. MOFs have been shown to adsorb various gases efficiently, making them ideal candidates for applications in adsorption-based cooling systems. By incorporating MOFs with metals or metal oxides, the adsorption–desorption properties of these composites can be optimized for rapid thermal response, a crucial feature for efficient heat exchangers [50]. This adaptability not only improves energy efficiency but also contributes to reducing the overall operational costs of cooling systems [51]. In addition to their thermal properties, MOF-based nanocomposites exhibit improved resistance to fouling and corrosion, common challenges in heat exchangers. For example, by integrating hydrophobic MOFs with anti-corrosive metals, researchers have developed nanocomposites that can minimize fouling from organic and inorganic deposits, thereby preserving heat transfer efficiency over time. This feature is especially beneficial in applications where maintenance and cleaning are difficult or costly [52].

The development of advanced materials for efficient thermal management is a critical focus in engineering, especially in the design of heat exchangers, which are central to systems such as HVAC, energy storage, and industrial processes. This review paper aims to examine the growing body of research on MOF-based nanocomposites as promising materials in heat exchangers. Recent advancements indicate that these composites, known for their high surface area, tunable thermal properties, and resilience in extreme environments, could significantly enhance heat exchanger efficiency. However, while MOF-based nanocomposites show potential, their practical application is challenged by factors such as thermal stability, material degradation, and scaling limitations. This review will explore these innovations and examine technical challenges to highlight the current state of MOF-based nanocomposites in heat exchanger applications. By addressing the gaps in existing research, this paper seeks to provide a comprehensive overview of the field and propose future research directions, including potential material modifications, synthesis techniques, and structural optimizations. This review is intended to serve as a resource for researchers and engineers, aiming to advance the practical applications of MOF-based nanocomposites in energy-efficient, sustainable heat exchanger designs.

The growing demand for energy-efficient thermal management systems, particularly in industries reliant on heat exchange processes, has driven interest in innovative materials like MOFs due to their exceptional thermal properties and tunable functionalities. However, despite significant advancements in the synthesis and application of MOF-based materials, there remains a fragmented understanding of their performance and practical integration into industrial heat exchangers. This review provides a comprehensive examination of the current state of MOFs in enhancing heat exchanger efficiency, addressing not only their potential but also the challenges in scaling these materials for real-world use. Section 2 delves into the impact of MOFs on heat exchanger performance, examining their role in improving heat transfer and reducing fouling. Section 3 explores recent innovations and ongoing challenges in integrating MOF-based materials into heat exchangers, highlighting advancements in synthesis techniques, surface functionalization, and industrial feasibility. Finally, Section 4 identifies key research gaps and outlines future directions for the development of MOF-based nanocomposites, aiming to guide future research and accelerate the adoption of MOF technologies in energy-efficient applications. By addressing these critical areas, this review aims to bridge the gap between academic research and practical implementation, supporting the development of more sustainable and effective heat exchange solutions.

2. MOF-Based Nanocomposites for Heat Exchanger Performance

2.1. High-Performance MOF Nanocomposites for Thermal Conductivity

Incorporating MOFs into metals and polymers enhances thermal conductivity in heat exchangers by improving thermal transport pathways and reducing interfacial thermal resistance, often termed as Kapitza resistance, which arises at interfaces between materials with differing thermal properties, particularly at the nanoscale [53]. This phenomenon arises due to the mismatch in acoustic phonon properties—vibrational energy carriers—of the materials on either side of the interface [54]. When heat flows across the interface, phonons encounter impedance due to differences in lattice structure, phonon density of states, or material stiffness. This impedance reduces the efficiency of thermal energy transfer, creating a bottleneck for heat conduction [55]. Kapitza resistance is particularly significant in systems involving nanoscale materials, such as nanocomposites or interfaces in electronic devices, where the surface-to-volume ratio is high [56]. The effect of Kapitza resistance is a reduction in overall heat transfer efficiency, which can limit the performance of thermal management systems. For example, in electronic devices, high Kapitza resistance at material interfaces can lead to localized overheating, affecting reliability and longevity. Minimizing Kapitza resistance through techniques such as interface engineering, use of intermediate materials to bridge phonon mismatch, or applying thermal interface materials can improve heat transfer efficiency across interfaces [57]. By embedding MOFs, which possess a highly porous structure and customizable surface chemistry, researchers can create a network of micro- and nanopores that facilitates smoother energy transfer across interfaces [58,59]. By filling gaps at the interface, MOFs create a more uniform thermal contact area, which decreases Kapitza resistance. The smooth energy transfer results from the mitigation of phonon scattering caused by mismatches in acoustic properties between materials [60]. The ordered structure of MOFs ensures that heat conduction is not only through direct contact but also facilitated by the framework's intrinsic thermal conductivity. This supports efficient energy propagation.

The enhancement in thermal conductivity from MOFs is due to improved phonon transport, as MOFs create ordered pathways that reduce scattering events at material interfaces. This effect is particularly beneficial in polymer composites, aligning thermal pathways and promoting efficient heat conduction [61]. For example, incorporating MOFs with high thermal conductivity into polymers has been shown to enhance their thermal conductivity, making these composites suitable for high-performance thermal applications [62]. Thermal conductivity remains a critical factor in emerging technologies and applications, including heat exchangers, aerospace, and mechanical instrumentation. Many advanced smart materials used in electronic devices generate heat during operation, leading to temperature increases that can degrade their performance. Elevated temperatures may also compromise the mechanical stability and functionality of polymeric components. To address this challenge, various strategies have been proposed to improve the thermal conductivity of polymers, with most focusing on the integration of thermally conductive nanoparticles [63].

Certain MOFs, such as MIL-101, UiO-66, and HKUST-1, have shown remarkable promise in enhancing thermal conductivity when incorporated into heat exchanger materials. MIL-101, with its large pore structure and high BET surface area of 3873 m^2/g, facilitates efficient heat transfer and has been widely studied for improving thermal transport in composite systems [64,65]. MIL-101 has thermal conductivity ranging from 0.83 to 0.86 W/m-K. UiO-66, known for its exceptional stability and tunability, enables strong thermal conductivity enhancements, particularly in polymer composites [66,67]. UiO-66 has a BET surface area of 1041 m^2/g. HKUST-1, composed of copper nodes and organic linkers, also demonstrates superior thermal properties, helping reduce interfacial thermal resistance in metal composites [68,69]. These MOFs contribute significantly to optimizing heat exchanger efficiency through enhanced thermal pathways. Several MOFs are used in adsorption heat pumps, and they have different water uptake. Some of these can be seen in Figure 5, for which the original data values have been extracted from [70–82].

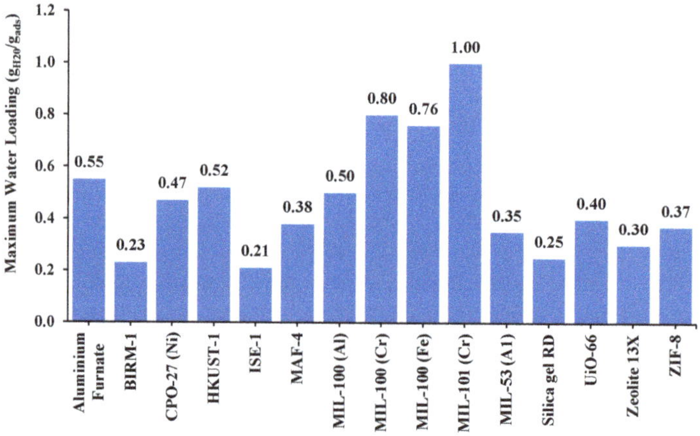

Figure 5. Maximum water vapor uptake (g_{H20}/g_{abs}) of various MOFs used in heat pumps.

Thermal conductivity is a critical property for materials used in heat transfer applications. The experimental measurement of MOF conductivity is highly influenced by pore size. A summary of some results on the thermal conductivity of MOFs is presented in Table 1. Similarly, Figure 6 illustrates that the Brunauer–Emmett–Teller (BET) surface area of MOFs significantly surpass those typically found in conventional porous materials like silica gel and some widely used metal oxides. The original data have been gathered from [83–97].

Table 1. Thermal conductivity values of various types of MOFs, as determined in different studies.

S. No.	MOF	Thermal Conductivity in W/(mK)	Main Findings	Ref.
1	Aluminum Fumarate	0.07	• Silres® was used as a binder in thicknesses from 140 to 610 μm, and it was extensively studied at 30–60 °C. • Aluminum sample plates coated with aluminum fumarate and a silicon binder demonstrated conduction heat transfer mechanisms within the coating layer (0.07 W/mK). • The samples exhibited consistently high performance.	[98]

Table 1. *Cont.*

S. No.	MOF	Thermal Conductivity in W/(mK)	Main Findings	Ref.
2	Cu-BTC UiO-66 UiO-67	0.39 0.11 0.19	• Increasing the pore size leads to higher thermal conductivity, as it reduces the number of resistive junctions per unit length. • The inter-particle contact conductance values for UiO-66, UiO-67, and Cu-BTC are 0.0117 μW/K, 0.09 μW/K, and 0.82 μW/K, respectively.	[99]
3	HKUST-1	0.44–0.73	• The thermal conductivity of HKUST-1 is reduced by 40–80%, influenced by the presence of adsorbates such as water, methanol, or ethanol.	[100]
4	Idealized MOF structures featuring various pore shapes, including cubic pores, triangular channels, and hexagonal channels	0.03	• MOFs with smaller pores exhibit lower thermal conductivity due to phonon scattering caused by interactions between gas molecules and the crystal structure. • Conversely, in larger pores (over 1.7 nm), the presence of adsorbed gas has minimal impact on thermal conductivity.	[101]
5	MIL-101/20% Few Layer Graphene (FLG)	0.8322–0.8603	• These observations were made at 30–130 °C. • Observed thermal conductivities were 5.79–6.54 times that of pure MIL-101.	[102]
6	MOF-5	0.1218–0.1477	• In this study, the transient plane source technique was used. • The specific heat capacities of pure MOF-5 at low temperatures were observed for a range between 93 and 313 K.	[103]
7	MOF-5	0.32	• The value was determined at 300 K. • The thermal conductivity increases with temperature in the range of 6–300 K, reaching a peak around 20 K.	[104]
8	MOF-5	0.33	• They observed that the extremely low thermal conductivity is attributed to intense phonon–phonon scattering caused by the dense and interwoven low-frequency phonons.	[105]
9	Ni_3(2,3,6,7,10,11-hexaiminotriphenylene)$_2$ ($Ni_3(HITP)_2$) a 2D MOF	0.21	• The steady-state method was used in this research study. • The ultralow thermal conductivity is attributed to several factors: (a) phonons, which represent lattice vibrations; (b) the heterogeneity of atomic masses and bond stiffness, which leads to phonon scattering; (c) the disordered arrangement of individual layers, resulting in additional phonon scattering; and (d) the very small particle size of crystallites and grain boundaries, which further contribute to phonon scattering.	[106]
10	Tetracyanoquinodimethane (TCNQ) @HKUST-1	0.23–0.31	• The incorporation of TCNQ enhances thermal conductivity by increasing the phonon density in the low-frequency/high-group-velocity range.	[107]
11	ZIF-8	0.165–0190	• With the temperature increment from 300 to 1000 K. • In ZIF-8, longitudinal vibrations account for 60% of thermal transport, while transverse vibrations contribute 40%.	[108]
12	ZIF-8(H)ZIF-8 (Cl)ZIF-8 (Br)ZIF-8 (CH3)	0.165 0.138 0.142 0.174	• The presence of functional groups such as -H, -CH3, -Cl, and -Br leads to a decrease in thermal conductivity, primarily due to the damping effect caused by acoustic mismatch.	[109]

Table 1. *Cont.*

S. No.	MOF	Thermal Conductivity in W/(mK)	Main Findings	Ref.
13	ZIF-8	0324–0.328	• In this study, the transient plane source technique was used. • Thermal conductivity remains nearly constant regardless of the ambient gas and pressure, ranging from atmospheric pressure to a vacuum.	[110]

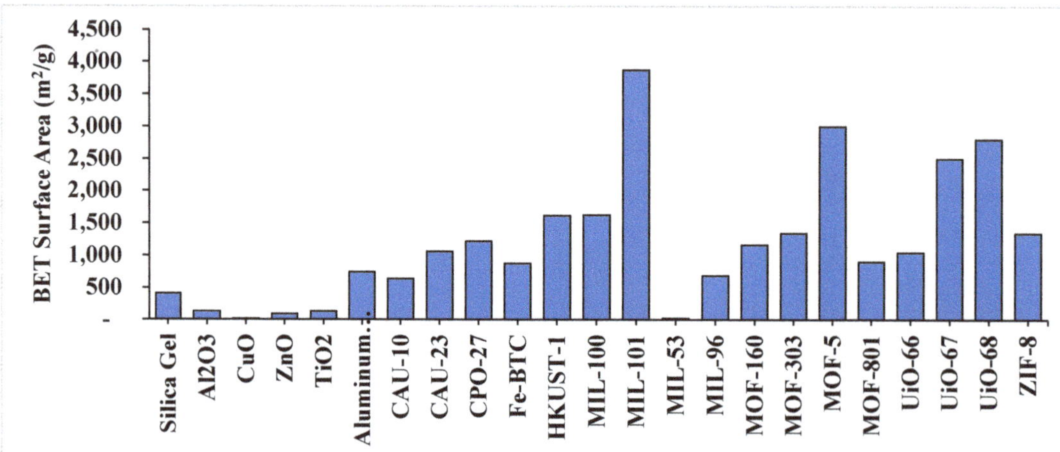

Figure 6. A schematic representation of the ultrahigh porosity of MOFs in heat transfer systems, highlighting their surface area in comparison with commonly used silica gel and commonly used metal oxides.

Embedding MOFs within metal and polymer matrices is an effective strategy to leverage the thermal properties, mechanical strength, and structural tunability of MOFs. These composites are especially valuable in heat exchanger applications, where both thermal conductivity and mechanical stability are critical. This section reviews key methods for embedding MOFs into metal and polymer matrices, highlighting techniques like in situ growth, physical blending, and surface functionalization. Figure 7 illustrates the MOF synthesis method over the years [111].

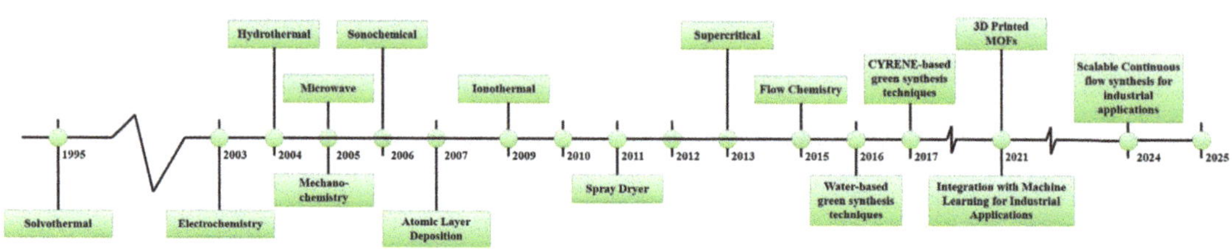

Figure 7. The development of MOF synthesis method over the years (updated and reproduced with permission from the Royal Society of Chemistry).

In situ growth is a widely used approach for embedding MOFs in both metal and polymer matrices. In this method, MOF crystals form directly within the matrix material, resulting in a composite with strong interfacial adhesion and uniform distribution.

For metal matrices, in situ growth involves immersing the metal substrate in a solution containing MOF precursors, allowing MOF crystals to nucleate and grow on the metal surface. This process promotes thermal conductivity, as it creates continuous thermal pathways between the MOF and metal, reducing thermal boundary resistance [112]. In polymer matrices, in situ growth can involve dissolving MOF precursors within a polymer solution, which leads to MOF crystallization as the polymer matrix solidifies. This method ensures a well-integrated composite structure with enhanced mechanical properties and homogeneity [113].

Physical blending involves mixing pre-synthesized MOF particles with metal or polymer matrices, often through mechanical blending or extrusion. While cost-effective and scalable, this method faces challenges in achieving uniform dispersion and strong interfacial bonding, which can limit thermal transport enhancement [114]. In polymer matrices, physical blending typically involves dispersing MOF particles in a polymer solution or melt, followed by casting or extrusion to form a composite. Advanced mixing techniques, such as ultrasonic dispersion or high-shear mixing, are often employed to improve the distribution of MOFs within the polymer, minimizing aggregation and enhancing the composite's thermal and mechanical properties [115]. Additionally, physical blending offers flexibility in choosing MOF–polymer combinations, allowing for fine-tuning of composite properties according to specific heat exchanger applications.

Surface functionalization enhances compatibility and bonding between MOFs and polymer matrices by introducing functional groups like -NH_2 or -COOH. This strengthens the interface, improves mechanical properties, and reduces thermal resistance, optimizing thermal transport in the composite [116]. For metal matrices, surface functionalization can involve coating the MOF particles with metal-compatible agents, which encourages better integration and reduces particle agglomeration within the metal host. This method has been shown to significantly enhance the thermal conductivity and stability of MOF–metal composites under high-temperature conditions. Studies have demonstrated that MOFs like MIL-101 and UiO-66, when functionalized and embedded in aluminum or copper matrices, exhibit superior thermal performance and robustness, making them highly suitable for heat exchanger applications [117].

Electrochemical deposition is a more specialized method used primarily for embedding MOFs in metal matrices. In this process, a metal or alloy is electrodeposited onto a surface along with MOF particles in a controlled environment, ensuring uniform distribution and integration. Electrochemical deposition creates a composite with excellent interfacial adhesion and controlled MOF loading, which is beneficial for enhancing thermal pathways within the material. This method is advantageous for applications requiring precise control over the composite's microstructure, and it has been successfully applied in embedding MOFs like ZIF-8 within copper and aluminum matrices for enhanced thermal conductivity in heat exchangers [118–120].

Each method for embedding MOFs in metal and polymer matrices has unique advantages and challenges. In situ growth ensures strong interfacial bonding but requires precise control of MOF growth conditions, while physical blending offers simplicity but may suffer from non-uniform distribution. Surface functionalization enhances compatibility and integration, especially in polymer matrices, but adds complexity in terms of processing. Electrochemical deposition, though specialized, allows for precise structural control in metal matrices. By carefully selecting and optimizing these embedding techniques, researchers can develop MOF-based composites that meet the rigorous thermal and mechanical demands of advanced heat exchangers [121–123].

2.2. MOF–Polymer and MOF–Metal Nanocomposites

2.2.1. Advantages and Applications of MOF–Polymer Composites

The integration of MOFs with polymers has shown substantial potential in enhancing the efficiency and performance of heat exchangers, particularly for compact systems like HVAC units. The combination leverages the inherent benefits of polymers, such as flexibility, lightweight structure, and mechanical robustness, alongside the superior surface area and thermal properties of MOFs [124]. This synergy not only improves thermal conductivity but also enhances moisture management, making these composites highly efficient for energy-intensive applications [125]. MOF–polymer composites, such as those involving ZIF-8 embedded within polyimides (PI), have been extensively studied for their enhanced heat transfer capabilities. ZIF-8 is known for its high porosity, stability, and large surface area, which allows for efficient heat dissipation. When ZIF-8 is incorporated into polyimide matrices, it significantly improves thermal conductivity while preserving the mechanical flexibility of the polymer [126]. Some of the prominent MOF–Polymer composites are listed in Table 2, along with their applications.

Table 2. Prominent MOF–Polymer Composites and their applications in different fields.

S. No.	Composite Name	Description	Applications	Ref.
1	ZIF-8/Polyimide (PI) Composite	ZIF-8 embedded in polyimide enhances thermal conductivity while maintaining the flexibility and robustness of the polymer.	HVAC systems for efficient heat transfer and moisture management, as well as in gas separation applications.	[127–129]
2	MIL-101(Cr)/Polysulfone (PSU) Composite	MIL-101(Cr) combined with PSU improves mechanical strength and thermal stability. The high surface area of MIL-101(Cr) contributes to superior adsorption capabilities.	Air purification, nano-filtration, and gas separation.	[130–132]
3	UiO-66/Polyethylene Oxide (PEO) Composite	UiO-66 incorporated into PEO enhances thermal conductivity and mechanical properties, along with increasing moisture resistance.	Lithium-ion batteries, proton exchange membranes, drug delivery, and water filtration systems.	[133–135]
4	MOF-5/Polymethyl Methacrylate (PMMA) Composite	MOF-5 embedded in PMMA improves the thermal and mechanical properties of the polymer, enhancing its use in structural applications.	Lightweight thermal insulation panels and coatings.	[136,137]
5	HKUST-1/PDMS Composite	HKUST-1 integrated with PDMS increases thermal conductivity while retaining the flexibility and elasticity of PDMS.	Heat exchangers, wearable electronics, and gas sensors.	[138,139]
6	MOF-801/Polyurethane (PU) Composite	MOF-801 embedded in PU improves thermal stability and adsorption properties, making it ideal for applications requiring flexibility and durability.	HVAC systems, energy-efficient coatings, and adsorptive cooling systems.	[140,141]

2.2.2. Applications and Advantages of MOF–Metal Composites

MOF–metal nanocomposites have gained significant attention for their enhanced thermal conductivity and mechanical durability, making them ideal for applications where efficient heat transfer is crucial. By combining MOFs with high-conductivity metals like copper and aluminum, researchers have developed composites that outperform traditional materials in heat exchanger and cooling systems [142,143].

Copper (Cu) is widely recognized for its excellent thermal conductivity (~400 W/m·K), making it an ideal candidate for integration with MOFs to improve heat transfer performance. Studies have shown that embedding MOFs like HKUST-1 (a copper-based MOF) into copper matrices results in composites with enhanced thermal transport properties [144]. The combination of aluminum with MOFs such as MIL-101(Cr) or UiO-66 results in materials that not only enhance heat transfer but also exhibit increased resistance to corrosion and thermal degradation [142]. This improvement is attributed to the high surface area and porosity of the MOF, which facilitates faster heat dissipation, especially in compact systems like HVAC units. Some of the relevant case studies related to heat transfer enhancement using MOFs are noted in Table 3.

Table 3. Relevant research studies in which MOFs are used for enhancement of heat transfer.

S. No.	MOF Used	Heat Exchanger Type	Main Findings	Ref.
1	Sodium polyacrylate	Single-row finned-tube	• Exchanger total fin area was 7.26 m^2. • Regeneration temperature of 50 °C was used. • Cycle Time in this study was 15 min. • Obtained Coefficient of Performance (COP) of 0.294, which was increased by 16.2%.	[145]
2	MIL-100 (Fe)	Rectangular finned heat sinks	• Exchanger total fin area was 1.452 m^2. • Regeneration temperature of 50 °C was used. • Cycle Time in this study was 15 min. • Obtained COP of 0.42, which is an almost 43% enhancement in comparison to [117].	[146]
3	Aluminum Fumarate MIL-100 (Fe) MIL-100 (Fe)/G MOF-303/G MOF-801 MOF-801/G	Wire-finned heat exchanger	• Exchanger total fin area was 0.9948 m^2. • Regeneration temperatures of 45, 60, and 70 °C were used. • Cycle Time in this study was 15 min. • Obtained COP of 0.62, 1.02, and 1.20 at regeneration temperatures of 45, 60, and 70 °C, respectively.	[147]
4	Aluminum Fumarate	Wire-finned heat exchanger	• Coated wire-finned heat exchanger produced Specific Cooling Power (SCP) of 682 W/kg. • Obtained COP of 0.32, which is almost 39% more than packed exchanger.	[13]
5	Aluminum Fumarate	Packed heat exchanger	• The optimal half-cycle time was observed at 250 s, with the temperatures of the evaporator, condenser, adsorption, and desorption water at 24 °C, 25 °C, 25 °C, and 85 °C, respectively. • Under these conditions, the SCP, Specific Desorption Water Production (SDWP), and COP were 226 W/kg, 8.66 L/kg/day, and 0.5, respectively.	[148]
6	MIL-100 (Fe)	Packed heat exchanger	• The MOF-coated heat exchangers delivered a cooling power density of 82 $W \cdot L^{-1}$. • The results show that the system achieved a high COP of up to 7.9 and reduced energy consumption by 36.1% compared to the conventional Vapor-Compression Air Conditioning (VCAC) system with reheating.	[51]
7	CAU-23 CAU-10 Co_2Cl_2 (BTDD)	Plate-type heat exchanger	• It was observed that MOF-based systems can achieve a COP that is 2.7 to 6 times higher than silica gel-based systems, with an optimal regeneration temperature range of 40 to 75 °C.	[149]
8	MIL-101 (Cr)	Cross-flow heat exchanger	• This study was carried out in the hot and humid climate of Qatar. • Throughout the entire cooling season, the hybrid system utilizing MIL-101 (Cr) was found to reduce thermal and electrical energy consumption by 17% and 48%, respectively, compared to the system using silica gel. • This resulted in a 27% decrease in operating costs, with a payback period of 11 years, based on the current market price of MIL-101 (Cr).	[150]

Table 3. *Cont.*

S. No.	MOF Used	Heat Exchanger Type	Main Findings	Ref.
9	CPO27 (NI) MIL100 (Fe) MIL-101 (Cr) Aluminum Fumarate silica gel	Cross-flow heat exchanger	• The results indicated that Aluminum Fumarate achieved the highest COP of 0.65, with a water removal rate of 12.65 g/kg dry air. • In contrast, MIL-101 (Cr) demonstrated the highest moisture removal rate of 15.99 g/kg dry air, but its COP was 0.44 when compared to silica gel and other MOF materials used in the study.	[151]
10	MOF–801 Aluminum Fumarate MIL-100 (Fe)	-	• The MOF–801/water pair can achieve a gross temperature lift of approximately 30 K with a COP of 0.4867, outperforming aluminum fumarate with a COP of 0.3903 and MIL-100 (Fe) with a COP of 0.0926.	[152]

2.3. MOF-Based Nanomaterials for Enhanced Fouling Resistance

Fouling is a persistent issue in heat exchangers, as the accumulation of unwanted deposits on surfaces reduces heat transfer efficiency, increases energy consumption, and leads to higher maintenance costs. Recent advancements in materials science have shown that coating heat exchanger surfaces with MOFs can mitigate fouling, especially in systems where water or other fluids are involved [153]. The unique structural properties of MOFs, such as their high surface area, tunable pore sizes, and selective adsorption capabilities, make them effective in minimizing fouling, thereby enhancing the efficiency and longevity of heat exchanger systems [154].

Hydrophobic MOFs, such as ZIF-8, have been particularly effective in reducing fouling by repelling water and other contaminants. The hydrophobic nature of ZIF-8 helps in preventing moisture-induced biofilm formation, which is a common problem in condenser units. Some of the studies related to fouling reduced by using MOFs are discussed in Table 4.

Table 4. Relevant research studies in which reduction in fouling is observed using MOFs.

S. No.	MOF Used	Main Findings	Ref.
1	ZIF-8@PVDF45 HKUST-1@PVDF45	• The coated membranes were experimentally evaluated in an air-to-air membrane enthalpy exchanger. • Results showed that the effectiveness increased from 44.4% for the original polyvinylidene fluoride (PVDF45) to 45.9% and 47.7% for the ZIF-8@PVDF45 and HKUST-1@PVDF45 membranes, respectively, at a flow rate of 0.24 L/s. • This improvement occurs due to improved hydrophilicity and a higher specific surface area, facilitating better moisture adsorption and transfer. These enhancements optimize the interaction between the air and the membrane, boosting the overall dehumidification and heat exchange efficiency. • This study highlights a simple and effective approach to improve the efficiency of polymer membranes in air-to-air enthalpy exchangers, potentially reducing energy use in buildings and lowering greenhouse gas emissions.	[129]
2	MIL-110	• MOF-based slippery surface infused with ionic liquid demonstrates exceptional antifouling capabilities. • The MOF-based slippery surface exhibited excellent antifouling performance, maintaining effectiveness in a 10-day trial and even under more rigorous conditions in a subsequent 21-day test. • The amount of lipopolysaccharide (LPS) adsorbed on the Slippery Lubricant-Infused Porous Surface (SLIPS) was reduced by 50% compared to a plain aluminum sheet.	[155]
3	CuZn-MOF-74 (PBMA)	• After submerging the coatings in the local seawater for two months, the untreated side was heavily covered with calcareous tubeworms and barnacles, whereas the prepared coatings showed significantly fewer macrofouling organisms. • A UV-curable poly butyl-methacrylate (PBMA) coating incorporating a low concentration of MOF-74 (e.g., 3 wt% to 6 wt%) has proven effective in preventing biofouling.	[156]

Table 4. *Cont.*

S. No.	MOF Used	Main Findings	Ref.
4	GO/ZIF-8 MOF	• Antifouling tests included Adsorption of Bovine Serum Albumin (BSA) and Anti-Biofouling Activity against Bacteria. • All the tests supported that GO/ZIF-8 MOF coating has enhanced antifouling characteristics.	[157]
5	PDMS@MOF@Cu	• Due to its super hydrophobic properties, the PDMS@MOF@Cu mesh demonstrates outstanding antifouling and self-cleaning capabilities, effectively resisting contamination from everyday drinks and dirt.	[158]

2.4. Applications in Specific Heat Exchanger Designs

2.4.1. Compact Plate Heat Exchanger

The compact design of plate heat exchangers (PHEs) is highly advantageous for applications demanding space efficiency, such as HVAC systems, automotive cooling, and chemical processing units. These exchangers consist of closely packed plates that maximize the surface area for heat transfer; enhancing the overall efficiency while minimizing space requirements remains a significant challenge, leading to decreased thermal performance and increased maintenance costs [159]. In recent years, utilization of MOFs in the design of heat exchangers has demonstrated promising results in enhancing heat transfer efficiency and reducing fouling [160]. PHEs consist of corrugated stainless steel plates sealed with gaskets to prevent fluid mixing and ensure efficient heat transfer via a counter-current flow design. Their modular construction allows thermal output adjustment, easy maintenance, and cleaning, making them ideal for industries like food processing, pharmaceuticals, shipbuilding, and oil and gas [159,161]. ZIF-8 MOFs have demonstrated a substantial improvement in thermal performance due to their ability to facilitate efficient thermal transport and fluid flow.

A study reported by Zhang et al. demonstrated ZIF-8/PI coated heat exchanger plates could achieve a significant enhancement in heat transfer rates while also reducing fouling due to the hydrophobic characteristics of the composite. Additionally, an experimental investigation revealed a compact PHE with a ZIF-8 composite led to a 12% increase in heat transfer efficiency compared to uncoated units. This improvement is attributed to the unique structure of ZIF-8 reducing surface energy, thus minimizing the adhesion of fouling agents [162]. Table 5 discusses some research studies, along with their key findings, related to PHE employing MOF as coating material.

Table 5. Research studies and their key findings related to PHE employing MOF as coating material.

S. No.	MOFs Used	Main Findings	Ref.
1	Aluminum Fumarate	• The research explored the absorption of refrigerants on solid absorbents, presenting a potentially eco-friendly solution for HVAC systems. • The study utilized frequency response analysis to assess the adsorption dynamics and equilibrium properties of water on MOF-based nanocoatings. • Aluminum plates coated with aluminum fumarate and a silicon binder demonstrated conduction-driven heat transfer mechanisms within the coating layer, achieving thermal conductivity close to 0.07 W/(m·K). • The calculated effective thermal resistance ranged from 1 to 4 $(m^2{\cdot}K)/kW$. • The decrease in thermal resistance is due to the highly porous structure and large surface area of MOFs, which enhance heat transfer by reducing thermal boundary resistance, while the silicon binder ensures uniform particle distribution and better thermal connectivity.	[98]

Table 5. *Cont.*

S. No.	MOFs Used	Main Findings	Ref.
2	CAU-23 CAU-10 Co_2Cl_2 (BTDD)	• MOF-based systems demonstrate a COP that is 2.7 to 6 times greater than traditional silica gel-based systems, depending on external environmental conditions. • These systems perform most efficiently at regeneration temperatures between 40 °C and 75 °C, allowing the utilization of low-temperature heat sources to enhance energy efficiency. • The improvement in COP with MOF compared to silica gel is due to MOFs' higher water adsorption capacity, faster adsorption kinetics, and the ability to regenerate efficiently at lower temperatures, reducing energy consumption.	[149]
3	MIL-100 (Fe)	• The humidity pump utilizing MIL-100(Fe) as the desiccant layer demonstrates significantly better dehumidification performance compared to silica gel-based systems. • Key performance metrics for the MOF-based pump include a dehumidification rate of 34.9 g/h, a moisture removal efficiency of 1.14 g/Wh, and a dehumidification COP of 0.46. These performance indicators are approximately double those achieved by silica gel-coated systems. • The results highlight the enhanced potential of MOF materials in applications requiring efficient humidity control. • The superior performance of MOF-based systems arises from their higher adsorption capacity, faster kinetics, and ability to regenerate at lower energy inputs. • MIL-100(Fe)'s tailored pore structure and hydrophilic properties allow it to adsorb and release water more efficiently, significantly boosting dehumidification rates, moisture removal efficiency, and dehumidification COP compared to silica gel.	[146]

2.4.2. Fin Tube Heat Exchangers

The integration of MOFs in fin tube heat exchanger designs offers considerable promise, especially in high-temperature and high-pressure environments. Fin tube exchangers are commonly used in industrial settings due to their ability to increase heat transfer surfaces, thereby enhancing efficiency. By embedding MOF–metal nanocomposites, such as those made with copper or aluminum, these heat exchangers can achieve even greater thermal performance while maintaining structural integrity under challenging conditions [160].

Sun et al. [163] carried out an experimental study under outdoor conditions in Singapore. The system achieved a moisture removal rate of 9.32 g/kg of dry air, with a thermal COP of 0.70 and a waste heat utilization efficiency of approximately 74.2%. To further boost energy efficiency, a heat recovery subsystem was implemented to capture waste heat from the exhausted regeneration air. By optimizing the pre-regeneration process to 3 min, the system's moisture removal was slightly reduced to 9.01 g/kg; however, this adjustment significantly enhanced the thermal COP and waste heat utilization to 1.34% and 86.5%, respectively. Similarly, an another study demonstrated that coating aluminum fins with CPO-27 (Ni) resulted in a 15% increase in heat transfer efficiency due to improved thermal conductivity and enhanced fluid flow across the heat exchanger surfaces. The porous nature of MOFs contributes to effective heat dissipation by reducing thermal resistance while maintaining excellent mechanical stability, even at high operating temperatures [164]. Table 6 discusses some research studies, along with their key findings, related to fin tube heat exchangers employing MOF as a coating material.

2.4.3. Wire-Finned Tube Heat Exchangers

The use of Metal–Organic Frameworks (MOFs) in wire-finned tube heat exchangers for HVAC and refrigeration systems has gained attention due to their exceptional moisture adsorption and desorption properties. MOFs are highly porous materials with large surface areas and tunable pore sizes, which make them highly effective in enhancing the efficiency of dehumidification processes, thus improving overall system performance [160]. Some of the studies found in the literature are discussed in Table 7, along with their key findings.

Table 6. Research studies and their key findings related to fin tube heat exchanger employing MOF as coating material.

S. No.	MOFs Used	Main Findings	Ref.
1	MIL-160 (Al)	• A computational fluid dynamics (CFD) approach was used to evaluate the performance of a MOF-based heat pump (MOF-HP), establishing interactions among humid air, desiccant material, and the heat sink. • The model incorporated fluid flow, heat transfer, and species transport subdomains and was validated against experimental data under various operating conditions. • A parametric analysis identified the optimal desiccant thickness for the system, recommending a range of 0.3 to 0.5 mm for improved geometry. • The optimized system configuration demonstrated an increase in the coefficient of performance (COP) by 1.51 to 1.59 times compared to the initial design. • The improved COP is attributed to the optimized desiccant thickness because it has minimal thermal resistance, which enhances heat and mass transfer efficiency, reducing energy losses and maximizing the adsorption–desorption cycle's effectiveness. • Experimental validation maintained discrepancies within 15%, confirming the accuracy and reliability of the developed model.	[165]
2	MIL-100 (Fe)	• The coating's hygroscopic performance was tested with varying binder types, binder mass concentrations, and MOF mass concentrations. • Optimal performance was achieved at 15 wt% PVP and 20 wt% MOF, resulting in a 51.71% improvement over silica coating. • The composite adsorbent's cost was reduced by 78.04% compared to pure MOF fabrication. • The moisture removal capacity (MRC) of the DCHE was 6.740 g/kg, 149.63% higher than that of silica coating. • The improvement in MRC is due to the synergistic effect of optimized binder and MOF concentrations, enhancing the material's water adsorption capacity and reducing structural limitations associated with silica coatings.	[166]
3	Aluminum Fumarate	• A full-scale functional heat exchanger coated with microporous aluminum fumarate achieved a maximum COP of 0.72 at an evaporator temperature of 22 °C. • For data center cooling, operating with heat rejection at 35 °C and delivering useful cold at 18 °C, the system demonstrated an equilibrium cooling COP of 0.69. • The increase in COP results from the MOF coating's enhanced adsorption–desorption efficiency, which improves heat transfer and cooling performance compared to conventional heat exchangers.	[167]
4	Aluminum Fumarate	• The binder solution used for coating comprised a 10% binder–solvent mixture and 90% aluminum fumarate as MOF. • Increasing the coating thickness from 0.05 mm to 0.2 mm resulted in an average enhancement of 113% in the dehumidification performance. • Extending the operating time from 150 s to 300 s led to a 131% improvement in the average dehumidification performance. • The enhancements are due to the increased MOF coating thickness and extended operating time, which maximize the material's adsorption capacity. Additionally, reducing space velocity increases the contact time between air and the MOF, further improving moisture adsorption efficiency.	[168]

Table 7. Key results of research studies on using MOFs in Wire-Finned Tube Heat Exchangers.

S. No.	MOFs Used	Main Findings	Ref.
1	Aluminum Fumarate MIL-100 (Fe) MIL-100 (Fe)/G MOF-303/G MOF-801 MOF-801/G	• Wire-finned tubes coated with MOF-303/G demonstrated superior performance compared to those coated with Aluminum fumarate, MIL-100(Fe), MIL-100(Fe)/G, MOF-801, and MOF-801/G, achieving a water uptake of 22%. • The COP of the multi-tube heat exchanger coated with MOF-303/G was experimentally evaluated and compared to other desiccant-coated heat exchangers. The results indicated that the COP ranged from 0.6264 to 1.02 to 1.2 at regeneration temperatures of 45 °C, 60 °C, and 70 °C, respectively. • MOF-303/G exhibited superior water uptake and COP due to its highly porous structure and exceptional hydrophilic properties, which enhance water adsorption efficiency. Additionally, its optimized thermal conductivity ensures effective heat transfer during regeneration, enabling higher COP at lower regeneration temperatures compared to other MOFs.	[147]

Table 7. *Cont.*

S. No.	MOFs Used	Main Findings	Ref.
2	Aluminum Fumarate	• The wire-finned heat exchanger with a specialized coating achieved a Specific Daily Water Production (SDWP) of 23.5 m^3/ton/day, outperforming the packed heat exchanger, which reached only 12.7 m^3/ton/day. • In adsorption cooling applications, the coated wire-finned heat exchanger demonstrated a SCP of 682 W/kg and a COP of 0.32, significantly higher than the packed heat exchanger, which recorded 318.5 W/kg and a COP of 0.23. • Aluminum fumarate demonstrated superior water uptake and COP due to its high surface area and favorable adsorption isotherms, which enhance moisture capture and energy efficiency. Its effective integration into the wire-finned heat exchanger ensures improved heat and mass transfer, resulting in higher SCP and SDWP compared to packed heat exchangers.	[13]
3	Aluminum Fumarate	• A heat-powered adsorption heat pump utilizing MOF adsorbent material was applied in a combined desalination and cooling system. • The cooling unit generated over 6.9 kW of cooling power, achieving a COP of 0.26. This performance was obtained using a coating thickness of 0.75 mm on the wire-finned tube heat exchanger adsorber beds. • Aluminum fumarate's high adsorption capacity and efficient heat transfer properties contributed to its enhanced performance.	[169]
4	MIL-101 (Cr)	• The validated model was employed to examine the impact of fin height, fin spacing, and tube diameter on SCP and SDWP using silica gel. • The results demonstrated that reducing the fin height significantly enhanced both SCP and SDWP, reaching peak values of 1.3226 kW/kg and 23 L/kg/day at a fin height of 5 mm. • Additionally, decreasing the fin spacing from 11.5 mm to 3.6 mm led to an improvement in SCP and SDWP by approximately 25%. • The improvements in SCP and SDWP occurred because reducing the fin height and fin spacing enhanced the heat and mass transfer rates. A lower fin height reduced thermal resistance for adsorption, while closer fin spacing improved airflow distribution, enabling more efficient moisture removal and cooling performance.	[170]

3. Innovations and Challenges in MOF-Based Heat Exchanger Applications

3.1. Innovations in Synthesis and Fabrication

In recent years, significant innovations in the synthesis of MOF-based nanocomposites have focused on improving scalability and cost-effectiveness, crucial for their industrial adoption.

3.1.1. Scalable Synthesis Method

Traditional synthesis methods, such as solvothermal processes, have been widely used to produce MOFs, but their scalability has often been limited by high energy consumption and lengthy reaction times [171]. To address these limitations, microwave-assisted synthesis has emerged as a promising alternative. This method accelerates reaction times and offers better control over particle size and morphology, resulting in high-quality MOFs with improved performance in heat exchanger applications [172,173].

Solvothermal methods, though widely employed for laboratory-scale production, have seen advancements in recent years to improve scalability. Researchers have developed optimized protocols to enhance the synthesis efficiency and yield, making solvothermal processes more suitable for large-scale production [174]. This advancement, coupled with the development of new MOF precursors and solvents, has allowed for a more controlled and reproducible production process, enabling MOFs to be produced at a commercial scale without compromising material quality [175,176].

Microwave-assisted synthesis has revolutionized the production of metal–organic frameworks (MOFs), making it more cost-effective and efficient, especially for industrial applications [177]. Traditionally, the synthesis of MOFs using solvothermal methods is time-consuming, often requiring several hours to days, along with significant energy consumption [178]. In contrast, microwave-assisted synthesis dramatically reduces reaction times to just a few minutes, while also lowering the energy input, making it a more sustainable and economically viable method for large-scale production provided by microwave radiation, which leads to uniform nucleation and accelerated crystallization rates, resulting in MOFs with consistent particle size and morphology, and enhanced crystalline characteristics. This uniformity is crucial for industrial applications where consistent material properties are needed to ensure reliable performance, especially in heat exchanger systems where thermal conductivity and adsorption capacity are critical [179,180]. Figure 8 represents the schematics for the operating principles of the widely used synthesis methods for nanoparticles.

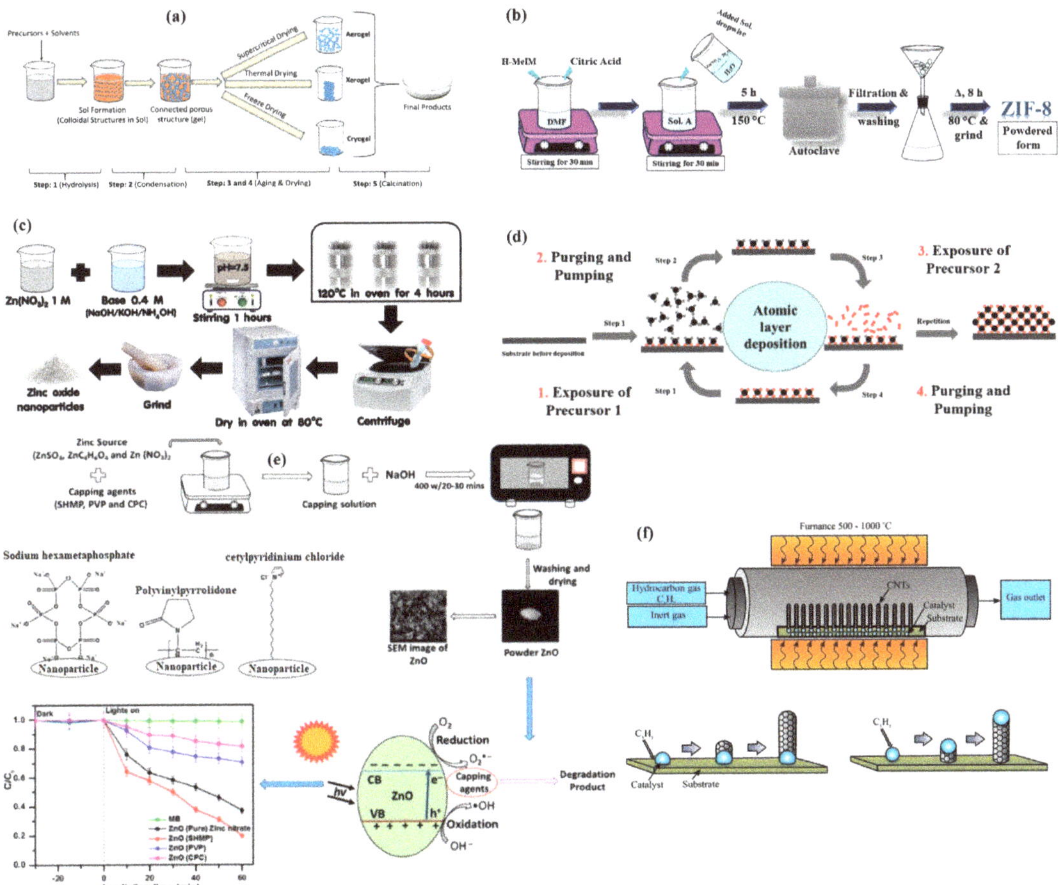

Figure 8. Operating principles for (**a**) sol–gel process to synthesize metal oxide nanoparticles [181], (**b**) solvothermal synthesis process for MOF (ZIF-8) [182], (**c**) hydrothermal process for synthesis of metal oxide (ZnO) nanoparticles [183], (**d**) atomic layer deposition (ALD) synthesis [184], (**e**) microwave-assisted synthesis of metal oxide nanoparticles [185], and (**f**) chemical vapor deposition (CVD) synthesis of CNTs along with their base and tip growth mechanism [186] (reprinted with permission from Springer Nature, Wiley, and MDPI).

3.1.2. Surface Functionalization Method

Surface functionalization of MOFs has gained significant attention as a means to enhance their properties for heat exchanger applications. By grafting specific functional groups onto the surface of MOFs, researchers can improve thermal conductivity, mechanical strength, and chemical stability, making these materials more suitable for demanding industrial environments [187]. Such techniques can be broadly categorized into structural modifications and post-synthetic functionalization techniques. Techniques such as ligand exchange, metal ion exchange, and annealing can improve stability by strengthening the coordination bonds or modifying the MOF's surface properties [188]. For instance, ligand exchange with hydrophobic molecules has been shown to protect MOFs against moisture degradation [189]. Figure 9 represents several strategies through which properties of MOFs can be tailored to attain stability.

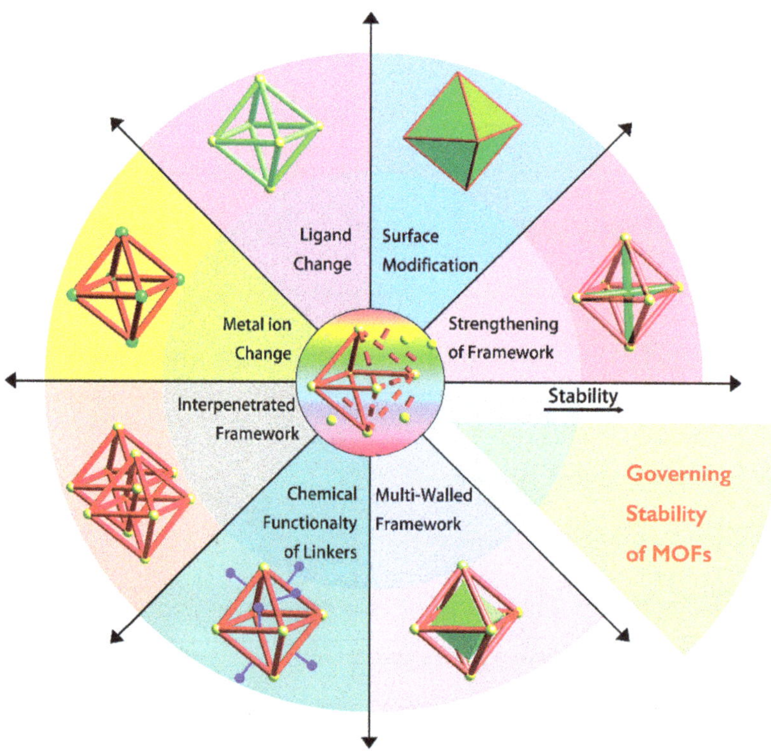

Figure 9. Various techniques used to attain stability of MOFs [190] (reprinted with permission from Royal Society of Chemistry).

Functionalization with alkyl or amine groups has been shown to enhance thermal conductivity by creating better interfaces between the MOF particles and the surrounding matrix. In composite material, ZIF-8 functionalized with amine groups exhibited better integration into polymer matrices, leading to increased mechanical strength and durability, which are essential for heat exchanger components that operate under variable thermal cycles [191]. Table 8 discusses some of the research studies carried out for enhancement in thermal properties of MOFs using the Surface Functionalization Method.

Table 8. Several research studies regarding Surface Functionalization Method to improve properties of MOFs.

S. No.	MOF Used	Function Group Added	Main Findings	Ref.
1	UiO-66 (Zr) MIL-125 (Ti)	amino (-NH_2)	• The amino-functionalized MOFs, H_2N-UiO-66 (Zr) and H_2N-MIL-125 (Ti), demonstrate high adsorption enthalpies (89.5 and 56.0 kJ/mol, respectively) and exhibit an exceptionally favorable water adsorption isotherm due to their increased hydrophilicity. • In particular, H_2N-MIL-125 shows a notably steep increase in its water adsorption isotherm, making it highly advantageous for heat pump applications.	[77]
2	MIL-101-NH_2	(Cr)	• A new fatty acid@MOF composite phase change material (PCM) has been developed, achieving shape stabilization for the first time. • This innovation addresses common challenges associated with traditional PCMs, such as leakage and significant volume changes during phase transitions. • The SA@Cr-MIL-101-NH_2 composite demonstrated excellent thermal stability and long-lasting durability, making it a promising candidate for renewable energy storage applications.	[192]
3	MIL-53	(Al)	• The modified MIL-53 (Al) MOFs were found to adjust both hydrophilicity and hydrophobicity characteristics, enhancing water adsorption capacity to as much as 0.9 g/g. • The ligand-extended MIL-53 (Al) variants demonstrated superior water transfer efficiency between humid (relative humidity of 80–90%) and regeneration conditions (RH of 30%). • They also achieved a substantial Thermal Energy Storage Density (TESD) of up to 1.54 MJ/L.	[193]
4	HKUST-1	(Cu)	• In this study, HKUST-1 was successfully grown on copper nanofibers using a homologous Cu self-transformation growth method, resulting in a uniform and tightly bonded MOF coating. • The composite retained a high surface area of 1473 m^2/g. • Consequently, the HK@Cu-NF composite, with an 80% loading of HKUST-1, demonstrated thermal diffusivity over three times greater than that of pure HKUST-1.	[194]
5	MIL-125 (Ti)	hydroxyl (-OH), amino (-NH^2), nitro (-NO_2), bromo (-Br), and pyridine (-C_5H_5N)	• The experimental findings indicate that (i) OH-MIL-125(Ti) exhibits adsorption and desorption kinetics that are two to three times faster than those of the unmodified MOF, and (ii) NH_2-MIL-125(Ti) demonstrates the most favorable water adsorption characteristics, particularly regarding maximum water capacity and efficient water transfer. • Additionally, the introduction of the NO_2 functional group increases the hydrophobic chain length, suggesting its suitability for desiccant-based dehumidification systems.	[195]

3.2. Challenges in Industrial Integration

3.2.1. Cost and Feasibility

Integrating MOFs into industrial applications, particularly in heat exchange systems, faces significant challenges, with cost being one of the most critical barriers. The synthesis, fabrication, and incorporation of MOFs into large-scale industrial processes remain expensive, largely due to the complex and energy-intensive methods traditionally used, such as solvothermal and hydrothermal synthesis. These methods require high temperatures and prolonged reaction times, leading to increased production costs [196]. Additionally, the precursor materials for MOFs, especially metals and organic linkers, are not always readily available with the desired stability, porosity, and mechanical robustness to withstand industrial conditions, such as high pressure and temperature, further driving up costs. For instance, integrating MOFs into polymer or metal matrices to enhance their thermal and mechanical properties involves additional steps like surface functionalization, which can increase both material and labor expenses [197]

Severino et al. [198] investigated the feasibility of using safe and economical batch production techniques to scale up the synthesis of MOF MIL-160(Al) for industrial applications. They developed a straightforward approach to estimate the production cost of this MOF, utilizing data gathered from laboratory pilot-scale experiments. This cost evaluation, for the first time, considered all critical process factors, including production scale, raw material expenses, recirculation, and washing procedures. According to their findings, the projected production costs could range from approximately USD 55 per kg for an annual output of 100 tons to about USD 29.5 per kg if scaled up to 1000 tons per year. Similarly, Shahvari et al. [199] explored the impact of substituting traditional solid sorbent materials like silica gel with various metal–organic frameworks (MOFs). Their study revealed that the top-performing MOFs among those tested—specifically, Co_2Cl_2 (BTDD) (Co), Aluminum fumarate, and CAU-23 (Al)—could be regenerated at significantly lower temperatures, ranging between 40 and 60 °C, depending on ambient conditions. This is a notable improvement over silica gel, which typically requires regeneration temperatures between 80 and 140 °C. Furthermore, the energy consumption for a MOF-based sorbent wheel was found to be substantially lower, ranging from half to as little as one-eighth of that needed for systems utilizing silica gel, varying with environmental conditions.

Cost–benefit analyses in the literature indicate that the initial capital expenditures are high, yet the long-term operational savings can justify the investment. For example, research discussed earlier reduced energy consumption up to 50%, resulting in substantial cost savings over time. As the technology scales and newer, more efficient synthesis methods, such as microwave-assisted synthesis, become more prevalent, the costs associated with MOF production are expected to decrease, potentially making industrial integration more feasible [200].

3.2.2. Long-Term Stability and Durability

Ensuring the long-term stability and durability of MOF-based nanocomposites under real-world operational conditions remains a significant challenge, particularly in harsh environments such as high temperatures, elevated pressures, and corrosive atmospheres. These conditions can degrade the structural integrity of MOFs, potentially reducing their performance over time. The inherent porous structures of MOFs, while advantageous for adsorption and heat transfer applications, are also prone to collapse or degradation when exposed to extreme conditions [201]. Addressing these challenges is essential for their practical use in industrial systems like heat exchangers and HVAC systems.

Recent studies have demonstrated efforts to enhance the stability of MOFs through surface modifications and structural tuning. Cai et al. found that the incorporation of stabilizing agents, such as graphene oxide, into MOF composites like MOF-808/GO resulted in enhanced resistance to thermal and mechanical stress. This hybrid structure improved the overall durability of the composite, particularly under conditions of fluctuating humidity and high heat [202]. Similarly, HKUST-1, while widely used due to its high surface area, suffers from instability in humid conditions, as water molecules can disrupt its metal–ligand bonds [203]. Research has proved that introducing hydrophobic functional groups or creating composite structures with polymers provides additional stability [204].

Despite these advancements, research is needed to develop MOFs that can withstand prolonged exposure to harsh conditions without compromising their adsorption capacity or thermal properties. There is ongoing work in the area of composite materials, where embedding MOFs into metal or polymer matrices helps in enhancing their structural integrity while retaining their desirable properties. These hybrid composites show promise in extending the operational life of MOFs in demanding applications, but scalability and cost remain barriers that must be addressed for widespread industrial adoption.

4. Research Gaps and Future Directions for MOF-Based Nanocomposites in Heat Exchangers

Advancements in material selection can lead to substantial improvements in performance across various applications. For Desiccant-Coated Heat Exchanger (DCHE) systems, the ideal desiccant should possess a high water adsorption isotherm, rapid water uptake kinetics, and regeneration temperatures close to or below room temperature. These properties help minimize the need for excess heat during desiccant regeneration, even after numerous operational cycles, ensuring energy efficiency in DCHE applications [205–207]. Moreover, the desiccant should be cost-effective, environmentally safe, and support recycling for sustainable use. Early materials like silica gel, activated alumina, and AC were widely used due to their high surface areas and effective water adsorption characteristics. To further enhance performance, composites were developed by using these porous materials as a foundational matrix, infused with hygroscopic salts, significantly boosting their adsorption capacity—often doubling or even quadrupling it [208,209]. However, increasing the salt content may lead to challenges such as corrosion and deliquescence, limiting the effectiveness of hygroscopic salt impregnation. This drawback did not hinder further research, which eventually led to the development of more advanced materials like MOFs. MOFs can be precisely engineered for specific properties, offering significant advantages. Yet, their high costs currently prevent widespread commercial use for large-scale applications. Nonetheless, advancements in the commercialization process and scaling up production could reduce costs, enabling efficient, high-speed manufacturing that would make MOFs more economically viable [196,210].

4.1. Multifunctional MOFs

The development of multifunctional MOF-based composites by integrating them with advanced materials, such as CNT and GO, represents a promising frontier in heat exchanger technology. These hybrid nanocomposites offer a combination of enhanced properties like superior heat transfer, efficient gas adsorption, and catalytic activity, making them ideal for next-generation heat exchangers. Due to their inherent characteristics of both MOFs and carbon-based materials, researchers aim to achieve optimal thermal conductivity, structural stability, and adsorption capacity [211].

For instance, incorporating CNTs into MOFs can significantly improve their thermal conductivity. A study demonstrated that combining CNTs with MOFs such as HKUST-1 resulted in enhanced heat transfer efficiency due to the high thermal conductivity of CNTs, which compensates for the low thermal conductivity typically associated with pure MOFs. Furthermore, MOF composites have shown superior thermal stability and mechanical strength, crucial for use in harsh industrial environments. Research involving MOF-5 combined with reduced graphene oxide (rGO) reported improved adsorption performance and heat transfer efficiency, making it suitable for energy-efficient HVAC systems [212]

These hybrid composites are multifunctional beyond heat transfer. For example, integrating MOF-74 with graphene oxide has been explored for applications that require both gas adsorption and thermal management [213]. This combination allows for effective heat dissipation while maintaining high adsorption capacity, which is advantageous in heat exchangers used for gas separation processes. Additionally, functionalizing MOFs with conductive materials enhances their electrical conductivity, which can be beneficial in thermoelectric applications [214]. Table 9 represents some multifunctional MOFs, along with their enhanced properties and applications.

Table 9. Enhancements in properties of multifunctional MOFs, along with their applications.

S. No.	Multifunctional MOFs	Enhancement in				Applications	Ref.
		Thermal Properties	Mechanical Stability	Electrical Conductivity	Adsorption Characteristics		
1	HKUST-1 with CNT	✔	✔	-	-	• Heat Exchangers • HVAC	[215]
2	MOF-5 with GO	-	✔	✔	✔	• Gas Separation • Electrocatalysis	[214]
3	MOF-74(Ni) with GO	-	✔	-	✔	• Gas Adsorption and Separation	[216]
4	MIL-101 (Cr) with AC	-	✔	-	✔	• Gas Separation • Carbon Capture	[217]
5	MIL-100 (Fe) with CNT	✔	✔	-	✔	• Gas Separation • Wastewater Decontamination	[218,219]
6	MOF-808 with CNT	-	✔	-	✔	• Gas Adsorption and Separation • Pharmaceuticals • Agrochemical Wastewater Treatment	[220,221]
7	MIL-53 with Graphene Oxide	✔	✔	-	✔	• Dye Adsorption • Photocatalyst	[222,223]
8	UiO-66 with CNT	-	✔	-	✔	• Photocatalyst • Membranes	[224,225]
9	Aluminum Fumarate	✔	✔	-	✔	• Dye Adsorption • Pharmaceuticals • Membranes	[226,227]
10	UiO-67 with Graphene Oxide	-	✔	-	✔	• Electrochemical Sensors • Energy Storage	[228]

Despite the promising performance of these multifunctional composites, challenges such as scalability, cost, and long-term stability under operational conditions remain. However, with continued research and the optimization of synthesis methods, the integration of MOFs with advanced materials could lead to the next generation of efficient and sustainable heat exchanger systems. As these hybrid materials continue to be tested and refined, they hold the promise of revolutionizing not only heat exchangers but also broader energy management and environmental technologies.

4.2. Bridging Gaps Between Research and Industry

Collaboration between academia and industry is crucial in advancing the field of MOFs research and ensuring that breakthroughs in laboratory settings translate into real-world applications. While academia excels in exploring the fundamental science behind MOFs, industry partners can bring the practical expertise needed for large-scale manufacturing, commercialization, and integration into existing systems. The synergy between these two sectors can accelerate the development of MOF technologies, making them more accessible and applicable for industrial processes like heat exchangers, gas storage, and separation.

One of BASF's recent innovations focuses on adsorptive dehumidification in air conditioning systems. Traditional air conditioners are tasked not only with cooling incoming air but also with controlling its humidity levels. Typically, these systems cool the air to a set temperature, causing moisture to condense. This dew pointing of air is highly energy-intensive and can account for up to 60% of the total energy usage of the unit, depending on environmental conditions. A more energy-efficient approach involves utilizing an adsorption-based system where moisture is selectively captured by a MOF. This method significantly lowers the electricity demand of air conditioning units. Under optimal conditions, using MOFs for air drying could reduce energy consumption by 50–60%, which translates to thousands of kWh savings annually for a standard office building. Compared to traditional materials like silica gel, MOFs offer 1.5 to 2 times higher volumetric energy density, allowing for more compact device designs, and provide a 27% improvement in the coefficient of performance. This not only enhances efficiency but also supports downsizing the equipment while achieving superior energy conservation [229]

However, adsorption systems come with their own energy costs. Once the MOF becomes saturated with moisture, it requires a desorption phase to regenerate the material and release the adsorbed water. This regeneration process can utilize thermal energy sourced from low-temperature heat supplies, such as solar thermal systems or waste heat produced by the air conditioner's own compressor. Such applications demonstrate that the distinctive characteristics of MOFs outperform those of current materials and technologies. As new use cases emerge, and production scales up, costs are expected to decline, making these materials more viable for broader applications. The continued reduction in expenses due to economies of scale will facilitate their integration into additional areas, further expanding their practical uses.

NuMat Technologies is a pioneer in the commercialization of MOF-based solutions, focusing on gas storage and separation technologies [230]. Their innovative systems utilize MOFs with high surface areas and tunable pore structures to store gases like methane and hydrogen efficiently. This makes them ideal for use in fuel cell vehicles and portable energy devices, where compact and lightweight storage is critical. NuMat Technologies has not only advanced MOF-based gas storage systems but has also addressed the challenges of cost-effectiveness and scalable production. By leveraging computational design and automation, NuMat reduces the time and expense associated with MOF synthesis, enabling large-scale production without compromising quality. Their manufacturing processes are optimized for industrial scalability, making MOFs more accessible for commercial applications. The cost-effectiveness of NuMat's technology is further highlighted in applications like hydrogen and methane storage for fuel cell vehicles, where lightweight and compact systems are essential. MOFs provide higher storage capacities compared to traditional materials, reducing the overall system size and cost. Additionally, NuMat's focus on modular production techniques ensures flexibility and adaptability for different industries [229]. These advancements underline how MOFs can transition from laboratory-scale research to practical, economically viable solutions, paving the way for widespread adoption in clean energy, gas separation, and other industrial applications.

MOFs could soon compete with established materials like activated carbons, zeolites, and silica in terms of cost and performance. At the very least, they may serve as crucial components in multi-material systems, such as polishing layers, depending on the specific application. Achieving this will require specialized expertise and robust supply chains to seize emerging opportunities effectively. In the long run, MOFs have the potential to become as widespread as plastics, becoming an essential part of various industries, products, and everyday life in the 21st century. Although it may take decades to see if this forecast fully materializes, it is clear that MOFs have already moved beyond purely academic research. The focus now is no longer on whether these materials will find commercial success, but rather on "what's next" and which players are best poised to leverage the emerging opportunities in this field.

4.3. Green and Pollution-Free Preparation Methods for MOFs

The development of environmentally friendly and sustainable synthesis methods for MOFs has gained increasing attention in recent years. Traditional MOF synthesis methods, such as solvothermal and hydrothermal processes, often involve the use of toxic solvents, high energy consumption, and significant waste generation [231]. To address these issues, researchers are exploring alternative green synthesis approaches that minimize environmental impact while maintaining material performance [232].

- One promising approach is solvent-free synthesis, which eliminates the use of hazardous organic solvents by employing mechanochemical methods [233]. In this process, the metal precursors and organic linkers are ground together in a ball mill, enabling the formation of MOFs under ambient conditions. This method not only reduces waste but also significantly lowers energy consumption, making it a highly sustainable alternative [234].
- Another green technique is the use of water as a reaction medium. Aqueous-based synthesis methods utilize water as a benign solvent, reducing the ecological footprint associated with solvent disposal [235]. For instance, the production of MIL-53 [236] and MIL-100/101 [237] has been successfully demonstrated using water as the primary solvent, achieving comparable performance to their traditionally synthesized counterparts.
- Microwave-assisted synthesis is another eco-friendly method that offers rapid reaction times and reduced energy requirements [172]. This technique utilizes microwave irradiation to heat the reaction mixture uniformly, leading to faster crystallization and higher yields [238]. Moreover, the process can often be conducted in water or ethanol, further reducing the environmental impact.
- Lastly, the adoption of waste valorization strategies, such as using industrial by-products as precursors for MOF synthesis, exemplifies a circular economy approach [239]. These strategies not only reduce the cost of raw materials but also contribute to waste reduction and resource conservation [240].

Incorporating these green synthesis methods into MOF production processes holds great promise for enhancing the sustainability of MOFs in heat exchanger applications, aligning with global efforts to reduce environmental pollution and promote sustainable industrial practices.

5. Conclusions

This review has provided an in-depth analysis of the current landscape, innovations, and future directions for the application of MOFs in heat exchangers. As highlighted throughout, MOFs hold immense promise for enhancing the efficiency and sustainability of heat exchange systems due to their tunable porosity, high surface area, and exceptional adsorption and thermal properties. However, realizing their full potential in real-world applications requires addressing several key challenges, from cost-effective synthesis to maintaining stability under harsh operational conditions. In this work, the authors have provided a comprehensive review on the applications, innovations, challenges, and future directions of MOFs in heat exchangers.

- MOFs integrated into a heat exchanger, such as wire-finned and shell-and-tube designs, have shown remarkable gains in heat transfer rates and moisture adsorption, and reduction in fouling resistance. Studies have proven that MOF-based coatings, like MOF-303/GO and HKUST-1 composites, enhance both water uptake, up to 22%, and thermal diffusivity, outperforming traditional materials such as silica gels and activated alumina. MOFs also outperformed traditional materials such as silica gel by enhancing COP as much as six times in comparison to it.
- This research also focused on the innovations and challenges surrounding MOF-based heat exchanger applications. Innovations in scalable synthesis methods, such as microwave-assisted synthesis and surface functionalization techniques, are key to reducing costs while ensuring consistent material quality. These advancements have the potential to make MOFs more economically viable for large-scale industrial applications. Furthermore, surface functionalization has been shown to enhance the thermal conductivity, mechanical stability, and chemical resilience of MOFs, making them more suitable for demanding environments. Yet, issues such as long-term stability under high temperatures, pressure, and corrosive conditions still need to be addressed, as shown by ongoing research into the durability of various MOF composites under operational stresses.
- The research gaps and future directions for MOF-based nanocomposites in heat exchanger systems are also studied in this research. The integration of multifunctional MOF hybrids, combining materials like GO, AC, and CNT, presents a promising path forward for developing heat exchangers with enhanced performance across multiple domains, such as heat transfer, gas adsorption, and catalysis. The multifunctional MOFs of CNT enhanced the thermal conductivity up to seven times in comparison to the actual MOFs. However, these hybrid solutions will require further collaboration between academia and industry to optimize material properties and develop scalable production processes.
- Real-world industrial case studies have also been analyzed such as BASF's innovation in adsorptive dehumidification, which leverages MOFs to reduce energy consumption in air conditioning systems by 50–60%, offering higher volumetric energy density and a 27% improvement in COP compared to silica gel. While MOFs require energy for regeneration, this can be sourced from low-temperature heat like solar or waste heat. Scaling production is expected to lower costs, broadening their applicability and enhancing energy efficiency. Similarly, NuMat Technologies exemplifies the transition of MOFs from research to industrial applications by developing cost-effective, scalable production methods for gas storage systems. Their innovations demonstrate the potential of MOFs in clean energy and industrial processes, emphasizing their role in advancing sustainable and economically viable solutions.
- Green synthesis methods for MOFs include solvent-free mechanochemical techniques, which eliminate hazardous solvents and reduce energy consumption by grinding metal precursors and linkers. Aqueous-based synthesis uses water as a benign solvent, lowering environmental impact. Methods for production of MIL-53 and MIL-100/101 using water as the primary solvent have displayed commendable performance in comparison to traditional synthesis counterparts. Microwave-assisted synthesis offers rapid, energy-efficient crystallization, often using eco-friendly solvents.

Overcoming challenges related to synthesis costs, material durability, and integration into existing industrial systems is essential to fully harness the potential of MOF-based technologies. Advances in scalable synthesis techniques, computational modeling, and collaborative efforts between academia and industry pave the way for MOFs to revolutionize heat exchanger efficiency, promoting energy conservation and sustainability. Ongoing research focused on improving stability and lowering production costs will be pivotal in establishing MOFs as a widely adopted solution for energy-efficient thermal management systems. Moreover, embedding MOF-based systems with IoT sensors and advanced monitoring tools can further optimize real-time performance and ensure efficient operation.

Author Contributions: Conceptualization, M.I. and T.B.N.; methodology, M.I. and T.B.N.; software, E.T. and T.B.N.; validation, M.I., T.B.N. and E.T.; formal analysis, E.T. and T.B.N.; investigation, T.B.N.; resources, M.I. and E.T.; data curation, T.B.N.; writing—original draft preparation, M.I. and T.B.N.; writing—review and editing, M.I., T.B.N. and E.T.; visualization, T.B.N.; supervision, M.I. and E.T.; project administration, M.I. and E.T.; funding acquisition, M.I. and E.T. All authors have read and agreed to the published version of the manuscript.

Funding: This research received no external funding.

Data Availability Statement: Not applicable.

Acknowledgments: The authors gratefully acknowledge the support provided by the Commonwealth Scholarship Commission (CSC) in the United Kingdom for funding the PhD research associated with this work. The technical and administrative assistance provided by the academic and research staff at Aston University is also highly appreciated.

Conflicts of Interest: The funders had no role in the design of the study; in the collection, analysis, or interpretation of data; in the writing of the manuscript; or in the decision to publish the results.

References

1. Hussain, A.R.J.; Alahyari, A.A.; Eastman, S.A.; Thibaud-Erkey, C.; Johnston, S.; Sobkowicz, M.J. Review of polymers for heat exchanger applications: Factors concerning thermal conductivity. *Appl. Therm. Eng.* **2017**, *113*, 1118–1127. [CrossRef]
2. Abdullah, S.; Zubir, M.N.B.M.; Muhamad, M.R.B.; Newaz, K.M.S.; Öztop, H.F.; Alam, M.S.; Shaikh, K. Technological development of evaporative cooling systems and its integration with air dehumidification processes: A review. *Energy Build.* **2023**, *283*, 112805. [CrossRef]
3. Sajjad, U.; Abbas, N.; Hamid, K.; Abbas, S.; Hussain, I.; Ammar, S.M.; Sultan, M.; Ali, H.M.; Hussain, M.; Rehman, T.U.; et al. A review of recent advances in indirect evaporative cooling technology. *Int. Commun. Heat Mass Transf.* **2021**, *122*, 105140. [CrossRef]
4. Salhein, K.; Kobus, C.J.; Zohdy, M.; Annekaa, A.M.; Alhawsawi, E.Y.; Salheen, S.A. Heat Transfer Performance Factors in a Vertical Ground Heat Exchanger for a Geothermal Heat Pump System. *Energies* **2024**, *17*, 5003. [CrossRef]
5. Ismail, M.; Yebiyo, M.; Chaer, I. A Review of Recent Advances in Emerging Alternative Heating and Cooling Technologies. *Energies* **2021**, *14*, 502. [CrossRef]
6. GaneshKumar, P.; VinothKumar, S.; Vigneswaran, V.S.; Kim, S.C.; Ramkumar, V. Advancing heat exchangers for energy storage: A comprehensive review of methods and techniques. *J. Energy Storage* **2024**, *99*, 113334. [CrossRef]
7. Shirmohammadi, R.; Gilani, N. Effectiveness enhancement and performance evaluation of indirect-direct evaporative cooling system for a wide variety of climates. *Environ. Prog. Sustain. Energy* **2019**, *38*, 13032. [CrossRef]
8. Kang, Z.; Zhou, X.; Wang, R.; Cheng, X.; Zhao, S. Research Status of Heat Recovery Device in a Train Station of Severe Cold Area. *Procedia Eng.* **2017**, *205*, 1705–1710. [CrossRef]
9. T'Joen, C.; Park, Y.; Wang, Q.; Sommers, A.; Han, X.; Jacobi, A. A review on polymer heat exchangers for HVAC&R applications. *Int. J. Refrig.* **2009**, *32*, 763–779.
10. Shah, R.K.; Sekulić, D.P. Classification of Heat Exchangers. In *Fundamentals of Heat Exchanger Design*; John Wiley & Sons, Inc.: Hoboken, NJ, USA, 2003; pp. 1–77.
11. Huu-Quan, D.; Rostami, A.M.; Rad, M.S.; Izadi, M.; Hajjar, A.; Xiong, Q. 3D numerical investigation of turbulent forced convection in a double-pipe heat exchanger with flat inner pipe. *Appl. Therm. Eng.* **2021**, *182*, 116106. [CrossRef]

12. Al Zahrani, S.; Islam, M.S.; Saha, S.C. Heat transfer enhancement of modified flat plate heat exchanger. *Appl. Therm. Eng.* **2021**, *186*, 116533. [CrossRef]
13. Saleh, M.M.; Elsayed, E.; Al-Dadh, R.; Mahmoud, S. Experimental testing of wire finned heat exchanger coated with aluminium fumarate MOF material for adsorption desalination application. *Therm. Sci. Eng. Prog.* **2022**, *28*, 101050. [CrossRef]
14. Patel, A. Heat Exchangers in Industrial Applications: Efficiency and Optimization Strategies. *Int. J. Eng. Res. Technol. (IJERT)* **2023**, *12*, 1–10.
15. Kaur, I.; Singh, P. State-of-the-art in heat exchanger additive manufacturing. *Int. J. Heat Mass Transf.* **2021**, *178*, 121600. [CrossRef]
16. Zhang, X.; Keramati, H.; Arie, M.; Singer, F.; Tiwari, R.; Shooshtari, A.; Ohadi, M. Recent developments in high temperature heat exchangers: A review. *Front. Heat Mass Transf.* **2018**, *11*, 1–14.
17. Soltan, Y.I.; Nasser, M.S.; Almomani, F.; Mahmoud, K.A.; Onaizi, S.A. Thermal conductivity of different materials nanofluids Nanofluids of MXenes, metal organic frameworks, and other Nanostructured materials in heat transfer applications: Review. *J. Mater. Res. Technol.* **2024**, *31*, 2723–2761. [CrossRef]
18. Kapustenko, P.; Klemeš, J.J.; Arsenyeva, O. Plate heat exchangers fouling mitigation effects in heating of water solutions: A review. *Renew. Sustain. Energy Rev.* **2023**, *179*, 113283. [CrossRef]
19. Shoar, Z.K.; Pourpasha, H.; Heris, S.Z.; Mousavi, S.B.; Mohammadpourfard, M. The effect of heat transfer characteristics of macromolecule fouling on heat exchanger surface: A dynamic simulation study. *Can. J. Chem. Eng.* **2023**, *101*, 5802–5817. [CrossRef]
20. Garcia, S.; Trueba, A. Fouling in Heat Exchanger. In *Inverse Heat Conduction and Heat Exchangers*; IntechOpen Limited: London, UK, 2020.
21. Müller-Steinhagen, H.; Malayeri, M.R.; Watkinson, A.P. Heat Exchanger Fouling: Mitigation and Cleaning Strategies. *Heat Transf. Eng.* **2011**, *32*, 189–196. [CrossRef]
22. Li, M.; Xiao, W.; Yin, Z.; Chen, Y.; Luo, Y.; Hong, Z.; Xue, M. Construction of a robust MOF-based superhydrophobic composite coating with the excellent performance in antifouling, drag reduction, and organic photodegradation. *Prog. Org. Coat.* **2024**, *186*, 108086. [CrossRef]
23. Al-Juboori, R.A.; Yusaf, T. Biofouling in RO system: Mechanisms, monitoring and controlling. *Desalination* **2012**, *302*, 1–23. [CrossRef]
24. Suckeveriene, R.Y. Grafting of polyaniline by a dynamic inverse emulsion polymerization technique onto reverse osmosis membranes as an antibiofouling agent. *Polym. Adv. Technol.* **2019**, *30*, 1759–1766. [CrossRef]
25. Farh, H.M.H.; Seghier, M.E.A.B.; Taiwo, R.; Zayed, T. Analysis and ranking of corrosion causes for water pipelines: A critical review. *NPJ Clean Water* **2023**, *6*, 65. [CrossRef]
26. Agala, A.; Khan, M.; Starr, A. Degradation mechanisms associated with metal pipes and the effective impact of LDMs and LLMs in water transport and distribution. *Proc. Inst. Mech. Eng. Part C J. Mech. Eng. Sci.* **2022**, *237*, 1855–1876. [CrossRef]
27. De Almeida, C.F.; Saget, M.; Delaplace, G.; Jimenez, M.; Fierro, V.; Celzard, A. Innovative fouling-resistant materials for industrial heat exchangers: A review. *Rev. Chem. Eng.* **2021**, *39*, 71–104. [CrossRef]
28. Weiss, S.; Ben-Shmuel, A.; Chajanovsky, I.; Mizrahi, D.M.; Suckeveriene, R.Y. Hybrid PANI-halamine design, synthesis and antibacterial activity. *J. Water Process Eng.* **2023**, *56*, 104539. [CrossRef]
29. Faes, W.; Lecompte, S.; Ahmed, Z.Y.; Van Bael, J.; Salenbien, R.; Verbeken, K.; De Paepe, M. Corrosion and corrosion prevention in heat exchangers. *Corros. Rev.* **2019**, *37*, 131–155. [CrossRef]
30. Viswanathan, R.; Stringer, J. Failure Mechanisms of High Temperature Components in Power Plants. *J. Eng. Mater. Technol.* **2000**, *122*, 246–255. [CrossRef]
31. Zhang, X.; Zhang, S.; Tang, Y.; Huang, X.; Pang, H. Recent advances and challenges of metal–organic framework/graphene-based composites. *Compos. Part B Eng.* **2022**, *230*, 109532. [CrossRef]
32. Liu, D.; Gu, W.; Zhou, L.; Wang, L.; Zhang, J.; Liu, Y.; Lei, J. Recent advances in MOF-derived carbon-based nanomaterials for environmental applications in adsorption and catalytic degradation. *Chem. Eng. J.* **2022**, *427*, 131503. [CrossRef]
33. Hayat, A.; Rauf, S.; Al Alwan, B.; El Jery, A.; Almuqati, N.; Melhi, S.; Amin, M.A.; Al-Hadeethi, Y.; Sohail, M.; Orooji, Y.; et al. Recent advance in MOFs and MOF-based composites: Synthesis, properties, and applications. *Mater. Today Energy* **2024**, *41*, 101542. [CrossRef]
34. Mohseni, M.M.; Jouyandeh, M.; Sajadi, S.M.; Hejna, A.; Habibzadeh, S.; Mohaddespour, A.; Rabiee, N.; Daneshgar, H.; Akhavan, O.; Asadnia, M.; et al. Metal-organic frameworks (MOF) based heat transfer: A comprehensive review. *Chem. Eng. J.* **2022**, *449*, 137700. [CrossRef]
35. Unnikrishnan, V.; Zabihi, O.; Ahmadi, M.; Li, Q.; Blanchard, P.; Kiziltas, A.; Naebe, M. Metal–organic framework structure–property relationships for high-performance multifunctional polymer nanocomposite applications. *J. Mater. Chem. A* **2021**, *9*, 4348–4378. [CrossRef]

36. Li, D.; Yadav, A.; Zhou, H.; Roy, K.; Thanasekaran, P.; Lee, C. Advances and Applications of Metal-Organic Frameworks (MOFs) in Emerging Technologies: A Comprehensive Review. *Glob. Chall.* **2024**, *8*, 2300244. [CrossRef]
37. Xu, K.; Zhang, S.; Zhuang, X.; Zhang, G.; Tang, Y.; Pang, H. Recent progress of MOF-functionalized nanocomposites: From structure to properties. *Adv. Colloid Interface Sci.* **2024**, *323*, 103050. [CrossRef] [PubMed]
38. Zhao, Y.; Song, Z.; Li, X.; Sun, Q.; Cheng, N.; Lawes, S.; Sun, X. Metal organic frameworks for energy storage and conversion. *Energy Storage Mater.* **2016**, *2*, 35–62. [CrossRef]
39. Vo, P.; Haranczyl, M. Insights into Thermal Conductivity at the MOF-Polymer Interface. *ACS Appl. Mater. Interfaces* **2024**, *16*, 56221–56231. [CrossRef]
40. Jayaramulu, K.; Mukherjee, S.; Marales, D.M.; Dubal, D.P.; Nanjudan, A.K.; Schneemann, A.; Masa, J.; Kment, S.; Schuhmann, W.; Otyepka, M.; et al. Graphene-Based Metal–Organic Framework Hybrids for Applications in Catalysis, Environmental, and Energy Technologies. *Chem. Rev.* **2022**, *122*, 17241–17338. [CrossRef]
41. Liu, X.; Shan, Y.; Zhang, S.; Kong, Q.; Pang, H. Application of metal organic framework in wastewater treatment. *Green Energy Environ.* **2023**, *8*, 698–721. [CrossRef]
42. He, X.; Deng, F.; Shen, T.; Yang, L.; Chen, D.; Luo, J.; Luo, X.; Min, X.; Wang, F. Exceptional adsorption of arsenic by zirconium metal-organic frameworks: Engineering exploration and mechanism insight. *J. Colloid Interface Sci.* **2019**, *539*, 223–234. [CrossRef]
43. Jrad, A.; Damacet, P.; Yaghi, Z.; Ahmad, M.; Hmadeh, M. Zr-Based Metal–Organic Framework Nanocrystals for Water Remediation. *ACS Appl. Nano Mater.* **2022**, *5*, 10795–10808. [CrossRef]
44. Kalauni, K.; Vedrtnam, A.; Wdowin, M.; Chaturvedi, S. ZIF for CO_2 Capture: Structure, Mechanism, Optimization, and Modeling. *Processes* **2022**, *10*, 2689. [CrossRef]
45. Mirkatuli, F.S.; Nosratinia, F.; Rohani, A.A.; Rashidi, A.M.; Ardjamand, M. Enhancing CO_2 adsorption with modified zeolite imidazolate frameworks (ZIF-8) functionalized with amine groups. *J. Environ. Chem. Eng.* **2024**, *12*, 114770. [CrossRef]
46. Elsaidi, S.K.; Mohamed, M.H.; Banerjee, D.; Thallapally, P.K. Flexibility in Metal–Organic Frameworks: A fundamental understanding. *Coord. Chem. Rev.* **2018**, *358*, 125–152. [CrossRef]
47. Zhang, X.; Xiao, L.; Zhang, G.; Xu, L.; Xu, Z.; Meng, Q. Preparation of stable multilayer PDMS composite pervaporation membrane incorporated with in-situ transformed metal organic frameworks for enhanced butanol recovery. *J. Membr. Sci.* **2024**, *700*, 122727. [CrossRef]
48. Rabiee, N.; Bagherzadeh, M.; Haris, M.H.; Ghadiri, A.M.; Moghaddam, F.M.; Fatahi, Y.; Dinarvand, R.; Jarahiyan, A.; Ahmadi, S.; Shokouhimehr, M. Polymer-Coated NH_2-UiO-66 for the Codelivery of DOX/pCRISPR. *ACS Appl. Mater. Interfaces* **2021**, *13*, 10796–10811. [CrossRef]
49. Rabiee, N.; Bagherzadeh, M.; Jouyandeh, M.; Zarrintaj, P.; Mozafari, M.; Shokouhimehr, M.; Varma, R.S. Natural Polymers Decorated MOF-MXene Nanocarriers for Co-delivery of Doxorubicin/pCRISPR. *ACS Appl. Bio Mater.* **2021**, *4*, 5106–5121. [CrossRef]
50. Karmakar, A.; Prabakaran, V.; Zhao, D.; Chua, K.J. A review of metal-organic frameworks (MOFs) as energy-efficient desiccants for adsorption driven heat-transformation applications. *Appl. Energy* **2020**, *269*, 115070. [CrossRef]
51. Cui, S.; Qin, M.; Marandi, A.; Steggles, V.; Wang, S.; Feng, X.; Nouar, F.; Serre, C. Metal-Organic Frameworks as advanced moisture sorbents for energy-efficient high temperature cooling. *Sci. Rep.* **2018**, *8*, 15284. [CrossRef]
52. Chen, H.; Wang, F.; Fan, H.; Hong, R.; Li, W. Construction of MOF-based superhydrophobic composite coating with excellent abrasion resistance and durability for self-cleaning, corrosion resistance, anti-icing, and loading-increasing research. *Chem. Eng. J.* **2021**, *408*, 127343. [CrossRef]
53. Xiao, C.; He, H.; Li, J.; Zhu, W. Kapitza resistance for nanoscale crystalline and amorphous silicon carbide. In Proceedings of the 19th International Conference on Thermal, Mechanical and Multi-Physics Simulation and Experiments in Microelectronics and Microsystems (EuroSimE), Toulouse, France, 15–18 April 2018.
54. Bebek, M.B.; Stanley, C.M.; Gibbons, T.M.; Estreicher, S.K. Temperature dependence of phonon-defect interactions: Phonon scattering vs. phonon trapping. *Sci. Rep.* **2016**, *6*, 32150. [CrossRef] [PubMed]
55. Cavigli, L.; Milanesi, A.; Khlebtsov, B.N.; Centi, S.; Ratto, F.; Khlebtsov, N.G.; Pini, R. Impact of Kapitza resistance on the stability and efficiency of photoacoustic conversion from gold nanorods. *J. Colloid Interface Sci.* **2020**, *578*, 358–365. [CrossRef] [PubMed]
56. He, B.; Mortazavi, B.; Zhuang, X.; Rabczuk, T. Modeling Kapitza resistance of two-phase composite material. *Compos. Struct.* **2016**, *152*, 939–946. [CrossRef]
57. Giri, A.; Hopkins, P.E.; Wessel, J.G.; Duda, J.C. Kapitza resistance and the thermal conductivity of amorphous superlattices. *J. Appl. Phys.* **2015**, *118*, 165303. [CrossRef]
58. Su, Y.; Li, J.J.; Weng, G.J. Theory of thermal conductivity of graphene-polymer nanocomposites with interfacial Kapitza resistance and graphene-graphene contact resistance. *Carbon* **2018**, *137*, 222–233. [CrossRef]

59. Chen, J.; Xu, X.; Zhou, J.; Li, B. Interfacial thermal resistance: Past, present, and future. *Rev. Mod. Phyics* **2022**, *94*, 025002. [CrossRef]
60. Schelling, P.K.; Phillpot, S.R.; Keblinski, P. Kapitza conductance and phonon scattering at grain boundaries by simulation. *J. Appl. Phys.* **2004**, *95*, 6082–6091. [CrossRef]
61. Fan, Y.; Liu, Z.; Chen, G. Recent Progress in Designing Thermoelectric Metal–Organic Frameworks. *Small* **2021**, *17*, 2100505. [CrossRef]
62. Ma, H.; Gao, B.; Wang, M.; Yuan, Z.; Shen, J.; Zhao, J.; Feng, Y. Strategies for enhancing thermal conductivity of polymer-based thermal interface materials: A review. *J. Mater. Sci.* **2020**, *56*, 1064–1086. [CrossRef]
63. Suckeveriene, R. Chapter 4: Polymers and Nanocomposites with High. In *Importance & Applications of Nanotechnology (Volume 7)*; MedDocs Publishers: Reno, NV, USA, 2021; pp. 26–32.
64. Ge, L.; Feng, Y.; Wu, J.; Wang, R.; Ge, T. Performance evaluation of MIL-101(Cr) based desiccant-coated heat exchangers for efficient dehumidification. *Energy* **2024**, *289*, 130049. [CrossRef]
65. Wee, M.G.V.; Chinnappan, A.; Shang, R.; Lee, P.S.; Ramakrishna, S. Enhanced moisture sorption through regulated MIL-101(Cr) synthesis and its integration onto heat exchangers. *J. Mater. Chem. A* **2024**, *12*, 824–839. [CrossRef]
66. Fan, Y.; Liu, Z.; Chen, G. Constructing flexible metal-organic framework/polymer/carbon nanotubes ternary composite films with enhanced thermoelectric properties for heat-to-electricity conversion. *Compos. Commun.* **2022**, *29*, 100997. [CrossRef]
67. Yang, S.; Karve, V.V.; Justin, A.; Kochetygov, I.; Espín, J.; Asgari, M.; Trukhina, O.; Sun, D.T.; Peng, L.; Queen, W.L. Enhancing MOF performance through the introduction of polymer guests. *Coord. Chem. Rev.* **2021**, *427*, 213525. [CrossRef]
68. Kim, H.K.; Yun, W.S.; Kim, M.-B.; Kim, J.Y.; Bae, Y.-S.; Lee, J.; Jeong, N.C. A Chemical Route to Activation of Open Metal Sites in the Copper-Based Metal–Organic Framework Materials HKUST-1 and Cu-MOF-2. *J. Am. Chem. Soc.* **2015**, *137*, 10009–10015. [CrossRef] [PubMed]
69. Singh, N.; Thakur, A. Applications of copper based metal organic frameworks. *Mater. Today Proc.* **2022**, *50*, 1906–1911. [CrossRef]
70. Henninger, S.K.; Habib, H.A.; Janiak, C. MOFs as Adsorbents for Low Temperature Heating and Cooling Applications. *J. Am. Chem. Soc.* **2009**, *131*, 2776–2777. [CrossRef]
71. Ehrenmann, J.; Henninger, S.K.; Janiak, C. Water Adsorption Characteristics of MIL-101 for Heat-Transformation Applications of MOFs. *Eur. J. Inorg. Chem.* **2011**, *2011*, 471–474. [CrossRef]
72. Henninger, S.K.; Jeremias, F.; Kummer, H.; Janiak, C. MOFs for Use in Adsorption Heat Pump Processes. *Eur. J. Inorg. Chem.* **2012**, *2012*, 2625–2634. [CrossRef]
73. Henninger, S.K.; Jeremias, F.; Kummer, H.; Schossig, P.; Henning, H.-M. Novel Sorption Materials for Solar Heating and Cooling. *Energy Procedia* **2012**, *30*, 279–288. [CrossRef]
74. Janiak, C.; Henninger, S.K. Porous Coordination Polymers as Novel Sorption Materials for Heat Transformation Processes. *CHIMIA* **2013**, *67*, 419. [CrossRef]
75. Jeremias, F.; Fröhlich, D.; Janiak, C.; Henninger, S.K. Advancement of sorption-based heat transformation by a metal coating of highly-stable, hydrophilic aluminium fumarate MOF. *RSC Adv.* **2014**, *4*, 24073. [CrossRef]
76. Jeremias, F.; Khutia, A.; Henninger, S.K.; Janiak, C. MIL-100 (Al, Fe) as water adsorbents for heat transformation purposes—A promising application. *J. Mater. Chem.* **2012**, *22*, 10148–10151. [CrossRef]
77. Jeremias, F.; Lozan, V.; Henninger, S.K.; Janiak, S.K. Programming MOFs for water sorption: Amino-functionalized MIL-125 and UiO-66 for heat transformation and heat storage applications. *Dalton Trans.* **2013**, *42*, 15967–15973. [CrossRef] [PubMed]
78. Khutia, A.; Rammelberg, H.U.; Schmidt, T.; Henninger, S.; Janiak, C. Water Sorption Cycle Measurements on Functionalized MIL-101Cr for Heat Transformation Application. *Chem. Mater.* **2013**, *25*, 790–798. [CrossRef]
79. Elsayed, E.; Saleh, M.M.; Al-Dadah, R.; Mahmoud, S.; Elsayed, A. Aluminium fumarate metal-organic framework coating for adsorption cooling application: Experimental study. *Int. J. Refrig.* **2021**, *130*, 288–304. [CrossRef]
80. Rezk, A.; Al-Dadah, R.; Mahmoud, S.; Elsayed, A. Experimental investigation of metal organic frameworks characteristics for water adsorption chillers. *Proc. Inst. Mech. Eng. Part C: J. Mech. Eng. Sci.* **2012**, *227*, 992–1005. [CrossRef]
81. Shi, B.; Al-Dadah, R.; Mahmoud, S.; Elsayed, A.; Elsayed, E. CPO-27(Ni) metal–organic framework based adsorption system for automotive air conditioning. *Appl. Therm. Eng.* **2016**, *106*, 325–333. [CrossRef]
82. Rezk, A.; Al-Dadah, R.; Mahmoud, S.; Elsayed, A. Characterisation of metal organic frameworks for adsorption cooling. *Int. J. Heat Mass Transf.* **2012**, *55*, 7366–7374. [CrossRef]
83. Sundari, S.S.K.; Rishwana, S.S.; Kotresh, T.M.; Ramani, R.; Shekar, R.I.; Vijayakumar, C.T. Effect of structural variation on the thermal degradation of nanoporous aluminum fumarate metal organic framework (MOF). *J. Therm. Anal. Calorim.* **2021**, *147*, 5067–5085. [CrossRef]
84. Tannert, N.; Jansen, C.; Nießing, S.; Janiak, C. Robust synthesis routes and porosity of the Al-based metal–organic frameworks Al-fumarate, CAU-10-H and MIL-160. *Dalton Trans.* **2019**, *48*, 2967–2976. [CrossRef]

85. Schlüsener, C.; Jordan, D.N.; Xhinovci, M.; Ntep, T.J.M.M.; Schmitz, A.; Giesen, B.; Janiak, C. Probing the limits of linker substitution in aluminum MOFs through water vapor sorption studies: Mixed-MOFs instead of mixed-linker CAU-23 and MIL-160 materials. *Dalton Trans.* **2020**, *49*, 7373–7383. [CrossRef] [PubMed]
86. Ghazy, M.; Askalany, A.A.; Ibrahim, E.M.M.; Mohamed, A.S.A.; Ali, E.S.; Al-Dadah, R. Solar powered adsorption desalination system employing CPO-27(Ni). *J. Energy Storage* **2022**, *53*, 105174. [CrossRef]
87. García, E.R.; Medina, R.L.; Lozano, M.M.; Pérez, I.H.; Valero, M.J.; Franco, A.M.M. Adsorption of Azo-Dye Orange II from Aqueous Solutions Using a Metal-Organic Framework Material: Iron- Benzenetricarboxylate. *Materials* **2014**, *7*, 8037–8057. [CrossRef] [PubMed]
88. Chen, Y.; Mu, X.; Lester, E.; Wu, T. High efficiency synthesis of HKUST-1 under mild conditions with high BET surface area and CO_2 uptake capacity. *Prog. Nat. Sci. Mater. Int.* **2018**, *28*, 584–589. [CrossRef]
89. Huo, S.-H.; Yan, X.-P. Metal–organic framework MIL-100(Fe) for the adsorption of malachite green from aqueous solution. *J. Mater. Chem.* **2012**, *22*, 7449–7455. [CrossRef]
90. Yılmaz, E.; Sert, E.; Atalay, F.S. Synthesis, characterization of a metal organic framework: MIL-53 (Fe) and adsorption mechanisms of methyl red onto MIL-53 (Fe). *J. Taiwan Inst. Chem. Eng.* **2016**, *65*, 323–330. [CrossRef]
91. Abid, H.R.; Rada, Z.H.; Shang, J.; Wang, S. Synthesis, characterization, and CO2 adsorption of three metal-organic frameworks (MOFs): MIL-53, MIL-96, and amino-MIL-53. *Polyhedron* **2016**, *120*, 103–111. [CrossRef]
92. Zheng, Z.; Nguyen, H.L.; Hanikel, N.; Li, K.K.-Y.; Zhou, Z.; Ma, T.; Yaghi, O.M. High-yield, green and scalable methods for producing MOF-303 for water harvesting from desert air. *Nat. Protoc.* **2023**, *18*, 136–156. [CrossRef]
93. Solovyeva, M.; Gordeeva, L.; Krieger, T.; Aristov, Y. MOF-801 as a promising material for adsorption cooling: Equilibrium and dynamics of water adsorption. *Energy Convers. Manag.* **2018**, *174*, 356–363. [CrossRef]
94. Luu, C.L.; Nguyen, T.T.V.; Nguyen, T.; Hoang, T.C. Synthesis, characterization and adsorption ability of UiO-66-NH_2. *Adv. Nat. Sci. Nanosci. Nanotechnol.* **2015**, *6*, 025004. [CrossRef]
95. Vahabi, A.H.; Norouzi, F.; Sheibani, E.; Rahimi-Nasrabadi, M. Functionalized Zr-UiO-67 metal-organic frameworks: Structural landscape and application. *Coord. Chem. Rev.* **2021**, *445*, 214050. [CrossRef]
96. Ye, X.; Liu, D. Metal–Organic Framework UiO-68 and Its Derivatives with Sufficiently Good Properties and Performance Show Promising Prospects in Potential Industrial Applications. *Cryst. Growth Des.* **2021**, *21*, 4780–4804. [CrossRef]
97. Butova, V.; Budnyk, A.; Bulanova, E.; Lamberti, C.; Soldatov, A. Hydrothermal synthesis of high surface area ZIF-8 with minimal use of TEA. *Solid State Sci.* **2017**, *69*, 13–21. [CrossRef]
98. Laurenz, E.; Füldner, G.; Velte, A.; Schnabel, L.; Schmidtz, G. Frequency response analysis for the determination of thermal conductivity and water transport in MOF adsorbent coatings for heat transformation. *Int. J. Heat Mass Transf.* **2021**, *169*, 120921. [CrossRef]
99. Huang, J.; Xia, X.; Hu, X.; Li, S.; Liu, K. A general method for measuring the thermal conductivity of MOF crystals. *Int. J. Heat Mass Transf.* **2019**, *138*, 11–16. [CrossRef]
100. Babaei, H.; DeCoster, M.E.; Jeong, M.; Hassan, Z.M.; Islamoglu, T.; Baumgart, H.; McGaughey, A.J.H.; Redel, E.; Farha, O.K.; Hopkins, P.E.; et al. Observation of reduced thermal conductivity in a metal-organic framework due to the presence of adsorbates. *Nat. Commun.* **2020**, *11*, 4010. [CrossRef]
101. Babaei, H.; McGhaughey, A.J.H.; Wilmer, C.E. Effect of pore size and shape on the thermal conductivity of metal-organic frameworks. *Chem. Sci.* **2017**, *8*, 583–589. [CrossRef]
102. Yin, Y.; Shao, J.; Zhang, L.; Zui, Q.; Wang, H. Experimental Study on Heat Transfer and Adsorption Cooling Performance of MIL-101/Few Layer Graphene Composite. *Energies* **2021**, *14*, 4970. [CrossRef]
103. Semelsberger, T.A.; Veenstra, M.; Dixon, C. Room temperature thermal conductivity measurements of neat MOF-5 compacts with high pressure hydrogen and helium. *Int. J. Hydrogen Energy* **2016**, *41*, 4690–4702. [CrossRef]
104. Huang, B.; Ni, Z.; Millward, A.; McGaughey, A.; Uher, C.; Kaviany, M.; Yaghi, O. Thermal conductivity of a metal-organic framework (MOF-5): Part II. Measurement. *Int. J. Heat Mass Transf.* **2007**, *50*, 405–411. [CrossRef]
105. Zhang, Y.; Sun, C.; Wang, C.; Jankovic, K.E.; Dong, Y. Lipids and Lipid Derivatives for RNA Delivery. *Chem. Rev.* **2021**, *121*, 12181–12277. [CrossRef] [PubMed]
106. Sun, L.; Liao, B.; Sheberla, D.; Kraemer, D.; Zhou, J.; Stach, E.A.; Zakharov, D.; Stavila, V.; Talin, A.A.; Ge, Y.; et al. A Microporous and Naturally Nanostructured Thermoelectric Metal-Organic Framework with Ultralow Thermal Conductivity. *Joule* **2017**, *1*, 168–177. [CrossRef]
107. Erickson, K.J.; Léonard, F.; Stavila, V.; Foster, M.E.; Spataru, C.D.; Jones, R.E.; Foley, B.M.; Hopkins, P.E.; Allendorf, M.D.; Talin, A.A. Thin Film Thermoelectric Metal–Organic Framework with High Seebeck Coefficient and Low Thermal Conductivity. *Adv. Mater.* **2015**, *27*, 3341–3459. [CrossRef] [PubMed]

108. Zhang, X.; Jiang, J. Thermal Conductivity of Zeolitic Imidazolate Framework-8: A Molecular Simulation Study. *J. Phys. Chem. C* **2013**, *117*, 18441–18447. [CrossRef]
109. Ying, P.; Zhang, J.; Zhang, X.; Zhong, Z. Impacts of Functional Group Substitution and Pressure on the Thermal Conductivity of ZIF-8. *J. Phys. Chem. C* **2020**, *124*, 6274–6283. [CrossRef]
110. Cui, B.; Audu, C.O.; Liao, Y.; Nguyen, S.T.; Farha, O.K.; Hupp, J.T.; Grayson, M. Thermal conductivity of ZIF-8 thin-film under ambient gas pressure. *ACS Appl. Mater. Interfaces* **2017**, *9*, 28139–28143. [CrossRef]
111. Rubio-Martinez, M.; Avci-Camur, C.; Thornton, A.W.; Imaz, I.; Maspoch, D.; Hill, M.R. New synthetic routes towards MOF production at scale. *Chem. Soc. Rev.* **2017**, *46*, 3453–3480. [CrossRef]
112. Ji, H.; Hwang, S.; Kim, K.; Kim, C.; Jeong, N.C. Direct in Situ Conversion of Metals into Metal–Organic Frameworks: A Strategy for the Rapid Growth of MOF Films on Metal Substrates. *ACS Appl. Mater. Interfaces* **2016**, *8*, 32414–32420. [CrossRef]
113. Peterson, G.W.; Lee, D.T.; Barton, H.F.; Epps, T.H.; Parsons, G.N. Fibre-based composites from the integration of metal–organic frameworks and polymers. *Nat. Rev. Mater.* **2021**, *6*, 605–621. [CrossRef]
114. Abednatanzi, S.; Derakhshandeh, P.G.; Depauw, H.; Coudert, F.X.; Vreilinck, H.; Voort, P.V.D.; Leus, K. Mixed-metal metal–organic frameworks. *Chem. Soc. Rev.* **2019**, *48*, 2535. [CrossRef]
115. Lin, R.; Hernandez, B.V.; Ge, L.; Zhu, Z. Metal organic framework based mixed matrix membranes: An overview on filler/polymer interfaces. *J. Mater. Chem. A* **2018**, *6*, 293–312. [CrossRef]
116. Elrasheedy, A.; Nady, N.; Bassyouni, M.; El-Shazly, A. Metal Organic Framework Based Polymer Mixed Matrix Membranes: Review on Applications in Water Purification. *Membranes* **2019**, *9*, 88. [CrossRef] [PubMed]
117. Han, B.; Chakraborty, A. Recent advances in metal-organic frameworks for adsorption heat transformations. *Renew. Sustain. Energy Rev.* **2024**, *198*, 114411. [CrossRef]
118. Babu, A.M.; Varghese, A. Electrochemical deposition for metal organic Frameworks: Advanced Energy, Catalysis, sensing and separation applications. *J. Electroanal. Chem.* **2023**, *937*, 117417. [CrossRef]
119. Campagnol, N.; Van Assche, T.R.C.; Li, M.; Stappers, L.; Dincă, M.; Denayer, J.F.M.; Binnemans, K.; De Vos, D.E.; Fransaer, J. On the electrochemical deposition of metal–organic frameworks. *J. Mater. Chem. A* **2016**, *4*, 3914–3925. [CrossRef]
120. Zhang, X.; Wan, K.; Subramanian, P.; Xu, M.; Luo, J.; Fransaer, J. Electrochemical deposition of metal–organic framework films and their applications. *J. Mater. Chem. A* **2020**, *8*, 7569. [CrossRef]
121. Dhakshinamoorthy, A.; Garcia, H. Catalysis by metal nanoparticles embedded on metal–organic frameworks. *Chem. Soc. Rev.* **2012**, *41*, 5262–5284. [CrossRef]
122. Muresan, L.M. Nanocomposite Coatings for Anti-Corrosion Properties of Metallic Substrates. *Materials* **2023**, *16*, 5092. [CrossRef]
123. Wei, L.; Dongxue, J.; Wenhui, L.; Yu, Z. Recent advances in the applications of metal-organic frameworks-based molecularly imprinted materials. *Chin. J. Chromatogr.* **2023**, *41*, 651–661.
124. Mahdi, E.M.; Tan, J.C. Mixed-matrix membranes of zeolitic imidazolate framework (ZIF-8)/Matrimid nanocomposite: Thermo-mechanical stability and viscoelasticity underpinning membrane separation performance. *J. Membr. Sci.* **2016**, *498*, 276–290. [CrossRef]
125. Kalaj, M.; Bentz, K.C.; Ayala, S., Jr.; Palomba, J.M.; Barcus, K.S.; Katayama, Y.; Cohen, S.M. MOF-Polymer Hybrid Materials: From Simple Composites to Tailored Architectures. *Chem. Rev.* **2020**, *120*, 8267–8302. [CrossRef] [PubMed]
126. Dai, H.; Yuan, X.; Jiang, L.; Wang, H.; Zhang, J.; Zhang, J.; Xiang, T. Recent advances on ZIF-8 composites for adsorption and photocatalytic wastewater pollutant removal: Fabrication, applications and perspective. *Coord. Chem. Rev.* **2021**, *441*, 213985. [CrossRef]
127. Carter, D.; Tezel, F.H.; Kruczek, B.; Kalipcilar, H. Investigation and comparison of mixed matrix membranes composed of polyimide matrimid with ZIF—8, silicalite, and SAPO—34. *J. Membr. Sci.* **2017**, *544*, 35–46. [CrossRef]
128. Li, T.T.; Cen, X.; Ren, H.T.; Wu, L.; Peng, H.K.; Wang, W.; Gao, B.; Lou, C.W.; Lin, J.H. Zeolitic Imidazolate Framework-8/Polypropylene–Polycarbonate Barklike Meltblown Fibrous Membranes by a Facile in Situ Growth Method for Efficient PM2.5 Capture. *ACS Appl. Mater. Interfaces* **2020**, *12*, 8730–8739. [CrossRef] [PubMed]
129. Albdoor, A.K.; Ma, Z.; Cooper, P.; Al-Ghazzawi, F.; Liu, J.; Richardson, C.; Wagner, P. Air-to-air enthalpy exchangers: Membrane modification using metal-organic frameworks, characterisation and performance assessment. *J. Clean. Prod.* **2021**, *293*, 126157. [CrossRef]
130. Chung, C.K. *Development of Dual-Layer Membrane and Metal Organic Framework for Gas Separation to Improve Indoor Air Quality*; Universiti Tunku Abdul Rahman (UTAR): Jaya, Malaysia, 2023.
131. Kim, E.; Patel, R. Recent Advances in Metal Organic Framework based Thin Film Nanocomposite Membrane for Nanofiltration. *Membr. J.* **2021**, *31*, 35–51. [CrossRef]

132. Iqbal, A.; Cevik, E.; Mustafa, A.; Qahtan, T.F.; Zeeshan, M.; Bozkurt, A. Emerging developments in polymeric nanocomposite membrane-based filtration for water purification: A concise overview of toxic metal removal. *Chem. Eng. J.* **2024**, *481*, 148760. [CrossRef]
133. Guo, F.; Li, B.; Ding, R.; Li, D.; Jiang, X.; He, G.; Xiao, W. A Novel Composite Material UiO-66@HNT/Pebax Mixed Matrix Membranes for Enhanced CO_2/N_2 Separation. *Membranes* **2021**, *11*, 693. [CrossRef]
134. Giliopoulus, D.; Zamboulis, A.; Giannakoudakis, D.; Bikiaris, D.; Triantafyllidis, K. Polymer/Metal Organic Framework (MOF) Nanocomposites for Biomedical Applications. *Molecules* **2019**, *25*, 185. [CrossRef]
135. Khosrojerdi, S.; Gholami, L.; Khazaei, M.; Hashemzadeh, A.; Darroudi, M.; Oskuee, R.K. Synthesis and evaluation of gene delivery vectors based on PEI-modified metal-organic framework (MOF) nanoparticles. *Iran. J. Basic Med. Sci.* **2023**, *27*, 203–213.
136. Singh, V.; Saini, I. Recent Developments in MOF-Polymer Composites. In *Metal-Organic Frameworks-Based Hybrid Materials for Environmental Sensing and Monitoring*; CRC Press: Boca Raton, FL, USA, 2022; pp. 20–32.
137. Yang, X.; Cheng, T.C.; Morris, A.J. Polymer-grafted metal–organic frameworks: Design, synthesis, and application. *J. Mater. Chem. C* **2024**, *12*, 4562–4592. [CrossRef]
138. Zornoza, B.; Tellez, C.; Coronas, J.; Gascon, J.; Kapteijn, F. Metal organic framework based mixed matrix membranes: An increasingly important field of research with a large application potential. *Microporous Mesoporous Mater.* **2013**, *166*, 67–78. [CrossRef]
139. Tian, N.; Gao, Y.; Wu, J.; Luo, S.; Dai, W. Water-resistant HKUST-1 functionalized with polydimethylsiloxane for efficient rubidium ion capture. *New J. Chem.* **2019**, *43*, 15339–15547. [CrossRef]
140. Liu, Q.; Wang, C.; Wang, H.; Li, L.; Zhang, X.; Zhao, R.; Han, J.; Wang, L. MOF-derived hollow octahedral CoxP/MOF-801 p-n heterojunction for efficient photocatalytic hydrogen production. *Int. J. Hydrogen Energy* **2024**, *81*, 66–74. [CrossRef]
141. Lenzen, D.; Zhao, J.; Ernst, S.-J.; Wahiduzzaman, M.; Inge, A.K.; Fröhlich, D.; Xu, H.; Bart, H.-J.; Janiak, C.; Henninger, S.; et al. A metal–organic framework for efficient water-based ultra-low-temperature-driven cooling. *Nat. Commun.* **2019**, *10*, 3025. [CrossRef] [PubMed]
142. Chaemchuen, S.; Xiao, X.; Klomkliang, N.; Yusubov, M.S.; Verpoort, F. Tunable Metal–Organic Frameworks for Heat Transformation Applications. *Nanomaterials* **2018**, *8*, 661. [CrossRef]
143. Rocchetti, A.; Lippi, M.; Socci, L.; Gullo, P.; Khorshidi, V.; Talluri, L. Metal-Organic Framework Adsorbent Materials in HVAC Systems: General Survey and Theoretical Assessment. *Energies* **2022**, *15*, 8908. [CrossRef]
144. Emam, H.E.; Abdelhameed, R.M.; Ahmed, H.B. Adsorptive Performance of MOFs and MOF Containing Composites for Clean Energy and Safe Environment. *J. Environ. Chem. Eng.* **2020**, *8*, 104386. [CrossRef]
145. Chang, C.; Luo, W.; Lu, C.; Cheng, Y.; Tsai, B.; Lin, Z. Effects of process air conditions and switching cycle period on dehumidification performance of desiccant-coated heat exchangers. *Sci. Technol. Built Environ.* **2017**, *23*, 81–90. [CrossRef]
146. Hou, P.; Zu, K.; Qin, M.; Cui, S. A novel metal-organic frameworks based humidity pump for indoor moisture control. *Build. Environ.* **2021**, *187*, 107396. [CrossRef]
147. Aziz, A.N.; Mahmoud, S.; Al-Dadah, R.; Taskin, A.; Ismail, M.A.; Fahmy, Y.M.; Rashid, M.M. Novel MOF-303/G coated wire-finned heat exchanger for dehumidification applications–Experimental investigation. *Energy* **2024**, *305*, 132333. [CrossRef]
148. Albaik, I.; Elsheniti, M.B.; Al-Dadah, R.; Mahmoud, S.; Solmaz, I. Numerical and experimental investigation of multiple heat exchanger. *Energy Convers. Manag.* **2022**, *251*, 114934. [CrossRef]
149. Shahvari, S.Z.; Kalkhorani, V.A.; Clark, J.D. Performance evaluation of a metal organic frameworks based combined dehumidification and indirect evaporative cooling system in different climates. *Int. J. Refrig.* **2022**, *140*, 186–197. [CrossRef]
150. Harrouz, J.P.; Katramiz, E.; Ghali, K.; Ouahrani, D.; Ghaddar, N. Life cycle assessment of desiccant—Dew point evaporative cooling systems with water reclamation for poultry houses in hot and humid climate. *Appl. Therm. Eng.* **2022**, *210*, 118419. [CrossRef]
151. Aziz, A.N.; Mahmoud, S.; Al-Dadah, R.; Ismail, M.A.; Al Mesfer, M.K. Numerical and experimental investigation of desiccant cooling system using metal organic framework materials. *Appl. Therm. Eng.* **2022**, *215*, 118940. [CrossRef]
152. Ye, L.; Rupam, T.H.; Islam, M.A.; Saha, B.B. Study of metal–organic framework (MOF)/water pairs for adsorption heat transformer applications. *Therm. Sci. Eng. Prog.* **2024**, *51*, 102595. [CrossRef]
153. Gholami, F.; Zinadini, S.; Zinatizadeh, A.; Abbasi, A.R. TMU-5 metal-organic frameworks (MOFs) as a novel nanofiller for flux increment and fouling mitigation in PES ultrafiltration membrane. *Sep. Purif. Technol.* **2018**, *194*, 272–280. [CrossRef]
154. Campagnol, N.; Assche, T.V.; Boudewijns, T.; Denayer, J.; Binnemans, K.; De Vos, D.; Fransaer, J. High pressure, high temperature electrochemical synthesis of metal–organic frameworks: Films of MIL-100 (Fe) and HKUST-1 in different morphologies. *J. Mater. Hemistry A* **2013**, *1*, 5827–5830. [CrossRef]
155. Li, H.; Yan, M.; Zhao, W. Designing a MOF-based slippery lubricant-infused porous surface with dual functional anti-fouling strategy. *J. Colloid Interface Sci.* **2022**, *607*, 1424–1435. [CrossRef]

156. Zhang, L.; Li, H.; Zahng, X.; Li, Q.; Zhang, G.; Pu, S.; Liu, F.Q. UV-curable PBMA coating containing CuZn-MOF-74 for fouling-resistance. *Microporous Mesoporous Mater.* **2024**, *368*, 113020. [CrossRef]
157. Moaness, M.; El-Sayed, S.A.M.; Beherei, H.H.; Mabrouk, M. Enhancing the Antifouling Properties of Alumina Nanoporous Membranes by GO/MOF Impregnated Polymer Coatings: In Vitro Studies. *J. Funct. Biom.* **2024**, *15*, 50. [CrossRef] [PubMed]
158. Wang, M.; Zi, Y.; Zhu, J.; Huang, W.; Zhang, Z.; Zhang, H. Construction of super-hydrophobic PDMS@MOF@Cu mesh for reduced drag, anti-fouling and self-cleaning towards marine vehicle applications. *Chem. Eng. J.* **2021**, *417*, 129265. [CrossRef]
159. Arsenyeva, O.; Tovazhnyanskyy, L.; Kapustenko, P.; Klemeš, J.J.; Varbanov, P.S. Review of Developments in Plate Heat Exchanger Heat Transfer Enhancement for Single-Phase Applications in Process Industries. *Energies* **2023**, *16*, 4976. [CrossRef]
160. Venegas, T.; Qu, M.; Nawaz, K.; Wang, L. Critical review and future prospects for desiccant coated heat exchangers: Materials, design, and manufacturing. *Renew. Sustain. Energy Rev.* **2021**, *151*, 111531. [CrossRef]
161. Edries, E.; Petrov, A. Types of heat exchangers in industry, their advantages and disadvantages, and the study of their parameters. In Proceedings of the IOP Conference Series: Materials Science and Engineering, Moscow, Russia, 8 September 2020.
162. Zhang, J.; Zhu, X.; Mondejar, M.E.; Haglind, F. A review of heat transfer enhancement techniques in plate heat exchangers. *Renew. Sustain. Energy Rev.* **2019**, *101*, 305–328. [CrossRef]
163. Sun, X.; Chen, J.; Zhao, Y.; Li, X.; Ge, T.; Wang, C.; Dai, Y. Experimental investigation on a dehumidification unit with heat recovery using desiccant coated heat exchanger in waste to energy system. *Appl. Therm. Eng.* **2021**, *185*, 116342. [CrossRef]
164. Pinheiro, J.M.; Salústio, S.; Geraldes, V.; Valente, A.A.; Silva, C.M. Copper foam coated with CPO-27(Ni) metal–organic framework for adsorption heat pump: Simulation study using OpenFOAM. *Appl. Therm. Eng.* **2020**, *178*, 115498. [CrossRef]
165. Zu, K.; Qin, M. Optimization of the hygrothermal performance of novel metal-organic framework (MOF) based humidity pump: A CFD approach. *Energy* **2022**, *259*, 125073. [CrossRef]
166. Hua, Z.; Cai, S.; Xu, H.; Yuan, W.; Li, S.; Tu, Z. Investigations of Silica/MOF composite coating and its dehumidification performance on a desiccant-coated heat exchanger. *Energy* **2024**, *307*, 132576. [CrossRef]
167. Kummer, H.; Jeremias, F.; Warlo, A.; Füldner, G.; Fröhlich, D.; Janiak, C.; Gläser, R.; Henninger, S.K. A Functional Full-Scale Heat Exchanger Coated with Aluminum Fumarate Metal–Organic Framework for Adsorption Heat Transformation. *Ind. Eng. Chem. Res.* **2017**, *56*, 8393–8398. [CrossRef]
168. Lee, J.-G.; Bae, K.J.; Kwon, O.K. Experimental investigation of the solid desiccant dehumidification system with metal organic frameworks. *Int. J. Refrig.* **2021**, *130*, 179–186. [CrossRef]
169. Albaik, I.; Diab, K.E.; Saleh, M.; Al-Dadah, R.; Mahmoud, S.; Elsheniti, M.B.; Solmaz, I.; Salama, E.; Hassan, H.S.; Elkadi, M.F. MOF based coated adsorption system for water desalination and cooling integrated with Pre-treatment unit. *Sustain. Energy Technol. Assess.* **2023**, *56*, 103006. [CrossRef]
170. Albaik, I.; Al-Dadah, R.; Mahmoud, S.; Solmaz, I. Non-equilibrium numerical modelling of finned tube heat exchanger for adsorption desalination/cooling system using segregated solution approach. *Appl. Therm. Eng.* **2021**, *183*, 116171. [CrossRef]
171. Sud, S.; Kaur, G. A comprehensive review on synthetic approaches for metal-organic frameworks: From traditional solvothermal to greener protocols. *Polyhedron* **2021**, *193*, 114897. [CrossRef]
172. Phan, P.T.; Hong, J.; Tran, N.; Le, T.H. The Properties of Microwave-Assisted Synthesis of Metal–Organic Frameworks and Their Applications. *Nanomaterials* **2023**, *13*, 352. [CrossRef]
173. Klinowski, J.; Paz, F.A.A.; Silva, P.; Rocha, J. Microwave-Assisted Synthesis of Metal–Organic Frameworks. *Dalton Trans.* **2011**, *321*, 40. [CrossRef]
174. Chakraborty, D.; Yurdusen, A.; Mouchaham, G.; Nouar, F.; Serre, C. Large-Scale Production of Metal–Organic Frameworks. *Adv. Funct. Mater.* **2023**, *34*, 2309089. [CrossRef]
175. Paul, T.; Juma, A.; Alqerem, R.; Karanikolos, G.; Arafat, H.A.; Dumée, L.F. Scale-up of metal-organic frameworks production: Engineering strategies and prospects towards sustainable manufacturing. *J. Environ. Chem. Eng.* **2023**, *11*, 111112. [CrossRef]
176. Julien, P.A.; Mottillo, C.; Friščić, T. Metal–organic frameworks meet scalable and sustainable synthesis. *Green Chem.* **2017**, *19*, 2729. [CrossRef]
177. Isaeva, V.I.; Kustov, L.M. Microwave activation as an alternative production of metal-organic frameworks. *Russ. Chem. Bull.* **2016**, *65*, 2103–2114. [CrossRef]
178. Ni, Z.; Masel, R.I. Rapid Production of Metal−Organic Frameworks via Microwave-Assisted Solvothermal Synthesis. *J. Am. Chem. Soc.* **2006**, *128*, 12394–12395. [CrossRef] [PubMed]
179. Chen, C.; Feng, X.; Zhu, Q.; Dong, R.; Yang, R.; Cheng, Y.; He, C. Microwave-Assisted Rapid Synthesis of Well-Shaped MOF-74 (Ni) for CO_2 Efficient Capture. *Inorg. Chem.* **2019**, *58*, 2717–2728. [CrossRef] [PubMed]
180. Rahaman, S.K.; Chatterjee, T.; Alam, S.M. Chapter 4—Microwave-assisted synthesis of metal–organic frameworks. In *Synthesis of Metal-Organic Frameworks Via Water-Based Routes—A Green and Sustainable Approach*; Elsevier Inc.: Amsterdam, The Netherlands, 2024; pp. 51–72.

181. Parashar, M.; Shukla, V.K.; Singh, R. Metal oxides nanoparticles via sol–gel method: A review on synthesis, characterization and applications. *J. Mater. Sci. Mater. Electron.* **2020**, *31*, 3729–3749. [CrossRef]
182. Sharma, S.; Chand, P. Effect of calcination temperature on electrochemical behavior of pristine zeolitic imidazolate framework-8. *Ionics* **2022**, *28*, 5395–5404. [CrossRef]
183. Wirunchit, S.; Koetniyom, W. ZnO Nanoparticles Synthesis and Characterization by Hydrothermal Process for Biological Applications. *Phys. Status Solidi A* **2022**, *220*, 2200364. [CrossRef]
184. Kim, S.Y.; Cha, B.J.; Saqlain, S.; Seo, H.O.; Kim, Y.D. Atomic Layer Deposition for Preparation of Highly Efficient Catalysts for Dry Reforming of Methane. *Catalysts* **2019**, *9*, 266. [CrossRef]
185. Mageswari, K.; Prabukanthan, P.; Madhavan, J. Microwave-assisted synthesis of ZnO nanoparticles using different capping agents and their photocatalytic application. *Environ. Sci. Pollut. Res.* **2023**, *30*, 40174–40188. [CrossRef]
186. Zaytseva, O.; Neumann, G. Carbon nanomaterials: Production, impact on plant development, agricultural and environmental applications. *Chem. Biol. Technol. Agric.* **2016**, *3*, 17. [CrossRef]
187. Mandal, S.; Natarajan, S.; Mani, P.; Pankajakshan, A. Post-Synthetic Modification of Metal–Organic Frameworks Toward Applications. *Adv. Funct. Mater.* **2020**, *31*, 2006291. [CrossRef]
188. Wu, M.X.; Yang, Y.W. Metal–Organic Framework (MOF)-Based Drug/Cargo Delivery and Cancer Therapy. *Adv. Mater.* **2017**, *29*, 1606134. [CrossRef]
189. Lal, S.; Singh, P.; Singhal, A.; Kumar, S.; Gahlot, A.P.S.; Gandhi, N.; Kumari, P. Advances in metal–organic frameworks for water remediation applications. *RSC Adv.* **2024**, *14*, 3413–3446. [CrossRef] [PubMed]
190. Li, N.; Xu, J.; Feng, R.; Hu, T.-L.; Bu, X.-H. Governing metal–organic frameworks towards high stability. *Chem. Commun.* **2016**, *52*, 8501–8513. [CrossRef] [PubMed]
191. James, J.B.; Lang, L.; Meng, L.; Lin, J.Y.S. Postsynthetic Modification of ZIF-8 Membranes via Membrane Surface Ligand Exchange for Light Hydrocarbon Gas Separation Enhancement. *ACS Appl. Mater. Interfaces* **2019**, *12*, 3893–3902. [CrossRef] [PubMed]
192. Luan, Y.; Yang, M.; Ma, Q.; Qi, Y.; Gao, H.; Wu, Z.; Wang, G. Introduction of an organic acid phase changing material into metal–organic frameworks and the study of its thermal properties. *J. Mater. Chem. A* **2016**, *4*, 7641. [CrossRef]
193. Han, B.; Chakraborty, A. Functionalization, protonation and ligand extension on MIL-53 (Al) MOFs to boost water adsorption and thermal energy storage for heat transformations. *Chem. Eng. J.* **2023**, *472*, 145137. [CrossRef]
194. Liu, L.; Li, J.; Subhan, S.; Yu, X.; Liu, Z.; Chen, R.; Deng, J.; Ji, H.; Zhao, Z.; Zhao, Z. Construction of HKUST-1@Cu nanofibers with thermal conductive adsorption sites for synchronous enhancement of toluene adsorption and desorption efficiency. *Sep. Purif. Technol.* **2024**, *339*, 126624. [CrossRef]
195. Han, B.; Chakraborty, A. Experimental investigation for water adsorption characteristics on functionalized MIL-125 (Ti) MOFs: Enhanced water transfer and kinetics for heat transformation systems. *Int. J. Heat Mass Transf.* **2022**, *186*, 122473. [CrossRef]
196. Huo, Y.; Xiu, S.; Meng, L.-Y.; Quan, B. Solvothermal synthesis and applications of micro/nano carbons: A review. *Chem. Eng. J.* **2023**, *451*, 138572. [CrossRef]
197. Begum, S.; Hassan, Z.; Bra, S.; Wo, C.; Tsotsalas, M. Metal–Organic Framework-Templated Biomaterials: Recent Progress in Synthesis, Functionalization, and Applications. *Acc. Chem. Res.* **2019**, *52*, 1598–1610. [CrossRef]
198. Severino, M.I.; Gkaniatsou, E.; Nouar, F.; Pinto, M.L.; Serre, C. MOFs industrialization: A complete assessment of production costs. *Faraday Discuss.* **2021**, *231*, 326–341. [CrossRef]
199. Shahvari, S.Z.; Kalkhorani, V.A.; Wade, C.R.; Clark, J.D. Benefits of metal–organic frameworks sorbents for sorbent wheels used in air conditioning systems. *Appl. Therm. Eng.* **2022**, *210*, 118407. [CrossRef]
200. Thomas-Hillman, I.; Laybourn, A.; Dodds, C.; Kingman, S.W. Realising the environmental benefits of metal–organic frameworks: Recent advances in microwave synthesis. *J. Mater. Chem. A* **2018**, *6*, 11564–11581. [CrossRef]
201. Ding, M.; Cai, X.; Jiang, H.-L. Improving MOF stability: Approaches and applications. *Chem. Sci.* **2019**, *10*, 10209–10230. [CrossRef]
202. Cai, Y.Y.; Wang, J.J.; Cai, Z.H.; Zhang, Q.G.; Zhu, A.M.; Liu, Q.L. Enhanced Performance of Sulfonated Poly(ether ether Ketone) Hybrid Membranes by Introducing Sulfated MOF-808/Graphene Oxide Composites. *ACS Appl. Energy Mater.* **2021**, *4*, 9664–9672. [CrossRef]
203. Álvarez, J.R.; Sánchez-González, E.; Pérez, E.; Schneider-Revueltas, E.; Martínez, A.; Tejeda-Cruz, A.; Islas-Jácome, A.; González-Zamora, E.; Ibarra, I.A. Structure stability of HKUST-1 towards water and ethanol and their effect on its CO_2 capture properties. *Dalton Trans.* **2017**, *46*, 9192–9200. [CrossRef] [PubMed]
204. Ding, M.; Jiang, H.-L. Improving Water Stability of Metal–Organic Frameworks by a General Surface Hydrophobic Polymerization. *CCS Chem.* **2021**, *3*, 2740–2748. [CrossRef]
205. Khouki, M.; Al Khatib, O. Performance of Desiccant-Based Cooling Systems in Hot-Humid Climates: A Review. *Energy Eng.* **2021**, *118*, 875–909. [CrossRef]

206. Zhao, L.H.; Wang, R.Z.; Ge, T.S. Desiccant coated heat exchanger and its applications. *Int. J. Refrig.* **2021**, *130*, 217–232. [CrossRef]
207. Zheng, X.; Ge, T.S.; Wang, R.Z. Recent progress on desiccant materials for solid desiccant cooling systems. *Energy* **2014**, *74*, 280–294. [CrossRef]
208. Mahmoodi, N.M.; Taghizadeh, M.; Taghizadeh, A. Activated carbon/metal-organic framework composite as a bio-based novel green adsorbent: Preparation and mathematical pollutant removal modeling. *J. Mol. Liq.* **2019**, *277*, 310–322. [CrossRef]
209. Hu, L.; Ge, T.; Jiang, Y.; Wang, R. Performance study on composite desiccant material coated fin-tube heat exchangers. *Int. J. Heat Mass Transf.* **2015**, *90*, 109–120. [CrossRef]
210. Cui, S.; Marandi, A.; Lebourleux, G.; Thimon, M.; Bourdon, M.; Chen, C.; Severino, M.I.; Steggles, V.; Nouar, F.; Serre, C. Heat properties of a hydrophilic carboxylate-based MOF for water adsorption applications. *Appl. Therm. Eng.* **2019**, *161*, 114135. [CrossRef]
211. Silva, P.; Vilela, S.M.F.; Tome, J.P.C. Multifunctional metal–organic frameworks: From academia to industrial applications. *Chem. Soc. Rev.* **2015**, *44*, 6774–6803. [CrossRef] [PubMed]
212. Qu, H.J.; Huang, L.J.; Han, Z.Y.; Wang, Y.W.; Zhang, Z.J.; Wang, Y.; Chang, Q.R.; Wei, N.; Kipper, M.J.; Tang, J.G. A review of graphene-oxide/metal–organic framework composites materials: Characteristics, preparation and applications. *J. Porous Mater.* **2021**, *28*, 1837–1865. [CrossRef]
213. Szczęśniak, B.; Choma, J.; Jaroniec, M. Gas adsorption properties of hybrid graphene-MOF materials. *J. Colloid Interface Sci.* **2018**, *514*, 801–813. [CrossRef] [PubMed]
214. Li, C.; Zhang, L.; Chen, J.; Li, X.; Sun, J.; Zhu, J.; Wang, X.; Fu, Y. Recent development and applications of electrical conductive MOFs. *Nanoscale* **2020**, *13*, 485–509. [CrossRef]
215. Atinafu, D.G.; Chang, S.J.; Berardi, U.; Kim, K.-H.; Kim, S. Potential utility of HKUST-1-graphite nanocomposite to endow alkane with high thermal properties and low electrical resistivity. *J. Hazard. Mater.* **2021**, *402*, 123695. [CrossRef]
216. Ghanbari, T.; Patah, M.F.A.; Wong, Y.H.; Abnisa, F.; Daud, W.M.A.W. Probing the capability of the MOF-74(Ni)@GrO composite for CO_2 adsorption and CO_2/N_2 separation: A combination of experimental and molecular dynamic simulation studies. *Fuel* **2024**, *372*, 131837. [CrossRef]
217. Yu, Z.; Deschamps, J.; Hamon, L.; Prabhakaran, P.K.; Pré, P. Hydrogen adsorption and kinetics in MIL-101(Cr) and hybrid activated carbon-MIL-101(Cr) materials. *Int. J. Hydrogen Energy* **2017**, *42*, 8021–8031. [CrossRef]
218. Qadir, N.U.; Said, S.A.M.; Mansour, R.B.; Mezghani, K.; Ul-Hamid, A. Synthesis, characterization, and water adsorption properties of a novel multi-walled carbon nanotube/MIL-100(Fe) composite. *Dalton Trans.* **2016**, *45*, 15621–15633. [CrossRef]
219. Fang, Y.; Yang, Z.; Li, H.; Liu, X. MIL-100(Fe) and its derivatives: From synthesis to application for wastewater decontamination. *Environ. Sci. Pollut. Res.* **2020**, *27*, 4703–4724. [CrossRef] [PubMed]
220. Samy, M.; Ibrahim, M.G.; Fujii, M.; Diab, K.E.; ElKady, M.; Alalm, M.G. CNTs/MOF-808 painted plates for extended treatment of pharmaceutical and agrochemical wastewaters in a novel photocatalytic reactor. *Chem. Eng. J.* **2021**, *406*, 127152. [CrossRef]
221. Kazemi, A.; Pordsari, M.A.; Tamtaji, M.; Afshari, M.H.; Keshavarz, S.; Zeinali, F.; Baesmat, H.; Zahiri, S.; Manteghi, F.; Ghaemi, A.; et al. Unveiling the power of defect engineering in MOF-808 to enhance efficient carbon dioxide adsorption and separation by harnessing the potential of DFT analysis. *Chem. Eng. J.* **2024**, *494*, 153049. [CrossRef]
222. Yang, Z.; Xu, X.; Liang, X.; Lei, C.; Wei, Y.; He, P.; Lv, B.; Ma, H.; Lei, Z. MIL-53(Fe)-graphene nanocomposites: Efficient visible-light photocatalysts for the selective oxidation of alcohols. *Appl. Catal. B Environ.* **2016**, *198*, 112–123. [CrossRef]
223. Serventi, D.R. MIL-53 (Al) and Graphene Oxide Nanocomposites for Dye Adsorption. 2020. Available online: http://hdl.handle.net/11122/11877 (accessed on 14 November 2024).
224. Abdi, J.; Banisharif, F.; Khataee, A. Amine-functionalized Zr-MOF/CNTs nanocomposite as an efficient and reusable photocatalyst for removing organic contaminants. *J. Mol. Liq.* **2021**, *334*, 116129. [CrossRef]
225. Wang, C.; Wang, W.; Wu, Y.; Li, Y.; Yan, Y.; Zhu, J. In-situ synthesis of CNT/UiO-66-NH_2-based molecularly imprinted nanocomposite membranes for selective recognition and separation of sulfamethoxazole: A synergistic promotion system. *Surf. Interfaces* **2022**, *31*, 101986. [CrossRef]
226. Mahmoodi, N.M.; Bakhtiari, M.; Oveisi, M.; Mahmoodi, B.; Hayati, B. Green synthesis of eco-friendly magnetic metal-organic framework nanocomposites (AlFum-graphene oxide) and pollutants (dye and pharmaceuticals) removal capacity from water. *Mater. Chem. Phys.* **2023**, *302*, 127720. [CrossRef]
227. Abdi, S.; Nasiri, M. Enhanced Hydrophilicity and Water Flux of Poly(ether sulfone) Membranes in the Presence of Aluminum Fumarate Metal–Organic Framework Nanoparticles: Preparation and Characterization. *ACS Appl. Mater. Interfaces* **2019**, *11*, 15060–15070. [CrossRef]
228. Ru, J.; Wang, X.; Zhao, J.; Yang, J.; Zhou, Z.; Du, X.; Lu, X. Evaluation and development of GO/UiO-67@PtNPs nanohybrid-based electrochemical sensor for invisible arsenic (III) in water samples. *Microchem. J.* **2022**, *181*, 107765. [CrossRef]

229. Chen, Z.; Wasson, M.C.; Drout, R.J.; Robison, L.; Idrees, K.B.; Knapp, J.G.; Son, F.A.; Zhang, X.; Hierse, W.; Kühn, C.; et al. The state of the field: From inception to commercialization of metal–organic frameworks. *Faraday Discuss.* **2021**, *225*, 9–69. [CrossRef]
230. Kong, X.J.; Li, J.R. An Overview of Metal–Organic Frameworks for Green Chemical Engineering. *Engineering* **2021**, *7*, 1115–1139. [CrossRef]
231. Zhang, Q.; Yan, S.; Yan, X.; Lv, Y. Recent advances in metal-organic frameworks: Synthesis, application and toxicity. *Sci. Total Environ.* **2023**, *902*, 165944. [CrossRef]
232. Jahan, I.; Rupam, T.H.; Palash, M.; Rocky, K.A.; Saha, B.B. Energy efficient green synthesized MOF-801 for adsorption cooling applications. *J. Mol. Liq.* **2022**, *345*, 117760. [CrossRef]
233. Klimakow, M.; Klobes, P.; Thünemann, A.F.; Rademann, K.; Emmerling, F. Mechanochemical Synthesis of Metal−Organic Frameworks: A Fast and Facile Approach toward Quantitative Yields and High Specific Surface Areas. *Chem. Mater.* **2010**, *22*, 5216–5221. [CrossRef]
234. Chen, D.; Zhao, J.; Zhang, P.; Dai, S. Mechanochemical synthesis of metal–organic frameworks. *Polyhedron* **2019**, *162*, 59–64. [CrossRef]
235. Almohasin, J.A.; Balag, J.; Miral, V.G.; Moreno, R.V.; Tongco, L.J.; Lopez, E.C.R. Green Solvents for Liquid–Liquid Extraction: Recent Advances and Future Trends. *Eng. Proc.* **2023**, *56*, 174. [CrossRef]
236. Qian, X.; Yadian, B.; Wu, R.; Long, Y.; Zhou, K.; Zhu, B.; Huang, Y. Structure stability of metal-organic framework MIL-53 (Al) in aqueous solutions. *Int. J. Hydrogen Energy* **2013**, *38*, 16710–16715. [CrossRef]
237. Alvares, E.; Tantoro, S.; Wijaya, C.J.; Cheng, K.C.; Soetaredjo, F.E.; Hsu, H.Y.; Angkawijaya, A.E.; Go, A.W.; Hsieh, C.W.; Santoso, S.P. Preparation of MIL100/MIL101-alginate composite beads for selective phosphate removal from aqueous solution. *Int. J. Biol. Macromol.* **2023**, *231*, 123322. [CrossRef] [PubMed]
238. Głowniak, S.; Szczęśniak, B.; Choma, J.; Jaroniec, M. Advances in Microwave Synthesis of Nanoporous Materials. *Adv. Mater.* **2021**, *33*, 2103477. [CrossRef] [PubMed]
239. Boukayouht, K.; Bazzi, L.; El Hankari, S. Sustainable synthesis of metal-organic frameworks and their derived materials from organic and inorganic wastes. *Coord. Chem. Rev.* **2023**, *478*, 214986. [CrossRef]
240. Pham, H.K.; Sim, Y.; Carboni, M.; Meyer, D.; Mathews, N. Generating metal-organic frameworks (MOFs) from photovoltaic modules for wastewater remediation. *J. Environ. Chem. Eng.* **2022**, *10*, 108346. [CrossRef]

nanomaterials

Article

M-Doped (M = Zn, Mn, Ni) Co-MOF-Derived Transition Metal Oxide Nanosheets on Carbon Fibers for Energy Storage Applications

Andrés González-Banciella [1,*], David Martinez-Diaz [1], Adrián de Hita [1], María Sánchez [1,2,*] and Alejandro Ureña [1,2]

[1] Materials Science and Engineering Area, Escuela Superior de Ciencias Experimentales y Tecnología, Universidad Rey Juan Carlos, C/Tulipán s/n, 28933 Móstoles, Spain

[2] Instituto de Investigación de Tecnologías para la Sostenibilidad, Universidad Rey Juan Carlos, C/Tulipán s/n, 28933 Móstoles, Spain

* Correspondence: andres.banciella@urjc.es (A.G.-B.); maria.sanchez@urjc.es (M.S.)

Abstract: Carbon fiber, with its strong mechanical properties and electrical conductivity, is ideal as a fiber electrode in wearable or structural energy storage devices. However, its energy storage capacity is limited, and coatings like transition metal oxides (TMOs) enhance its electrochemical performance. Metal–organic frameworks (MOFs) are commonly used to grow TMOs on carbon fibers, increasing the surface area for better energy storage. Despite this, TMOs have limited electrical conductivity, so ion exchange is often used to dope them with additional cations, improving both conductivity and energy storage capacity. This study compares different ion-exchange cations in ZIF-L-derived TMO coatings on carbon fiber. Testing both supercapacitor and Li-ion battery applications, Ni-doped samples showed superior results, attributed to their higher exchange ratio with cobalt. As a supercapacitor electrode, the Ni-doped material achieved 13.3 F/g at 50 mA/g—66% higher than undoped samples. For Li-ion battery anodes, it reached a specific capacity of 410.5 mAh/g at 25 mA/g, outperforming undoped samples by 21.4%.

Keywords: metal–organic frameworks; supercapacitors; Li-ion batteries; transition metal oxides; carbon fiber

1. Introduction

The development of all-solid-state energy storage devices has emerged as one of the most important challenges for the achievement of sustainable development goals [1]. The replacement of conventional flammable electrolytes with their new solid counterparts is not only a safety issue [2] but also an opportunity to achieve new functionalities [3–5]. All-solid-state batteries and supercapacitors could be wearable or even present mechanical properties, which allow their use not only as energy storage components but also as structural ones [6–8]. This multifunctionality could open the way for a new generation of structures with energy storage capability, which would be more efficient and sustainable [9]. Nevertheless, this goal is still far from being achieved due to the difficulty of finding materials that show a good relationship between electrochemical and mechanical performance [1,9]. In this context, carbon fiber (CF) fabrics have been proposed as a promising substrate for battery or supercapacitor electrodes due to their high elastic modulus, high mechanical resistance, excellent electrical conductivity, and low density [10,11]. So much so that even the bare carbon fiber fabric has been commonly used as an all-solid-state battery anode [12,13]. However, to improve the electrochemical performance of the device, the design of anodes that display higher capacities and capacitances is essential [10]. In fact, for this purpose, several coatings with other materials, which display higher energy storage capability, have been widely reported in the last years [14–20]. Transition metal oxides

(TMOs) are known to be widely reported as battery anodes [21,22]. These materials are abundant, their cost is low, and, what is even more relevant, their specific capacity is about three times higher than that of conventional graphitic anodes [23]. In addition, TMOs are widely reported as supercapacitor (SC) electrode materials due to their chemical stability and larger capacitances due to pseudocapacitive reactions [24–26]. However, TMOs show two main disadvantages for these purposes. The first one is that TMOs present a relatively low electrical conductivity, which limits the electrode's rate capability. Thus, bimetallic and bivalence TMOs have attracted more attention than their counterparts, which involve only a single-valence metal. So, the presence of several metals or metals in more than one oxidation state normally improves the electrical conductivity due to the activation energy for the electron transfer between cations being lower [27,28]. The second disadvantage is the poor cyclability of TMOs related to the volume expansion during the lithiation–delithiation process [21,23], which damages the material, worsening the electrical contact. In order to solve this problem, the use of porous TMOs has been demonstrated to be key. The synthesis of porous TMOs has been widely researched, with the obtention through metal–organic frameworks (MOFs) calcination a common way to achieve it [29–31]. MOFs exhibit large mesoporosity, which partially remains after the calcination [32]. Moreover, the porosity associated with the MOF-derived obtention improves the capacitance of TMOs as supercapacitor electrodes by the surface area [18]. Thus, this synthesis strategy has turned into an attractive way to improve TMO performance as both battery anodes [31,33] and supercapacitor electrodes [34]. In the last years, some examples of MOF-derived TMO coatings on carbon fiber have been reported [35–40]. The majority of works are based on cobalt and/or zinc zeolitic imidazole frameworks as the MOF precursor due to their demonstrated easy synthesis on carbon fiber and the excellent electrochemical performance of the derived Co_3O_4 and $ZnCo_2O_4$ spinels [35–53]. In this way, these materials have been demonstrated to be promising electrodes for all-solid-state Li-ion batteries [42,49–53] and supercapacitors [36–39,41,45,48,54]. Nevertheless, derived compounds obtained by the ion exchange of the MOF precursor with different cations have exhibited better performance [41,44–46]. In this work, Co-ZIF-L synthesized over carbon fiber fabric has been submitted to an ion exchange with Ni^{2+}, Zn^{2+}, and Mn^{2+} cations in order to obtain M-doped Co_3O_4 (M = Ni, Mn, Zn) coatings and compare their electrochemical performance as both Li-ion battery anode and supercapacitor electrode for all-solid-state devices, studying the physicochemical reasons behind the differences.

2. Materials and Methods

2.1. Materials

Plain weaving (1 × 1) (193 g/m^2) of 3K AS4 GP carbon fiber supplied by Hexcel® (Stamford, EUUU). Nitric acid (HNO_3) 65% v/v and zinc (II) nitrate hexahydrate ($Zn(NO_3)_2{\cdot}6H_2O$) ≥ 99% supplied by Sigma-Aldrich (Burlington, EEUU). 2-Methylimidazole (2-mIM) < 98% supplied by TCI (Tokyo, Japan). Cobalt (II) nitrate hexahydrate ($Co(NO_3)_2{\cdot}6H_2O$) 98.0–102.0%, Manganese (II) nitrate tetrahydrate ($Mn(NO_3)_2{\cdot}4H_2O$) 98%, and Nickel (II) nitrate hexahydrate ($Ni(NO_3)_2{\cdot}6H_2O$) 98% supplied by Thermo Fisher scientific (Waltham, EEUU).

2.2. TMO Synthesis on Carbon Fiber Fabric

First, ZIF-L MOF was synthesized on the carbon fiber fabrics to be used as a precursor for the later obtention of the TMO materials. For this purpose, CF fabrics of 4 × 4 cm were cut, and then the original coating of the fibers (commonly called sizing) was removed through immersion in acetone for 48 h. After that, CF fabrics were immersed in HNO_3 for 1 h to generate polar groups on the surface through partial oxidation of the fibers [55]. Subsequently, fabrics were washed with distilled water and immersed in a $Co(NO_3)_2$ solution of 0.05 M for 10 min to favor the MOF growth on the CF surface through electrostatic interactions between the generated polar groups and Co^{2+} cations [42,56]. Having passed this time, CF pieces were removed from the solution, and 2-methylimidazole (0.4 M) was added to the previous solution. Finally, after mixing both solutions for 1 min, the carbon fiber fabrics were introduced into the mixture for 1 h, washed with distilled water, and dried overnight at 60 °C. Moreover, the synthesized MOF powder particles, which did not adhere to the CF or did not begin to grow on it, were filtered, washed, and dried using the same method as the CF fabrics. The next step involved ion exchange. CF samples were individually introduced into 0.01 M solutions of $M(NO_3)_2$ (M = Zn, Mn, and Ni) in ethanol for 10 min. After drying, all the samples were submitted to a heat treatment to obtain the corresponding TMOs from the MOF coatings. To evaluate the effect on the final performance of the three proposed different ion-exchange steps, samples without ion exchange were also subjected to the same thermal treatment. More in detail, the heat treatment consisted of two steps. The first one was annealing at 500 °C for 1 h in the Ar atmosphere to generate oxygen vacancies, which improve the electrical conductivity [40]. The second one involved a calcination step at 350 °C for 30 min with the aim of transforming the MOF material into the final TMO coating. The samples were named Zn-doped TMO, Mn-doped TMO, Ni-doped TMO, and undoped TMO. As references, powders collected in the synthesis underwent the same ion exchange in ethanol solutions with $M(NO_3)_2$ (M = Zn, Mn, and Ni), followed by the same heat treatments.

2.3. Material Characterization

Scanning electron microscopy (SEM) images from MOF precursor coatings on carbon fiber were taken by S-3400 N by Hitachi (Chiyoda, Japan), while SEM images of TMO coatings on the carbon fiber were taken using a PRISMA-E by Thermo Fisher (Waltham, EEUU). Transmission electron microscopy (TEM) images were acquired using STEM F200 by JEOL (Tokyo, Japan). X-ray diffraction (XRD) patterns were acquired from the reference powders by X'PERT diffractometer by PHILIPS (Amsterdam, Netherlands), using Cu Kα radiation. Moreover, X-ray photoelectronic spectra (XPS) were collected by VersaProbe II by PHI (Chanhassen, EEUU) using Al 1486.6 eV mono at 47.3 W as an X-ray source.

2.4. Supercapacitor Electrode Characterization

The electrochemical performance of the samples as a supercapacitor electrode was evaluated in a three-electrode system in which the counter-electrode was a Pt electrode, the reference electrode an Ag/AgCl electrode, and the electrolyte was a KOH 2 M solution. Cyclic voltammetry (CV) and galvanostatic charge–discharge (GCD) measurements were acquired with an Autolab PGSTAT302N potentiostat, being the potential window between 0 and 0.45 V. The studied current densities were 50, 100, 150, 200 and 250 mA/g. Additionally, the stability of the electrodes was tested by measuring the specific capacitance after 5000 GCD cycles at 150 mA/g of current density using a NEWARE battery testing system BTS4000-5V10M. The specific capacitance values (C_s) were calculated from CV as:

$$C_s = \frac{\int_{V_{min}}^{V_{max}} i dV}{2\, \nu\, V_0}, \quad (1)$$

where V_0 is the potential window, ν the scan rate, and i the current density. On the other hand, the specific capacitance was also calculated from GCD results as:

$$C_s = \frac{it}{V_0}, \quad (2)$$

where t is the time of the discharge. The masses of the electrodes were 0.0021 g for the Undoped sample, 0.0019 g for the Zn-doped sample, 0.0019 g for the Mn-doped sample, and 0.0019 g for the Ni-doped one.

2.5. Li-Ion Battery Anode Characterization

Electrochemical measurements of TMO coatings on CF for Li-ion anodes were performed in half-cell configuration coins 2025. Lithium metal served as counter-electrode, and glass microfiber filters with a diameter of 150 mm, supplied by Whatman, were used as separators. Coins were assembled into an Ar-filled glovebox to avoid O_2 and H_2O presence. A 1 M $LiPF_6$ electrolyte in ethylene carbonate and dimethyl carbonate (1:1 v/v) supplied by Merck was used. Galvanostatic charge–discharge (GCD) tests were performed using a NEWARE battery testing system BTS4000-5V10MA. To determine the specific capacity and evaluate the rate capability and capacity retention, tests consisting of five cycles at different current densities were performed (25, 50, 100, 250, 500, and 25 mA/g). Additionally, to fully evaluate the capacity retention, 100 GCD cycle tests were carried out at 100 mA/g to evaluate the long-term performance of the batteries. Additionally, cyclic voltammetry (CV) and electrochemical impedance spectroscopy (EIS) tests were conducted using an Autolab PGSTAT302N potentiostat. The potential window of CV tests was from 0 to 3 V, while EIS measures were taken at 0.01 V between 0.1 and 105 Hz.

In this case, the specific capacity values were calculated from GCD results as follows:

$$C_s = it \quad (3)$$

The masses of the electrodes were 0.0022 g for the undoped sample, 0.0019 g for the Zn-doped sample, 0.0020 g for the Mn-doped sample, and 0.0017 g for the Ni-doped one.

3. Results

3.1. Coating Characterization

In Figure 1, the obtained coating morphology was examined by SEM. Figure 1a shows a general image of the ZIF-L MOF coating on the CFs, demonstrating both its homogeneity and the high density achieved. As previously detailed, this MOF coating will be used as a precursor for the later formation of the different TMO materials. Furthermore, it can be observed how the MOF coats each CF instead of being a continuous layer over the tows. This kind of coating could be advantageous in ensuring a good interface between the CF fabric electrode and the solid polymer electrolyte. Figure 1b shows a more detailed image of the ZIF-L coating, revealing triangular nanosheets with nanometric thickness, around 1 μm in size. The resulting TMO without ion exchange (undoped TMO) after the heat treatment is shown in Figure 1c, where some similarities and differences in comparison with the ZIF-L precursor coating can be analyzed. The coating retains its high density, homogeneity, and nanosheet morphology, which provide a high surface area that is essential for good electrochemical performance, especially for supercapacitor applications. Nevertheless, the thickness is lower, and the nanosheets appear less defined after the heat treatment. Finally, Figure 1d–f show the resulting TMO coating with the different proposed ion exchanges: Zn-doped, Mn-doped, and Ni-doped, respectively. As can be observed, no significant differences in the morphology of the coating were detected between the undoped and doped TMO coatings.

In order to make sure that ion exchange has occurred satisfactorily and to analyze the composition of each sample, XPS and EDS spectra were collected and presented in Figure 2a and Figure S1, respectively. Both EDS spectra and XPS surveys demonstrated the presence of the dopant cations (Ni^{2+}, Zn^{2+}, and Mn^{2+}) in the corresponding samples. Figure 2b shows the Co 2p core level of the undoped sample. Two main peaks, located about 790 and 795.6 eV, correspond to $2p_{3/2}$ and $2p_{1/2}$, respectively, while the wider peaks, located at 786 and 802.3 eV, are the corresponding satellites of these levels. However, the main peaks can be deconvoluted into the other two peaks, each one obtaining four peaks at 779.8, 794.8, 781.2, and 797.0 eV. The first two peaks are attributed to Co^{3+}, and the other two are associated with Co^{2+} [36,57]. The presence of the two oxidation states suggests that the composition of the undoped TMO could be Co_3O_4. Figure 2c–e show the Zn 2p, Mn 2p, and Ni 2p core levels of the ion-exchanged samples Zn-doped, Mn-doped, and Ni-doped, respectively. In these core levels, it is realizable that all dopant cations show a fully or very majority 2+ oxidation state. The cation atomic contribution was quantified, and the results are summarized in Figure 2f. Ni exchanged to Co by 25.3%, being the highest among the dopants, possibly due to its size similarity to Co, which better suits

its positions than the others. Moreover, Zn exchanged to Co by 21.1% while Mn only by 16.9% due to its larger size, which makes it difficult not only to exchange in the crystalline structure but also the intercalation. Having demonstrated that the ion exchange occurs for all dopant cations, XRD analysis was performed and summarized in Figure 2g to determine the crystalline structure and how it is affected by the ion exchange. In regard to undoped TMO, the XRD spectra match the expected XRD for Co_3O_4 spinel, while the doped-TMOs present slight differences with respect to the undoped one. Thus, XRD spectra indicate the same crystalline structure for all obtained TMOs. However, some conclusions can be drawn from them. Taking into account the fact that, according to XPS analysis, the doping in neither case reaches 33% of total cations and XRD results, the obtained TMOs are Co_3O_4 for the undoped sample, $ZnCo_2O_4@Co_3O_4$ for the Zn-doped sample, $MnCo_2O_4@Co_3O_4$ for the Mn-doped sample, and $NiCo_2O_4@Co_3O_4$ for the Ni-doped one. Dopant cations distort spinel unit cells, so crystallinity is affected by the ion exchange. As a result of the great difference in size between Co^{2+} and Zn^{2+}, the Zn-doped sample presents wider peaks, which indicate lower crystallinity. Nevertheless, in spite of being Ni^{2+}, the more similar cation to Co^{2+} in size, the Mn-doped sample presents more crystallinity due to the low exchange degree.

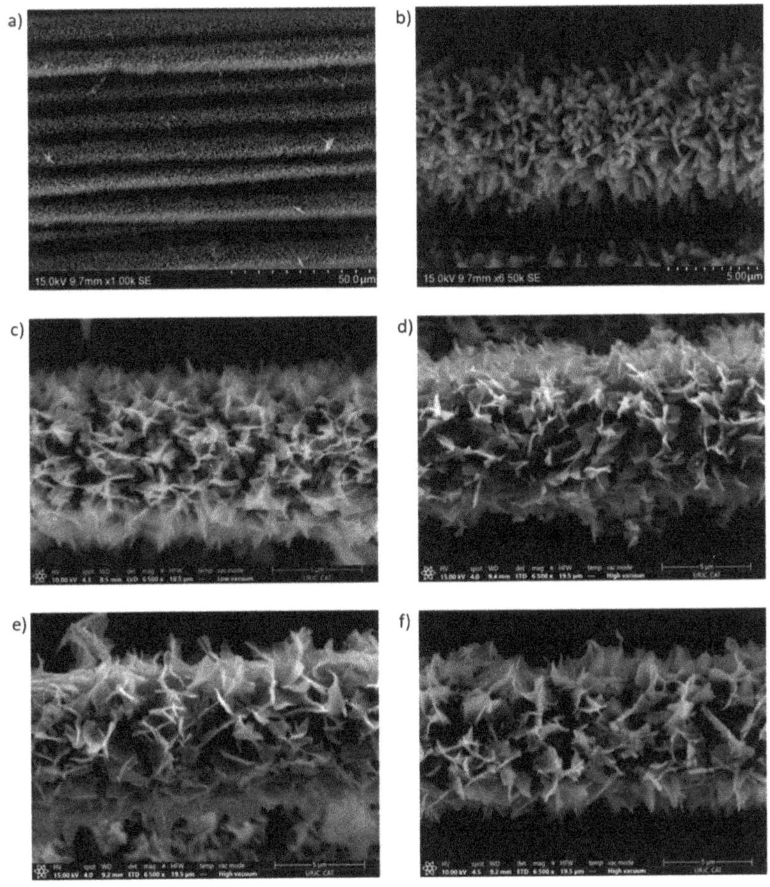

Figure 1. SEM images of ZIF-L coating on carbon fiber fabric at (**a**) lower magnification and (**b**) higher magnification, (**c**) undoped TMO sample, (**d**) Zn-doped TMO sample, (**e**) Mn-doped TMO sample, and (**f**) Ni-doped TMO sample.

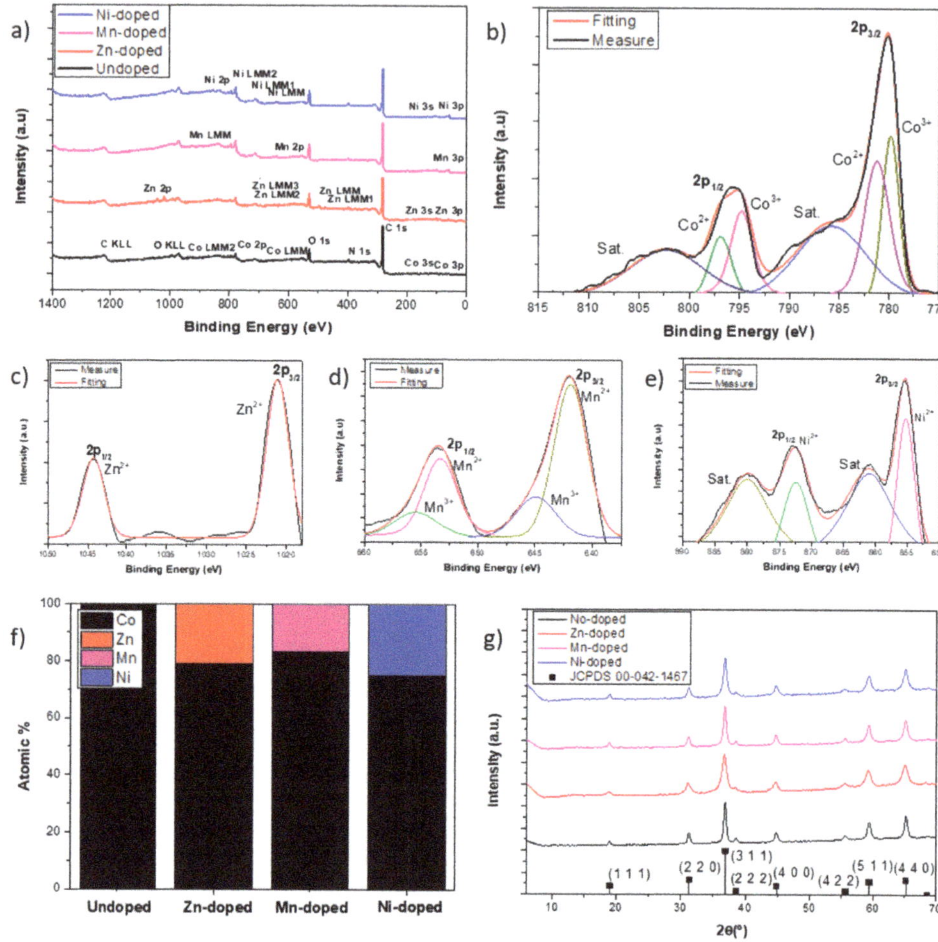

Figure 2. (**a**) XPS surveys. (**b**) XPS Co 2p core level of the undoped sample. (**c**) XPS Zn 2p core level of Zn-doped sample. (**d**) XPS Mn 2p core level of Mn-doped sample. (**e**) XPS Ni 2p core level of Ni-doped sample. (**f**) The atomic percentage of each cation in each sample. (**g**) XRD spectra of TMO powder samples.

Additionally, in order to study the microstructure, TEM images of the undoped TMO are presented in Figure 3. It is possible to observe that the nanosheets are composed of little nanocrystals with a diameter of around 10 nm, displaying high crystallinity. Although diffraction patterns were not possible to obtain due to the small grain size, some interplanar distances were measured in real space. One of them was 2.31 Å and the other 2.41 Å. Both could correspond to Co_3O_4 spinel structure, the first one to the plane (2 2 2) and the second one to the plane (3 1 1), confirming XRD crystal structure results.

3.2. Characterization as Supercapacitor Electrode

After compositional and morphological characterization of the synthesized TMO coatings, CV tests were performed to compare the electrochemical processes of the samples acting as supercapacitor electrodes. Figure 4a shows a comparison of CV curves for all samples at a scan rate of 20 mV/s, revealing that all samples displayed an oxidation peak at 0.4 V and another reduction peak at 0.36 V, indicating a pseudocapacitance energy storage mechanism. These peaks can be associated with the reversible faradaic reactions (4) and (5) [58,59]:

$$MCo_2O_4 + H_2O + OH^- \leftrightarrow MOOH + 2CoOOH + e^- \quad (4)$$

$$CoOOH + OH^- \leftrightarrow CoO_2 + H_2O + e^- \quad (5)$$

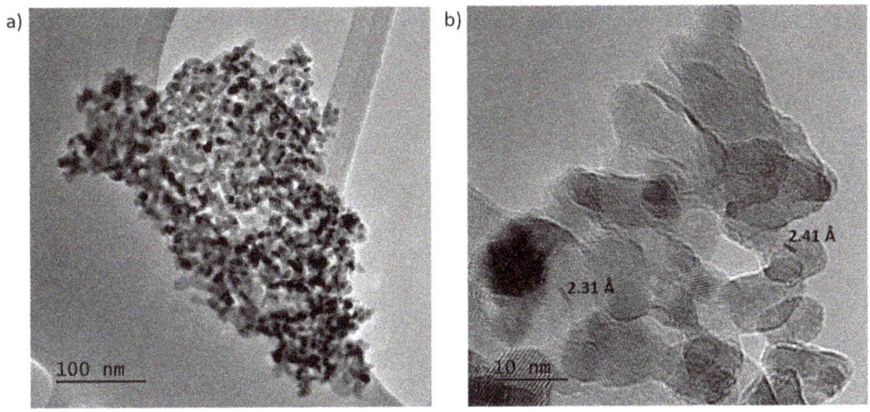

Figure 3. TEM images of undoped TMO samples at (**a**) lower and (**b**) higher magnification.

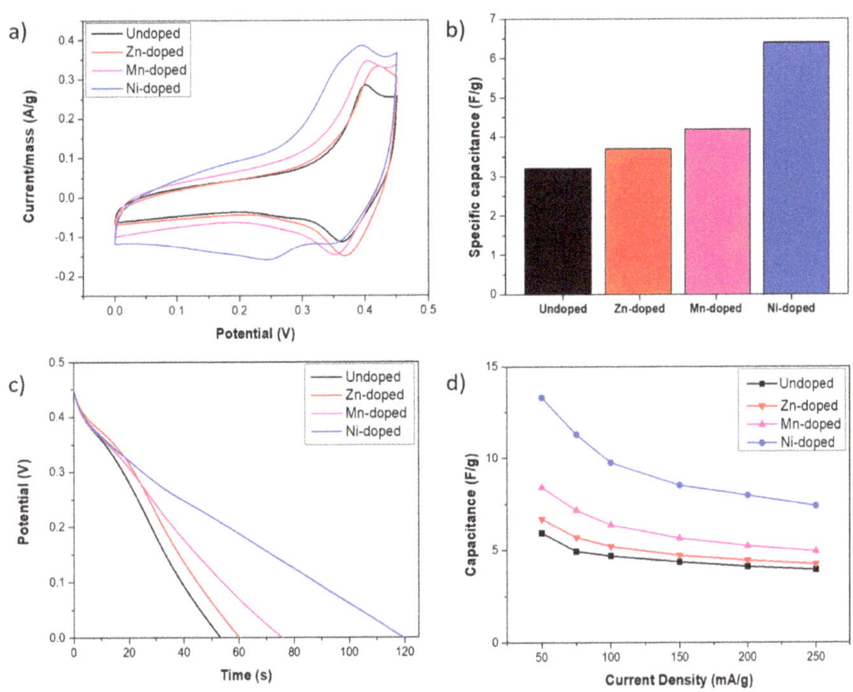

Figure 4. (**a**) CV curves at a scan rate of 20 mV/s. (**b**) Specific capacitance values are calculated from CV at 20 mV/s. (**c**) GCD test discharges at 50 mA/g. (**d**) Specific capacitance calculated from GCD tests at different current densities.

The material that exhibited the lower current at these peaks was the undoped one due to the presence of other cations on cobaltite spinels, resulting in a lattice distortion that improves the electrochemical activity [60], being the undoped material the only one that does not exhibit this effect. Moreover, in the Ni-doped sample, the oxidation peak appears to consist of two different peaks, one at 3.95 V and the other at 3.6 V, while another reduction peak at 0.24 V, hardly visible on the other sample curve, is evident. These peaks are related to the reaction (6) [61–63].

$$NiOOH + OH^{-} \leftrightarrow NiO + H_2O + e^{-} \quad (6)$$

Figure 4b summarizes the calculated specific capacitance displayed for each sample from CV curves at 20 mV/s, considering the mass as the whole mass of the electrode. Because of the aforementioned additional reactions, the Ni-doped sample displayed a larger specific capacitance, achieving 6.4 F/g per gram of the whole electrode. Nevertheless, all doped samples exhibited larger specific capacitance than the undoped one, which only displayed 3.2 F/g, while the Zn-doped and Mn-doped displayed 3.7 F/g and 4.6 F/g, respectively. In order to evaluate the electrochemical performance of the samples as supercapacitor electrodes, GCD tests were carried out and represented in Figure S2. Figure 4c depicts only discharges at 50 mA/g, facilitating the comparison. The above-discussed reduction peak at 0.36 V is observable as a plateau in the GCD discharge in all cases. Nonetheless, the plateau at 0.24 V is also observable in the Ni-doped sample curve, as was expected from CV results. In addition, it is possible to obtain information about the internal resistance from GCD discharge curves that are associated with the IR drop. Analyzing the discharge curve of the CGD test (Figure S3), performed from 50 to 250 mA/g, it can be concluded that the IR drops at 250 mA/g are more evident. These IR drops and their associated internal resistance are summarized in Table 1. The results suggest that the Zn-doped sample displays the lower electrical resistance, followed by the Ni-doped, Mn-doped, and undoped ones, respectively. Moreover, the specific capacitances were also evaluated by GCD tests, and the results are summarized in Figure 4d. Inconsistent with CV tests, Ni-doped samples exhibited the highest capacitance values at all studied current densities. For instance, at a current density of 50 mA/g, it reached 13.3 F/g, while the Mn-doped, Zn-doped, and undoped samples achieved 8.4, 6.7, and 5.9 F/g, respectively. This represents a 37%, 50%, and 66% decrease compared to the Ni-doped sample. Even at a current density of 250 mA/g, the Ni-doped sample maintained a specific capacitance of 7.4 F/g, which is 56% of the value observed at 50 mA/g. Thus, Ni-doped sample electrochemical performance as a supercapacitor electrode was further characterized due to its superior performance in this application.

Table 1. IR drops at a current density of 250 mA/g.

	Undoped	Zn-Doped	Mn-Doped	Ni-Doped
IR drop	67 mV	55 mV	64 mV	58 mV
Internal Resistance	0.134 Ω	0.110 Ω	0.128 Ω	0.116 Ω

Kinetic information about the electrode can be obtained from CV tests. The current of a peak (i) increases proportionally to the scan rate (v) when the process is totally capacitive-controlled, while it increases proportionally to the square root of the scan rate when the process is totally diffusion-controlled. Then, it is possible to understand the dependence of the current peak with the scan rate as detailed in Equation (7):

$$i = av^{b} \quad (7)$$

where a and b are coefficients. The b coefficient is between 0.5 and 1; a b value close to 0.5 indicates a more diffusion-controlled process, while a b close to 1 is a more capacitive-controlled process [64]. Figure 5a represents the CV curves of the Ni-doped sample at different scan rates, and Figure 5b illustrates the $log(i)$ versus $log(\nu^{1/2})$, whose slope is the coefficient b. The obtained b value was 0.85, suggesting a more capacitive-controlled process. Moreover, the current peak can be understood as the sum of two contributions [30] following Equation (2): one related to capacitive-controlled processes, which depends on the scan rate (ν), and the other to diffusion-controlled processes, which depends on the square root of the scan rate ($\nu^{1/2}$).

$$i = i_{cap} + i_{dif} = k_1 v + k_2 v^{1/2} \quad (8)$$

where the parameters k_1 and k_2 can be calculated by the lineal regression obtained from the $i/\nu^{1/2}$ versus the $\nu^{1/2}$ in order to calculate the contribution of each process at different scan rates, as it is represented in Figure 5c for the Ni-doped electrode. As was expected because of the calculated b value, the behavior is mainly controlled by capacitive processes rather than diffusion processes. However, as the scan rate rises, the contribution of capacitive-controlled processes increases further, primarily due to the kinetic inhibition of diffusion at high scan rates. On the other hand, the Ni-doped specific capacitance and the coulombic efficiency along 5000 GCD cycles at 150 mA/g are shown in Figure 5d to evaluate the capacitance retention. The reached coulombic efficiency was 94%, indicating the redox process was partially irreversible. Because of that, the initial specific capacitance, which was 9 F/g, was reduced by 32% after 5000 GCD cycles. Nevertheless, this value appears to stabilize, and it is still higher than the specific capacitances displayed by the undoped, Zn-doped, and Mn-doped samples in only one GCD test.

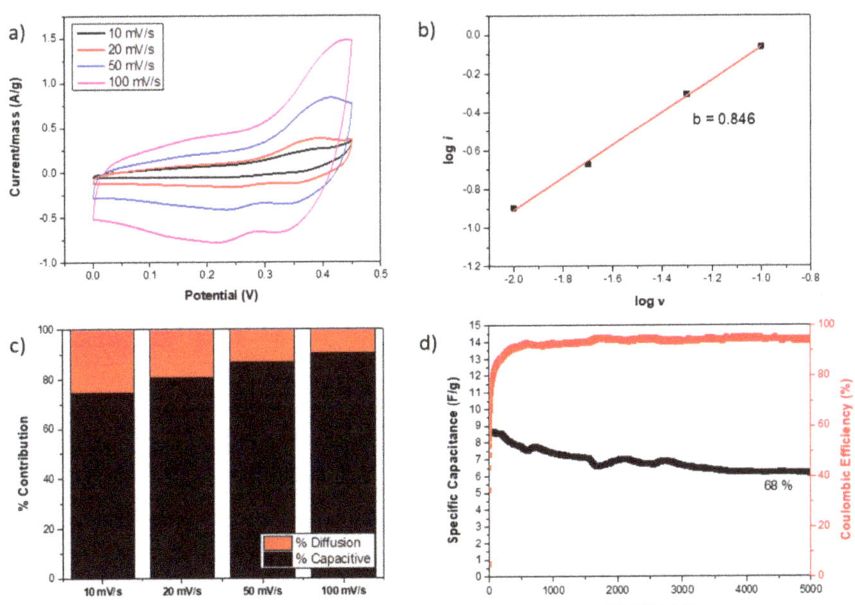

Figure 5. (**a**) CV curve of Ni-doped sample at different scan rates and (**b**) calculated b value in the inset. (**c**) Diffusion and capacitive-controlled processes contribute to different scan rates for Ni-doped samples. (**d**) Specific capacitance and coulombic efficiency of Ni-doped sample after 5000 GCD cycles at 150 mA/g.

3.3. Characterization as Li-Ion Battery Anode

After evaluating the behavior of the different TMOs as an electrode for supercapacitors, their performance as an anode in Li-ion batteries was also investigated to explore the versatility of the synthesized TMO materials for diverse energy storage devices. In order to study the TMO coating electrochemical reactions, CV tests were carried out. Figure 6a shows the current normalized CV curves, enabling a comparative analysis of the coating peaks. Peaks associated with $MCo_2O_4@Co_3O_4$ coatings are those located around 1.3 V in the cathodic process and around 1.6 and 2.2 V in the anodic process, which are being expanded in the figure, while the others, presented in the inset, can be attributed to the carbon fiber [44,51–53]. In the cathodic process, the peak around 1 V is related to the reduction in MCo_2O_4 and Co_3O_4 to metallic M and Co. In the anodic process, the peak at around 1.6 V is related to the oxidation of M and Co to M^{2+} and Co^{2+}, and the peak around 2.2 V to the oxidation of Co^{2+} to Co^{3+}. Thus, the involved electrochemical reactions during the charge–discharge process are the following:

$$MCo_2O_4 + 8Li^+ \rightarrow M + 2Co + 8Li_2O \tag{9}$$

$$Co_3O_4 + 8Li^+ \rightarrow 3Co + 4Li_2O \tag{10}$$

$$M + Li_2O \rightarrow MO + 2Li^+ \tag{11}$$

$$Co + Li_2O \rightarrow CoO + 2Li^+ \tag{12}$$

$$2CoO + MO + Li_2O \rightarrow MCo_2O_4 + 2Li^+ \tag{13}$$

$$3CoO + Li_2O \rightarrow Co_3O_4 + 2Li^+ \tag{14}$$

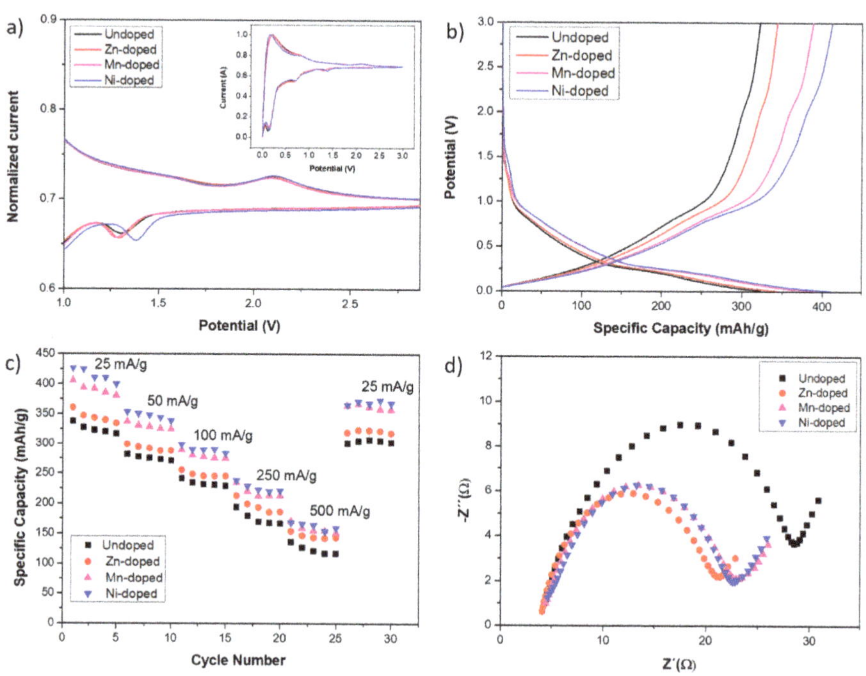

Figure 6. (**a**) Normalized third-cycle CV of the samples. (**b**) GCD curves at 25 mA/g of the samples. (**c**) GCD calculated capacities at different current densities. (**d**) Nyquist plots of the samples.

The main difference among CV curves of the diverse samples is related to the cathodic peak located around 1.3 V. This peak is narrower in CVs obtained from doped samples in comparison to undoped ones, indicating a better electrical conductivity induced by the cation substitution. These CV curves can be correlated to GCD curves, where each electrochemical reaction, manifested as a peak in CV, is represented in the GCD curve as a plateau. Figure 6b shows the GCD curves at a current density of 25 mA/g collected from the samples. The anodic reaction at 1.6 V is hardly observable as a change in the curvature during the charge process, while the reaction at 2.2 V is evident. The cathodic reaction at 1.3 V is observable during the discharge process. Figure 6c shows the calculated capacities, where it can be observed how the Ni-doped sample displayed the highest capacity value across all cases; for instance, at a current density of 25 mA/g, it reached 410.5 mAh/g, whereas the Mn-doped sample exhibited 392 mAh/g, Zn-doped 346.3 mAh/g, and undoped 322.7 mAh/g. These values are lower by 4.5%, 15.7%, and 21.4% compared to the Ni-doped sample, respectively. Assessing the rate capability, i.e., the retention of the capacity as the current density increases, the undoped sample retains 37.6% of its capacity, while the Zn-doped one retains 41.8%, the Mn-doped retains 39.4%, and the Ni-doped retains 40%, as the current density increases from 25 to 500 mA/g. This parameter is related to the electrical conductivity, suggesting that the most electrically conductive sample is Zn-doped, followed by Ni-doped and Mn-doped, whereas the undoped sample, being the monometallic TMO, exhibits the least conductivity. Nevertheless, it is important to realize that even at a current density of 500 mA/g, the Ni-doped sample capacity is higher than Mn-doped, Zn-doped, and undoped samples by 5.9%, 12.5%, and 26.1%, respectively. Moreover, the capacity retention can also be evaluated from the results of Figure 6c, relating the initial capacity at 25 mA/g with the capacity at the same current density after the samples have been tested at the other current densities. The undoped sample retains the most capacity, 95.1%, suggesting that the doping with other cations decreases the cyclability, which could be a consequence of the distortions in the crystalline structure. However, smaller cations like Zn^{2+} decrease cyclability by only 6% compared to larger ones like Mn^{2+}, which, despite being less exchanged, lose only 7.7% of the initial capacity. Owing to the large degree of exchange, the Ni-doped sample exhibits the least capacity retention, losing 9.3%. Nevertheless, again, it is important to realize that despite these results, even after all these tests, the Ni-doped sample still demonstrates the highest specific capacity.

Finally, EIS tests were carried out to evaluate the charge transfer resistance. The Nyquist plots of each sample are presented in Figure 6d, while the results are summarized in Table 2. Typically, the charge transfer process between electrode and electrolyte is related to the first semicircle [35], and its resistance can be calculated as its width. The charge transfer resistance value was 28.5 Ω for the undoped sample, 19.4 Ω for the Zn-doped sample, 20.6 Ω for the Mn-doped sample, and 20.2 Ω for the Ni-doped sample. These results agree with the rate capability discussion presented above, wherein lower electrical resistance was expected for the Zn-doped sample and higher resistance for the undoped one, while Ni-doped and Mn-doped are expected to present similar values.

Table 2. Summary of the EIS calculated resistances from different samples.

	Undoped	Zn-Doped	Mn-Doped	Ni-Doped
Charge-transfer Resistance (R_{ct})	28.5 Ω	19.4 Ω	20.6 Ω	20.2 Ω
Serial Resistance (R_s)	4.0 Ω	3.9 Ω	3.8 Ω	3.6 Ω

Owing to its better performance as a Li-ion battery anode, the Ni-doped sample was more extensively characterized. Figure 7a shows Ni-doped sample CV at different scan rates, focusing on $NiCo_2O_4$@Co_3O_4 characteristic peaks and the b value in the inset, which was 0.78. This *b* value indicates a half behavior between capacitive and diffusion-controlled processes. In Figure 7b, capacitive and diffusion-controlled contributions are represented for all tested scan rates. It is observable how the capacitive contribution is higher as the scan rate increases due to high scan rates kinetically preventing diffusion. On the other hand, the cyclability of the Ni-doped electrode was evaluated through GCD 100 cycles at 100 mA/g of current density, being the specific capacity evolution shown in Figure 7c, just as the coulombic efficiency. The prominent decrease after the first cycle, where capacity is reduced by 29%, is attributed to the formation of the solid electrolyte interphase (SEI) layer. Following the next five cycles, the capacity rapidly decreases by another 8%. After the 6th cycle, the decrease is continuous until reaching 76% of the value at the second cycle by the 100th cycle. Additionally, the coulombic efficiency was about 100% at all cycles except at the first, so the SEI formation is an irreversible process. EIS tests were also carried out to study the evolution of the charge transfer resistance as the cycle number increased. Figure 7d shows the Nyquist plot of the fresh sample after the first cycle and after 100 cycles; the result is summarized in Table 3. Attending to the charge transfer resistances, measured values were 20.2 Ω, 32.4 Ω, and 74.8 Ω, respectively. The first increment of the resistance is due to the SEI formation, while the increment after that is related to the degradation of the TMO during Li intercalation–deintercalation processes [65].

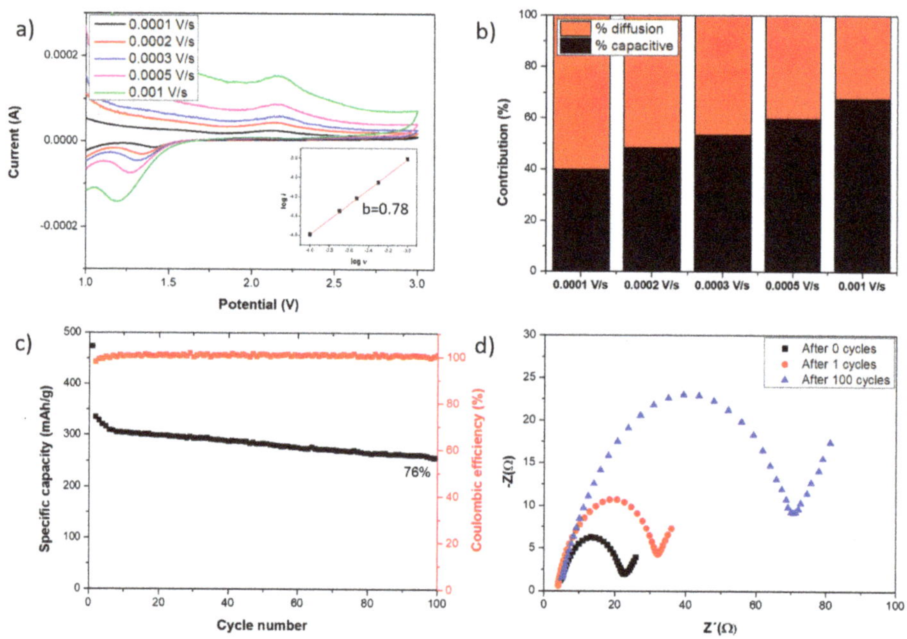

Figure 7. (**a**) Ni-doped sample CV at different scan rates and calculated b value in the inset. (**b**) Percentage of each contribution at different scan rates in the Ni-doped sample. (**c**) Specific capacity of Ni-doped sample after 100 GCD cycles at 100 mA/g. (**d**) Nyquist plot of Ni-doped sample after several number of cycles.

Table 3. Summary of the EIS calculated resistances of the Ni-doped sample after different number of cycles.

	Fresh	After the First Cycle	After 100 Cycles
Charge-transfer resistance (R_{ct})	20.2 Ω	32.4 Ω	74.8 Ω
Serial resistance (R_s)	3.6 Ω	4.0 Ω	3.7 Ω

4. Conclusions

In this work, homogeneous and dense coatings with different TMOs have been developed on the carbon fiber through a Co-ZIF-L MOF precursor and ion exchange with Zn^{2+}, Mn^{2+}, and Ni^{2+} as dopant cations. All doped samples have improved undoped TMO coating electrochemical performance, both as supercapacitor electrodes and Li-ion battery anodes. This enhancement can be attributed to the increased electrical conductivity caused by the bimetallic composition and the improved electrochemical activity as a consequence of the lattice distortion. As a supercapacitor electrode, the Ni-doped sample exhibited extra peaks with respect to other samples, attributed to the redox couple NiOOH/NiO, which is possibly due to its higher exchange degree. These extra peaks provided a Ni-doped sample with higher specific capacitance than its counterparts, 13.3 F/g of the electrode in the GCD test at 50 mA/g. This value was 37%, 50%, and 66% higher than Mn-doped, Zn-doped, and undoped samples, respectively. On the other hand, Ni-doped electrochemical performance was also higher than their counterparts as Li-ion battery anodes. The specific capacity of this sample was 410.5 mAh/g at a current density of 25 mA/g, which means 4.5%, 15.7%, and 21.4% more than Mn-doped, Zn-doped, and undoped samples, respectively. Thus, the results of this work highlight Ni^{2+} as the most suitable dopant cation for achieving enhanced electrochemical performance in both applications, emerging this Ni-doped cobaltite ZIF-L-derived as a promising coating on the carbon fiber fabrics to develop supercapacitor electrodes and Li-ion battery anodes for structural energy storage devices.

Supplementary Materials: The following supporting information can be downloaded at: https://www.mdpi.com/article/10.3390/nano14221846/s1, Figure S1: EDS spectra of (a) no-doped sample, (b) Zn-doped sample, (c) Mn-doped sample, and (d) Ni-doped sample; Figure S2: GCD test at different current densities of (a) undoped sample, (b) Zn-doped sample, (c) Mn-doped sample, and (d) Ni-doped sample; Figure S3: Galvanostatic discharges at different current densities of (a) undoped sample, (b) Zn-doped sample, (c) Mn-doped sample, and (d) Ni-doped sample.

Author Contributions: A.G.-B.: Conceptualization, methodology, formal analysis, investigation, writing—original draft preparation, writing—review and editing, and visualization. D.M.-D.: Conceptualization, methodology, formal analysis, investigation, writing—original draft preparation, writing—review and editing, and visualization. A.d.H.: Investigation, formal analysis. M.S.: Conceptualization, writing—review and editing, supervision, and funding acquisition. A.U.: Conceptualization, writing—review and editing, supervision, and funding acquisition. All authors have read and agreed to the published version of the manuscript.

Funding: This work was supported by the Agencia Estatal de Investigación of the Spanish Government (Project MULTISENS PID2022-136636OB-I00) and (PRE2020-094255).

Data Availability Statement: Data will be made available on request.

Conflicts of Interest: The authors declare no conflicts of interest.

References

1. Danzi, F.; Salgado, R.M.; Oliveira, J.E.; Arteiro, A.; Camanho, P.P.; Braga, M.H. Structural batteries: A review. *Molecules* **2021**, *26*, 2203. [CrossRef] [PubMed]
2. Kalnaus, S.; Asp, L.E.; Li, J.; Veith, G.M.; Nanda, J.; Daniel, C.; Chen, X.C.; Westover, A.; Dudney, N.J. Multifunctional approaches for safe structural batteries. *J. Energy Storage* **2021**, *40*, 102747. [CrossRef]
3. Kalita, G.; Endo, T.; Nishi, T. Recent development on low temperature synthesis of cubic-phase LLZO electrolyte particles for application in all-solid-state batteries. *J. Alloy Compd.* **2023**, *969*, 102747. [CrossRef]
4. Tuo, K.; Sun, C.; Liu, S. Recent Progress in and Perspectives on Emerging Halide Superionic Conductors for All-Solid-State Batteries. *Electrochem. Energy Rev.* **2023**, *6*, 17. [CrossRef]
5. Yang, X.; Yin, Q.; Wang, C.; Doyle-Davis, K.; Sun, X.; Li, X. Towards practically accessible high-voltage solid-state lithium batteries: From fundamental understanding to engineering design. *Prog. Mater. Sci.* **2023**, *140*, 102747. [CrossRef]
6. Snyder, J.; Gienger, E.; Wetzel, E. Performance metrics for structural composites with electrochemical multifunctionality. *J. Compos. Mater.* **2015**, *49*, 1835–1848. [CrossRef]
7. González, C.; Vilatela, J.; Molina-Aldareguía, J.; Lopes, C.; Llorca, J. Structural composites for multifunctional applications: Current challenges and future trends. *Prog. Mater. Sci.* **2017**, *89*, 194–251. [CrossRef]
8. Greenhalgh, E.S.; Nguyen, S.; Valkova, M.; Shirshova, N.; Shaffer, M.S.; Kucernak, A. A critical review of structural supercapacitors and outlook on future research challenges. *Compos. Sci. Technol.* **2023**, *235*, 102747. [CrossRef]
9. Jin, T.; Singer, G.; Liang, K.; Yang, Y. Structural batteries: Advances, challenges and perspectives. *Mater. Today* **2023**, *62*, 151–167. [CrossRef]

10. Zhang, S.; Xiao, S.; Li, D.; Liao, J.; Ji, F.; Liu, H.; Ci, L. Commercial carbon cloth: An emerging substrate for practical lithium metal batteries. *Energy Storage Mater.* **2022**, *48*, 172–190. [CrossRef]
11. Gulzar, U.; Goriparti, S.; Miele, E.; Li, T.; Maidecchi, G.; Toma, A.; De Angelis, F.; Capiglia, C.; Zaccaria, R.P. Next-generation textiles: From embedded supercapacitors to lithium ion batteries. *J. Mater. Chem. A* **2016**, *4*, 16771–16800. [CrossRef]
12. Asp, L.E.; Bouton, K.; Carlstedt, D.; Duan, S.; Harnden, R.; Johannisson, W.; Johansen, M.; Johansson, M.K.G.; Lindbergh, G.; Liu, F.; et al. A Structural Battery and its Multifunctional Performance. *Adv. Energy Sustain. Res.* **2021**, *2*, 2000093. [CrossRef]
13. Moyer, K.; Meng, C.; Marshall, B.; Assal, O.; Eaves, J.; Perez, D.; Karkkainen, R.; Roberson, L.; Pint, C.L. Carbon fiber reinforced structural lithium-ion battery composite: Multifunctional power integration for CubeSats. *Energy Storage Mater.* **2020**, *24*, 676–681. [CrossRef]
14. Yao, S.; Zhang, G.; Zhang, X.; Shi, Z. Mace-like carbon fibers@Fe_3O_4@carbon composites as anode materials for lithium-ion batteries. *Ionics* **2020**, *26*, 5923–5934. [CrossRef]
15. Huang, Y.; Yang, H.; Xiong, T.; Adekoya, D.; Qiu, W.; Wang, Z.; Zhang, S.; Balogun, M.-S. Adsorption energy engineering of nickel oxide hybrid nanosheets for high areal capacity flexible lithium-ion batteries. *Energy Storage Mater.* **2020**, *25*, 41–51. [CrossRef]
16. Han, Q.; Zhang, W.; Han, Z.; Wang, F.; Geng, D.; Li, X.; Li, Y.; Zhang, X. Preparation of PAN-based carbon fiber@MnO2 composite as an anode material for structural lithium-ion batteries. *J. Mater. Sci.* **2019**, *54*, 11972–11982. [CrossRef]
17. Subhani, K.; Hameed, N.; Al-Qatatsheh, A.; Ince, J.; Mahon, P.J.; Lau, A.; Salim, N.V. Multifunctional structural composite supercapacitors based on MnO2-nanowhiskers decorated carbon fibers. *J. Energy Storage* **2022**, *56*, 105936. [CrossRef]
18. Artigas-Arnaudas, J.; Sánchez-Romate, X.F.; Sánchez, M.; Ureña, A. Effect of electrode surface treatment on carbon fiber based structural supercapacitors: Electrochemical analysis, mechanical performance and proof-of-concept. *J. Energy Storage* **2023**, *59*, 106599. [CrossRef]
19. Cen, T.; Chen, L.; Zhang, X.; Tian, Y.; Fan, X. A novel fiber-shaped asymmetric supercapacitor prepared by twisting carbon fiber/carbon nanotube/MnO_2 and carbon fiber/carbon nanotube/polypyrrole electrodes+. *Electrochim. Acta* **2021**, *367*, 137488. [CrossRef]
20. Huang, R.; Zhang, J.; Dong, Z.; Lin, H.; Han, S. Flexible carbon fiber/reduced-TiO_2 composites for constructing remarkable performance supercapacitors. *J. Power Sources* **2022**, *550*, 232169. [CrossRef]
21. Saravanan, R.S.A.; Bejigo, K.S.; Kim, S.-J. Scope and significance of transition metal oxide nanomaterials for next-generation Li-ion batteries. *Mater. Chem. Front.* **2023**, *7*, 4613–4634. [CrossRef]
22. Ayyanusamy, P.; Alphonse, R.; Minakshi, M.; Sivasubramanian, R. Synthesis of amorphous nickel-cobalt hydroxides for Ni−Zn batteries. *Chem. A Eur. J.* **2024**, *30*, e202402325. [CrossRef] [PubMed]
23. Zhu, J.; Ding, Y.; Ma, Z.; Tang, W.; Chen, X.; Lu, Y. Recent Progress on Nanostructured Transition Metal Oxides As Anode Materials for Lithium-Ion Batteries. *J. Electron. Mater.* **2022**, *51*, 3391–3417. [CrossRef]
24. Yadav, S.; Sharma, A. Importance and challenges of hydrothermal technique for synthesis of transition metal oxides and composites as supercapacitor electrode materials. *J. Energy Storage* **2021**, *44*, 103295. [CrossRef]
25. Zhu, X. Recent advances of transition metal oxides and chalcogenides in pseudo-capacitors and hybrid capacitors: A review of structures, synthetic strategies, and mechanism studies. *J. Energy Storage* **2022**, *49*, 104148. [CrossRef]
26. Haripriya, M.; Manimekala, T.; Dharmalingam, G.; Minakshi, M.; Sivasubramanian, R. Asymmetric Supercapacitors Based on $ZnCo_2O_4$ Nanohexagons and Orange Peel Derived Activated Carbon Electrodes. *Chem.–Asian J.* **2024**, *19*, e202400202. [CrossRef]
27. Yuan, C.; Bin Wu, H.; Xie, Y.; Lou, X.W. Mixed transition-metal oxides: Design, synthesis, and energy-related applications. *Angew. Chem. Int. Ed.* **2014**, *53*, 1488–1504. [CrossRef]
28. Li, M.; Meng, Z.; Feng, R.; Zhu, K.; Zhao, F.; Wang, C.; Wang, J.; Wang, L.; Chu, P.K. Fabrication of bimetallic oxides (MCo_2O_4: M=Cu, Mn) on ordered microchannel electro-conductive plate for high-performance hybrid supercapacitors. *Sustainability* **2021**, *13*, 9896. [CrossRef]
29. Xu, Y.; Chu, K.; Li, Z.; Xu, S.; Yao, G.; Niu, P.; Zheng, F. Porous CuO@C composite as high-performance anode materials for lithium-ion batteries. *Dalton Trans.* **2020**, *49*, 11597–11604. [CrossRef]
30. Zhang, X.; Du, W.; Lin, Z.; Tan, X.; Li, Y.; Ou, G.; Xu, X.; Lin, X.; Wu, Y.; Zeb, A.; et al. Templated formation of Mn2O3 derived from metal-organic frameworks with different organic ligands as anode materials for enhanced lithium-ion storage. *J. Alloy. Compd.* **2022**, *927*, 166977. [CrossRef]
31. Tan, X.; Wu, Y.; Lin, X.; Zeb, A.; Xu, X.; Luo, Y.; Liu, J. Application of MOF-derived transition metal oxides and composites as anodes for lithium-ion batteries. *Inorg. Chem. Front.* **2020**, *7*, 4939–4955. [CrossRef]
32. Zhou, J.; Yang, Q.; Xie, Q.; Ou, H.; Lin, X.; Zeb, A.; Hu, L.; Wu, Y.; Ma, G. Recent progress in Co–based metal–organic framework derivatives for advanced batteries. *J. Mater. Sci. Technol.* **2022**, *96*, 262–284. [CrossRef]
33. Wang, Y.; Li, B.; Zhang, B.; Tian, S.; Yang, X.; Ye, H.; Xia, Z.; Zheng, G. Application of MOFs-derived mixed metal oxides in energy storage. *J. Electroanal. Chem.* **2020**, *878*, 114576. [CrossRef]
34. Vanaraj, R.; Daniel, S.; Haldhar, R.; Asrafali, S.P.; Kim, S.C. Direct growth of TiO_2–MoO_2/MnO_2–MoO_2 on plasma-treated carbon-cloth surface for high-performance supercapacitor and oxygen evolution reaction. *Electrochim. Acta* **2023**, *440*, 141705. [CrossRef]
35. Zhang, S.; Dai, P.; Liu, H.; Yan, L.; Song, H.; Liu, D.; Zhao, X. Metal-organic framework derived porous flakes of cobalt chalcogenides (CoX, X = O, S, Se and Te) rooted in carbon fibers as flexible electrode materials for pseudocapacitive energy storage. *Electrochim. Acta* **2021**, *369*, 137681. [CrossRef]

36. Gong, H.; Bie, S.; Zhang, J.; Ke, X.; Wang, X.; Liang, J.; Wu, N.; Zhang, Q.; Luo, C.; Jia, Y. In Situ Construction of ZIF-67-Derived Hybrid Tricobalt Tetraoxide@Carbon for Supercapacitor. *Nanomaterials* **2022**, *12*, 1571. [CrossRef]
37. Guan, C.; Zhao, W.; Hu, Y.; Lai, Z.; Li, X.; Sun, S.; Zhang, H.; Cheetham, A.K.; Wang, J. Cobalt oxide and N-doped carbon nanosheets derived from a single two-dimensional metal–organic framework precursor and their application in flexible asymmetric supercapacitors. *Nanoscale Horiz.* **2017**, *2*, 99–105. [CrossRef]
38. Liu, S.; Kang, L.; Zhang, J.; Jung, E.; Lee, S.; Jun, S.C. Structural engineering and surface modification of MOF-derived cobalt-based hybrid nanosheets for flexible solid-state supercapacitors. *Energy Storage Mater.* **2020**, *32*, 167–177. [CrossRef]
39. Dai, S.; Han, F.; Tang, J.; Tang, W. MOF-derived Co3O4 nanosheets rich in oxygen vacancies for efficient all-solid-state symmetric supercapacitors. *Electrochim. Acta* **2019**, *328*, 135103. [CrossRef]
40. Lim, G.J.; Liu, X.; Guan, C.; Wang, J. Co/Zn bimetallic oxides derived from metal organic frameworks for high performance electrochemical energy storage. *Electrochim. Acta* **2018**, *291*, 177–187. [CrossRef]
41. Yang, Q.; Liu, Y.; Yan, M.; Lei, Y.; Shi, W. MOF-derived hierarchical nanosheet arrays constructed by interconnected NiCo-alloy@NiCo-sulfide core-shell nanoparticles for high-performance asymmetric supercapacitors. *Chem. Eng. J.* **2019**, *370*, 666–676. [CrossRef]
42. Wang, F.; Han, Q.; Yi, Z.; Geng, D.; Li, X.; Wang, Z.; Wang, L. Synthesis and performances of carbon fiber@Co3O4 based on metal organic frameworks as anode materials for structural lithium-ion battery. *J. Electroanal. Chem.* **2017**, *807*, 196–202. [CrossRef]
43. Han, Q.; Li, X.; Wang, F.; Han, Z.; Geng, D.; Zhang, W.; Li, Y.; Deng, Y.; Zhang, J.; Niu, S.; et al. Carbon fiber@ pore-ZnO composite as anode materials for structural lithium-ion batteries. *J. Electroanal. Chem.* **2019**, *833*, 39–46. [CrossRef]
44. Huang, T.; Lou, Z.; Lu, Y.; Li, R.; Jiang, Y.; Shen, G.; Chen, D. Metal-Organic-Framework-Derived MCo_2O_4 (M=Mn and Zn) Nanosheet Arrays on Carbon Cloth as Integrated Anodes for Energy Storage Applications. *ChemElectroChem* **2019**, *6*, 5836–5843. [CrossRef]
45. Guan, C.; Liu, X.; Ren, W.; Li, X.; Cheng, C.; Wang, J. Rational Design of Metal-Organic Framework Derived Hollow $NiCo_2O_4$ Arrays for Flexible Supercapacitor and Electrocatalysis. *Adv. Energy Mater.* **2017**, *7*, 1602391. [CrossRef]
46. Javed, M.S.; Aslam, M.K.; Asim, S.; Batool, S.; Idrees, M.; Hussain, S.; Shah, S.S.A.; Saleem, M.; Mai, W.; Hu, C. High-performance flexible hybrid-supercapacitor enabled by pairing binder-free ultrathin Ni–Co–O nanosheets and metal-organic framework derived N-doped carbon nanosheets. *Electrochim. Acta* **2020**, *349*, 136384. [CrossRef]
47. Chen, S.; Wu, J.; Zhou, R.; Chen, Y.; Song, Y.; Wang, L. Controllable growth of $NiCo_2O_4$ nanoarrays on carbon fiber cloth and its anodic performance for lithium-ion batteries. *RSC Adv.* **2015**, *5*, 104433–104440. [CrossRef]
48. Dai, S.; Yuan, Y.; Yu, J.; Tang, J.; Zhou, J.; Tang, W. Metal–organic framework-templated synthesis of sulfur-doped core–sheath nanoarrays and nanoporous carbon for flexible all-solid-state asymmetric supercapacitors. *Nanoscale* **2018**, *10*, 15454–15461. [CrossRef]
49. Fu, Y.; Zhou, H.; Hu, Z.; Yin, S.; Zhou, L. Temperature-induced microstructure optimization of Co_3O_4 for the achievement of a high-areal-capacity carbon cloth-based lithium ion battery anode. *Compos. Commun.* **2020**, *22*, 100446. [CrossRef]
50. Fang, G.; Zhou, J.; Liang, C.; Pan, A.; Zhang, C.; Tang, Y.; Tan, X.; Liu, J.; Liang, S. MOFs nanosheets derived porous metal oxide-coated three-dimensional substrates for lithium-ion battery applications. *Nano Energy* **2016**, *26*, 57–65. [CrossRef]
51. Liu, T.; Wang, W.; Yi, M.; Chen, Q.; Xu, C.; Cai, D.; Zhan, H. Metal-organic framework derived porous ternary $ZnCo_2O_4$ nanoplate arrays grown on carbon cloth as binder-free electrodes for lithium-ion batteries. *Chem. Eng. J.* **2018**, *354*, 454–462. [CrossRef]
52. Dai, Z.; Long, Z.; Li, R.; Shi, C.; Qiao, H.; Wang, K.; Liu, K. Metal–Organic Framework-Structured Porous $ZnCo_2O_4$/C Composite Nanofibers for High-Rate Lithium-Ion Batteries. *ACS Appl. Energy Mater.* **2020**, *3*, 12378–12384. [CrossRef]
53. Li, H.; Wang, S.; Feng, M.; Yang, J.; Zhang, B. MOF-derived $ZnCo_2O_4$/C wrapped on carbon fiber as anode materials for structural lithium-ion batteries. *Chin. Chem. Lett.* **2019**, *30*, 529–532. [CrossRef]
54. Wu, X.; Meng, L.; Wang, Q.; Zhang, W.; Wang, Y. Highly flexible and large areal/volumetric capacitances for asymmetric supercapacitor based on $ZnCo_2O_4$ nanorods arrays and polypyrrole on carbon cloth as binder-free electrodes. *Mater. Lett.* **2019**, *234*, 1–4. [CrossRef]
55. Feng, M.; Wang, S.; Yu, Y.; Feng, Q.; Yang, J.; Zhang, B. Carboxyl functionalized carbon fibers with preserved tensile strength and electrochemical performance used as anodes of structural lithium-ion batteries. *Appl. Surf. Sci.* **2017**, *392*, 27–35. [CrossRef]
56. Gholampour, N.; Ahmadian-Yazdi, M.-R. Investigation of zeolitic imidazolate frameworks–derived carbon nanotubes thin film in solar vapor generation. *J. Porous Mater.* **2021**, *28*, 1105–1113. [CrossRef]
57. Zhang, Y.; Chen, Z.; Tian, J.; Sun, M.; Yuan, D.; Zhang, L. Nitrogen doped $CuCo_2O_4$ nanoparticles anchored on beaded-like carbon nanofibers as an efficient bifunctional oxygen catalyst toward zinc-air battery. *J. Colloid Interface Sci.* **2022**, *608*, 1105–1115. [CrossRef]
58. Mary, A.J.C.; Bose, A.C. Incorporating $Mn^{2+}/Ni^{2+}/Cu^{2+}/Zn^{2+}$ in the Co_3O_4 Nanorod: To Investigate the Effect of Structural Modification in the Co_3O_4 Nanorod and Its Electrochemical Performance. *ChemistrySelect* **2019**, *4*, 160–170. [CrossRef]
59. Kavinkumar, T.; Vinodgopal, K.; Neppolian, B. Development of nanohybrids based on porous spinel MCo_2O_4 (M = Zn, Cu, Ni and Mn)/reduced graphene oxide/carbon nanotube as promising electrodes for high performance energy storage devices. *Appl. Surf. Sci.* **2020**, *513*, 145781. [CrossRef]
60. Asghari, A.; Kazemi, S.H.; Khanmohammadi, M. Facile and binder-free synthesis of N-doped carbon/$ZnCo_2O_4$ hybrid nanostructures on nickel foam for high-performance solid-state asymmetric supercapacitor. *J. Mater. Sci. Mater. Electron.* **2020**, *31*, 4354–4363. [CrossRef]

61. Xiao, J.; Yang, S. Sequential crystallization of sea urchin-like bimetallic (Ni, Co) carbonate hydroxide and its morphology conserved conversion to porous $NiCo_2O_4$ spinel for pseudocapacitors. *RSC Adv.* **2011**, *1*, 588–595. [CrossRef]
62. Kamble, G.P.; Kashale, A.A.; Rasal, A.S.; Mane, S.A.; Chavan, R.A.; Chang, J.-Y.; Ling, Y.-C.; Kolekar, S.S.; Ghule, A.V. Marigold micro-flower like $NiCo_2O_4$ grown on flexible stainless-steel mesh as an electrode for supercapacitors. *RSC Adv.* **2021**, *11*, 3666–3672. [CrossRef] [PubMed]
63. Luo, Y.; Zhang, H.; Guo, D.; Ma, J.; Li, Q.; Chen, L.; Wang, T. Porous $NiCo_2O_4$-reduced graphene oxide (rGO) composite with superior capacitance retention for supercapacitors. *Electrochim. Acta* **2014**, *132*, 332–337. [CrossRef]
64. Patil, D.R.; Koteswararao, B.; Begari, K.; Yogi, A.; Moussa, M.; Dubal, D.P. Cobalt Cyclotetraphosphate ($Co_2P_4O_{12}$): A New High-Performance Electrode Material for Supercapacitors. *ACS Appl. Energy Mater.* **2019**, *2*, 2972–2981. [CrossRef]
65. Wilson, M.K.; Saikrishna, V.; Mannayil, J.; Sreeja, E.M.; Abhilash, A.; Antony, A.; Jayaraj, M.K.; Jayalekshmi, S. Exploring the potential of iron oxide nanoparticle embedded carbon nanotube/polyaniline composite as anode material for Li-ion cells. *J. Mater. Sci. Mater. Electron.* **2023**, *34*, 1689. [CrossRef]

nanomaterials MDPI

Article

Water Sorption on Isoreticular CPO-27-Type MOFs: From Discrete Sorption Sites to Water-Bridge-Mediated Pore Condensation

Marvin Kloß, Lara Schäfers, Zhenyu Zhao, Christian Weinberger, Hans Egold and Michael Tiemann *

Department of Chemistry, Paderborn University, 33098 Paderborn, Germany
* Correspondence: michael.tiemann@upb.de

Abstract: Pore engineering is commonly used to alter the properties of metal–organic frameworks. This is achieved by incorporating different linker molecules (*L*) into the structure, generating isoreticular frameworks. CPO-27, also named MOF-74, is a prototypical material for this approach, offering the potential to modify the size of its one-dimensional pore channels and the hydrophobicity of pore walls using various linker ligands during synthesis. Thermal activation of these materials yields accessible open metal sites (i.e., under-coordinated metal centers) at the pore walls, thus acting as strong primary binding sites for guest molecules, including water. We study the effect of the pore size and linker hydrophobicity within a series of Ni^{2+}-based isoreticular frameworks (i.e., Ni_2L, L = dhtp, dhip, dondc, bpp, bpm, tpp), analyzing their water sorption behavior and the water interactions in the confined pore space. For this purpose, we apply water vapor sorption analysis and Fourier transform infrared spectroscopy. In addition, defect degrees of all compounds are determined by thermogravimetric analysis and solution ^{1}H nuclear magnetic resonance spectroscopy. We find that larger defect degrees affect the preferential sorption sites in Ni_2dhtp, while no such indication is found for the other materials in our study. Instead, strong evidence is found for the formation of water bridges/chains between coordinating water molecules, as previously observed for hydrophobic porous carbons and mesoporous silica. This suggests similar sorption energies for additional water molecules in materials with larger pore sizes after saturation of the primary binding sites, resulting in more bulk-like water arrangements. Consequently, the sorption mechanism is driven by classical pore condensation through H-bonding anchor sites instead of sorption at discrete sites.

Keywords: metal–organic frameworks; MOF-74; IRMOF-74; open metal sites; water vapor sorption; hydrophobicity

1. Introduction

The concept of reticular synthesis [1] has proven to be one of the most fundamental concepts within the field of metal–organic frameworks (MOFs), a class of porous inorganic–organic hybrid materials. On the one hand, it provides a systematic approach to reduce the complexity of numerous MOFs [2,3] by simplifying their extended structures, reducing them to simple geometric building blocks, so-called secondary building units (SBUs). On the other hand, it guides chemists to rationally design new frameworks adopting the same framework connectivity, i.e., network topology, referring to the structure and connectivity of the underlying periodic network [4,5]. Substitution of geometrically identical units allows creation of precisely tailored pore environments by inclusion of new functionalities [6–10], pore size modification [6,10–13] or the use of different metal cations [14].

One of the most extensively studied examples of the applicability of this concept is CPO-27-*M* [15] (i.e., M_2dhtp, dhtp = 2,5-dihydroxyterephthalate, also known as dobdc), also well-known as MOF-74-*M* [16]. Substitution of the dhtp-linker molecule [11,17–19] generates three-dimensional, isoreticular MOFs (i.e., IRMOF-74) with a honeycomb-like cross section and one-dimensional, cylindrical pores. This pore engineering concept resulted in multiple compounds with different pore sizes and functionalities such as M_2dondc [20] (dondc = 1,5-dihydroxynaphthalene-2,6-dicarboxylate), M_2bpp [11,17–19,21] (bpp = 3,3′-dihydroxy-[1,1′-biphenyl]-4,4′-dicarboxylate, also known as dobpdc) and M_2tpp [11,17,19,21] (tpp = 3,3″-dihydroxy-2′,5′-dimethyl-[1,1′:4′,1″-terphenyl]-4,4″-dicarboxylate). Further studies showed that modification of the linker substitution pattern (i.e., from para- to meta-arrangement of carboxylic acid groups) results in frameworks with bended, 'banana-like' pore walls as in the cases of M_2dhip [22] (dhip = 4,6-dihydroxyisophthalate, also known as *m*-dobdc) and M_2bpm [21–24] (bpm = 4,4′-Dihydroxy [1,1′-biphenyl]-3,3′-dicarboxylate). All IRMOF-74 frameworks have accessible open metal sites [25] (i.e., under-coordinated metal centers) created by removing coordinating solvent molecules. These sites are exposed to the pores, acting as attractive sorption sites for various guest molecules such as CO_2 [26–29] or water [28,30–32]. Therefore, these isostructural frameworks offer unique possibilities for sorption-based applications.

In order to evaluate the material performance, a deep understanding of the water sorption properties is needed, since applications either heavily rely on the absence (e.g., battery technology [33] and carbon capture [34,35]) or presence of water (e.g., proton conduction [36–40] and water harvesting [12,41]). Hence, a comprehensive knowledge of the interactions of water molecules inside the pores of the host framework is necessary. One powerful method to study the adsorption/desorption behavior of MOFs is water vapor sorption analysis. The isothermal water uptake of the sample is measured, typically applying either gravimetric methods (microbalance) [42,43] or manometric techniques [44,45]. For the latter method, we recently demonstrated the potential provided by high accuracy in the low-pressure range, i.e., for small amounts of water, studying the sorption mechanism of M_2dhtp (M^{2+} = Co, Cu, Mg, Mn, Ni and Zn) [46,47], unveiling distinct differences of the sorption mechanisms, depending on the respective type of metal center.

In general, different sorption sites and unique sorption properties in MOFs [48,49] are a consequence of the chemical nature of the surface of these porous frameworks, gathered by a complex interplay of the hydrophilic inorganic building units and the (mostly) hydrophobic organic domains (i.e., linker molecules). However, discrete information on the polarity of the framework surface is obtained by determination of the inflection point α, i.e., the point at which half of the maximum adsorption is observed, of water adsorption steps. For polar materials, this point is shifted toward lower relative pressures (p/p_0), when compared to less polar compounds [1,50]. Many MOF materials tend to lack sufficient hydrolytic stability [51,52], which is an immediate consequence of the dynamic nature of the coordinative bonds in competition with water molecules [53]. Nevertheless, the above-discussed isoreticular approaches to modulate the pore properties in MOFs make them interesting materials for detailed water sorption studies. Qian et al. [10] systematically studied UiO-66(Zr) by independently analyzing the effect of the pore wall polarity (via linker functionalization) and pore size (via incorporation of larger linker moieties) on its water sorption behavior. Linkers with higher hydrophilicity revealed a pronounced shift of the observed sorption step toward lower relative pressures, suggesting stronger water–framework interactions. Similar observations were also made for other MOFs [13]. However, incorporation of highly polar linkers drastically altered the water sorption behavior, changing the shape of the observed water sorption isotherm (from type IV to type I [54]), suggesting highly attractive conditions for water adsorption, as similarly observed for porous carbon [55]. In contrast, hydrophobic groups only led to severely reduced water uptake without further visible changes. Further, pore size-dependent shifts of the sorption step toward higher pressures were observed, indicating that micropore filling is highly affected by the given pore diameter and the increased size of hydrophobic domains, since pore condensation relies on water cluster formation, as similarly observed for other MOFs [56–58].

We present a comprehensive study on the effects of pore size and linker substitution pattern on the water sorption mechanism and water arrangement within the one-dimensional, cylindrical pores of a series of isoreticular MOFs (i.e., Ni_2L; L = dhtp, dhip, dondc, bpp, bpm and tpp). We utilize thermogravimetric analysis (TGA) and Fourier transform infrared (FTIR) spectroscopy to monitor the interactions of water with its environment in the confined space. In addition, we apply solution ^{1}H nuclear magnetic resonance (NMR) spectroscopy to monitor the number of defects from missing linker molecules. Finally, manometric water vapor sorption analysis is used to study the adsorption/desorption behavior of the IRMOF-74 series.

2. Materials and Methods

2.1. Linker and MOF Synthesis

All M_2L MOF materials (i.e., Ni_2dhtp [59], Ni_2dhip [22], Ni_2dondc [20], Ni_2bpp [11], Ni_2bpm [11], Ni_2tpp [11]) were synthesized under solvothermal conditions according to modified reported procedures (details are given in the Supplementary Materials). Solvent was removed by heating under dynamic vacuum conditions to obtain desolvated samples, which were stored in a glove box under Ar atmosphere until further use. All products were characterized by powder X-ray diffraction (XRD) and compared with literature results, ensuring structural integrity with no evidence of crystalline impurities.

H_4dhtp [46,60], H_4dondc [20], H_4bpp [17,61] and H_4tpp [17,61] linker molecules were synthesized according to modified literature procedures (details are given in the Supplementary Materials). All other chemicals (including H_4dhip and H_4bpm) and solvents were purchased from commercial suppliers and used without further purification. All air- or water-sensitive reactions were carried out under standard Schlenk techniques using a dry argon atmosphere.

2.2. Preparation of Hydrated Samples

The desolvated samples were placed in a closed vessel with a separate water reservoir (1 mL per sample, ca. 10 to 15 mg) to load them via the gas phase at 40 °C for 24 h. Afterwards, we refer to them as hydrated samples (Ni_2L-hyd).

2.3. General Characterization Techniques

Powder X-ray diffraction (XRD) data were collected on a Bruker D8 Advance diffractometer (Bruker, Karlsruhe, Germany) with a step size of 0.02° and a counting time of 3 s per step. Patterns are normalized (data range from 0, 1) to the most intense reflection for better comparison of relative intensities. ^{1}H and ^{13}C nuclear magnetic resonance (NMR) spectra were recorded using Bruker Advance 500 and Bruker Ascent 700 spectrometers (Bruker, Ettlingen, Germany). Chemical shifts were calibrated to the resonance of residual non-deuterated solvent. For defect analysis, small amounts of the samples (ca. 3 to 5 mg) were diluted overnight in 600 µL of a solvent mixture prepared from dmso-d_6 (10 mL) and deuterium chloride (20 wt% in D_2O; 0.5 mL). Thermogravimetric analysis (TGA) was performed using a TGA/DSC1 STAR System thermobalance from Mettler-Toledo (Gießen, Germany). Samples were placed in a 70 µL corundum crucible. Measurements were performed under a constant nitrogen gas flow (50 mL min^{-1}, purity 5.0) in a temperature range from 40 °C to 800 °C using a heating rate of 10 °C min^{-1}.

2.4. Fourier Transform Infrared (FTIR) Spectroscopy

FTIR spectroscopy was performed using a Bruker Vertex 70 spectrometer (Bruker, Ettlingen, Germany) in annual total reflection (ATR) mode utilizing the Platinum ATR unit A225 with a diamond ATR crystal. Powdered samples were pressed on the crystal during the measurement. After the measurement of hydrated samples, an automatic baseline correction was applied using the OPUS 7.2.1 software package. The correction was performed using a concave rubberband correction, with 10 iterations and 32 baseline points. Gaussian least square fits of the water stretching band were performed individually using the peak analyzer function of the Origin23b software package. First, the data within the range from 4500 to 2400 cm^{-1} were normalized (data range from 0.1) to the maximum absorbance. For the fitting procedure, the data from 3900 to 2400 cm^{-1} were selected.

Baseline correction was performed, using a linear baseline. Then, three peaks were set with the starting positions near 3560 cm^{-1} (multimer water), 3380 cm^{-1} (intermediate water) and 3210 cm^{-1} (network water). Only positive peak areas were allowed, while the previously corrected baseline was set constant. No further restrictions were used. It was noted that the fit results possess a small error, caused by the overlap of the aromatic C-H vibration of the linker molecules with the water stretching bands.

2.5. Sorption Analysis

2.5.1. Nitrogen Physisorption Measurements

N_2 physisorption analysis was performed with a Quantachrome Autosorb 6B (Quantechrome Instruments, Boca Raton, FL, USA) at 77 K. The relative pressure range suitable for the BET area calculation was determined using the Rouquerol [62] criteria. Total pore volumes were determined from the uptake at $p/p_0 \approx 0.9$. Activation of the samples was performed in a stepwise manner. The desolvated samples were transferred to the measurement cell and dispersed in methanol twice for 45 min. After each step, the solvent was removed under dynamic vacuum. The pre-dried samples were degassed for approximately 17 h applying the following procedure: Samples were heated from room temperature to 60 °C with a heating rate of 2 °C min^{-1} after which the temperature was held for two hours. Then, the sample was heated to 100 °C with the same heating rate (2 °C min^{-1}) and the temperature was held for another two hours. Finally, the sample was heated (2 °C min^{-1}) to 150 °C and the temperature held for an additional 12 h, after which the sample was allowed to cool to room temperature. Pore size analysis was accomplished with the supplied AS-Multistation 2.01 software package using the NLDFT data-based method (N_2 at 77 K on silica, cylind. pore, NLDFT adsorption branch data kernel), treating the samples as silica.

2.5.2. Water Vapor Sorption Measurements

Water vapor sorption experiments were performed on a 3Flex instrument (Micromeritics, Unterschleißheim, Germany) at 298 K (25 °C) with double distilled (and degassed) water. Activation of the samples was performed in a stepwise manner, applying the same procedure as mentioned in the N_2 physisorption section.

3. Results and Discussion

3.1. General Characterization

All members of our nickel-based frameworks, i.e., Ni_2L (L = linker molecule), were prepared using six linker molecules, each presenting either a different substitution pattern (para- vs. meta-position of carboxylate groups) or different sizes, resulting in isoreticular materials with tunable size and pore wall polarities (Figure 1).

Powder X-ray diffraction (XRD) data of the prepared frameworks (i.e., Ni_2L) are in agreement with literature data [11,17,20,21], confirming the periodic, three-dimensional framework structure with a honeycomb-like cross section, with no evidence of crystalline impurities in any of the prepared samples (Figure 2a). The respective (110) lattice plane distances, characteristic for the perpendicular arrangement of the one-dimensional cylindrical pores, indicate the expected pore diameters in the range 13 ... 23 Å (Table 1). We note that the broad reflections for most materials indicate small crystallite size, while the signal-to-noise ratio is attributed to sample preparation and the crystallinity of the frameworks (longer measurement times were avoided to minimize the influence of humidity).

N_2 physisorption analysis enabled us to confirm the permanent porosity of all products, revealing type I sorption isotherms [64] for most frameworks in our series, consistent with the existence of micropores (Figure 2b). In contrast, Ni_2tpp reveals a type IV(b) isotherm shape typical of mesoporous materials with pore diameters below 40 Å. The suitable pressure range for BET area determination was selected using the Rouquerol [62] criteria. The high specific surface areas (560 ... 2320 m^2 g^{-1}) and large specific pore volumes (0.38 ... 1.03 cm^3 g^{-1}, at $p/p_0 \approx 0.9$) of the networks mostly align with literature data (reference data only for Mn_2dondc) [17,20–23]. We note that the comparable low surface areas for some materials hint toward the existence of amorphous byproduct, while the increase in the adsorbed volume at high relative pressures generally indicates interparticle porosity as a consequence of small crystallite size Observed differences for Ni_2dhip and Ni_2bpm are slightly larger. Pore sizes analysis (Figure 2c) is in good agreement with previous studies [21] and our XRD results, verifying that the pore size within our series approximately doubles from Ni_2dhip to Ni_2tpp (see Table 1, details are shown in the Supplementary Materials; NLDFT-based fits displayed in Figures S6–S11). However, the uniformity of the pore sizes is seemingly affected by the average pore diameters, resulting in a broadening of the distribution curve, in agreement with the expectation that larger inorganic moieties negatively influence the frameworks crystallinity.

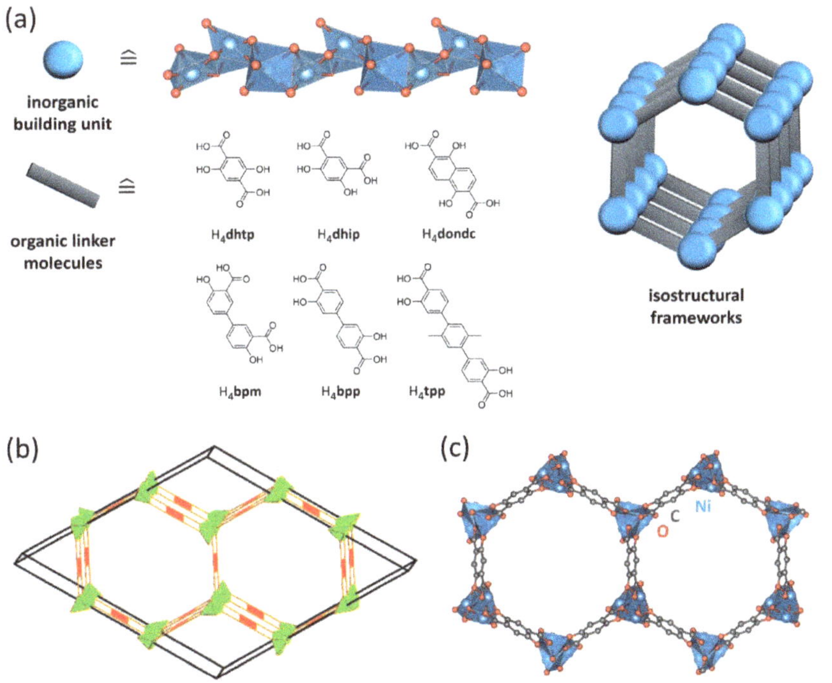

Figure 1. (**a**) Illustration of the inorganic building units and used linker molecules in this research, forming isostructural Ni_2L frameworks. (**b**) Network topology of Ni_2dhtp (msf, reprinted from ref. [3]) and (**c**) representation of the honeycomb-like cross section along the crystallographic *c*-axis, showing Ni_2dhtp as an example.

Table 1. Determined lattice plane distance d_{110} calculated from the Bragg equitation as well as surface areas S_{BET}, pore volumes V_{Pore} and pore sizes $d_{Pore,NLDFT}$ of Ni_2L materials determined from nitrogen sorption data at 77 K.

	d_{110} (Å)	S_{BET} ($m^2\ g^{-1}$)	V_{Pore} ($cm^3\ g^{-1}$)	$d_{Pore,NLDFT}$ (Å)
Ni_2dhtp	12.93 ± 0.02	1240 ± 80	0.55 ± 0.07	12.73 ± 0.22
Ni_2dhip	12.97 ± 0.06	560 ± 120	0.38 ± 0.09	11.93 ± 0.30
Ni_2dondc	14.57 ± 0.10	1010 ± 10	0.55 ± 0.03	13.63 ± 0.42
Ni_2bpp	18.36 ± 0.04	1660 ± 230	0.74 ± 0.02	16.68 ± 0.57
Ni_2bpm	18.87 ± 0.00	1200 ± 50	0.58 ± 0.02	18.52 ± 0.52
Ni_2tpp	23.05 ± 0.06	2320 ± 140	1.03 ± 0.06	25.09 ± 0.67

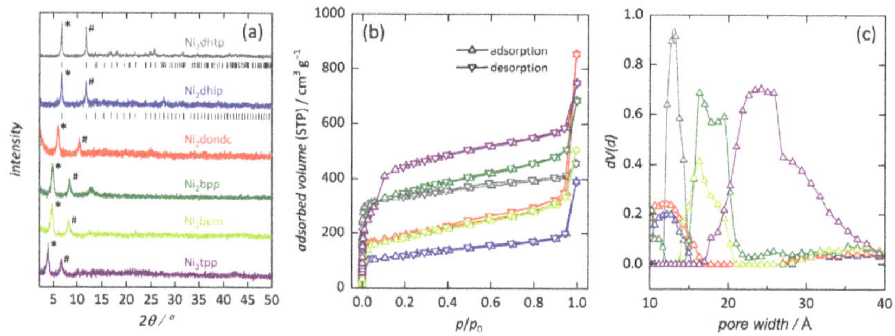

Figure 2. XRD patterns of Ni_2L frameworks (**a**). The most intense reflections represent the (110) and (300) lattice planes (marked with * and # respectively), being shifted toward lower angles with increasing pore diameter. Patterns are normalized and compared to literature data where accessible [22,63]. N_2 physisorption isotherms of Ni_2L (**b**) measured at 77 K and their respective NLDFT-based pore size plots (**c**).

Recent studies analyzed formate defects in Ni_2dhtp stemming from decomposed DMF (DMF = *N*,*N*-dimethylformamide) solvent molecules. They found that the number of defect sites increases with the metal-to-linker ratio [65]. Therefore, we applied ^{1}H NMR spectroscopy to analyze potential defects within our series. Samples were diluted in a solvent mixture of dmso-d_6 and deuterium chloride (20 wt% in D_2O). We note that acid-caused decomposition of formate anions [66] is a possible error source. However, we suggest that the impact of this is comparably low due to the reasonably low acid concentration used for decomposition. NMR spectra of diluted frameworks are shown in the Supplementary Materials (Figures S12–S17). Respective ratios of the linker molecules and formate anions (For^-), or acetate (OAc^-) anions for Ni_2dondc, were calculated considering the different charges of the coordinating ligands, i.e., one linker (L^{4-}) must be replaced by four For^-/OAc^- ions (capping ligand) to maintain charge balance. This enabled us to determine the respective sum formula for our materials (see Table 2). We found that even for low metal-to-linker ratios (e.g., Ni_2dhtp) the amount of incorporated formate ions is significant (i.e., 0.62, suggesting 0.16 dhtp^{4-} molecules are replaced). However, we see the reported trend of increased defect sites with increasing metal-to-linker ratios for all materials in our series. The precise amount of the capping ligands depends on the nature of the linker molecule (acidity, solubility, . . .) and the precise reaction conditions. Ni_2dhip revealed the largest differences between individually synthesized samples, indicating that its synthesis is more sensitive to minor changes compared to the other materials in this study. Further, we found formate ions in diluted samples of Ni_2dondc (Figure S18), which we attribute to thermally decomposed DMF used for the replacement of *N*-methyl-2-pyrrolidone (NMP) solvent molecules. We suggest that these post-synthetically generated formate ions do not affect the defect degree of the pristine framework but coordinate to the open metal sites. Hence, these molecules can be replaced by additional solvent exchange procedures, as performed prior to sorption analysis. However, this suggests that the determined formate ion concentration in diluted samples is slightly overestimated as a consequence of residual DMF solvent molecules. Nevertheless, the amount of incorporated capping ligands clearly is another contributing factor for the broadening of the pore size distribution (see above), in agreement with the negative effect of defect sites on the materials' crystallinity. For the sake of simplicity, all materials are referred to Ni_2L, and presented defects are discussed when necessary.

Table 2. Determined linker and formate (For^-) or acetate (OAc^-) capping ligand concentrations of Ni_2L materials derived from solution ^1H NMR and their respective sum formulas, considering charge balance. (Sum formula for a defect free material would be Ni_2L.).

	Linker Molecules per Formula Unit	For^-/OAc^- Ligands per Formula Unit	Sum Formula
Ni_2dhtp	0.84 ± 0.02	0.62 ± 0.10	$Ni_2(dhtp)_{0.84}(For)_{0.62}$
Ni_2dhip	0.65 ± 0.09	1.41 ± 0.35	$Ni_2(dhip)_{0.65}(For)_{1.41}$
Ni_2dondc	0.65 ± 0.01	1.39 ± 0.00	$Ni_2(dondc)_{0.65}(OAc)_{1.39}$
Ni_2bpp	0.49 ± 0.06	2.04 ± 0.25	$Ni_2(bpp)_{0.49}(For)_{2.04}$
Ni_2bpm	0.61 ± 0.06	1.54 ± 0.25	$Ni_2(bpm)_{0.61}(For)_{1.54}$
Ni_2tpp	0.54 ± 0.04	1.85 ± 0.16	$Ni_2(tpp)_{0.54}(For)_{1.85}$

3.2. Thermal Analysis

To evaluate the thermal stability of our frameworks and provide an ideal starting point for our hydration/dehydration experiments, we performed thermogravimetric analysis (TGA) of the hydrated frameworks (Ni_2L-hyd, Figure 3). Samples were prepared by loading the desolvated materials with water through the gas phase. Therefore, samples were placed in a closed vessel with a separate water reservoir to load them with water above room temperature (i.e., 40 °C) for 24 h; for more details, see Section 2.2.

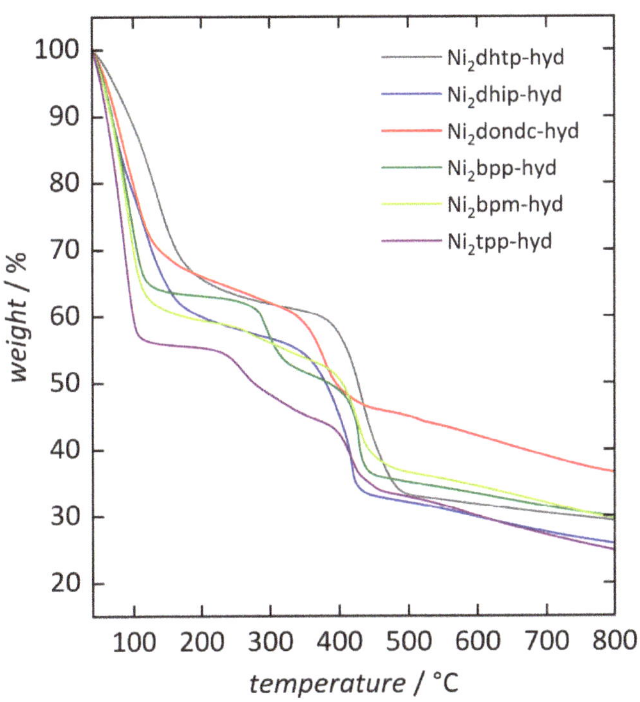

Figure 3. Thermogravimetric analysis of the hydrated materials (Ni_2L-hyd) from 40 to 800 °C under N_2 atmosphere (heating rate: 10 °C min^{-1}).

In general, framework decomposition is observed in the range 350 . . . 450 °C, comparable with previous reports [17,21]. However, the above-mentioned formate/acetate defects affect the thermal stability of the frameworks. Prior to framework decomposition, an additional smaller mass loss step around 250 °C (Ni_2tpp) and 300 °C (Ni_2bpp) is observed for some frameworks. This is in accordance with previous studies on defect engineering in Ni_2dhtp, revealing additional decomposition steps, becoming more pronounced with higher defect degrees [67]. Our results on Ni_2bpp and Ni_2tpp are comparable to previous studies, although the here-observed steps are more pronounced and, for Ni_2tpp, shifted toward lower temperatures [17,21]. These differences are likely to arise from slightly different synthetic approaches (here: mostly DMF [59] vs. DMF/EtOH/water (1/1/1) mixture with lower Ni-to-*L* ratios), resulting in varying defect sites [65]. TGA data further indicate that the mass stability, i.e., formation of a stable plateau, depends on the defect degree, as increasing amounts of For^-/OAc^- lead to the loss of a stable mass plateau after the water loss step and finally, an additional pronounced decomposition step. We note that our results for Ni_2dhtp deviate from our previous studies [46], where larger amounts of OAc^- ions were incorporated (stemming from the metal source), visible in an additional decomposition step prior to the collapse of the framework. This suggests that the herein-found formate defects do not significantly alter the thermal stability of Ni_2dhtp, agreeing with the above-referenced studies.

Next, we determined the amount of adsorbed water in all hydrated Ni_2L-hyd materials by considering the first mass loss step between 40 and 250 °C. (We note that the adsorbed amount of water is generally underestimated due to its strong bonding to the Ni^{2+} sites, being slowly desorbed over a broad temperature range as well as during framework collapse [32,46]). All materials reveal similar values (21 ... 25% weight loss, Supplementary Materials Table S1), slightly increasing with pore size. However, Ni_2tpp deviates from this trend, indicating that some loosely bonded water molecules could desorb during sample preparation, as indicated by a less stable sample mass during measurement setup. In general, the dehydration step shifts toward lower temperatures with increasing pore diameter, indicating water–framework interactions become less attractive and the adsorbed water becomes more comparable to bulk water.

3.3. FTIR Spectra of Hydrated Samples

Our results presented above indicate the expected effect of the channel size on water–framework interactions and hint toward different water arrangements in the one-dimensional channels. Thus far, infrared spectroscopy is frequently used to analyze the interactions of water molecules at surfaces (e.g., inverse micelles [68,69]) or within pores (e.g., mesoporous silica [45,70] and MOFs [46,71]). This possibility arises from the deconvolution of the O–H stretching vibration, a superposition of vibrational bands, typically located in the range of 2800 to 3700 cm^{-1} [72]. Least squares fitting of three Gaussian profiles enables us to determine the respective contributions (peak areas) of the different vibrational bands assigned to water in a particular H-bonding environment [45,68,69]. Brubach et al. [72] designated these water molecules as different types with respect to their coordination numbers (CN), i.e., numbers of H-bonding partners. These three types are termed 'network water' (NW), 'intermediate water' (IW) and 'multimer water' (MW), as their respective stretching frequencies reveal a red shift, i.e., toward lower wavenumbers, with an increasing number of intermolecular interactions. NW resembles highly H-bonded water molecules interacting strongly with their respective environment, forming roughly four H-bonds (CN $\geq$ 4) with adjacent atoms. These structures are often referred to as 'ice-like' arrangements, possessing oscillation frequencies around 3260 cm^{-1}. As the number of interactions decreases, IW water populations form, showing frequencies around 3460 cm^{-1}. These populations offer a 'bulk-like' water arrangement of more dynamic water molecules. Lastly, MW resembles poorly connected water molecules, often found at interfaces, forming only few H-bonds (as found in dimeric/trimeric structures). These MW water molecules are also referred to 'dangling' water molecules, being a consequence of their isolation, i.e., low connectivity, to other water molecules. MW typically shows vibration frequencies located around 3600 cm^{-1}.

We carried out Fourier transform infrared (FTIR) spectroscopic measurements in attenuated total reflection (ATR) mode on hydrated samples (i.e., Ni_2L-hyd) to analyze the effect of the pore size and linker substitution pattern on the arrangement and interactions of adsorbed water molecules. The spectra reveal the integrity of the organic linker molecules as well as the presence of three of the characteristic vibration bands of water, i.e., the stretching band (3750 ... 2800 cm^{-1}), the bending band (1700 ... 1650 cm^{-1}), and the libration band (1000 ... 400 cm^{-1}) (see Supplementary Materials Figures S19–S24) [72]. Next, least squares fitting was performed for a pure liquid water film located at the surface of the ATR crystal (Supplementary Materials Figure S25), as a reference to quantify the relative contributions of the different water vibrations. We observed a contribution of 13% from multimer water (3535 cm^{-1}), stemming from water in the liquid film that is located at either of the two interfaces (ATR diamond crystal or air) [46]. Further, contributions of 18% from intermediate water (3405 cm^{-1}) and 69% from network water (3251 cm^{-1}) indicate that most of the water molecules not located at the interfaces form multiple H-bonds with surrounding molecules. All Ni_2*L*-hyd samples possess significantly lower contributions for the multimer peak (5–6%) (Figure 4 and Supplementary Materials Table S2) when compared to the pure water film. This suggests that, similar to M_2dhtp [46], water molecules in these confined environments show a higher tendency for H-bonding than interface-near water molecules of a pure liquid water film. This further suggests that additional water molecules interact strongly with the coordinating water molecules, located at the primary binding sites, and with oxygen atoms of the inorganic building unit, giving rise to further H-bonding possibilities. Further, all Ni_2*L*-hyd networks show slightly higher contributions from network water (73–78%) as well as similar contributions from intermediate water (17–22%) when compared to the pure water film, suggesting that multiple water molecules form a highly connected H-bonding network.

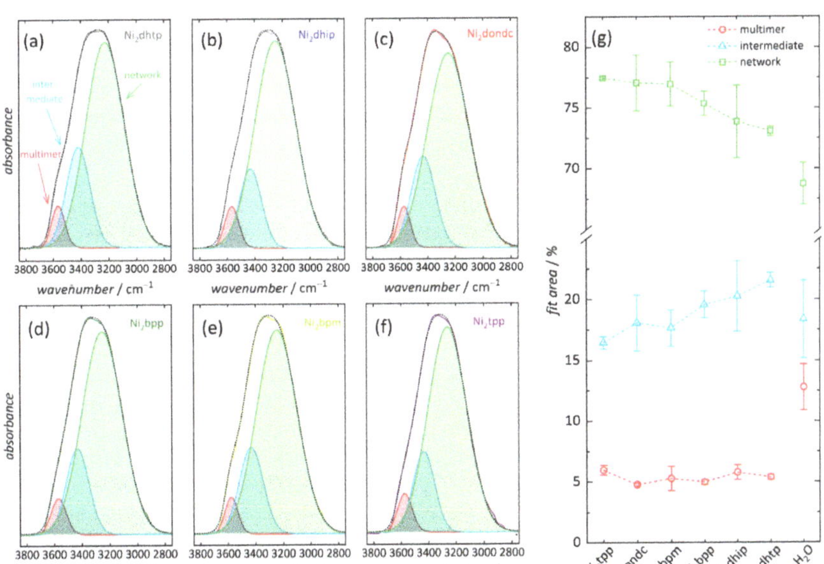

Figure 4. Deconvolution of the FTIR O–H stretching vibration band of water in hydrated Ni_2L-hyd samples (**a**–**f**) by least squares fitting (Dotted lines: sums of the three Gaussians). Relative contributions of multimer, intermediate and network water in comparison with a pure liquid water film (**g**).

We note that the observed relative contributions of the three vibrational bands in Ni_2dhtp-hyd differ from earlier results [46]. As mentioned above, this is explained by different synthetic procedures, resulting in more defect sites caused by significant incorporation of acetate anions (from the metal source). These capping ligands alter the pore surface, resulting in a less uniform pore wall structure, causing less attractive interactions of water with the framework because of reduced H-bonding possibilities. Further, the precise adsorption mechanism is affected (as discussed below). As stated above, the Ni_2dhtp formed here possess fewer defects, which is typical for reactions applying small metal to linker rations [65]. Consequently, a more uniform pore wall structure and stronger interactions with more H-bonding opportunities are found. Expecting crystallographically defined water positions in Ni_2dondc, similarly for Mn_2dondc [20], we found different MW (lowest observed) and NW (second highest) contributions, suggesting a similar number of water molecules within the hydrated framework (ca. five molecules per Ni^{2+}). However, defects within the other frameworks seem to play a minor role for these interactions, since most water molecules form multiple H-bonds even in the presence of significant amounts of capping ligands. This indicates that potential binding sites near the pore walls are significantly less attractive for frameworks with larger hydrophobic domains. We hypothesize that the high NW contributions found therein stem from inner pore, more bulk-like water, rather than from interfacial water (i.e., MOF–water interactions as found for M_2dhtp [46]).

Consideration of the respective wavenumbers $\widetilde{v}$ (peak positions, see Supplementary Materials Table S3) of the three Gaussian contributions allows us to gain more information on the O–H bond situation of confined water. When compared to the pure water film, all Ni_2L-hyd samples show a blue shift (i.e., toward higher $\widetilde{v}$) of the multimer and intermediate water peak. Observed shifts are comparable for all materials except for Ni_2dhtp-hyd, which reveals a less pronounced blue shift, generally indicating O–H bond strengthening upon hydration. In addition, most compounds reveal no significant changes of the network water peak. However, Ni_2dhtp-hyd reveals a red shift (toward lower $\widetilde{v}$), indicating weakening of the O–H bonds in NW water stemming from highly confined water molecules. For Ni_2tpp-hyd, the opposite, i.e., blue shift, is found, being a consequence of less confined water molecules.

The local H-bonding situation inside the one-dimensional pores of hydrated frameworks is another aspect to bear in mind to monitor pore-size-dependent differences. Information on this is derived from the respective differences between the peak center positions x and y of different vibrational bands, i.e., from the wavenumber splitting $\Delta\widetilde{v}_{xy}$ (Table 3). Smaller $\Delta\widetilde{v}$ values suggest a higher symmetry in the local H-bonding of water, indicating more mobile water molecules with high rotational degrees of freedom and vice versa [73,74]. We determined higher values for all Ni_2L-hyd materials than found for a pure water film. This indicates a higher degree of H-bonding asymmetry, which is consistent with the expectation of confined water molecules. Despite the similar trend, the precise $\Delta\widetilde{v}$ values differ for any given compound. For Ni_2L-hyd materials with smaller pore diameters, i.e., L = dhtp, dhip and dondc, generally higher wavenumber splitting is found. In the particular cases of Ni_2dhtp [32] and Ni_2dondc [20], this is in accordance with the expectation of crystallographically defined water positions in a fixed, highly asymmetric local H-bonding situation. However, frameworks with larger pore diameters, i.e., L = bpp, bpm and tpp, show the opposite trend, revealing lower $\Delta\widetilde{v}$ values. H-bonding becomes more symmetric, and thus are more comparable with bulk water because of the less pronounced confinement effects. These observations are in accordance with our TGA experiments and previous results on mesoporous silica revealing similar effects of the pore size [70], highlighting a more bulk-like behavior with more flexible water molecules. Considering these results, the members with larger pores (Ni_2L, L = bpp, bpm and tpp) should reveal a water sorption behavior that mostly resembles classical pore condensation (i.e., water cluster formation), rather than the occupation of discrete positions and confinement effects (as expected for the smaller pores).

Table 3. Wavenumber splitting $\Delta\widetilde{v}$ of the stretching vibration bands of water in the Ni_2L-hyd materials and of pure water.

	$\Delta\widetilde{v}_{multimer\text{-}network}$ (cm^{-1})	$\Delta\widetilde{v}_{multimer\text{-}intermed.}$ (cm^{-1})	$\Delta\widetilde{v}_{network\text{-}intermed.}$ (cm^{-1})
Ni_2dhtp	336.2 ± 2.5	146.1 ± 1.1	190.1 ± 2.5
Ni_2dhip	325.4 ± 1.8	142.2 ± 1.2	183.2 ± 1.8
Ni_2dondc	337.0 ± 2.5	141.3 ± 1.2	181.7 ± 1.7
Ni_2bpp	321.0 ± 1.9	140.9 ± 1.5	182.9 ± 1.6
Ni_2bpm	320.6 ± 1.7	142.6 ± 1.4	178.8 ± 1.5
Ni_2tpp	309.1 ± 2.2	140.0 ± 2.0	169.1 ± 1.8
water	284.4 ± 1.5	130.2 ± 2.9	154.2 ± 1.5

3.4. Water Vapor Sorption

To prove our hypothesis regarding the effect of the pore size on the sorption mechanism, we analyzed the hydration/dehydration behavior of the Ni_2L with manometric water vapor sorption measurements at 25 °C. We carried out two consecutive adsorption/desorption cycles without additional thermal activation in between the two cycles (i.e., samples were not removed from the device), since our previous water sorption studies on M_2dhtp [46,47] highlighted that coordination of water to the open metal site during the first sorption isotherm results in different starting conditions ascribed to strongly bound (chemisorbed) water molecules. These cannot be removed during the desorption process without additional thermal activation.

We observed an overall higher total uptake during the second adsorption/desorption cycle when compared to the first cycle for Ni_2dhtp. This is explained by a partial degradation of the frameworks resulting in the formation of additional open metal sites caused by cleavage of metal–linker bonds [42,46,47] during the time-consuming measurements (several days). Results on Ni_2dhip indicate a poor hydrolytically stability since we observe a decrease in water uptake during the first desorption isotherm, reflected by the crossing of the adsorption and desorption branch and large equilibrium times, making it impossible to further evaluate its water sorption behavior in detail. On the other hand, Ni_2L frameworks containing larger linker moieties (i.e., L = dondc, bpp, bpm and tpp), and thus possessing larger pores, reveal a severely decreased total water uptake during the second cycle, indicating that water molecules affect the framework's longevity. All observations are consistent with the decreased crystallinity of all samples after water sorption experiments (Supplementary Materials Figure S26), while most materials maintain a larger degree of their initial pore ordering (except for Ni_2dondc, which reveals complete loss of crystallinity). Since all M_2L frameworks adsorb water at low relative pressure ($p/p_0 < 0.1$) due to highly attractive interactions with vacant coordination sites, we displayed all isotherms, both in a linear and in a semi-logarithmic representation, to analyze differences in the sorption behavior. In addition, first derivatives of the isotherms ($\delta V/\delta(p/p_0)$) were calculated to highlight the distinct water sorption steps in all isotherms (see Supplementary Materials Figures S27–S32) (Missing data points below p/p_0 = ca. 0.002, except in the first adsorption isotherms, are due to technical experimental restrictions).

The water sorption isotherms of Ni_2dhtp (Figure 5a,b), reveal similar characteristics as observed in previous studies on M_2dhtp [46,47], revealing a type I [54] water sorption isotherm. However, differences arise as both materials are compared with each other. Accordingly, uptake of the first 20% (with respect to the total uptake) occurs at particularly low pressures ($p/p_0 < 0.003$), attributed to the coordination of water to open meal sites, even though the desorption data are inconclusive with respect to the irreversibility of this process, i.e., missing low-pressure data due to technical limitations. The further course of the isotherms suggests a slightly different sorption mechanism: The observed sorption step at $p/p_0 = 0.03 \ldots 0.05$ in this study corresponds to the uptake of one rather than two water molecules in our previous study (ca. 20%, resulting in 60% of the total water uptake vs. 40%, resulting in 80%) and presumably results in the formation of a uniform water cluster along the pore walls of the framework, being in agreement with the water sorption mechanisms found for most other metals within the M_2dhtp series (except M = Cu) [46,47]. As mentioned above, these findings result from varying defect degrees. Significant incorporation of acetate anions (and presumably capping ligands in general), results in less uniform pore walls within the Ni_2dhtp framework, and thus less attractive interactions of water with the pore walls, i.e., via H-bonding with the phenylene backbone at the proposed third adsorption site. Consequently, the adsorption mechanism is altered, making the adsorption of two water molecules more attractive to create a metastable state (i.e., preliminary pore saturation). However, the further cause of our isotherm suggests that the here observed formation of a continuous water layer along the pore walls seems to be a meta-stable state, since gradual absorption of additional water molecules occurs afterwards until complete hydration is observed, without any sign of additional sorption steps. We attribute this to stronger Ni–O interactions in the inorganic SBU when compared to the other members of the M_2dhtp series [47], affecting atomic charges on the metal center, thus preferring a square planer coordination environment of Ni^{2+}. Consequently, chemisorption of water molecules to the metal center as well as H-bonding with framework O atoms becomes less attractive, i.e., larger H-bonding distances, causing similar adsorption energies of both sorption sites. Subsequent occupation of the third site (near the phenylene backbone) results in more water molecules arranged in the confined environment. Hence, preliminary pore saturation is observed at lower relative pressures when compared to M_2dhtp [46,47]. Then, uptake of additional water molecules with increasing vapor pressure (i.e., during adsorption) is enabled by relaxation of previously adsorbed molecules, providing additional space while decreasing their respective mobility. This phenomenon creates a hysteresis because of an increased density of the adsorbed water, which is irreversible during the desorption, i.e., with decreasing pressure. We previously ascribed this effect to the alternating hydrophobic (phenylene) and hydrophilic (coordinated metal centers) domains within M_2dhtp [46,47], which is in accordance with similar effects found for other porous materials possessing larger and/or more hydrophobic pores [45,75–84]. The total water uptake after the first adsorption branch was determined to be 0.51 ± 0.06 g g^{-1}, which is in agreement with earlier studies [17,85]. Our findings on Ni_2dhtp indicate that the precise sorption mechanism can be tailored by selective incorporation of defects, giving rise to tailor-made water sorption properties [59,65,67,86]. We will carry out further experiments in the future to validate this hypothesis and confirm if this is a general property of the M_2dhtp framework series.

Continuing with Ni_2dhip, the water sorption isotherms reveal a similar shape compared to Ni_2dhtp (Figure 5c,d). Slight differences within the low-pressure range $p/p_0 < 0.05$ suggest that the less uniform structure of the pore walls because of the different substitution patterns (i.e., para- vs. meta-arrangements of the carboxylate groups) has only a minor influence on the possible sorption sites within the one-dimensional pores. In addition, the total water uptake is slightly lower (0.41 ± 0.06 g g^{-1}), which is probably due to the decreased void spaces, affecting especially sorption sites near the pore center. We note that these effects could also stem from the observed higher defect-degrees within Ni_2dhip when compared to Ni_2dhtp. As mentioned above, the framework decomposes during the first desorption branch, making it impossible to analyze any effects of the chemisorbed water on the second cycle.

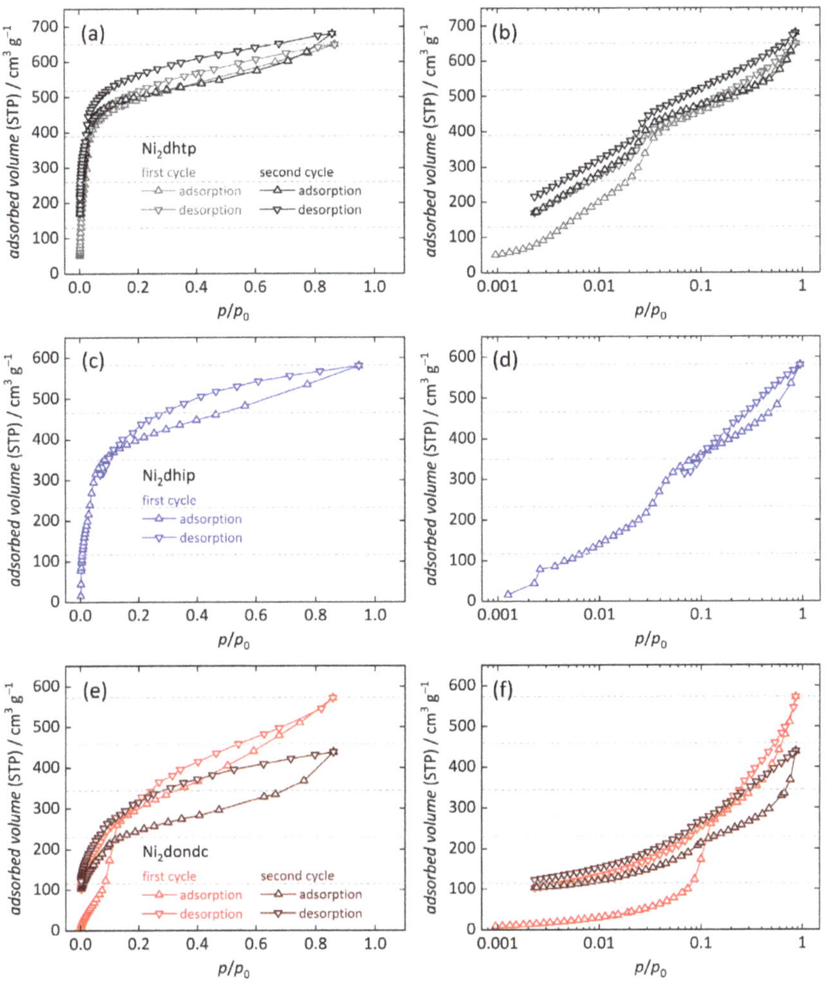

Figure 5. Water vapor sorption isotherms (25 °C) of Ni_2dhtp, Ni_2dhip and Ni_2dondc (two consecutive adsorption/desorption cycles). Data are shown at linear scale (**a**,**c**,**e**) and in semi-logarithmic representation (**b**,**d**,**f**). Horizontal lines mark 20%, 40%, 60%, 80% and 100% of the total water uptake during the first cycle.

The water sorption isotherms of Ni_2dondc (Figure 5e,f) offer fewer similarities to the previously discussed Ni_2dhtp, displaying a type IV [54] water sorption isotherm. Here, water adsorption proceeds less at low relative pressures ($p/p_0 < 0.1$) and is attributed to the irreversible saturation of the open metal sites, i.e., chemisorption to the primary binding sites (ca. 20% of the total uptake remain adsorbed after completed desorption). Following this, a pronounced adsorption step at $p/p_0 = 0.1$ is detected, accountable for a majority of the water uptake. This indicates less favorable conditions for water adsorption, i.e., similar sorption energies of additional water molecules after chemisorption, because of the larger pore diameter and hydrophobic moieties. Therefore, water cluster formation is the driving force for pore condensation, initiated by previously adsorbed water molecules coordinating to the metal centers [56,57]. After this step, more water molecules are gradually adsorbed over a large pressure range. Desorption proceeds gradually as no pronounced desorption steps are visible. This results in a hysteresis between the first adsorption and desorption branch, again suggesting the formation of a meta-stable state during the adsorption process. The total water uptake after the first adsorption branch (0.43 ± 0.03 g g^{-1}) is slightly smaller than for Ni_2dhtp, which is attributed to the higher molecular mass of the used dondc^{4-} ligand, resulting in a larger molecular mass. However, our results are in agreement with single crystal data on as-synthesized Mn_2dondc [20], suggesting that, once completely hydrated, water occupies similar positions in fully hydrated M_2dhtp (i.e., five distinct positions) (We note that the authors mentioned significant disorder of the water molecules due to residual *N*-Methyl-2-pyrolidone solvent molecules). However, unlike for Ni_2dhtp, these distinct positions do not stem from the occupation of preferred sorption sites during the adsorption but rather from an attractive configuration of adsorbed water molecules in hydrated pores. Continuing with the second cycle, a lower overall uptake is observed after the adsorption branch, suggesting decomposition of the framework during the measurement, applying a similar mechanism as discussed for Ni_2dhtp [42,46,47], i.e., with partial pore blockage, resulting in decreased water uptake.

Finally, we investigated the adsorption/desorption behavior of the remaining Ni_2L materials (i.e., L = bpp, bpm and tpp). These compounds offer even larger pores than Ni_2dondc, but reveal distinct similarities, showing type IV [54] water sorption isotherms. All these Ni_2L materials reveal only minor adsorption below a pressure of $p/p_0 < 0.2$, which is again attributed to the irreversible adsorption to the primary sorption sites (Ni^{2+} metal sites, Figure 6a–f). The amount of irreversibly bond water molecules decreases as shown by 13% (Ni_2bpp) > 11% (Ni_2bpm) > 8% (Ni_2tpp), suggesting a smaller amount of water strongly interacting with the framework, resulting in a more bulk-like state, agreeing with our FTIR experiments. Slight differences in the trend presumably stem from the significantly higher incorporation of For$^-$ within Ni_2bpp, resulting in lower hydrolytic stability. As the relative water pressure increases, one large adsorption step is visible in all isotherms, which is shifted toward higher pressures with increasing pore size: $p/p_0 \approx 0.21$ for (Ni_2bpp and Ni_2bpm) < $p/p_0 \approx 0.46$ for (Ni_2tpp). Similar dependencies of the condensation of water on the pore size were reported for other MOFs [10] and carbon nanotubes [81,87,88]. These pronounced sorption steps are accountable for the majority of the total water uptake, indicating that for Ni_2dondc and other MOFs [56,57], water adsorption is driven by water cluster formation initiated by chemisorbed water molecules, facilitating pore condensation. Over the remaining pressure ranges, water is gradually adsorbed until complete hydration is observed. During desorption, a small hysteresis is observed, attributed to the above-stated preliminary pore condensation. However, a rather unusual behavior is found for Ni_2bpp and Ni_2bpm, as the first desorption branch crosses the first adsorption branch within the pressure range $p/p_0 = 0.25 \ldots 0.41$ (Figure 6a–d). This is assigned to structural degradation of the framework [89], adopting a similar decomposition mechanism as reported for Ni_2dhtp [42,47]. There, additional open metal sites are formed with individual pores tilting against each other, resulting in decreased porosity. Further evidence for this is found as the second adsorption/desorption cycle is considered, revealing a significantly reduced water uptake after the second adsorption branch when compared to the first adsorption branch. We note that a similar phenomenon might occur for Ni_2tpp (see data range from $p/p_0 = 0.5 \ldots 0.7$) although data are inconclusive in this respect (Figure 6e,f). In addition to this crossing, the second cycles are mostly identical to the first ones, clearly indicating that this effect is caused by framework degradation rather than unique material properties. The maximum water uptake after the first sorption branch increases with pore size, 0.57 ± 0.02 g g^{-1} (Ni_2bpp) < 0.65 ± 0.07 g g^{-1} (Ni_2bpm) < 0.80 ± 0.01 g g^{-1} (Ni_2tpp). Except for the observed crossing of the branches, our results on Ni_2bpp and Ni_2tpp are generally in agreement with previous studies, revealing a slightly lower total water uptake [17]. Differences are attributed to either different measurement settings (fewer data points in the respective crossing regions) or lower defect amounts (not analyzed therein).

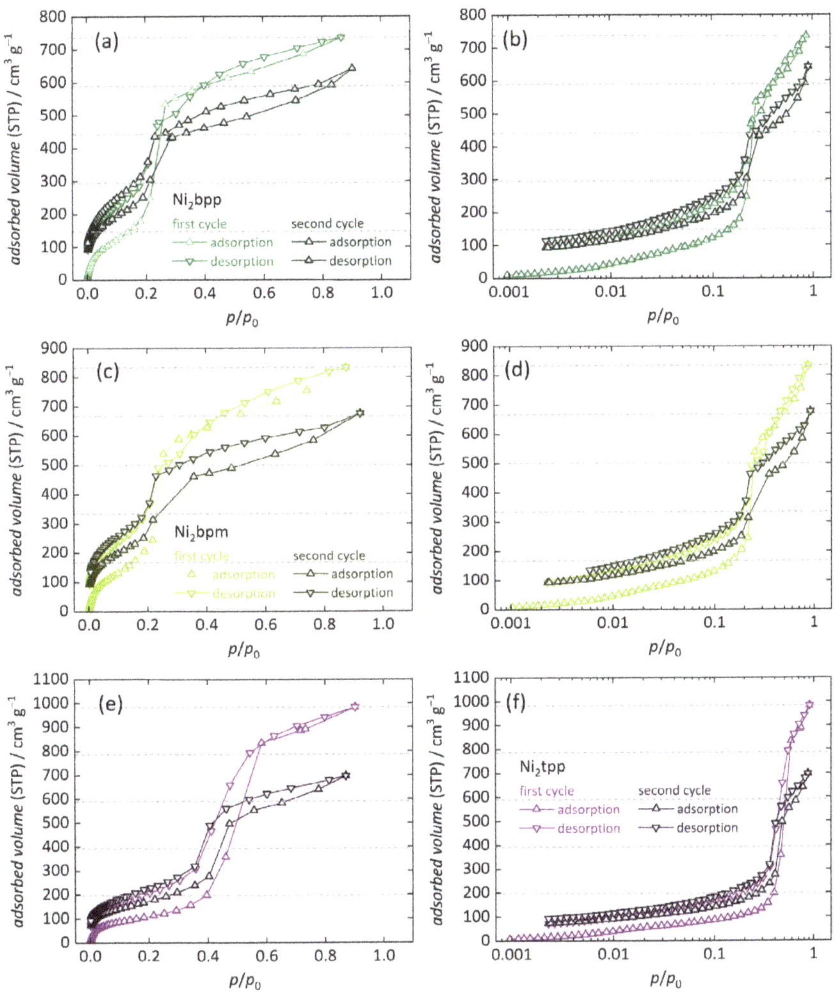

Figure 6. Water vapor sorption isotherms (25 °C) of Ni_2bpp, Ni_2bpm and Ni_2tpp (two consecutive adsorption/desorption cycles). Data are shown at linear scale (**a**,**c**,**e**) and in semi-logarithmic representation (**b**,**d**,**f**). Horizontal lines mark 20%, 40%, 60%, 80% and 100% of the total water uptake during the first cycle (Missing data points for Ni_2bpm below ca. p/p_0 = 0.006 stem from technical difficulties).

When compared to our TGA results, the overall water uptake determined by water vapor sorption experiments (see Supplementary Materials Table S4) is higher for all frameworks (23 . . . 48 mmol g^{-1}). Differences become larger with increasing pore size. As stated above, our TGA analysis underestimates the amount of adsorbed water because the chemisorbed water is strongly bound and gradually released over a larger temperature range and during decomposition. Hence, it is not considered in the initial dehydration step. While this is sufficient to explain the difference for Ni_2L frameworks with a smaller pore apparatus, i.e., L = dhtp, dhip and dondc, additional factors for larger pores must be considered. We hypothesize that some water molecules in these frameworks, i.e., L = bpp, bpm, tpp, are loosely bond and could leave prior to the TGA measurements during sample preparation. Further, we note that we do not see a clear trend of the maximum adsorbed volumes in our nitrogen sorption and water sorption experiments (see Table S6), as found in other studies [82]. Nevertheless, our water vapor sorption analysis indicates that additional hydrophilic sorption sites for H-bonding, besides the open metal sites, i.e., near the inorganic SBUs as present in Ni_2dhtp, are optimal for high water sorption at low relative pressures (Figure 7a). Those additional sorption sites are not found in the other members of the Ni_2L series, which is an immediate consequence of the larger hydrophobic domains. We suggest that water adsorption within the more porous Ni_2L frameworks (L = dondc, bpp, bpm, tpp), is dominated by water cluster formation and follows a similar mechanism in the low-pressure region, as previously reported for other MOFs with strong sorption sites [56–58]. First, the open metal sites are saturated before additional, low-entropy water molecules adsorb thereon, forming distinct water clusters driven by large hydration enthalpies (We note that this could also be a simultaneous process). With increasing pressures, previously adsorbed water molecules initiate the formation of water bridges/chains between adjacent chemisorbed water molecules, acting as hydrophilic anchor sites (Figure 7b–d), similar to studies previously reported for mesoporous silica [83,90] and porous carbons [91] possessing similar hydrophobic domain size. As soon as enough water molecules are present in the gas phase to connect enough adjacent anchor sites, the formation of a continuous H-bonding network is initiated, resulting in pore condensation with minimal framework–water interactions. This process is shifted toward higher relative pressures p/p_0 with increasing hydrophobic domain size, i.e., larger distances of adjacent chemisorbed water molecules, revealing a direct correlation of the inflection point α and the linker polarity (i.e., octanol–water partition coefficients [92,93], see Table S5). Further, this suggests that the pore wall surface polarity is mostly controlled by the inorganic domains, even at small domain sizes. These effects are a direct consequence of the Gibbs hydration energy $\Delta G_{\text{hydration}}$ being dominated by the enthalpy term for larger hydrophobic domains, with water molecules avoiding these units. Conversely, smaller units are viewed as discrete perturbations, forcing a high-entropy water arrangement near the hydrophobic surface, resulting in dominant entropic contributions as suggested for Ni_2dhtp (i.e., dominant surface–water interactions) [94–96]. The transition length for the reversal of entropy/enthalpy-driven hydration lies around 10 Å (or 1 nm), which is within the same range as water–water distances for coordination water molecules in Ni_2bpp and Ni_2bpm. Following the condensation step, the relaxation of adsorbed water molecules provides additional room for further water molecules until complete hydration is observed. Consequently, a more bulk-like water arrangement with increased H-bonding symmetry is found, as suggested by our FTIR experiments. Hence, the observed H-bonding in NW water is dominated by water–water interactions for larger pores, which is additionally supported by the observation of a decreasing amount of irreversible bond water molecules with increasing pore size. For Ni_2dondc, this relaxation process initiates the formation of more entropic water molecules with more asymmetric H-bonding forced to interact with the pore walls. Eventually, this results in the occupation of discrete positions, as suggested by XRD data [20], even though the initial hydration mechanism, i.e., pore condensation, is enthalpy-driven. This is supported by our spectroscopic results, highlighting a more asymmetric H-bonding situation.

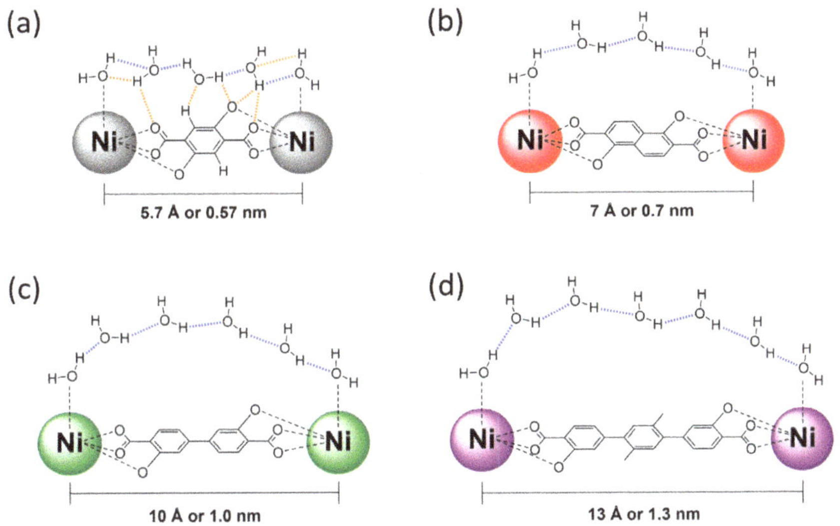

Figure 7. Illustration of the different arrangements of water molecules near the framework pore walls in hydrated Ni_2L frameworks. (**a**) Ni_2dhtp reveals multiple attractive H-bonding interactions with framework atoms (dashed orange lines) and surrounding water molecules (dashed blue lines), located at distinct positions. Formed water bridges, avoiding the hydrophobic domains in (**b**) Ni_2dondc, (**c**) Ni_2bpp and (**d**) Ni_2tpp, suggest that the size of the water bridges/chains increases with larger distances of adjacent coordinating water molecules, i.e., H-bonding anchors (distances estimated from XRD data [11,20,97]).

4. Conclusions

Our study on the isoreticular Ni_2L framework series enabled us to analyze the effects of the pore size and linker hydrophobicity on the water sorption mechanism and the arrangement of confined water. We found evidence that defects in Ni_2dhtp play an important role on the given sorption mechanism. Higher amounts of formate defects result in a less uniform pore wall, reducing the attractiveness of the third sorption site (near the phenylene backbone), offering less H-bonding possibilities. Conversely, incorporation of fewer defect sites results in a sorption mechanism, more comparable to those discussed for most other M_2dhtp frameworks. This suggests that selective incorporation of defects enables one to precisely tune the water sorption properties in the M_2dhtp series. On the other hand, these effects play a minor role for larger pores (i.e., Ni_2L; L = dondc, bpp, bpm, tpp). Within the low-pressure range, water predominantly adsorbs to the open metal sites while additional molecules adsorb thereon as a result of large hydration enthalpies, forming distinct water clusters. Water bridges between adjacent chemisorbed water molecules, located at the primary binding sites, form thereafter, providing additional H-bonding opportunities. Consequently, the sorption process is facilitated, resulting in classical pore condensation characterized by a continuous H-bonding network. The observed sorption step is governed by the pore size, with the pressure range additionally depending on the linker hydrophobicity/hydrophobic domain size. We found that H-bonding is generally more asymmetric in smaller pores, consistent with the expectation of defined water positions in fully hydrated Ni_2dhtp and Ni_2dondc, offering multiple attractive water–framework interactions. Larger pores offer a more symmetric H-bonding situation, suggesting a more bulk-like water arrangement under less confinement, dominated by water–water interactions, as indicated by decreasing dehydration temperatures with increasing pore size.

Supplementary Materials: The following supporting information can be downloaded at: https://www.mdpi.com/article/10.3390/nano14221791/s1, Synthetic procedures (linker and MOF synthesis), Figures S1–S5: NMR spectra of organic compounds; Figures S6–S11: Pore size distribution fits by NLDFT methods; Figures S12–S18: NMR spectra of digested MOFs; Figures S19–S24: FTIR spectra of MOFs in comparison with the used linkers; Figure S25: Deconvolution of the O-H stretching vibration of liquid water; Figure S26: XRD data of re-isolated MOF samples after water vapor sorption analysis; Figures S27–S32: Normalized first derivatives ($\delta V/\delta(p/p_0)$) of water vapor sorption data; Table S1: Total water uptake of MOF samples, determined by TGA analysis; Table S2: Results of least square fits of hydrated MOF samples in comparison with pure water; Table S3: Peak center position of the O-H stretching vibration of water in hydrated MOFs; Table S4: Total water uptake of MOF samples, determined by water vapor sorption analysis; Table S5: Octanol–water partition coefficients of protonated linker molecules; Table S6: Comparison of the maximum adsorption of the nitrogen and water sorption isotherms.

Author Contributions: Conceptualization, M.K. and M.T.; methodology, M.K.; validation, M.K., L.S., Z.Z., C.W. and M.T.; formal analysis, M.K., L.S., Z.Z., C.W. and H.E.; investigation, M.K. and L.S.; resources, M.T.; data curation, M.K.; writing—original draft preparation, M.K., C.W. and M.T.; writing—review and editing, M.K., C.W. and M.T.; visualization, M.K. and Z.Z.; supervision, M.T.; project administration, M.T. All authors have read and agreed to the published version of the manuscript.

Funding: This research received no external funding.

Data Availability Statement: Dataset available on request from the authors.

Acknowledgments: The authors thank Michael Reimer and Tamay Meric from the working group for Plastics Technology and Plastics Processing of the Faculty of Mechanical Engineering of Paderborn University for performing the TGA experiments.

Conflicts of Interest: The authors declare no conflicts of interest.

References

1. Yaghi, O.M.; O'Keeffe, M.; Ockwig, N.W.; Chae, H.K.; Eddaoudi, M.; Kim, J. Reticular synthesis and the design of new materials. *Nature* **2003**, *423*, 705–714. [CrossRef] [PubMed]
2. Janiak, C.; Vieth, J.K. MOFs, MILs and more: Concepts, properties and applications for porous coordination networks (PCNs). *New J. Chem.* **2010**, *34*, 2366. [CrossRef]
3. Schoedel, A.; Li, M.; Li, D.; O'Keeffe, M.; Yaghi, O.M. Structures of Metal–Organic Frameworks with Rod Secondary Building. Units. *Chem. Rev.* **2016**, *116*, 12466–12535. [CrossRef]
4. Keeffe, M.O.; Peskov, M.A.; Ramsden, S.J.; Yaghi, O.M. The Reticular Chemistry Structure Resource (RCSR) Database of, and Symbols for, Crystal Nets. *Acc. Chem. Res.* **2008**, *41*, 1782–1789. [CrossRef]
5. Zhao, D.; Timmons, D.J.; Yuan, D.; Zhou, H.-C. Tuning the Topology and Functionality of Metal−Organic Frameworks by Ligand Design. *Acc. Chem. Res.* **2011**, *44*, 123–133. [CrossRef]
6. Eddaoudi, M.; Kim, J.; Rosi, N.; Vodak, D.; Wachter, J.; O'Keeffe, M.; Yaghi, O.M. Systematic Design of Pore Size and Functionality in Isoreticular MOFs and Their Application in Methane Storage. *Science* **2002**, *295*, 469–472. [CrossRef]
7. Pallach, R.; Keupp, J.; Terlinden, K.; Frentzel-Beyme, L.; Kloß, M.; Machalica, A.; Kotschy, J.; Vasa, S.K.; Chater, P.A.; Sternemann, C.; et al. Frustrated flexibility in metal-organic frameworks. *Nat. Commun.* **2021**, *12*, 4097. [CrossRef]
8. Frentzel-Beyme, L.; Kloß, M.; Kolodzeiski, P.; Pallach, R.; Henke, S. Meltable Mixed-Linker Zeolitic Imidazolate Frameworks and Their Microporous Glasses: From Melting Point Engineering to Selective Hydrocarbon Sorption. *J. Am. Chem. Soc.* **2019**, *141*, 12362–12371. [CrossRef]
9. Henke, S.; Schneemann, A.; Wütscher, A.; Fischer, R.A. Directing the Breathing Behavior of Pillared-Layered Metal–Organic Frameworks via a Systematic Library of Functionalized Linkers Bearing Flexible Substituents. *J. Am. Chem. Soc.* **2012**, *134*, 9464–9474. [CrossRef]
10. Lu, F.-F.; Gu, X.-W.; Wu, E.; Li, B.; Qian, G. Systematic evaluation of water adsorption in isoreticular UiO-type metal–organic frameworks. *J. Mater. Chem. A* **2023**, *11*, 1246–1255. [CrossRef]
11. Deng, H.; Grunder, S.; Cordova, K.E.; Valente, C.; Furukawa, H.; Hmadeh, M.; Gándara, F.; Whalley, A.C.; Liu, Z.; Asahina, S.; et al. Large-Pore Apertures in a Series of Metal-Organic Frameworks. *Science* **2012**, *336*, 1018–1023. [CrossRef] [PubMed]
12. Kalmutzki, M.J.; Diercks, C.S.; Yaghi, O.M. Metal–Organic Frameworks for Water Harvesting from Air. *Adv. Mater.* **2018**, *30*, 1704304. [CrossRef]
13. Akiyama, G.; Matsuda, R.; Sato, H.; Hori, A.; Takata, M.; Kitagawa, S. Effect of functional groups in MIL-101 on water sorption behavior. *Microporous Mesoporous Mater.* **2012**, *157*, 89–93. [CrossRef]

14. Frentzel-Beyme, L.; Kloß, M.; Pallach, R.; Salamon, S.; Moldenhauer, H.; Landers, J.; Wende, H.; Debus, J.; Henke, S. Porous purple glass—A cobalt imidazolate glass with accessible porosity from a meltable cobalt imidazolate framework. *J. Mater. Chem. A* **2019**, *7*, 985–990. [CrossRef]
15. Dietzel, P.D.C.; Morita, Y.; Blom, R.; Fjellvåg, H. An In Situ High-Temperature Single-Crystal Investigation of a Dehydrated Metal–Organic Framework Compound and Field-Induced Magnetization of One-Dimensional Metal–Oxygen Chains. *Angew. Chem. Int. Ed.* **2005**, *44*, 6354–6358. [CrossRef]
16. Rosi, N.L.; Kim, J.; Eddaoudi, M.; Chen, B.; O'Keeffe, M.; Yaghi, O.M. Rod Packings and Metal−Organic Frameworks Constructed from Rod-Shaped Secondary Building Units. *J. Am. Chem. Soc.* **2005**, *127*, 1504–1508. [CrossRef] [PubMed]
17. Zheng, J.; Vemuri, R.S.; Estevez, L.; Koech, P.K.; Varga, T.; Camaioni, D.M.; Blake, T.A.; Mcgrail, B.P.; Motkuri, R.K. Pore-Engineered Metal–Organic Frameworks with Excellent Adsorption of Water and Fluorocarbon Refrigerant for Cooling Applications. *J. Am. Chem. Soc.* **2017**, *139*, 10601–10604. [CrossRef]
18. Gheorghe, A.; Reus, S.; Koenis, M.; Dubbeldam, D.; Woutersen, S.; Tanase, S. Role of additives and solvents in the synthesis of chiral isoreticular MOF-74 topologies. *Dalt. Trans.* **2021**, *50*, 12159–12167. [CrossRef]
19. Peng, S.; Bie, B.; Jia, H.; Tang, H.; Zhang, X.; Sun, Y.; Wei, Q.; Wu, F.; Yuan, Y.; Deng, H.; et al. Efficient Separation of Nucleic Acids with Different Secondary Structures by Metal–Organic Frameworks. *J. Am. Chem. Soc.* **2020**, *142*, 5049–5059. [CrossRef]
20. Dietzel, P.D.C.; Georgiev, P.A.; Frøseth, M.; Johnsen, R.E.; Fjellvåg, H.; Blom, R. Effect of Larger Pore Size on the Sorption Properties of Isoreticular Metal–Organic Frameworks with High Number of Open Metal Sites. *Chem. A Eur. J.* **2020**, *26*, 13523–13531. [CrossRef]
21. Zheng, J.; Barpaga, D.; Trump, B.A.; Shetty, M.; Fan, Y.; Bhattacharya, P.; Jenks, J.J.; Su, C.Y.; Brown, C.M.; Maurin, G.; et al. Molecular Insight into Fluorocarbon Adsorption in Pore Expanded Metal–Organic Framework Analogs. *J. Am. Chem. Soc.* **2020**, *142*, 3002–3012. [CrossRef]
22. Kapelewski, M.T.; Geier, S.J.; Hudson, M.R.; Stück, D.; Mason, J.A.; Nelson, J.N.; Xiao, D.J.; Hulvey, Z.; Gilmour, E.; Fitzgerald, S.A.; et al. M_2(*m*-dobdc) (M = Mg, Mn, Fe, Co, Ni) Metal–Organic Frameworks Exhibiting Increased Charge Density and Enhanced H_2 Binding at the Open Metal Sites. *J. Am. Chem. Soc.* **2014**, *136*, 12119–12129. [CrossRef] [PubMed]
23. Gygi, D.; Bloch, E.D.; Mason, J.A.; Hudson, M.R.; Gonzalez, M.I.; Siegelman, R.L.; Darwish, T.A.; Queen, W.L.; Brown, C.M.; Long, J.R. Hydrogen Storage in the Expanded Pore Metal–Organic Frameworks M_2(dobpdc) (M = Mg, Mn, Fe, Co, Ni, Zn). *Chem. Mater.* **2016**, *28*, 1128–1138. [CrossRef]
24. Forse, A.C.; Gonzalez, M.I.; Siegelman, R.L.; Witherspoon, V.J.; Jawahery, S.; Mercado, R.; Milner, P.J.; Martell, J.D.; Smit, B.; Blümich, B.; et al. Unexpected Diffusion Anisotropy of Carbon Dioxide in the Metal–Organic Framework Zn_2(dobpdc). *J. Am. Chem. Soc.* **2018**, *140*, 1663–1673. [CrossRef]
25. Kökçam-Demir, Ü.; Goldman, A.; Esrafili, L.; Gharib, M.; Morsali, A.; Weingart, O.; Janiak, C. Coordinatively unsaturated metal sites (open metal sites) in metal–organic frameworks: Design and applications. *Chem. Soc. Rev.* **2020**, *49*, 2751. [CrossRef]
26. Dietzel, P.D.C.; Besikiotis, V.; Blom, R. Application of metal–organic frameworks with coordinatively unsaturated metal sites in storage and separation of methane and carbon dioxide. *J. Mater. Chem.* **2009**, *19*, 7362–7370. [CrossRef]
27. Pato-Doldán, B.; Rosnes, M.H.; Dietzel, P.D.C. An In-Depth Structural Study of the Carbon Dioxide Adsorption Process in the Porous Metal–Organic Frameworks CPO-27-M. *ChemSusChem* **2017**, *10*, 1710–1719. [CrossRef] [PubMed]
28. Fetisov, E.O.; Shah, M.S.; Long, J.R.; Tsapatsis, M.; Siepmann, J.I. First principles Monte Carlo simulations of unary and binary adsorption: CO_2, N_2, and H_2O in Mg-MOF-74. *Chem. Commun.* **2018**, *54*, 10816–10819. [CrossRef]
29. Queen, W.L.; Hudson, M.R.; Bloch, E.D.; Mason, J.A.; Gonzalez, M.I.; Lee, J.S.; Gygi, D.; Howe, J.D.; Lee, K.; Darwish, T.A.; et al. Comprehensive study of carbon dioxide adsorption in the metal–organic frameworks M_2(dobdc) (M = Mg, Mn, Fe, Co, Ni, Cu, Zn). *Chem. Sci.* **2014**, *5*, 4569–4581. [CrossRef]
30. Drisdell, W.S.; Poloni, R.; McDonald, T.M.; Long, J.R.; Smit, B.; Neaton, J.B.; Prendergast, D.; Kortright, J.B. Probing Adsorption Interactions in Metal–Organic Frameworks using X-ray Spectroscopy. *J. Am. Chem. Soc.* **2013**, *135*, 18183–18190. [CrossRef]
31. Strauss, I.; Mundstock, A.; Hinrichs, D.; Himstedt, R.; Knebel, A.; Reinhardt, C.; Dorfs, D.; Caro, J. The Interaction of Guest Molecules with Co-MOF-74: A Vis/NIR and Raman Approach. *Angew. Chemie Int. Ed.* **2018**, *57*, 7434–7439. [CrossRef] [PubMed]
32. Rosnes, M.H.; Pato-Doldán, B.; Johnsen, R.E.; Mundstock, A.; Caro, J.; Dietzel, P.D.C. Role of the metal cation in the dehydration of the microporous metal–organic frameworks CPO-27-M. *Microporous Mesoporous Mater.* **2020**, *309*, 110503. [CrossRef]
33. Chen, T.; Chen, S.; Chen, Y.; Zhao, M.; Losic, D.; Zhang, S. Metal-organic frameworks containing solid-state electrolytes for lithium metal batteries and beyond. *Mater. Chem. Front.* **2021**, *5*, 1771–1794. [CrossRef]
34. McDonald, T.M.; Lee, W.R.; Mason, J.A.; Wiers, B.M.; Hong, C.S.; Long, J.R. Capture of Carbon Dioxide from Air and Flue Gas in the Alkylamine-Appended Metal–Organic Framework mmen-Mg_2(dobpdc). *J. Am. Chem. Soc.* **2012**, *134*, 7056–7065. [CrossRef] [PubMed]
35. Trickett, C.A.; Helal, A.; Al-Maythalony, B.A.; Yamani, Z.H.; Cordova, K.E.; Yaghi, O.M. The chemistry of metal–organic frameworks for CO_2 capture, regeneration and conversion. *Nat. Rev. Mater.* **2017**, *2*, 17045. [CrossRef]
36. Solís, C.; Palaci, D.; Llabres i Xamena, F.X.; Serra, J.M. Proton Transport through Robust CPO-27-type Metal Organic Frameworks. *J. Phys. Chem. C* **2014**, *118*, 21663–21670. [CrossRef]
37. Hwang, S.; Lee, E.J.; Song, D.; Jeong, N.C. High Proton Mobility with High Directionality in Isolated Channels of MOF-74. *ACS Appl. Mater. Interfaces* **2018**, *10*, 35354–35360. [CrossRef]

38. Javed, A.; Strauss, I.; Bunzen, H.; Caro, J.; Tiemann, M. Humidity-Mediated Anisotropic Proton Conductivity through the 1D Channels of Co-MOF-74. *Nanomaterials* **2020**, *10*, 1263. [CrossRef]
39. Khalil, I.E.; Fonseca, J.; Reithofer, M.R.; Eder, T.; Chin, J.M. Tackling orientation of metal-organic frameworks (MOFs): The quest to enhance MOF performance. *Coord. Chem. Rev.* **2023**, *481*, 215043. [CrossRef]
40. Liu, Y.R.; Chen, Y.Y.; Zhao, H.Y.; Li, G. Exploration of single-crystal proton conduction in ordered networks. *Coord. Chem. Rev.* **2024**, *499*, 215516. [CrossRef]
41. Hanikel, N.; Prévot, M.S.; Yaghi, O.M. MOF water harvesters. *Nat. Nanotechnol.* **2020**, *15*, 348. [CrossRef] [PubMed]
42. DeCoste, J.B.; Peterson, G.W.; Schindler, B.J.; Killops, K.L.; Browe, M.A.; Mahle, J.J. The effect of water adsorption on the structure of the carboxylate containing metal–organic frameworks Cu-BTC, Mg-MOF-74, and UiO-66. *J. Mater. Chem. A* **2013**, *1*, 11922–11932. [CrossRef]
43. Schoenecker, P.M.; Carson, C.G.; Jasuja, H.; Flemming, C.J.J.; Walton, K.S. Effect of Water Adsorption on Retention of Structure and Surface Area of Metal–Organic Frameworks. *Ind. Eng. Chem. Res.* **2012**, *51*, 6513–6519. [CrossRef]
44. Metrane, A.; Delhali, A.; Ouikhalfan, M.; Assen, A.H.; Belmabkhout, Y. Water Vapor Adsorption by Porous Materials: From Chemistry to Practical Applications. *J. Chem. Eng. Data* **2022**, *67*, 1617–1653. [CrossRef]
45. Weinberger, C.; Zysk, F.; Hartmann, M.; Kaliannan, N.K.; Keil, W.; Kühne, T.D.; Tiemann, M. The Structure of Water in Silica Mesopores—Influence of the Pore Wall Polarity. *Adv. Mater. Interfaces* **2022**, *9*, 2200245. [CrossRef]
46. Kloß, M.; Beerbaum, M.; Baier, D.; Weinberger, C.; Zysk, F.; Elgabarty, H.; Kühne, T.D.; Tiemann, M. Understanding Hydration in CPO-27 Metal-Organic Frameworks: Strong Impact of the Chemical Nature of the Metal (Cu, Zn). *Adv. Mater. Interfaces* **2024**, 2400476. [CrossRef]
47. Kloß, M.; Weinberger, C.; Tiemann, M. Water in the Micropores of CPO-27 Metal-Organic Frameworks: A Comprehensive Study. *Microporous Mesoporous Mater.* **2024**, *381*, 113352. [CrossRef]
48. Furukawa, H.; Gándara, F.; Zhang, Y.B.; Jiang, J.; Queen, W.L.; Hudson, M.R.; Yaghi, O.M. Water Adsorption in Porous Metal–Organic Frameworks and Related Materials. *J. Am. Chem. Soc.* **2014**, *136*, 4369–4381. [CrossRef]
49. Zhang, B.; Zhu, Z.; Wang, X.; Liu, X.; Kapteijn, F. Water Adsorption in MOFs: Structures and Applications. *Adv. Funct. Mater.* **2023**, *34*, 2304788. [CrossRef]
50. Yaghi, O.M.; Kalmutzki, M.J.; Diercks, C.S. Water Sorption Applications of MOFs. In *Introduction to Reticular Chemistry: Metal-Organic Frameworks and Covalent Organic Frameworks*; Wiley-VCH: Weinheim, Germany, 2019; pp. 359–427. [CrossRef]
51. Tan, K.; Nijem, N.; Canepa, P.; Gong, Q.; Li, J.; Thonhauser, T.; Chabal, Y.J. Stability and Hydrolyzation of Metal Organic Frameworks with Paddle-Wheel SBUs upon Hydration. *Chem. Mater.* **2012**, *24*, 3153–3167. [CrossRef]
52. Tan, K.; Nijem, N.; Gao, Y.; Zuluaga, S.; Li, J.; Thonhauser, T.; Chabal, Y.J. Water interactions in metal organic frameworks. *CrystEngComm* **2015**, *17*, 247–260. [CrossRef]
53. Lincoln, S.F. Mechanistic Studies of Metal Aqua Ions: A Semi-Historical Perspective. *Helv. Chim. Acta* **2005**, *88*, 523–545. [CrossRef]
54. Ng, E.-P.; Mintova, S. Nanoporous materials with enhanced hydrophilicity and high water sorption capacity. *Microporous Mesoporous Mater.* **2008**, *114*, 1–26. [CrossRef]
55. Liu, L.; Zeng, W.; Tan, S.J.; Liu, M.; Do, D.D. Microscopic insights into water adsorption in carbon nanopores—The role of acidic and basic functional groups and their configurations. *Phys. Chem. Chem. Phys.* **2021**, *23*, 18369–18377. [CrossRef]
56. Mileo, P.G.M.; Cho, K.H.; Park, J.; Devautour-Vinot, S.; Chang, J.S.; Maurin, G. Unraveling the Water Adsorption Mechanism in the Mesoporous MIL-100(Fe) Metal−Organic Framework. *J. Phys. Chem. C* **2019**, *123*, 23014–23025. [CrossRef]
57. Ho, C.-H.; Valentine, M.L.; Chen, Z.; Xie, H.; Farha, O.; Xiong, W.; Paesani, F. Structure and thermodynamics of water adsorption in NU-1500-Cr. *Commun. Chem.* **2023**, *6*, 70. [CrossRef]
58. Hanikel, N.; Pei, X.; Chheda, S.; Lyu, H.; Jeong, W.S.; Sauer, J.; Gagliardi, L.; Yaghi, O.M. Evolution of water structures in metal-organic frameworks for improved atmospheric water harvesting. *Science* **2021**, *374*, 454–459. [CrossRef]
59. Heidary, N.; Chartrand, D.; Guiet, A.; Kornienko, N. Rational incorporation of defects within metal–organic frameworks generates highly active electrocatalytic sites. *Chem. Sci.* **2021**, *12*, 7324–7333. [CrossRef] [PubMed]
60. Song, Q.; Li, Y.; Cao, Z.; Liu, H.; Tian, C.; Yang, Z.; Qiang, X.; Tan, Z.; Deng, Y. Discovery of novel 2,5-dihydroxyterephthalamide derivatives as multifunctional agents for the treatment of Alzheimer's disease. *Bioorg. Med. Chem.* **2018**, *26*, 6115–6127. [CrossRef]
61. Schukraft, G.E.M.; Ayala, S.; Dick, B.L.; Cohen, S.M. Isoreticular expansion of polyMOFs achieves high surface area materials. *Chem. Commun.* **2017**, *53*, 10684–10687. [CrossRef]
62. Rouquerol, J.; Llewellyn, P.; Rouquerol, F. Is the bet equation applicable to microporous adsorbents? *Stud. Surf. Sci. Catal.* **2007**, *160*, 49–56. [CrossRef]
63. Kim, H.; Sohail, M.; Yim, K.; Park, Y.C.; Chun, D.H.; Kim, H.J.; Han, S.O.; Moon, J.-H. Effective CO_2 and CO Separation Using [M_2(DOBDC)] (M = Mg, Co, Ni) with Unsaturated Metal Sites and Excavation of Their Adsorption Sites. *ACS Appl. Mater. Interfaces* **2019**, *11*, 7014–7021. [CrossRef] [PubMed]
64. Thommes, M.; Kaneko, K.; Neimark, A.V.; Olivier, J.P.; Rodriguez-Reinoso, F.; Rouquerol, J.; Sing, K.S.W. Physisorption of gases, with special reference to the evaluation of surface area and pore size distribution (IUPAC Technical Report) Pure. *Appl. Chem.* **2015**, *87*, 1051. [CrossRef]
65. Fu, Y.; Yao, Y.; Forse, A.C.; Li, J.; Mochizuki, K.; Long, J.R.; Reimer, J.A.; De Paëpe, G.; Kong, X. Solvent-derived defects suppress adsorption in MOF-74. *Nat. Commun.* **2023**, *14*, 2386. [CrossRef] [PubMed]

66. Yasaka, Y.; Yoshida, K.; Wakai, C.; Matubayasi, N.; Nakahara, M. Kinetic and Equilibrium Study on Formic Acid Decomposition in Relation to the Water-Gas-Shift Reaction. *J. Phys. Chem. A* **2006**, *110*, 11082–11090. [CrossRef]
67. Lim, J.; Lee, S.; Sharma, A.; Seong, J.; Baek, S.B.; Lah, M.S. Ligand functionalization of defect-engineered Ni-MOF-74. *RSC Adv.* **2022**, *12*, 31451–31455. [CrossRef]
68. Onori, G.; Santucci, A. IR investigations of water structure in Aerosol OT reverse micellar aggregates. *J. Phys. Chem.* **1993**, *97*, 5430. [CrossRef]
69. MacDonald, H.; Bedwell, B.; Gulari, E. FTIR spectroscopy of microemulsion structure. *Langmuir* **1986**, *2*, 704. [CrossRef]
70. Knight, A.W.; Kalugin, N.G.; Coker, E.; Ilgen, A.G. Water properties under nano-scale confinement. *Sci. Rep.* **2019**, *9*, 8246. [CrossRef]
71. Wagner, J.C.; Hunter, K.M.; Paesani, F.; Xiong, W. Water Capture Mechanisms at Zeolitic Imidazolate Framework Interfaces. *J. Am. Chem. Soc.* **2021**, *143*, 21189–21194. [CrossRef]
72. Brubach, J.-B.; Mermet, A.; Filabozzi, A.; Gerschel, A.; Roy, P. Signatures of the hydrogen bonding in the infrared bands of water. *J. Chem. Phys.* **2005**, *122*, 184509. [CrossRef] [PubMed]
73. Zhang, C.; Khaliullin, R.Z.; Bovi, D.; Guidoni, L.; Kühne, T.D. Vibrational Signature of Water Molecules in Asymmetric Hydrogen Bonding Environments. *J. Phys. Chem. Lett.* **2013**, *4*, 3245–3250. [CrossRef]
74. Zhang, C.; Guidoni, L.; Kühne, T.D. Competing factors on the frequency separation between the OH stretching modes in water. *J. Mol. Liq.* **2015**, *205*, 42–45. [CrossRef]
75. Iiyama, T.; Ruike, M.; Kaneko, K. Structural mechanism of water adsorption in hydrophobic micropores from in situ small angle X-ray scattering. *Chem. Phys. Lett.* **2000**, *331*, 359–364. [CrossRef]
76. Alcañiz-Monge, J.; Linares-Solano, A.; Rand, B. Mechanism of Adsorption of Water in Carbon Micropores As Revealed by a Study of Activated Carbon Fibers. *J. Phys. Chem. B* **2002**, *106*, 3209–3216. [CrossRef]
77. Brovchenko, I.; Geiger, A.; Oleinikova, A. Water in nanopores: II. The liquid–vapour phase transition near hydrophobic surfaces. *J. Phys. Condens. Matter* **2004**, *16*, S5345–S5370. [CrossRef]
78. Ohba, T.; Kanoh, H.; Kaneko, K. Cluster-Growth-Induced Water Adsorption in Hydrophobic Carbon Nanopores. *J. Phys. Chem. B* **2004**, *108*, 14964–14969. [CrossRef]
79. Ohba, T.; Kanoh, H.; Kaneko, K. Structures and Stability of Water Nanoclusters in Hydrophobic Nanospaces. *Nano Lett.* **2005**, *5*, 227–230. [CrossRef] [PubMed]
80. Gallo, P.; Rovere, M.; Chen, S.-H. Anomalous dynamics of water confined in MCM-41 at different hydrations. *J. Phys. Condens. Matter* **2010**, *22*, 284102. [CrossRef]
81. Sadeghi, M.; Parsafar, G.A. Density-induced molecular arrangements of water inside carbon nanotubes. *Phys. Chem. Chem. Phys.* **2013**, *15*, 7379–7388. [CrossRef]
82. Thommes, M.; Morell, J.; Cychosz, K.A.; Fröba, M. Combining Nitrogen, Argon, and Water Adsorption for Advanced Characterization of Ordered Mesoporous Carbons (CMKs) and Periodic Mesoporous Organosilicas (PMOs). *Langmuir* **2013**, *29*, 14893–14902. [CrossRef] [PubMed]
83. Mietner, J.B.; Brieler, F.J.; Lee, Y.J.; Fröba, M. Properties of Water Confined in Periodic Mesoporous Organosilicas: Nanoimprinting the Local Structure. *Angew. Chem. Int. Ed.* **2017**, *56*, 12348–12351. [CrossRef] [PubMed]
84. Malfait, B.; Moréac, A.; Jani, A.; Lefort, R.; Huber, P.; Fröba, M.; Morineau, D. Structure of Water at Hydrophilic and Hydrophobic Interfaces: Raman Spectroscopy of Water Confined in Periodic Mesoporous (Organo)Silicas. *J. Phys. Chem. C* **2022**, *126*, 3520–3531. [CrossRef]
85. Liu, J.; Wang, Y.; Benin, A.I.; Jakubczak, P.; Willis, R.R.; LeVan, M.D. CO_2/H_2O Adsorption Equilibrium and Rates on Metal−Organic Frameworks: HKUST-1 and Ni/DOBDC. *Langmuir* **2010**, *26*, 14301–14307. [CrossRef]
86. Villajos, J.A.; Jagorel, N.; Reinsch, S.; Emmerling, F. Increasing Exposed Metal Site Accessibility in a Co-MOF-74 Material with Induced Structure-Defects. *Front. Mater.* **2019**, *6*, 230. [CrossRef]
87. Dalla Bernardina, S.; Paineau, E.; Brubach, J.-B.; Judeinstein, P.; Rouzière, S.; Launois, P.; Roy, P. Water in Carbon Nanotubes: The Peculiar Hydrogen Bond Network Revealed by Infrared Spectroscopy. *J. Am. Chem. Soc.* **2016**, *138*, 10437–10443. [CrossRef] [PubMed]
88. Hanasaki, I.; Nakatani, A. Hydrogen bond dynamics and microscopic structure of confined water inside carbon nanotubes. *J. Chem. Phys.* **2006**, *124*, 144708. [CrossRef]
89. Küsgens, P.; Rose, M.; Senkovska, I.; Fröde, H.; Henschel, A.; Siegle, S.; Kaskel, S. Characterization of metal-organic frameworks by water adsorption. *Microporous Mesoporous Mater.* **2009**, *120*, 325–330. [CrossRef]
90. Moon, H.; Collanton, R.P.; Monroe, J.I.; Casey, T.M.; Shell, M.S.; Han, S.; Scott, S.L. Evidence for Entropically Controlled Interfacial Hydration in Mesoporous. Organosilicas. *J. Am. Chem. Soc.* **2022**, *144*, 1766–1777. [CrossRef]
91. Wang, C.; Xing, Y.; Lei, Y.; Xia, Y.; Zhang, C.; Zhang, R.; Wang, S.; Chen, P.; Zhu, S.; Li, J.; et al. Adsorption of water on carbon materials: The formation of "water bridge" and its effect on water adsorption. *Colloids Surf. A Physicochem. Eng. Asp.* **2021**, *631*, 127719. [CrossRef]
92. Harris, M.F.; Logan, J.L. Determination of log K_{ow} Values for Four Drugs. *J. Chem. Educ.* **2014**, *91*, 915–918. [CrossRef]
93. Paschke, A.; Neitzel, P.L.; Walther, W.; Schüürmann, G. Octanol/Water Partition Coefficient of Selected Herbicides: Determination Using Shake-Flask Method and Reversed-Phase High-Performance Liquid Chromatography. *J. Chem. Eng. Data* **2004**, *49*, 1639–1642. [CrossRef]

94. Lum, K.; Chandler, D.; Weeks, J.D. Hydrophobicity at Small and Large Length Scales. *J. Phys. Chem. B* **1999**, *103*, 4570–4577. [CrossRef]
95. Huang, D.M.; Chandler, D. Temperature and length scale dependence of hydrophobic effects and their possible implications for protein folding. *Proc. Natl. Acad. Sci. USA* **2000**, *97*, 8324–8327. [CrossRef]
96. Huang, D.M.; Geissler, P.L.; Chandler, D. Scaling of Hydrophobic Solvation Free Energies. *J. Phys. Chem. B* **2001**, *105*, 6704–6709. [CrossRef]
97. Wong-Ng, W.; Kaduk, J.A.; Wu, H.; Suchomel, M. Synchrotron X-ray studies of metal-organic framework M_2(2,5-dihydroxyterephthalate), M = (Mn, Co, Ni, Zn) (MOF74). *Powder Diffr.* **2012**, *27*, 256–262. [CrossRef]

nanomaterials

Article

Data-Driven and Machine Learning to Screen Metal–Organic Frameworks for the Efficient Separation of Methane

Yafang Guan [1], Xiaoshan Huang [1], Fangyi Xu [1], Wenfei Wang [1], Huilin Li [1], Lingtao Gong [1], Yue Zhao [2], Shuya Guo [1,*], Hong Liang [1,*] and Zhiwei Qiao [1,*]

1 Guangzhou Key Laboratory for New Energy and Green Catalysis, School of Chemistry and Chemical Engineering, Guangzhou University, Guangzhou 510006, China; 2005200075@e.gzhu.edu.cn (Y.G.); xshuang@e.gzhu.edu.cn (X.H.); 32105200048@e.gzhu.edu.cn (F.X.); 2112105103@e.gzhu.edu.cn (W.W.); lihuilin1@e.gzhu.edu.cn (H.L.); 32305200102@e.gzhu.edu.cn (L.G.)

2 State Key Laboratory of NBC Protection for Civilian, Beijing 100191, China; sa11226532@mail.ustc.edu.cn

* Correspondence: 2112105073@e.gzhu.edu.cn (S.G.); lhong@gzhu.edu.cn (H.L.); zqiao@gzhu.edu.cn (Z.Q.)

Abstract: With the rapid growth of the economy, people are increasingly reliant on energy sources. However, in recent years, the energy crisis has gradually intensified. As a clean energy source, methane has garnered widespread attention for its development and utilization. This study employed both large-scale computational screening and machine learning to investigate the adsorption and diffusion properties of thousands of metal–organic frameworks (MOFs) in six gas binary mixtures of CH_4 (H_2/CH_4, N_2/CH_4, O_2/CH_4, CO_2/CH_4, H_2S/CH_4, He/CH_4) for methane purification. Firstly, a univariate analysis was conducted to discuss the relationships between the performance indicators of adsorbents and their characteristic descriptors. Subsequently, four machine learning methods were utilized to predict the diffusivity/selectivity of gas, with the light gradient boosting machine (LGBM) algorithm emerging as the optimal one, yielding R^2 values of 0.954 for the diffusivity and 0.931 for the selectivity. Furthermore, the LGBM algorithm was combined with the SHapley Additive exPlanation (SHAP) technique to quantitatively analyze the relative importance of each MOF descriptor, revealing that the pore limiting diameter (PLD) was the most critical structural descriptor affecting molecular diffusivity. Finally, for each system of CH_4 mixture, three high-performance MOFs were identified, and the commonalities among high-performance MOFs were analyzed, leading to the proposals of three design principles involving changes only to the metal centers, organic linkers, or topological structures. Thus, this work reveals microscopic insights into the separation mechanisms of CH_4 from different binary mixtures in MOFs.

Keywords: methane; metal–organic frameworks; gas separation; molecular simulation; machine learning; diffusion

1. Introduction

Coal mine methane (CMM) [1] is methane released during the coal mining process, typically mixed with gases such as nitrogen, oxygen, and carbon dioxide. CMM is primarily stored in the coal matrix through physical adsorption, where the coal matrix contains a large number of micropores and mesopores; the high specific surface area of micropores makes them crucial for methane adsorption [2]. In abandoned coal mines, microbial methane production is a new source of methane, indicating that, even after coal mining ceases, methane generation and emission remains a concern that needs attention [3]. Accurately predicting the emission of coal mine methane [4] is essential for understanding its potential environmental impact and planning the use of methane as an energy source. Methane in nature often contains impurities, such as nitrogen, carbon dioxide, and hydrogen sulfide, which can reduce the combustion efficiency of methane, corrode pipelines, and increase the environmental burden. Wu [5] and others developed a new amine ion-exchanged zeolite Y by introducing amine cations CH^+ and TWA^+ to the original zeolite Y and confirmed that they had good separation performance in the CH_4/N_2 system. The global warming potential of each mole of methane is 3.7 times that of carbon dioxide, and carbon dioxide emissions account for 80% of the contribution of current greenhouse gas emissions to global warming [6]. Although methane is a greenhouse gas, through effective capture and utilization, methane can also serve as an efficient and clean energy source, playing a positive role in global energy transition and climate change mitigation [7]. Biogas, formed by mixing methane with an appropriate amount of air, is an ideal gas fuel. Purifying biogas for use as vehicle fuel can further improve the utilization rate of biogas [8]. The pressure swing adsorption (PSA) purification method utilizes the selective adsorption of carbon dioxide by the adsorbent [9], that is, carbon dioxide has a higher separation coefficient relative to other gaseous components on the adsorbent, to achieve the purpose of removing carbon dioxide from biogas and purifying methane [10]. Purifying methane in biogas using physical or chemical methods, such as pressure swing adsorption (PSA) or chemical absorption, usually involves repeated pressurization and depressurization processes, which can lead to high energy consumption and production costs, affecting the economic benefits and sustainable development of enterprises [11].

In recent years, a new type of adsorbent material, metal–organic framework (MOF), has emerged as a novel class of porous material. Formed through self-assembling metal ions or clusters with organic ligands, MOFs boast high porosity, large surface area, and tunable pore sizes. They have been extensively studied and applied in various fields, including gas adsorption and separation [12], storage [13], catalysis [14], drug delivery [15], and sensing applications [16]. The structure of MOFs can be regulated by selecting different metal ions and organic ligands, thereby achieving precise control over pore size, shape, and chemical environment [17]. Understanding multi-component adsorption equilibria on MOFs is essential for designing adsorption-based separation processes [18]. Although a large number of MOFs have been synthesized and reported to date, given the vast variety of metal ions and organic linkers and diverse connection sequences, manually sifting through a vast database to select MOFs for a specific system using only experimental methods is undoubtedly inefficient.

Therefore, molecular simulation, high-throughput computing combined with machine learning (ML), and molecular fingerprints (MFs) are used to replace traditional research methods in searching for and designing suitable new MOF absorbents [19]. Anbia [20] and others studied the adsorption characteristics of MOF-235 for CH_4, H_2, and CO_2 by measuring volume, finding that the absolute amount of adsorption was in the order of $CH_4 > H_2 > CO_2$. The selectivity of MOF-235 for CH_4 was higher than that for CO_2 (14.7) and H_2 (8.3), indicating that MOF-235 is a potential adsorbent for CH_4 separation in gas mixtures. Niu [21] and others reported a methane nano-trap, discovering that an alkyl-MOF-based methane nano-trap has a high methane adsorption rate and CH_4/N_2 selectivity at 298 K and 1 bar. Yang [22] and others synthesized a microporous hetero metal–organic framework (MOF) material named CuIn(Ina)$_4$, which showed excellent adsorption performance in separating and purifying CH_4 from a ternary mixed system containing C_2H_2 and CO_2. Mulu [23] and others determined that original wood ash, after calcination treatment, can effectively adsorb and remove CO_2 and H_2S from biogas, thereby increasing the purity of methane in biogas. Wang [24] and others designed and synthesized a series of porous coordination polymers (PCPs) with hourglass-shaped nanochannels, one of which, the porous coordination polymer NTU-30, showed excellent separation efficiency in C_2H_6/CH_4 and C_2H_4/CH_4 gas systems. Experiments proved that NTU-30 has weaker adsorption for CH_4, but stronger adsorption capacity for C_2H_6 and C_2H_4, making it an adsorbent material for natural gas purification and increasing CH_4 purity. Yang [25] and others studied the field of low-concentration coal mine methane and used the VPSA (vacuum pressure swing adsorption) method with modified carbon molecular sieves as adsorbents for methane adsorption. The results showed that this adsorption technology is particularly suitable for purifying raw gases containing low concentrations of methane. Tu [26] and others reported a nickel-based metal–organic framework Ni-BPZ, which can effectively achieve methane separation and purification, including capturing methane in coal mine gas, and revealed its efficient separation mechanism through molecular simulation. Kang [27] found in his research that, compared with other gas molecules, the MOFJUC-150 membrane has a significant preference for H_2 permeation, and at room temperature, the selectivity factor of this membrane for H_2/CH_4 is 26.3.

The purpose of this work is to employ large-scale computational screening and ML to study the adsorption and diffusion properties of 6013 MOF adsorbents in six gas mixtures (H_2/CH_4, N_2/CH_4, O_2/CH_4, CO_2/CH_4, H_2S/CH_4, He/CH_4). In Section 2, the atomic models of CoRE-MOF and gases, simulation methods, and principles of machine learning are described. In Section 3, the relationship between the performance indicators of the adsorbents and their characteristic descriptors is first discussed through univariate analysis, and then the best ML model for a specific performance indicator for the six systems is identified using four ML techniques, followed by the application of SHAP for a quantitative analysis of the relative importance of each CoRE-MOF descriptor. Finally, high-performance MOF materials for different systems are screened, and the commonalities among high-performance MOFs are analyzed to propose design principles.

2. Model and Methods

2.1. Molecular Model

In this work, Chung [28] and colleagues conducted a series of experiments to screen CoRE-MOFs by removing free solvent molecules using high-throughput molecular simulation and established a database containing MOF crystal models. Five structural descriptors of MOF were calculated, including pore-limiting diameter (PLD (Å)), large cavity diameter (LCD (Å)), volumetric surface area (VSA (m^2/cm^3)), void fraction (ϕ), and MOF density (ρ (kg/m^3)). The LCD and PLD were calculated using zeo++ software (version 0.3) [29]. The VSA and ϕ were computed using the RASPA software (version 1.9.15) [30] with He with a diameter of 2.58 Å and N_2 with a diameter of 3.64 Å as probes, respectively. In the grand canonical Monte Carlo (GCMC) simulations, the Lennard-Jones (LJ) potential function was used to describe the interactions between atoms in CoRE-MOFs, allowing particles to freely enter and exit the simulation system, thereby simulating the adsorption behavior of the system under different chemical potentials.

$$u_{LJ+elec}(r) = \sum 4\varepsilon_{ij}\left[\left(\frac{\sigma_{ij}}{r_{ij}}\right)^{12} - \left(\frac{\sigma_{ij}}{r_{ij}}\right)^{6}\right] + \sum \frac{q_i q_j}{4\pi\varepsilon_0 r_{ij}} \tag{1}$$

In Equation (1), r_{ij} represents the distance between two atoms; q_i, q_j denote the charge quantities of the two atoms, respectively; $\varepsilon_0 = 8.8542 \times 10^{-12}\ C^2\ N^{-1}m^{-2}$ is the vacuum permittivity, which is used to describe the Coulombic interactions between atoms. The Lennard-Jones (LJ) potential parameters for MOFs are derived from the universal force field (UFF) [31] and are listed in Table S1.

We constructed a database for seven gas components (CH_4, N_2, H_2S, O_2, CO_2, H_2, He). The force field parameters for these seven gas components listed in Table S2 are derived from the transferable potential of the Trappe force field [32]. CH_4 is a united atom model; N_2 is a three-site model with an N-N bond length of 1.10 Å; H_2S is a four-site model with an S-H bond length of 1.13 Å, featuring L-J potentials on the S and H atoms. Additionally, there is a dummy atom near the S atom, with partial charges on the H atoms and the dummy atom, while the S atom is uncharged. O_2 is a three-site model, and for CO_2, the C-O bond length is 1.16 Å and an ∠OCO was 180°. The He atom has a diameter of 2.58 Å.

2.2. Molecular Simulations

In this study, the temperature was set to 298 K and the pressure to 1 bar. The diffusion coefficients D of CH_4, N_2, H_2S, O_2, CO_2, H_2, and He in the MOF were calculated using MD simulation [33]. Each MOF underwent an independent MD simulation to individually assess its diffusion characteristics for gas molecules. The time step for the MD simulation was 1 fs, which was the smallest time interval used in the simulation to calculate atomic motion. The duration of each MOF's MD was 5 ns, with the first 3 ns used for the system to reach an equilibrium state, and the last 2 ns for production. The data from the production phase will be used for the final analysis and results. The diffusion coefficient in the MD simulation was calculated from the mean square displacement using the Einstein equation. All MD simulations were performed using the RASPA software package. The simulation using this method showed better consistency with experiments (see Figure S1 for details). To quantify the diffusion separation of two gases, the diffusion selectivity between ideal binary gas pairs was estimated using Equation (2).

$$S_{diff(i/j)} = \frac{D_i}{D_j} \tag{2}$$

In this context, D_i and D_j represent the diffusion coefficients of gas molecules i or j within CoRE-MOFs, where i stands for N_2, H_2S, O_2, CO_2, H_2, He, and j stands for CH_4. The pairs i/j represent different gas combinations, listed in Table S3.

The interactions between CoRE-MOFs and adsorbate molecules were calculated using the Lorentz–Berthelot rules. Periodic boundary conditions (PBC) are a standard simulation technique, enabling the representation of an infinitely large material within a small repeating unit. In a 3D system with PBC, particles that exit one side of the simulation box re-enter from the opposite side, ensuring continuity. The simulation cell, which contains the particles, is expanded to a minimum of 24 Å in every dimension to ensure accurate modeling. To calculate the Lennard-Jones (LJ) interactions, a long-range correction spherical cutoff radius of 12 Å was set.

2.3. Machine Learning

In this work, four algorithms were used to predict two target values—diffusion coefficient (D) and diffusion selectivity (S_{diff}). In machine learning, constructing a diverse training library is crucial as it helps ensure the model has generalization capabilities, that is, it can accurately predict new unseen data. This training library includes gas molecules of different sizes, shapes, and polarities (CH_4, N_2, H_2S, O_2, CO_2, H_2, He), and CoRE-MOFs with different topological structures and chemical compositions. A total of 6013 MOFs with different topological structures and chemical compositions were sourced from the 2019 CoRE-MOF database. Additionally, to facilitate the exploration of diffusion similarities of different molecules within MOFs, the dataset was arranged longitudinally.

Firstly, molecular dynamics simulations (MD) were used to obtain the diffusion coefficients (D) of seven gas molecules (CH_4, N_2, H_2S, O_2, CO_2, H_2, He), resulting in 42,092 preprocessed data entries. The feature variables consisted of MOF structural descriptors (PLD, LCD, VSA, ϕ, ρ) and gas physical properties (see attachment Table S4) (kinetic diameter (*Dia*), quadrupole moment (*Qua*), dipole moment (*Dip*), and polarizability (*Pol*)). The diffusion selectivity (S_{diff}) for six ideal binary gas mixtures was derived using Equation (2), resulting in 36,078 preprocessed data entries. The feature variables consisted of MOF's five structural descriptors (PLD, LCD, VSA, ϕ, ρ), the physical properties of the separated gases (kinetic diameter (*Dia*), quadrupole moment (Qua_i), dipole moment (Dip_i), polarizability (Pol_i)), and differences in the physical properties of the separated gases (ΔDia, ΔPol, ΔDip, ΔQua) [34]. The data then proceeded to the next step of machine learning.

To explore the importance of geometric/energy descriptors (LCD, PLD, VSA, ρ, ϕ) in MOFs for their performance, four machine learning (ML) algorithms were used for data mining, including random forest (RF) (version 1.2.2) [35], gradient boosting regression trees (GBRT) in scikit-learn (version 1.0.2) [36], extreme gradient boosting (XGB) (version 1.1.2) [37], and lightGBM (LGBM) (version 3.3.2) [38] for comparison. All algorithms were executed in scikit-learn for Python [39], and detailed information on the principles of ML algorithms can be found in Supporting Information Section S3. All version numbers of the Python packages used during training are listed in Table S5. ML algorithm parameters can be found in detail in Table S6. Before training the ML models, the dataset was first randomly divided into training and testing sets at a ratio of 7:3, followed by standardization processing. During training, k-fold cross-validation was used (k = 5 in this work) to improve the stability and accuracy of the model; the principle can be found in Figure S6. In machine learning and statistical modeling, evaluating the performance of the model is a very important step. The coefficient of determination R^2, mean absolute error (MAE), and root mean square error (RMSE) are three commonly used evaluation metrics. A higher R^2 indicates greater predictive accuracy, while lower MAE and RMSE values signify enhanced stability in the model's predictions.

In this work, SHapley Additive exPlanation (SHAP) technology was combined with machine learning models for the analysis of the relative importance of features. SHAP (version 0.40.4) [40] is a popular interpretive tool for machine learning that helps us understand the specific impact of each feature of the model on the model's predictions. It can be calculated through the TreeExplainer algorithm developed by Lundberg [41] and others. TreeExplainer is a method for calculating SHAP values to interpret tree-based models. For each sample in the dataset, TreeExplainer individually calculates the result of that sample in each decision tree of the model. It then assesses the contribution of each feature to the sample's predicted outcome in each decision tree. By analyzing the SHAP values across all samples, the global behavior of the model can be interpreted.

3. Results and Discussion

3.1. Statistical Analysis

To explore the impact of MOF structure on the diffusion performance of seven gases during the diffusion process, we first analyzed the relationship between diffusivity (D) and five descriptors. As shown in Figure 1a–d, D increases with the increase in PLD, LCD, ϕ, and VSA, eventually tending to stabilize, as they are all descriptors of pore size or pore volume. ρ is negatively correlated with diffusion performance, showing a downward trend as ρ increases, which may be because ρ is not only related to the material's pore channels but also to its mass. When ρ > 3000 kg/m^3, the dispersion of D is very high, as shown in Figure 1e. MOFs typically have a high specific surface area and adjustable pore structure, which makes them excellent for gas storage (such as hydrogen, methane, and carbon dioxide). However, ρ may vary greatly due to the type and quantity of the constituent elements and may include heavier metal atoms such as gold (Au), platinum (Pt), or uranium (U), which can significantly increase the overall density of the structure. Although some MOFs may have a larger free space (high porosity), which can provide more gas storage capacity, the density of MOFs may still be high. Therefore, ρ is not easy to represent the performance of MOFs.

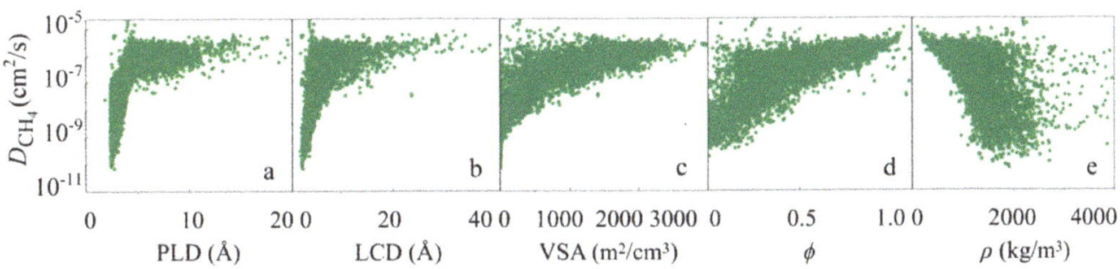

Figure 1. The variation of CH_4 diffusion performance with (**a**) PLD, (**b**) LCD, (**c**) VSA, (**d**) ϕ, (**e**) ρ.

To further analyze, the relationship between D and PLD for the seven gases (CH_4, N_2, H_2S, O_2, CO_2, H_2, He) was explored. The kinetic diameters of these seven molecules, sorted from smallest to largest, were He (2.6 Å) < H_2 (2.89 Å) < CO_2 (3.3 Å) < O_2 (3.46 Å) < H_2S (3.62 Å) < N_2 (3.64 Å) < CH_4 (3.8 Å). As can be seen from Figure 2a,b, within the range of PLD < 3.7 Å, the diffusion coefficient of molecules increases with the increase in PLD, showing a high positive correlation. After that, as the pore size increases and all gases enter the interior of MOFs, when PLD > 5 Å, diffusivity decreases and tends to stabilize.

Secondly, based on the diffusion coefficients of CH_4 and N_2 in the infinite dilution state, a quantitative relationship with PLD was established as $\log D = a\ PLD - b$, as shown in Figure S7. We found that the molecules affected by PLD in descending order were $CH_4 > N_2 > H_2S > O_2 > H_2 > CO_2 >$ He. Since CH_4 has the largest molecular diameter among the four gases and He has the smallest, larger kinetic diameters slow gas molecule diffusion, enabling methane capture and gas separation, which is consistent with the conclusions reported previously in the literature [42].

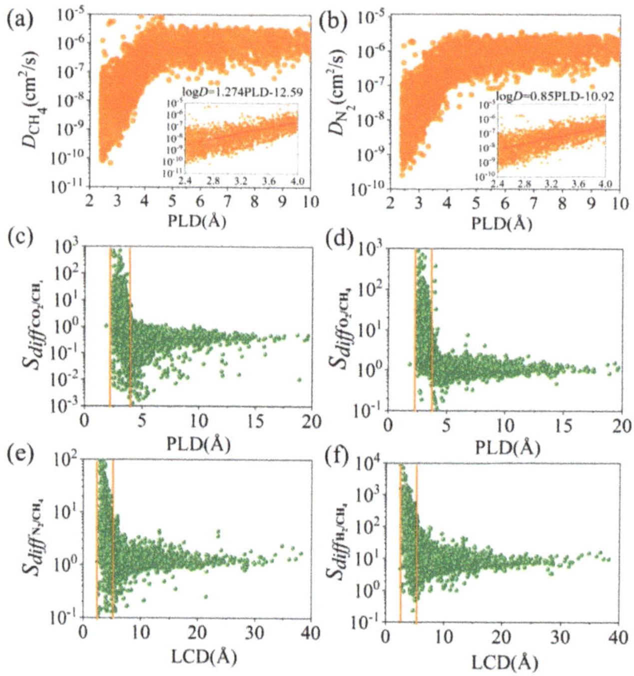

Figure 2. The effect of MOFs' PLD and LCD on the diffusion properties of gases for (**a**) CH_4—PLD, (**b**) N_2—PLD, (**c**) CO_2/CH_4—PLD, (**d**) O_2/CH_4—PLD, (**e**) N_2/CH_4—LCD, and (**f**) H_2/CH_4—LCD. (The orange line represents the range of optimal MOFs).

We then analyzed the relationship between selectivity (x = N_2, H_2S, O_2, CO_2, H_2, He) and PLD and LCD. Most MOFs with larger diffusion selectivity have relatively smaller PLD and LCD. From Figure 2c, it can be seen that when the PLD is between 2.4~4.2 Å, the adsorption selectivity for CO_2/CH_4 decreases. When PLD > 5 Å, both gas molecules can pass through freely, and the adsorption selectivity gradually stabilizes and tends towards 1. This is because CO_2 has a strong quadrupole moment and is preferentially adsorbed over CH_4 even in infinitely large pores, thus achieving the separation of CO_2 from CH_4. The pore size of MOFs can be precisely controlled, allowing only specific-sized molecules to pass through. Smaller pore sizes can prevent larger molecules from passing while allowing smaller molecules to diffuse freely. The molecular kinetic diameter of CO_2 is about 3.3 Å, also smaller than that of the methane molecule. Therefore, designing MOFs with a pore size smaller than the kinetic diameter of the methane molecule can promote the permeation of CO_2 molecules while hindering CH_4 molecules, achieving CO_2/CH_4 separation. Similarly, the kinetic diameter of O_2 is about 3.46 Å, and that of CH_4 is about 3.8 Å. If the pore size of MOFs is designed to be smaller than 3.5 Å, it will be more conducive to the rapid diffusion of oxygen molecules and restrict the passage of methane molecules, achieving O_2/CH_4 separation. Other gases also show the same trend, as shown in Figure S8. Therefore, by finely tuning the pore size and channel geometry, a certain degree of methane purification can be achieved.

S_{diff} decreases and then tends to flatten as the four descriptors (LCD, ϕ, VSA, PLD) increase, as shown in Figure 2. This is also due to the gradual release of strong steric hindrance. When the pore size is small, the pore walls of different framework materials have a strong or weak adsorption effect on gases, resulting in high selectivity differences. When ϕ is high, gas molecules have more space for free diffusion, so the diffusion capacity S_{diff} is larger. When VSA is close to 0, that is, the specific surface area of MOFs is very small, and the pores are very limited, the diffusion area S_{diff} of gas molecules is the largest. This is because at this time, MOF molecules either cannot pass any molecules or can only pass a small number of molecules other than methane. As VSA increases, that is, the specific surface area increases, the pores become more tortuous and complex, leading to a longer diffusion path for gas molecules, thereby reducing the diffusion capacity S_{diff}. Finally, because further increasing VSA has a limited effect on increasing the diffusion path, or the pore structure is already complex enough that the diffusion capacity of gas molecules no longer decreases significantly, the diffusion selectivity equals 1, that is, methane purification cannot be achieved. The complexity of the pore structure of MOFs may have an important impact on the diffusion of gas molecules. As shown in Figure 2e,f, PLD also affects the diffusion of gas molecules. A smaller pore size may limit the diffusion of gas molecules, while a larger pore size may help to improve the diffusion rate. Other gases also show the same trend, as shown in Figure S9.

The results show that PLD is a key indicator of x/CH_4 separation performance, but not a perfect one. The above univariate analysis can only preliminarily determine the relationship between a single parameter and performance. To screen structural variables that have a strong impact on all gas components and further predict the performance of MOF structures, we will further use ML algorithms for systematic multivariate analysis of the comprehensive structure–performance relationship of CoRE-MOFs.

3.2. Machine Learning

ML (machine learning) was used to model these complex large-scale data, and the prediction results were effective. Therefore, in this work, ensemble algorithms were employed in machine learning (ML) to enhance the performance of the predictive model, especially in forecasting diffusivity and diffusion selectivity. Details of R^2, MAE, and RMSE are presented in Table 1. It can be seen that the computational accuracy of the RF algorithm is relatively low, and its stability is poor. Relatively speaking, this algorithm has lower predictive performance for methane purification. GBRT usually reduces prediction errors step by step in the iterative process of multiple decision trees, with each tree trying to correct the errors of the previous one. This strategy makes GBRT slightly better in prediction accuracy than RF.

Table 1. Comparative prediction of R^2, MAE, and RMSE for D and S performance by four machine learning algorithms.

Indicators	Algorithm	R^2	MAE	RMSE
D	RF	0.934	0.151	0.247
	LGBM	0.954	0.127	0.207
	XGB	0.957	0.119	0.199
	GBRT	0.947	0.129	0.222
S	RF	0.887	0.155	0.264
	LGBM	0.931	0.127	0.207
	XGB	0.928	0.123	0.210
	GBRT	0.907	0.135	0.240

Figure 3 shows that the prediction results of XGB and LGBM are very close, with considerable accuracy and stability. The R^2 value of XGB is 0.957, and the R^2 value of LGBM is 0.954; both are close to 1, indicating that both models fit the training data very well. The RMSE of XGB is 0.199, and the RMSE of LGBM is 0.207. Given the similar R^2 and RMSE outcomes, we can deduce that both algorithms exhibit comparable accuracy.

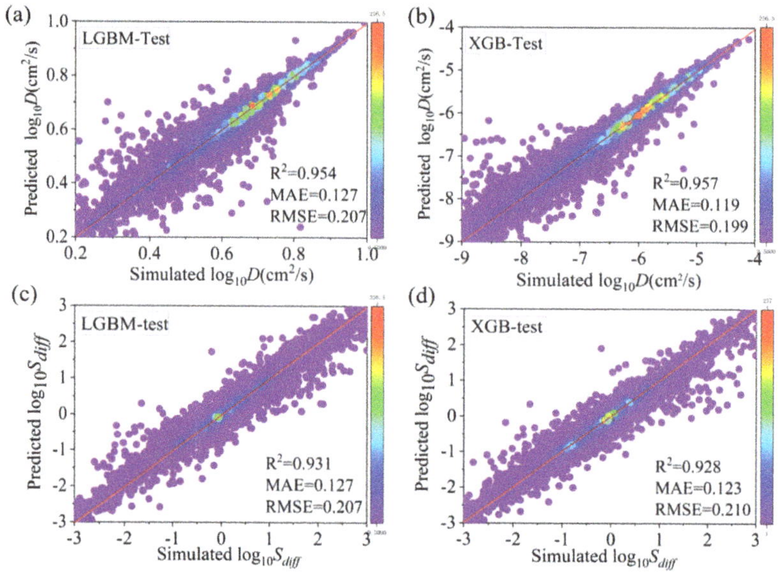

Figure 3. Prediction of machine learning for diffusivity and diffusion selectivity with (**a**) the predictions of D by LGBM, (**b**) the predictions of D by XGB, (**c**) the predictions of S_{diff} by LGBM, and (**d**) the predictions of S_{diff} by XGB. (The color of the dots represents the quantity).

Figure 4b shows the time performance of different machine learning algorithms in predicting diffusivity and selectivity, with the selectivity prediction time being shorter than that of diffusivity. Although the R^2 value of the XGB algorithm is slightly larger than that of LGBM, the computing time of XGB is over 300 s. The CPU computation time of LGBM is significantly reduced; especially, when predicting selectivity, it reached an impressive speed of 23 s. Therefore, comparing the four algorithms comprehensively, it can be concluded that LGBM > XGB > GBRT > RF, with LGBM having the highest fitting ability, stability, and efficiency. The excellent predictive effect of LGBM may be attributed to its difference in tree structure from other tree model algorithms. Traditional tree models use a depth-first strategy when growing, that is, building one branch at a time. LGBM adopts leaf-wise growth, which can lead to the asymmetry of the tree but usually finds better split points more quickly. The tree structure built from leaf growth may better reflect the interaction between MOF description and performance than trees grown horizontally. Another optimization of LGBM is histogram subtraction acceleration. With this method, after constructing a histogram of a leaf, the histogram of its sibling leaf can be obtained at a very low cost, which can double the speed. Therefore, LGBM is the fastest among the four algorithms.

To avoid the impact of the training sample on the coverage of molecules in the test set, we calculated each of the six systems separately, as shown in Table S7. As the training set sample size increases, the model is exposed to more data and can better learn the underlying patterns. Consequently, its predictive accuracy and generalization ability typically improve. Ideally, the improvement in accuracy observed on the training set should also be reflected in an independent test set, indicating that the model not only performs well on the training data but also generalizes effectively. For datasets with a small amount of data, the training score of LGBM is higher than the test score. Providing more data for training will enhance the model's generalization ability and reduce overfitting. Surprisingly, the accuracy of the LGBM model for predicting small sample sizes is still very high (R^2 = 0.910). Therefore, when the algorithm focuses on predicting within a certain range (for example, parts of materials with excellent diffusivity performance), the LGBM model still has a high degree of accuracy and can also be applied to predict the diffusivity performance of new gases or new MOF materials.

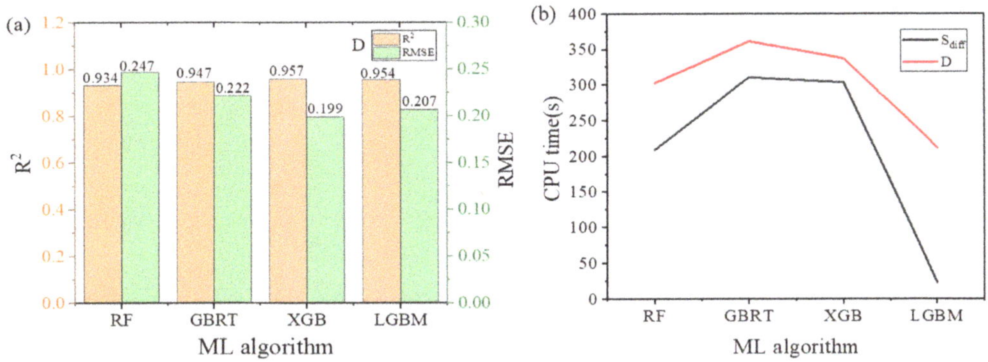

Figure 4. Comparison of the four algorithms. (**a**) R^2 and RMSE, (**b**) D and S.

3.3. SHAP Analysis

To further explore the impact of MOF descriptors on diffusion performance, in this work, SHAP was utilized to interpret their relationships. Calculating SHAP values aims to quantify the impact of features on performance in terms of magnitude (significant or insignificant) and direction (positive or negative correlation). As shown in Figure 5a, the SHAP values based on the LGBM algorithm illustrate the impact on diffusivity. The influence is ranked as PLD > *pol* > *Dia* > VSA > LCD > LCD/PLD > ϕ > ρ > *Qua* > *Dip*. PLD is positively correlated with gas diffusion performance, meaning that an increase in PLD usually leads to an improvement in diffusion performance. The size of PLD can determine the MOF's screening ability for gas molecules of different sizes, allowing smaller gas molecules to pass through quickly while hindering the diffusion of larger molecules, and achieving size-selective separation. A larger PLD provides more space for gas molecules to enter and diffuse, reducing intermolecular collisions and the diffusion resistance within the adsorbent, possessing a more open pore structure, and providing more diffusion channels for gas molecules. Therefore, the degree of match between the gas molecule's kinetic diameter and PLD is an important factor in determining diffusion performance. This is consistent with previous research results [43], further verifying the importance of PLD in MOF design. *Pol*, as the second most important influencing factor, shows a significant negative correlation, meaning that an increase in the *pol* descriptor will reduce gas diffusion performance. *Dia*, the descriptor for gas molecules, mostly has a negative impact on gas diffusion, but there are also some positive correlations. This may indicate that the impact of *Dia* on diffusion performance depends on the specific MOF structure and other factors. An increase in VSA and LCD is found to promote gas diffusion. This may be because a larger VSA and LCD provide more adsorption sites and more open diffusion channels. Different MOF descriptors may interact with each other, and they collectively determine the MOF's diffusion capability. By identifying and quantifying the impact of these descriptors, strategies can be provided for the optimization of MOF materials, such as adjusting pore size and surface chemical properties to improve the diffusion performance of target gases.

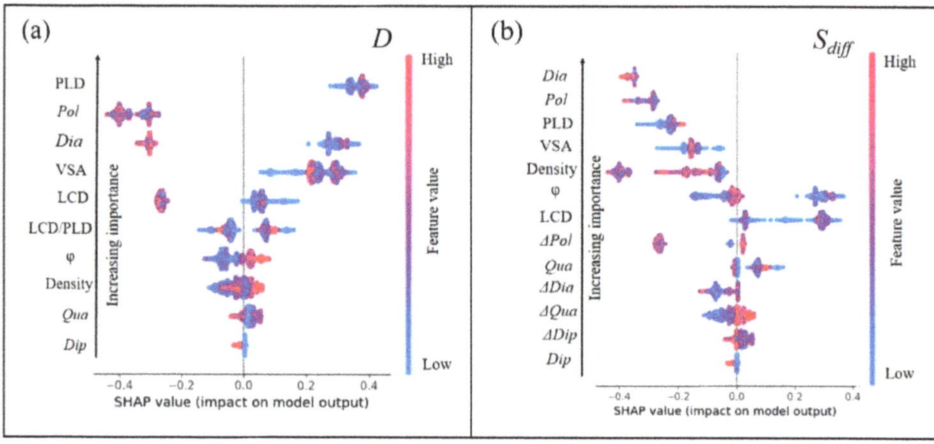

Figure 5. Relative importance analysis chart of structural descriptors for (**a**) *D* and (**b**) S_{diff}.

As shown in Figure 5b, the SHAP values based on the LGBM algorithm illustrate the impact on diffusion selectivity. The influence is ranked as *Dia* > *pol* > PLD > VSA > ρ > ϕ > LCD > Δ*Pol* > LCD/PLD > *Qua* > Δ*Dia* > Δ*Qua* > Δ*Dip* > *Dip*. *Dia* (as the most significant influencing factor)'s increase will have a negative correlation effect on diffusion selectivity, meaning that a larger *Dia* value will reduce the MOF's diffusion selectivity, which has been mentioned in previous studies [44]. *Pol*, as the second influencing factor, also inhibits diffusion selectivity, indicating that stronger intermolecular forces will reduce the MOF's separation efficiency. These two factors are the physical properties of the gas molecules themselves, indicating that the physical properties of the gas molecules have a significant negative impact on MOF diffusion selectivity [45]. In addition, PLD, VSA, and ρ each have an inhibitory effect on diffusion selectivity to varying degrees. An increase in PLD inhibits diffusion selectivity, possibly because an overly large PLD allows non-target molecules to pass through, reducing selectivity. An increase in VSA and ρ leads to a longer diffusion path for gas molecules within the MOF, increasing the resistance to mass transfer. Meanwhile, an increase in ϕ and LCD is positively correlated with diffusion selectivity, indicating that optimizing these parameters can improve the MOF's diffusion selectivity.

3.4. Top-Performing Metal–Organic Frameworks (MOFs)

Further research has found that, when the LCD value is smaller, the LCD/PLD ratio is lower, the channels appear more uniform, and the diffusion performance is better. When the LCD/PLD ratio is around 1, the diffusion rate rises sharply, reaching a peak in diffusion. As the LCD/PLD value increases, the diffusion rate shows a downward trend, indicating that more uniform channels are conducive to diffusion. For low PLD (below 5 Å) with smaller LCD, ϕ, and VSA, the SHAP value is greater than 0, having a positive impact. When the MOFs' PLD is between 2.4 and 5.0 Å, LCD is within 5 Å, ϕ is near 0.1, and the specific surface area is around 500, there will be a peak with a SHAP value greater than 0, which will have a positive effect on the target values of our ML model. Therefore, MOFs with higher diffusion selectivity can be obtained within this range.

To select CoRE-MOFs with excellent performance, we set limiting conditions according to the above range, as shown in Table S10. In the selection process, D_i and $S_{diff(i/j)}$ are the two factors we jointly consider. As can be seen from Figure S12, the specific strategy involves screening for MOF materials that exhibit high diffusivity and significant selectivity for the target gas *i* to be separated. In Table 2, three optimal MOFs are selected for each system, and most MOFs have a larger $S_{diff(i/j)}$. Among the materials selected in the He/CH_4 system, ELUQIM04 has the largest diffusion selectivity, $S_{diff(He/CH4)}$ = 61,406.36. Among the materials selected in the H_2/CH_4 system, ELUQIM05 has the largest diffusion selectivity,

$S_{diff(H2/CH4)}$ = 40,173.07. Figure 6a,b are the structural diagrams of the optimal MOFs, and Figure 6c,d are their corresponding pore diagrams. It can be seen that their channel structures are all circular, and it is speculated that the circular channels make the gas diffusion more uniform. In the selection of the six optimal MOFs for the system, the ranges of PLD and LCD are very concentrated, with PLD between 2.40 and 3.25 Å and LCD concentrated between 2.47 and 3.84 Å. This is because, within this range of PLD, CH_4 cannot enter the channels due to its large molecular kinetic diameter, which enhances the separation performance of the six gases. It can be seen from the SHAP relative importance analysis that PLD is an important descriptor affecting gas separation. PLD close to the kinetic diameter of the separated gas is the key condition to achieve CH_4 purification with good separation effects. This will provide effective theoretical guidance for the design of new MOFs with excellent performance.

Table 2. The best-performing MOFs selected from six systems.

Gas Mixture i/j	CSD Cord	LCD [Å]	ϕ	PLD [Å]	ρ [kg/m^3]	D_i [cm^2/s]	D_j [cm^2/s]	$S_{diff(i/j)}$
He/CH_4	ELUQIM04	2.92	0.04	2.44	1764.65	9.03×10^{-6}	1.47×10^{-10}	61,406.36
	ELUQIM05	2.90	0.04	2.43	1773.83	8.42×10^{-6}	1.76×10^{-10}	47,815.36
	ELUQIM06	2.89	0.04	2.41	1779.43	9.74×10^{-6}	8.11×10^{-10}	12,015.02
H_2/CH_4	ELUQIM05	2.90	0.04	2.43	1773.83	7.08×10^{-6}	1.76×10^{-10}	40,173.07
	ELUQIM04	2.92	0.04	2.44	1764.65	4.14×10^{-6}	1.47×10^{-10}	28,148.27
	FAPYEA04	2.47	0.00	2.40	1583.54	2.83×10^{-6}	3.01×10^{-10}	9397.11
CO_2/CH_4	XEKDUO	2.98	0.02	2.75	1903.55	1.70×10^{-7}	6.92×10^{-11}	2463.01
	ELUQIM05	2.90	0.04	2.43	1773.83	3.91×10^{-7}	1.76×10^{-10}	2220.35
	HIQPEE	3.84	0.15	3.12	1440.14	1.24×10^{-6}	5.69×10^{-10}	2185.18
O_2/CH_4	FAPYEA04	2.47	0.00	2.40	1583.54	2.92×10^{-7}	3.01×10^{-10}	968.16
	ELUQIM05	2.90	0.04	2.43	1773.83	1.58×10^{-7}	1.76×10^{-10}	898.78
	GUXQAS	2.79	0.02	2.52	1598.28	1.16×10^{-7}	1.30×10^{-10}	893.53
H_2S/CH_4	GUXQAS	2.79	0.02	2.52	1598.28	1.70×10^{-9}	1.30×10^{-10}	13.04
	RUPZIM	3.48	0.11	3.25	1549.49	1.51×10^{-7}	1.39×10^{-8}	10.87
	GUXPUL	2.79	0.02	2.58	1595.02	1.18×10^{-9}	1.10×10^{-10}	10.70
N_2/CH_4	FAPYEA04	2.47	0.00	2.40	1583.54	1.11×10^{-7}	3.01×10^{-10}	369.05
	PARFOF	2.77	0.05	2.46	1541.02	4.78×10^{-7}	2.16×10^{-9}	221.18
	HIWXER01	3.29	0.13	2.76	2533.00	1.80×10^{-7}	1.27×10^{-9}	141.34

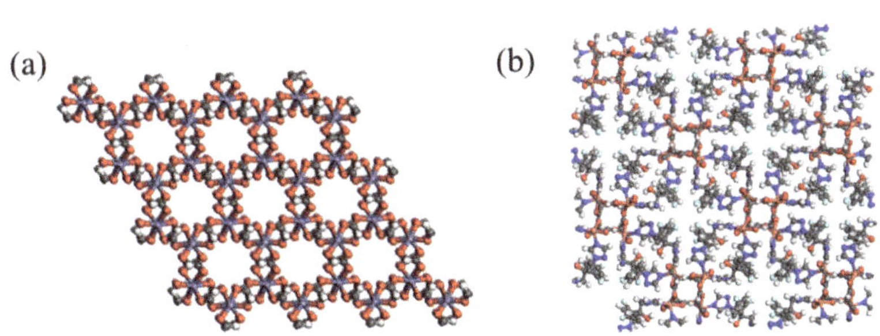

Figure 6. *Cont.*

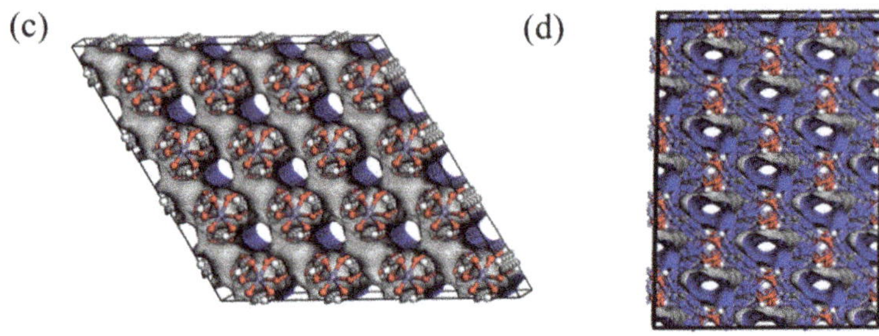

Figure 6. The structure of MOF for (**a**) ELUQIM04 and (**b**) GUXQAS; the pore channel diagram of MOF for (**c**) ELUQIM04 and (**d**) GUXQAS. (Color code: O atoms of MOFs: red; C atoms: gray; H atoms: white; N atoms: blue).

3.5. Design Strategies of MOFs with High Performances

To further guide experiments and design novel MOF materials with excellent methane purification and diffusion selectivity, this section discusses a series of design strategies for hypothetical MOF materials. Each pair comprises one MOF with superior diffusion selectivity and another with inferior performance, differentiated by a single compositional variation, such as the metal center, organic linker, or topology. Details of MOF composition are shown in Table 3. Based on this, we propose three strategies to diffuse gases from the six systems to achieve efficient methane purification, similar to previous studies [46].

Table 3. Comparison of the structures of different pairs of MOFs.

Gas Mixture (i/j)	NO.	CSD Cord	Metal Center	Organic Links	Top Structure	$S_{diff(i/j)}$
H_2/CH_4	a	GIRDUI	Co	MGFJDEHFNMWYBD	pcu	7.93
		GIRGUL	Co	MTAVBTGOXNGCJR	pcu	643.48
	b	CIMTAV	La	BVKZGUZCCUSVTD	llj	4.58
		CIMTEZ	Nd	BVKZGUZCCUSVTD	llj	12.97
	c	EBELUU	NiZn	JEVCWSUVFOYBFI	fsc	5.36
		EBEMEF	NiZn	JEVCWSUVFOYBFI	pts	147.93
CO_2/CH_4	d	CAJQEL	Cu	GRYHAGOZZMMYAO	scu	0.31
		CAJQIP	Cu	DUKMDOUQAIDJRW	scu	2.24
	e	HEBTEP	Zn	ABMFBCRYHDZLRD	lim	21.93
		SETFUT	Cd	ABMFBCRYHDZLRD	lim	1.22
H_2S/CH_4	f	ISIKIF	Co	GEBVRXNOWAYDCP	dia	16.07
		ISIKOL	Co	GEBVRXNOWAYDCP	bbf	0.97

In Table 3a, the alteration of the organic linker in this group of MOFs resulted in a dramatic increase in selectivity, jumping from 7.93 to 643.48—a remarkable enhancement of 80.15 times. As shown in Table 3e, in this group of MOFs, the sole modification was the metal center, with the higher performing MOF featuring Zn. Zn's atomic radius of 0.142 nm was more compact than Cd (0.151 nm), a difference that influenced the pore structure and dimensions of the material. Additionally, metal activity significantly influenced adsorption, with Zn noted for its greater reactivity compared with Cd, the selectivity was improved from 1.22 to 21.93, an increase of 16.97 times. In Figure 7c, a modification was made solely to the MOF's topology. By substituting the fsc topology with pst, the selectivity saw a remarkable enhancement, soaring from 5.36 to 147.93—a 26.6-fold improvement in performance. Variations in separation efficacy stem from the distinct pore dimensions and shapes created by the MOFs' topological networks. Ideally, in gas separation processes, pore sizes are meticulously regulated to permit the passage of molecules of a particular size while excluding others, ensuring selective gas permeation. It is evident that by adjusting topologies, organic linkers, or metal centers, one can obtain or design superior MOF adsorbents tailored for various systems.

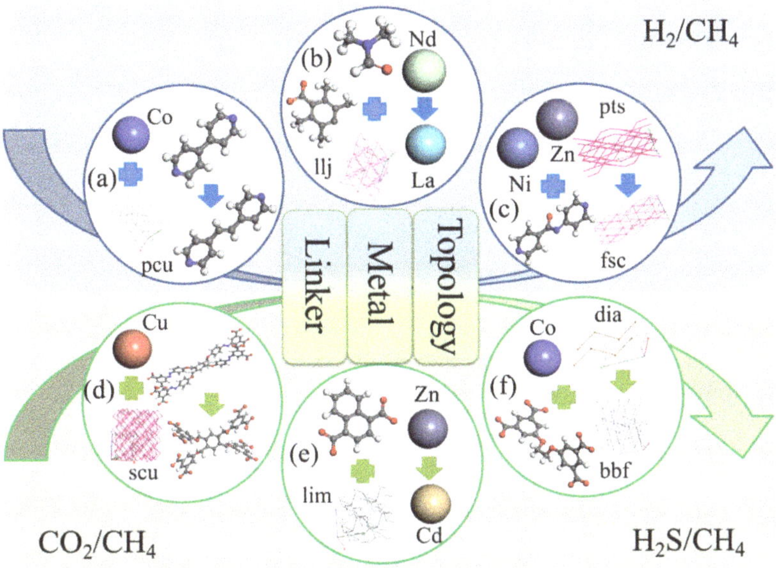

Figure 7. Enhancement of gas diffusion selectivity performance by three different MOF design strategies. The different pairs of MOFs for (**a**) GIRDUI and GIRGUL; (**b**) CIMTAV and CIMTEZ; (**c**) EBELUU and EBEMEF; (**d**) CAJQEL and CAJQIP; (**e**) HEBTEP and SETFUT; (**f**) ISIKIF and ISIKOL.

4. Conclusions

In this study, big data and machine learning were combined to simulate the diffusion and selectivity for methane purification in 6013 CoRE-MOFs within six gas mixtures (H_2/CH_4, N_2/CH_4, O_2/CH_4, CO_2/CH_4, H_2S/CH_4, He/CH_4). Preliminary analysis indicated a close relationship between D and PLD, establishing a quantitative relationship of log D = a PLD − b. Subsequently, four machine learning algorithms were applied to predict the diffusion and selectivity of MOFs in the six systems. The result analysis showed significant performance differences among the models, ranked by excellence as LGBM > XGB > GBRT > RF. Further SHAP relative importance analysis using the LGBM algorithm indicated that PLD is the most important structural descriptor affecting diffusivity, and *Dia* is the most important descriptor affecting selectivity. Additionally, three optimal MOFs were selected for each of the six systems, further verifying that the diffusion properties of gases are key to separating other gases from methane for its purification. Finally, three design strategies were proposed to effectively purify methane for specific systems by replacing different metal nodes, adjusting various topologies, and changing multiple organic linkers. This research provides effective guidance for experimentalists to select high-performance MOFs to separate other gases from methane.

Supplementary Materials: The following supporting information can be downloaded at: https://www.mdpi.com/article/10.3390/nano14131074/s1, Table S1. Lennard-Jones parameters of MOF; Table S2. Lennard-Jones parameters and charges of adsorbates; Table S3. Differences of kinetic diameter, polarizability, dipole moment, and quadruple moment between binary gas mixtures; Table S4. Physical properties of gas molecules; Table S5. The version information of tool packages used for building ML model; Table S6. Hyperparameters set in machine learning methods; Table S7. Evaluation of four algorithms for D and S_{diff}; Table S8. Importance ranking of features (based on D); Table S9. Importance ranking of features (based on S_{diff}); Table S10. Benchmark of D_i and $S_{diff\,(i/j)}$ for six gas mixtures; Figure S1. Comparison of simulated gas diffusivities and the experimental data for various MOF; Figure S2. Random forest model; Figure S3. GBRT model; Figure S4. XGBoost model; Figure S5. LightGBM model; Figure S6. 10-fold cross-validation; Figure S7. Gas diffusivity changes with PLD; Figure S8. Gas diffusion selectivity varies with PLD; Figure S9. Gas diffusion selectivity varies with LCD; Figure S10. Predicted results of D by RF, XGB, and GBRT ML algorithm models versus simulated results of CoRE-MOFs on the testing set; Figure S11. Predicted results of S_{diff} by RF, XGB, and GBRT ML algorithm models versus simulated results of CoRE-MOFs on the testing set. Figure S12. Example of optimal material screening strategy.

Author Contributions: Y.G.: Conceptualization, Methodology, Data Curation, Software, Writing—Original Draft, Validation. X.H.: Writing—Review and Editing, Validation. F.X.: Writing—Review and Editing, Supervision. W.W.: Writing—Review and Editing, Validation. H.L. (Huilin Li): Investigation, Supervision. L.G.: Writing—Review and Editing, Supervision. Y.Z.: Investigation, Supervision. S.G.: Resources, Supervision. H.L. (Hong Liang): Resources, Supervision, Funding Acquisition. Z.Q.: Writing—Review and Editing, Formal Analysis, Resources, Funding Acquisition. All authors have read and agreed to the published version of the manuscript.

Funding: The authors gratefully thank the National Natural Science Foundation of China (21978058), the Pearl River Talent Recruitment Program (2019QN01L255), the Natural Science Foundation of Guangdong Province (2022A1515011446, 2023A1515240076), Guangzhou University National Student Innovation Training Program (202311078015, 202211078209), the Special Fund for Scientific and Technological Innovation Strategy of Guangdong province (pdjh2023a0406) for their financial support.

Data Availability Statement: Data are contained within the article and Supplementary Materials.

Conflicts of Interest: The authors declare no conflicts of interest.

References

1. Warmuzinski, K. Harnessing methane emissions from coal mining. *Process Saf. Environ. Prot.* **2008**, *86*, 315–320. [CrossRef]
2. Moore, T.A. Coalbed methane: A review. *Int. J. Coal Geol.* **2012**, *101*, 36–81. [CrossRef]
3. Thielemann, T.; Cramer, B.; Schippers, A. Coalbed methane in the Ruhr Basin, Germany: A renewable energy resource. *Org. Geochem.* **2004**, *35*, 1537–1549. [CrossRef]
4. Tutak, M.; Brodny, J. Forecasting methane emissions from hard coal mines including the methane drainage process. *Energies* **2019**, *12*, 3840. [CrossRef]
5. Wu, Y.; Yuan, D.; Zeng, S.; Yang, L.; Dong, X.; Zhang, Q.; Xu, Y.; Liu, Z. Significant enhancement in CH_4/N_2 separation with amine-modified zeolite Y. *Fuel* **2021**, *301*, 121077. [CrossRef]
6. Lashof, D.A.; Ahuja, D.R. Relative contributions of greenhouse gas emissions to global warming. *Nature* **1990**, *344*, 529–531. [CrossRef]
7. Yin, Z.; Linga, P. Methane hydrates: A future clean energy resource. *Chin. J. Chem. Eng.* **2019**, *27*, 2026–2036. [CrossRef]
8. Kapoor, R.; Ghosh, P.; Tyagi, B.; Vijay, V.K.; Vijay, V.; Thakur, I.S.; Kamyab, H.; Nguyen, D.D.; Kumar, A. Advances in biogas valorization and utilization systems: A comprehensive review. *J. Clean. Prod.* **2020**, *273*, 123052. [CrossRef]
9. Gomes, V.G.; Yee, K.W.K. Pressure swing adsorption for carbon dioxide sequestration from exhaust gases. *Sep. Purif. Technol.* **2002**, *28*, 161–171. [CrossRef]
10. Wiheeb, A.D.; Helwani, Z.; Kim, J.; Othman, M.R. Pressure swing adsorption technologies for carbon dioxide capture. *Sep. Purif. Rev.* **2016**, *45*, 108–121. [CrossRef]
11. Mel, M.; Ibrahim MM, A.; Setyobudi, R.H. Preliminary study of biogas upgrading and purification by pressure swing adsorption. *Proc. AIP Conf.* **2016**, *1755*, 130010.
12. Lin, R.B.; Xiang, S.; Zhou, W.; Chen, B. Microporous Metal-Organic Framework Materials for Gas Separation. *Chem* **2020**, *6*, 337–363. [CrossRef]
13. Li, H.; Li, L.; Lin, R.B.; Zhou, W.; Zhang, Z.; Xiang, S.; Chen, B. Porous metal-organic frameworks for gas storage and separation: Status and challenges. *EnergyChem* **2019**, *1*, 100006. [CrossRef] [PubMed]
14. Wang, C.; An, B.; Lin, W. Metal–organic frameworks in solid–gas phase catalysis. *ACS Catal.* **2018**, *9*, 130–146. [CrossRef]

15. Cai, M.; Chen, G.; Qin, L.; Qu, C.; Dong, X.; Ni, J.; Yin, X. Metal organic frameworks as drug targeting delivery vehicles in the treatment of cancer. *Pharmaceutics* **2020**, *12*, 232. [CrossRef] [PubMed]
16. Yang, G.L.; Jiang, X.L.; Xu, H.; Zhao, B. Applications of MOF as luminescent sensors for environmental pollutants. *Small* **2021**, *17*, 2005327. [CrossRef] [PubMed]
17. Yin, H.Q.; Yin, X.B. Metal–organic frameworks with multiple luminescence emissions: Designs and applications. *Acc. Chem. Res.* **2020**, *53*, 485–495. [CrossRef] [PubMed]
18. Wang, L.; Huang, H.; Zhang, X.; Zhao, H.; Li, F.; Gu, Y. Designed metal-organic frameworks with potential for multi-component hydrocarbon separation. *Coord. Chem. Rev.* **2023**, *484*, 215111. [CrossRef]
19. Shi, Z.; Yang, W.; Deng, X.; Cai, C.; Yan, Y.; Liang, H.; Liu, Z.; Qiao, Z. Machine-learning-assisted high-throughput computational screening of high performance metal–organic frameworks. *Mol. Syst. Des. Eng.* **2020**, *5*, 725–742. [CrossRef]
20. Anbia, M.; Hoseini, V.; Sheykhi, S. Sorption of methane, hydrogen and carbon dioxide on metal-organic framework, iron terephthalate (MOF-235). *J. Ind. Eng. Chem.* **2012**, *18*, 1149–1152. [CrossRef]
21. Niu, Z.; Cui, X.; Pham, T.; Lan, P.C.; Xing, H.; Forrest, K.A.; Wojtas, L.; Space, B.; Ma, S. A metal–organic framework based methane Nano-trap for the capture of coal-mine methane. *Angew. Chem. Int. Ed.* **2019**, *58*, 10138–10141. [CrossRef] [PubMed]
22. Yang, L.; Gao, Q.; Li, Y.N.; Yang, L.; Gao, Q.; Li, Y.N.; Liu, Q.N.; Liu, X.T.; Wang, S.; Chen, L.Z. Efficient purification of CH_4 from ternary mixtures by a microporous heterometal-organic framework. *Sep. Purif. Technol.* **2024**, *335*, 126235. [CrossRef]
23. Mulu, E.; M'Arimi, M.M.; Ramkat, R.C.; Mecha, A.C. Potential of wood ash in purification of biogas. *Energy Sustain. Dev.* **2021**, *65*, 45–52. [CrossRef]
24. Wang, Y.; Dong, Q.; Cao, H.; Ji, W.; Duan, J.; Jing, S.; Jin, W. Finely Tuned Porous Coordination Polymers To Boost Methane Separation Efficiency. *Chem. Eur. J.* **2019**, *25*, 8790–8796. [CrossRef] [PubMed]
25. Yang, B.; Xu, E.L.; Li, M. Purification of coal mine methane on carbon molecular sieve by vacuum pressure swing adsorption. *Sep. Sci. Technol.* **2016**, *51*, 909–916. [CrossRef]
26. Tu, S.; Yu, L.; Lin, D.; Chen, Y.; Wu, Y.; Zhou, X.; Li, Z.; Xia, Q. Robust nickel-based metal–organic framework for highly efficient methane purification and capture. *ACS Appl. Mater. Interfaces* **2022**, *14*, 4242–4250. [CrossRef] [PubMed]
27. Kang, Z.; Xue, M.; Fan, L.; Huang, L.; Guo, L.; Wei, G.; Chen, B.; Qiu, S. Highly selective sieving of small gas molecules by using an ultra-microporous metal—Organic framework membrane. *Energy Environ. Sci.* **2014**, *7*, 4053–4060. [CrossRef]
28. Chung, Y.G.; Camp, J.; Haranczyk, M.; Sikora, B.J.; Bury, W.; Krungleviciute, V.; Yildirim, T.; Sholl, D.S.; Snurr, R.Q. Computation-ready, experimental metal—Organic frameworks: A tool to enable high-throughput screening of nanoporous crystals. *Chem. Mater.* **2014**, *26*, 6185–6192. [CrossRef]
29. Willems, T.F.; Rycroft, C.H.; Kazi, M.; Meza, J.C.; Haranczyk, M. Algorithms and tools for high-throughput geometry-based analysis of crystalline porous materials. *Microporous Mesoporous Mater.* **2012**, *149*, 134–141. [CrossRef]
30. Dubbeldam, D.; Calero, S.; Ellis, D.E.; Snurr, R.Q. RASPA: Molecular simulation software for adsorption and diffusion in flexible nanoporous materials. *Mol. Simul.* **2016**, *42*, 81–101. [CrossRef]
31. Rappé, A.K.; Casewit, C.J.; Colwell, K.S.; Goddard, W.A., III; Skiff, W.M. UFF, a full periodic table force field for molecular mechanics and molecular dynamics simulations. *J. Am. Chem. Soc.* **1992**, *114*, 10024–10035. [CrossRef]
32. Potoff, J.J.; Siepmann, J.I. Vapor–liquid equilibria of mixtures containing alkanes, carbon dioxide, and nitrogen. *AIChE J.* **2001**, *47*, 1676–1682. [CrossRef]
33. Chokbunpiam, T.; Chanajaree, R.; Caro, J.; Janke, W.; Remsungnen, T.; Hannongbua, S.; Fritzsche, S. Separation of nitrogen dioxide from the gas mixture with nitrogen by use of ZIF materials; computer simulation studies. *Comput. Mater. Sci.* **2019**, *168*, 246–252. [CrossRef]
34. Li, J.R.; Kuppler, R.J.; Zhou, H.C. Selective gas adsorption and separation in metal–organic frameworks. *Chem. Soc. Rev.* **2009**, *38*, 1477–1504. [CrossRef] [PubMed]
35. Breiman, L. Random forests. *Mach. Learn.* **2001**, *45*, 5–32. [CrossRef]
36. Natekin, A.; Knoll, A. Gradient boosting machines, a tutorial. *Front. Neurorobot.* **2013**, *7*, 21. [CrossRef] [PubMed]
37. Liang, W.; Luo, S.; Zhao, G.; Wu, H. Predicting hard rock pillar stability using GBDT, XGBoost, and LightGBM algorithms. *Mathematics* **2020**, *8*, 765. [CrossRef]
38. Ke, G.; Meng, Q.; Finley, T.; Wang, T.; Chen, W.; Ma, W.; Ye, Q.; Liu, T.Y. Lightgbm: A highly efficient gradient boosting decision tree. *Adv. Neural Inf. Process. Syst.* **2017**, *3149–3157*, 3149–3157.
39. Bergstra, J.; Komer, B.; Eliasmith, C.; Yamins, D.; Cox, D.D. Hyperopt: A python library for model selection and hyperparameter optimization. *Comput. Sci. Discov.* **2015**, *8*, 014008. [CrossRef]
40. Lundberg, S.M.; Lee, S.I. A unified approach to interpreting model predictions. *Adv. Neural Inf. Process. Syst.* **2017**, *4768–4777*, 4768–4777.
41. Lundberg, S.M.; Erion, G.; Chen, H.; DeGrave, A.; Prutkin, J.M.; Nair, B.; Lee, S.I. From local explanations to global understanding with explainable AI for trees. *Nat. Mach. Intell.* **2020**, *2*, 56–67. [CrossRef] [PubMed]
42. Zhou, M.; Vassallo, A.; Wu, J. Toward the inverse design of MOF membranes for efficient D2/H2 separation by combination of physics-based and data-driven modeling. *J. Membr. Sci.* **2020**, *598*, 117675. [CrossRef]
43. Qiao, Z.; Peng, C.; Zhou, J.; Jiang, J. High-throughput computational screening of 137953 metal–organic frameworks for membrane separation of a $CO_2/N_2/CH_4$ mixture. *J. Mater. Chem. A* **2016**, *4*, 15904–15912. [CrossRef]

44. Yu, S.; Bo, J.; Meijun, Q. Molecular dynamic simulation of self-and transport diffusion for $CO_2/CH_4/N_2$ in low-rank coal vitrinite. *Energy Fuels* **2018**, *32*, 3085–3096. [CrossRef]
45. Li, L.; Zhang, T.; Duan, Y.; Wei, Y.; Dong, C.; Ding, L.; Qiao, Z.; Wang, H. Selective gas diffusion in two-dimensional MXene lamellar membranes: Insights from molecular dynamics simulations. *J. Mater. Chem. A* **2018**, *6*, 11734–11742. [CrossRef]
46. Tang, H.; Jiang, J. In silico screening and design strategies of ethane-selective metal–organic frameworks for ethane/ethylene separation. *AIChE J.* **2021**, *67*, e17025. [CrossRef]

nanomaterials MDPI

Article

Enhancing the Green Synthesis of Glycerol Carbonate: Carboxylation of Glycerol with CO_2 Catalyzed by Metal Nanoparticles Encapsulated in Cerium Metal–Organic Frameworks

Simon Lukato, Michał Wójcik, Agnieszka Krogul-Sobczak * and Grzegorz Litwinienko *

Faculty of Chemistry, University of Warsaw, Pasteura 1, 02-093 Warsaw, Poland; s.lukato@uw.edu.pl (S.L.)
* Correspondence: akrogul@chem.uw.edu.pl (A.K.-S.); litwin@chem.uw.edu.pl (G.L.)

Abstract: The reaction of glycerol with CO_2 to produce glycerol carbonate was performed successfully in the presence of gold nanoparticles (AuNPs) supported by a metal–organic framework (MOF) constructed from mixed carboxylate (terephthalic acid and 1,3,5-benzenetricarboxylic acid). The most efficient were two AuNPs@MOF catalysts prepared from pre-synthesized MOF impregnated with Au^{3+} salt and subsequently reduced to AuNPs using H_2 (catalyst 4%Au(H_2)@MOF1) or reduced with $NaBH_4$ (catalyst 4%Au@PEI-MOF1). Compared to existing catalysts, AuNPs@MOFs require simple preparation and operate under mild and sustainable conditions, i.e., a much lower temperature and the lowest CO_2 overpressure ever reported, with $MgCO_3$ having been found to be the optimal dehydrating agent. Although the yield of the process is still not competitive with previously developed systems, the most promising advantage is the highest TOF (78 h^{-1}) ever reported for this reaction. The optimal parameters observed for AuNPs were also tested on AgNPs and CuNPs with promising results, suggesting their great potential for industrial application. The catalysts were characterized by XRD, TEM, SEM-EDS, ICP-MS, XPS, and porosity measurements, confirming that AuNPs are present in low concentration, uniformly distributed, and confined to the cavities of the MOF.

Keywords: glycerol carbonate; carbon dioxide; glycerol; metal nanoparticles; metal organic frameworks; NPs-CeMOFs

1. Introduction

The increasing release of carbon dioxide (CO_2) is a matter of global environmental concern, with intensive efforts aimed at recovering a global carbon balance by using energy generated from renewable sources in preference to traditional fossil fuels. Today, transport consumes most of the world's petroleum. Although electric vehicles are becoming standard, many regions described as the Global South depend on traditional fuels, with increasing use of biofuels (biodiesel and bioethanol) as alternatives to petroleum. Unfortunately, biodiesel production results in the formation of glycerol (GL) as a by-product of transesterification, with at least 9 kg of GL accumulating per 100 kg of biodiesel produced [1]. The scale of GL production exceeds any reasonable demand: the chemical and pharmaceutical industries utilize less than 20% of the generated glycerol, and the rest is not even purified because the costs of purification exceed the price of the final pure product. The problem of environmental significance and other challenges connected with excess bioglycerol and market dynamics have been highlighted and discussed in several reports and reviews [1–4]. The most sustainable method for glycerol utilization is to capture CO_2 by GL and convert both wastes into added-value products. Glycerol carbonate (GLC) appears to be the most versatile valuable product from GL [2,4,5], with a high dielectric constant (111.5), low vapor pressure (0.008 bar at 177 °C), low melting point (−69 °C), low flammability/high flash point (190 °C), negligible toxicity, and high miscibility, among other features.

GLC is a clean substitute for alkyl carbonates, and this green solvent [6] has many applications (as a solvent for electrolytes in batteries; for lubricants and plasticizers; for the production of polycarbonates and polyurethanes, membranes, coatings, paints, and surfactants). On an industrial scale, GLC is produced from toxic (and expensive) phosgene (see Scheme 1A). The other two technologies presented in Scheme 1A are also of limited applicability. Transesterification is expensive due to the high cost of organic carbonates used [7,8]. The glycerolysis of urea could be competitive but is not sufficiently sustainable because the industrial production of urea relies on the combustion of large amounts of methane to obtain ammonia. The reaction is shifted to the products by removing generated ammonia under reduced pressure. Therefore, the direct catalytic reaction of GL with CO_2 to form GLC (Scheme 1B) appears to be the greenest option and has attracted the attention of many research groups [4,9–11]. The efforts focused on applying metal and metal-oxide-based catalysts have resulted in relatively small progress [12–15] as the proposed catalytic materials exhibited TOF parameters are far from satisfactory (see Scheme 1B).

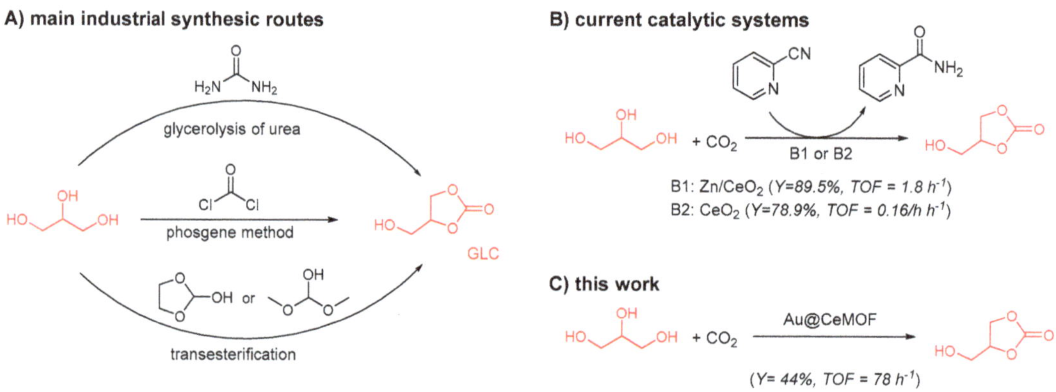

Scheme 1. Synthetic pathways to GLC. The TOFs in (**B**) are calculated based on reported data [16,17].

As presented in Scheme 1B, methods B1 and B2 [16,17] are thermodynamically uphill reactions driven forward by the irreversible loss of water due to the use of a stoichiometric excess of expensive 2-cyanopyridine as a dehydrating agent. In both methods, the catalysts are recovered by calcination, and such thermal treatment (at 600 and 400 °C, respectively) increases the overall costs of the process. Furthermore, a hazardous by-product (pyridine-2-carboxamide) is formed that is not readily biodegradable, has a low flash point (89 °C), and during calcination decomposes, causing the release of CO, CO_2 and nitrogen oxides (NO_x). The limited attractiveness of these methods caused attention to turn to other kinds of catalytic systems. A different approach to tackling the problem of low catalytic efficiency includes the application of metal nanoparticles (MNPs) combined with metal–organic frameworks (MOFs) and the use of an inert dehydrating agent in order to remove water and avoid the formation of other products, for example, by replacing toxic and expensive 2-cyanopirydine with $MgCO_3$. When water is removed physically by $MgCO_3$, no toxic by-product is formed and the dehydrating agent might be recovered together with the catalyst, activated by drying, and reused. Surprisingly, this direction is not represented in the rather broad literature on GL/CO_2 chemistry. MNP-based catalysts are commonly used in both organic and inorganic transformations [18], whereas MOFs are versatile materials in many fields, including catalysis; the sorption of gases [19,20], with catalytic reactions of CO_2 with olefins [21] or epoxides (obtaining up to 96% yield of propylene carbonate) [22], and the direct cyclodimerization of epichlorohydrin to produce 1,4-dioxane 2,5-bis-chloromethyl [23]. The catalytic applicability of MOFs is usually limited because of their moderate stability and low number of catalytic centers. However, the first weakness can be overcome by improving the architecture of MOFs aimed to increase in chemical and thermal stability (which are usually compromised), and the second drawback can be solved by enriching MOF with more active centers, like MNPs as guests. On the other hand, the idea of combining of MOFs with MNPs can also improve the catalytic activity of MNPs as a convenient protection against agglomeration, oxidation, leaching from the support material [24,25], or poor accessibility to active centers when MNPs are coated with stabilizing surfactants or ligands [26–28]. The accessibility/contact of MNPs with reactants could be enhanced by embedding MNPs within the cavities of an MOF [28,29], and such a combination might be facilitated by a proper design of the MOF architecture or its post-synthetic modifications.

Here, we used Cu, Ag, and Au NPs stabilized by a cerium-mixed carboxylate MOF as catalysts for the reaction of glycerol with carbon dioxide to form glycerol carbonate. AuNPs, with their high chemical and physical stability, well-defined morphology [30], and high catalytic activity in numerous transformations, including oxidative carboxylation [31], are frequently used as model NPs to develop new catalytic systems to be further extended on cheaper and more available metals or bimetallic catalysts. Cerium-based catalysts demonstrate exceptional abilities in activating small molecules, including CO_2 [16,32,33]. Moreover, Ce-based MOFs are inexpensive and more resistant to the presence of oxygen and moisture.

2. Materials and Methods

2.1. Chemicals and Instrumentation

Detailed information on the chemicals used, their origins, and detailed analytical protocols are given in the Supplementary Materials. UV–vis spectra were recorded using a Thermo Scientific Genesis 50 UV/Vis spectrophotometer (Waltham, MA, USA) in the 200–800 nm range using glass quartz cells (10 mm path length). Infrared spectra (in KBr) were obtained on a Shimadzu FTIR-8400S spectrophotometer (Kyoto, Japan). Using Cu radiation, the crystal structures were characterized using powder X-ray diffraction on a Bruker AXS D8 X-ray diffractometer (Billerica, MA, USA). The diffraction patterns were processed in the TOPAS application and compared with the literature data about MOF-808(Ce), UiO-66(Ce), and CeBTC. N_2 sorption isotherms were analyzed at 77 K with Autosorb iQ Station 1 (Quantachrome Instruments, Boynton Beach, FL, USA). Before analysis, the samples were degassed at 120 °C for 12 h. The surface area and pore properties were obtained using BET and BHJ methods, respectively. X-ray photo-electron spectroscopy analysis (XPS) was performed at 150 W. High-resolution spectra were obtained using a scan step of 0.2 eV with a pass energy of 200 eV for Au 4f and Ce 3d. The neutralizator was set at 1 eV/1 μA. The catalyst (fine powder) was placed on a conductive carbon tape. The beam was monochromated using an Al Kα (1486.7 eV) monocromator. Curve fitting of the XPS data was carried out using the casaXPS program (version 2.3.15) and Shirley background. For Ce 3d spectrum E Tougard background was used because the energy interval was too weak for reliable Shirley background use.

Scanning electron microscopy (SEM) images and energy-dispersive X-ray spectroscopy (EDS) measurements (SEM-EDS) were performed using three microscopes: a scanning electron microscope JEOL-JSM-5600 (Tokyo, Japan) equipped with an energy-dispersive X-ray spectrometer OXFORD Link-ISIS (Oxford, UK); a Zeiss Merlin FE-SEM scanning electron microscope (Oberkochen, Germany) equipped with a Bruker Quantax XFlash EDS detector 5010 (Billerica, MA, USA) operated at 3 kV and 30 pA for imaging and 15 kV and 500 μA for EDS analysis; and a Jeol JSM—6380LA SEM (Tokyo, Japan) operated at 20 kV with BSE detector, W.D = 16 mm, an aperture size of 120 μm, and a Bruker XFlash 6110 (Berlin, Germany) as the EDS probe. Transmission Electron Microscopy (TEM) images were taken on a Zeiss Libra 120 Microscope (Oberkochen, Germany) operated at 120 kV and on a JEOL 1400 electron microscope (Tokyo, Japan). Inductively coupled plasma-mass spectrometry data were acquired on digested samples. For 12 h, 20 mg of the catalyst was aged at room temperature in 5 mL of acid mixture (0.30 mL of conc HCl in 2 mL 65% HNO_3). Then the resultant mixture was digested at 180 °C with 2 mL of the acid mixture every 30 min until the solution became transparent and colorless (3 h) and came to a total of 25 mL with the HNO_3. Samples for analysis were further diluted with 3% HNO_3. Metal contents were determined on a NexION 300D spectrometer (Perkin Elmer, Waltham, MA, USA) equipped with a liquid sample introduction system. The ICP-MS apparatus was calibrated by measuring a series of reference solutions with concentrations ranging from 10 μgL^{-1} to 200 μgL^{-1} (obtained by diluting ICP standard solutions). Dilution was made using ultra-pure water. The measurement was obtained as a count of ions with a mass-to-charge ratio of appropriate isotopes. The apparatus was purged with a neutral solution for 60 s after each sample and was allowed to equilibrate for 40 s before measurement. Three measurements were obtained for each sample and averaged. The limit of quantification was calculated from the results obtained for the digested blank solutions. The reaction products were analyzed by gas chromatography (GC) using an Agilent 7820A (Santa Clara, CA, USA) chromatograph equipped with an HP-5 capillary column (30 m long × 0.32 mm internal diameter × 0.25 mm filling) and a flame ionization detector (FID).

2.2. Synthesis of Nanoparticles Au@PEI and Au@DBU

The preparation of the nanoparticles was based on the synthesis of gold nanoclusters in aqueous solutions using branched polyethylenimine (PEI) as proposed by Tao et al. [34] with further modification: briefly, a solution containing $HAuCl_4 \cdot 3H_2O$ (0.08 g, 0.2 mmol) and PEI (2 g) in ultrapure water (60 mL) was adjusted to about 7 °C in an ice bath. Then, a cold aqueous solution of freshly prepared sodium borohydride (2 mmol) was added with constant stirring (1200 rpm), and the mixture was allowed to react for 1 h. The resultant solution was reddish brown, indicating the formation of AuNPs designated as Au@PEI NPs. In a separate synthesis, Au@DBU NPs were synthesized using 1,8-diazabicyclo [5.4.0]undec-7-ene (DBU) (2 mL, 13 mmol) as a stabilizing ligand instead of PEI. The nanoparticles were precipitated with acetone after a reduction in the solvent volume under reduced pressure, then isolated by centrifugation (15,000 rpm, 20 min), dried, and stored for further use. For TEM measurements, a drop of NP solution in water or ethanol was placed on a carbon-coated copper grid and allowed to dry under ambient conditions before TEM measurements were conducted.

2.3. Pre-Synthesized MOF

The MOF was constructed from two organic components, terephthalic acid (BDC) and 1,3,5-benzenetricarboxylic acid (BTC), using the hydrothermal/solvothermal method [35]. BTC (11.2 mg, 0.05 mmol) and BDC (5.8 mg, 0.1 mmol) were ultrasonically dissolved in 3 mL of DMF and mixed with a solution of ammonium cerium (IV) nitrate (13.7 mg, 0.25 mmol) in ultra-pure water (1 mL) under stirring for 5 min. Then formic acid (250 µL) was added under constant mixing. The mixture was sealed in a Teflon-lined steel autoclave, heated to 70 °C, and kept at this temperature for 20 min. The mixture was allowed to cool down and the precipitated MOF was separated by centrifugation, washed twice with DMF and two times with methanol, and afterwards soaked in methanol for 12 h, with the methanol portions being replaced after 6 h. Finally, it was washed once with acetone and allowed to dry in the air (fume-hood) for 1 h. The crystalline powder was dried at 80 °C in a vacuum for 12 h and thermally activated (150 °C for 12 h) before use and characterization. The synthesis of MOF was scaled up with 10-fold higher amounts of reactants.

2.4. Synthesis of Catalysts

Three synthetic strategies were applied to combine the MOF and NPs into one catalytic system. **Method 1** is based on the generation of NPs inside the already pre-synthesized MOF. The catalyst obtained by this method is denoted as NP-MOF1, where NP is a nanoparticle of nanoclusters of Au, Ag, and Cu. **Method 2** involves the buildup of MOF around the already pre-synthesized MNPs, and these catalysts are abbreviated as NP-MOF2. The third strategy (**Method 3**) is a one-pot in situ formation of a catalyst from precursors of both MNPs and MOF; such catalysts are described as NP-MOF3.

Method 1. Synthesis of catalysts using pre-synthesized MOF

The NPs or nanoclusters (NCs) of Au@DBU and Au@PEI were embedded in the MOF using simple precursor impregnation method followed by reduction [36]. Pre-synthesized MOF material (75 mg) was ultrasonically dispersed in methanol (15 mL). Measures of 6 mg (0.015 mmol) of $HAuCl_4 \cdot 3H_2O$ and 100 μL DBU were dissolved in 1 mL of water and added, and the mixture was stirred for 6 h and then placed in an ice bath at about 7 °C before a cold aqueous solution of $NaBH_4$ (0.15 mmol, 2 mL) was added. After 30 min of stirring, the resulting solid product was removed by centrifugation at 8000 rpm for 10 min, washed twice with water and once with methanol, dried for 1 h at RT, and then dried overnight at 80 °C to produce approximately 72.5 mg of powder, which was finally reduced/activated in a flow of H_2/He (1:2 v/v) at 210 °C (quartz tube) in a horizontal furnace for 1 h before use and characterization. This catalyst is denoted as 4%Au@DBU-MOF1 and contains 4% of Au (calculated with respect to Ce in $(NH_4)_2Ce(NO_3)_6$ used, see the analysis of the catalysts in the Results and Discussion). By changing the amount of $HAuCl_4$ to 4 mg and 12 mg, two other catalytic systems were obtained: 3%Au@DBU-MOF1) and 8%Au@DBU-MOF1, respectively. The catalyst based on polyethylenimine (4%Au@PEI-MOF1) was synthesized following the same procedure except that 200 mg of PEI was used instead of DBU. Catalysts with silver and copper nanoparticles (2%Ag@PEI-MOF1 and 2%Cu@PEI-MOF1) were synthesized using an appropriate amount of $AgNO_3$ or $CuSO_4$ instead of Au precursor.

A variation of this procedure, but without surfactants and with gold reduced directly with H_2 (instead of $NaBH_4$), was used for the preparation of the catalyst Au(H_2)@MOF1, following the double solvent method [37]: 250 μL of aqueous solution of $HAuCl_4 \cdot 3H_2O$ (6 mg, 0.01 mmol) was added dropwise to ultrasonically dispersed pre-synthesized MOF (85 mg) in 15 mL of dry n-hexane. The reaction vessel was removed from the sonic bath and the mixture was stirred until all the solid was deposed on the walls of the flask to leave colorless n-hexane, which was allowed to stand for 3 h. The solvent was gently removed, and the MOF impregnated with $HAuCl_4$ was vacuum dried under ambient conditions, then dried at 70 °C under vacuum for 12 h. Finally, the Au^{3+} impregnated in the MOF was reduced with mixed H_2/He (approximately 1: 1 v/v) at 210 °C for 2 h to obtain the final catalysts with approximately 3.4 mol% Au (ICP-MS) based on Ce, which was labeled 4%Au(H_2)@MOF1.

Method 2. Synthesis of 4%Au@DBU-MOF2

A solution of sodium borohydride (0.06 g, 1.5 mmol) in DMF (5 mL) was added dropwise to a solution containing $HAuCl_4 \cdot 3H_2O$ (0.06 g, 0.15 mmol) and 1,8-diazabicyclo [5.4.0]undec-7-ene (DBU) (1 mL, 6.55 mmol) in DMF (20 mL), immersed in an ice bath. The reaction mixture was stirred and allowed to react for 1 h. The resultant mixture was red, which indicated the formation of AuNPs. The mixture was dispersed in a solution containing 1, 3, 5-benzenetricarboxylic acid (BTC) (0.263 g, 1 mmol) and terephthalic acid (BDC) (0.166 g, 1.2 mmol) in DMF (30 mL) and ammonium cerium(IV) nitrate (3.5 mmol, 10 mL H_2O) under ultra-sonication and then allowed to react at 120 °C for 1 h. The solid product was recovered by centrifugation and washed with DMF and then with methanol. Afterward, the product (4%Au@DBU-MOF2) was dried at 80 °C in a vacuum oven for 12 h and activated at 150 °C with H_2.

Method 3. One-pot synthesis of the catalyst

A solution containing $HAuCl_4 \cdot 3H_2O$ (0.06 g, 0.15 mmol) and 1,8-diazabicyclo [5.4.0] undec-7-ene (DBU) (1 mL, 6.55 mmol) in DMF (20 mL) was stirred for 5 min, and the resultant solution was mixed with a solution containing 0.263 g (1 mmol) of 1, 3, 5-benzenetricarboxylic acid (BTC) and 0.166 g (1.2 mmol) of terephthalic acid (BDC) in 30 mL of DMF and 6 mL (3.5 mmol) aqueous solution of ammonium cerium (IV) nitrate. This was then sonicated, sealed, and allowed to react at 120 °C for 12 h. The solid was recovered by centrifugation and washed with DMF and then with methanol. Afterward, the product was dried overnight at 80 °C in a vacuum. The dried catalyst was activated in H_2 gas at 150 °C for 1 h. The final material was marked as 4%Au@DBU-MOF3.

2.5. Carboxylation of Glycerol with Carbon Dioxide Using the Catalysts

Measures of 10 mL of DMF, 1.15 g (12.5 mmol) of GL, 2.5 mL of methanol, and the dehydrating agent (1.0 g of $MgCO_3$, or 2 mL of adiponitrile, or 2.5 mL of acetonitrile) were placed in a stainless-steel autoclave equipped with a Teflon vessel and a magnetic stirring bar. For crude glycerol, 3 mL of the crude GL were diluted to 10 mL with DMF without the addition of methanol or the dehydrating agent (1.0 g of $MgCO_3$ was added, see Supplementary Material for chemicals and reagents for information on how crude GL was obtained). The catalyst (0.01, 0.05, 0.10, or 0.15 g) was loaded, and the vessel was purged with CO_2, pressurized with CO_2, and placed in a preheated oil bath on a magnetic stirrer. After the required time, the reactor was cooled down and carefully depressurized. The solids were separated from the reaction mixture by centrifugation at 4000 rpm for 3 min and the supernatant was derivatized (silylated) and analyzed by gas chromatography. Derivatization was carried out as follows: 0.50 g of the supernatant from the carboxylation reaction was shaken with approximately 2 g of the silylating reagent (a mixture of pyridine, hexamethyldisilazane, and trimethylchlorosilane at molar ratio of 9:3:1, respectively) in a vial. After 0.5 h, the formed precipitate was separated by centrifugation at 4000 rpm for 2 min, and the supernatant was analyzed by gas chromatography. The conversion of GL and yield of GLC were determined from the integrated peaks for GL and GLC on chromatograms, calculated with respect to the initial amount of GL using the internal standard method and expressed in %. See Supplementary Materials for details on the calculation of GLC yield, GL conversion, TON, and TOF (Equations (S1)–(S6)). Please note that TON/TOF parameters were calculated based on moles of Au embedded in MOF.

3. Results and Discussion

3.1. Synthesis and Characterization of the Nanoparticles

Good stabilization and protection of NPs against agglomeration in water was obtained due to the strong interaction with nitrogen-containing ligands PEI [34] and DBU (see Figure 1). The UV–vis absorption spectra (Figure 1B) recorded for Au@DBU indicate a clearly visible localized surface plasmon resonance (LSPR) band at around 530 nm, which suggests that the average size of formed AuNPs is greater than 2 nm. On the other hand, a weak LSPR at 530 nm for Au@PEI indicates the presence of few large nanoparticles, while a strong band with a maximum at ca. 365 nm can be assigned to Au nanoclusters [38,39].

These observations were supported by the TEM measurements presented in Figure 1. AuNPs are spherical and have a very narrow size distribution, with a mean diameter of 3.2 nm for Au@DBU and 2.6 nm for Au@PEI (Figures 1C and 1D, respectively).

3.2. Analysis of Pre-Synthesized MOFs

Among the rare earth metals, cerium and its compounds are the most accessible, with costs comparable to those of metals like copper. Ce- and CeO_2-containing catalysts have exceptional abilities to activate small molecules such as CO_2 [40]; therefore, we decided to prepare cerium-containing MOFs as part of the catalytic system. Chemically and thermally stable MOF-808(Ce) and UiO-66(Ce) and their Zr predecessors can be produced with the BDC and BTC linkers [41], respectively, and we applied this convenient, non-expensive, and environmentally friendly method to prepare Ce-MOF based on both organic components assembled within one framework (see Section 2—Materials and Methods). The SEM, TEM, and XRD images of the synthesized Ce-MOFs are given in Figure 2.

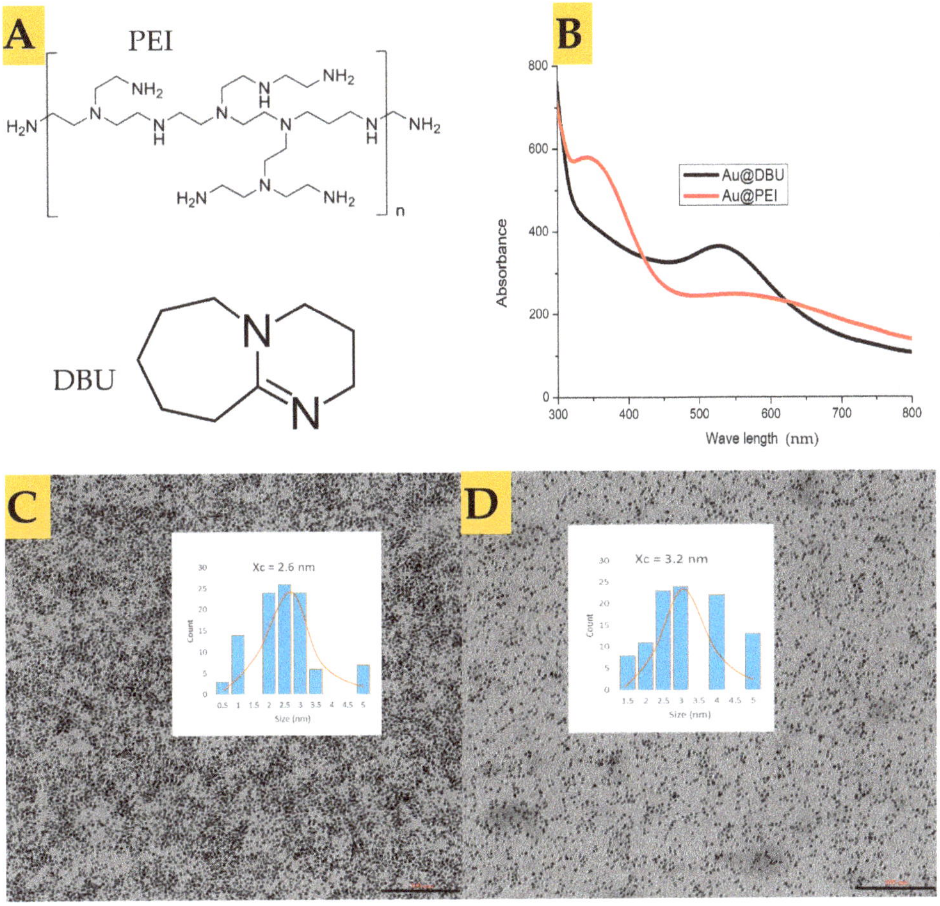

Figure 1. (**A**) Structures of 1,8-diazabicyclo [5.4.0]undec-7-ene (DBU) and polyethylenimine (PEI). (**B**) UV–Vis of the NPs at concentration 1.3 mM (calculated with respect to Au) in water. Lower panels: TEM images with histograms for AuNPs stabilized with PEI (**C**) and DBU (**D**).

The pure MOF is shown in Figure 2A,B (SEM and TEM, respectively). We expected to obtain a material with a new crystal-like morphology as a result of cerium bridges formed between dicarboxylate and tricarboxylate anions. However, the morphology and size of the synthesized powder particles indicate a mixture of MOF-808(Ce) and UiO-66(Ce) [42], suggesting that two-valent (BDC) and trivalent (BTC) anions do not assemble in one hybrid system, but each of them forms a separate MOF with cerium cations. The TEM image is typical for UiO-66 and MOF-808 with aggregates, and the sizes of the pristine MOFs range from 40 nm to ca. 560 nm (note that the small MOF NPs appear as aggregates in the bigger picture, see Figure S1). The MOF presented in Figure 2A contains a larger number of aggregated crystals that cover the regular octahedral crystals, which is common with UiO-66(Ce), as recently reported [43,44]. Based on this similarity, we suggest that our product contains more UiO-66(Ce) MOF than MOF-808(Ce). This is supported by the comparison of powder X-ray diffraction patterns for mixed MOF and for the two catalysts 4%Ag@PEI-MOF1 and 4%Au(H_2)@MOF1 presented in Figure 3E, whose XRD results are similar to the diffraction pattern obtained for UiO-66(Ce) [45–47].

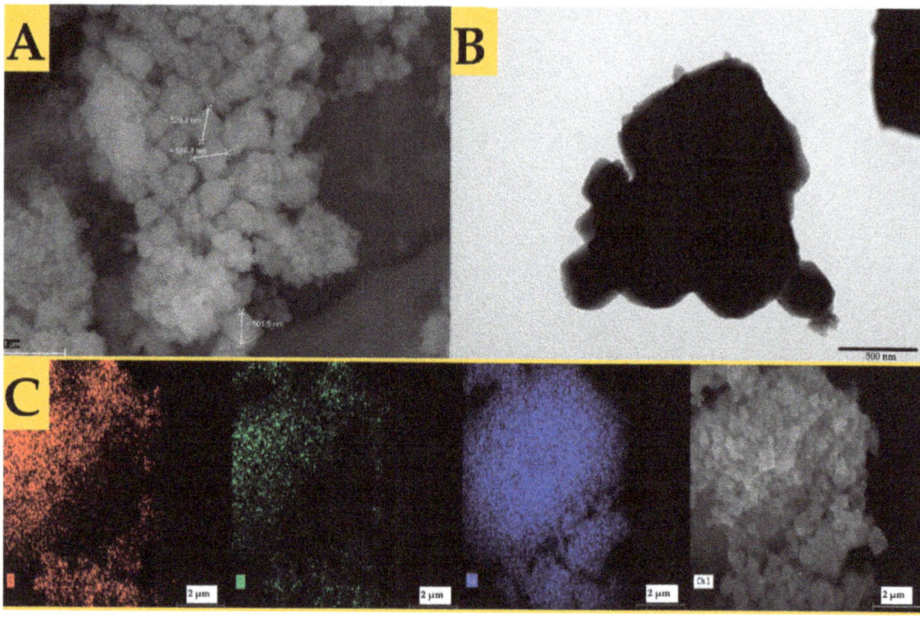

Figure 2. SEM (**A**), TEM (**B**), and SEM-EDS (**C**) elemental maps of O, C, and Ce, respectively, and far right is the image of the prepared Ce-MOF containing BTC and BDC.

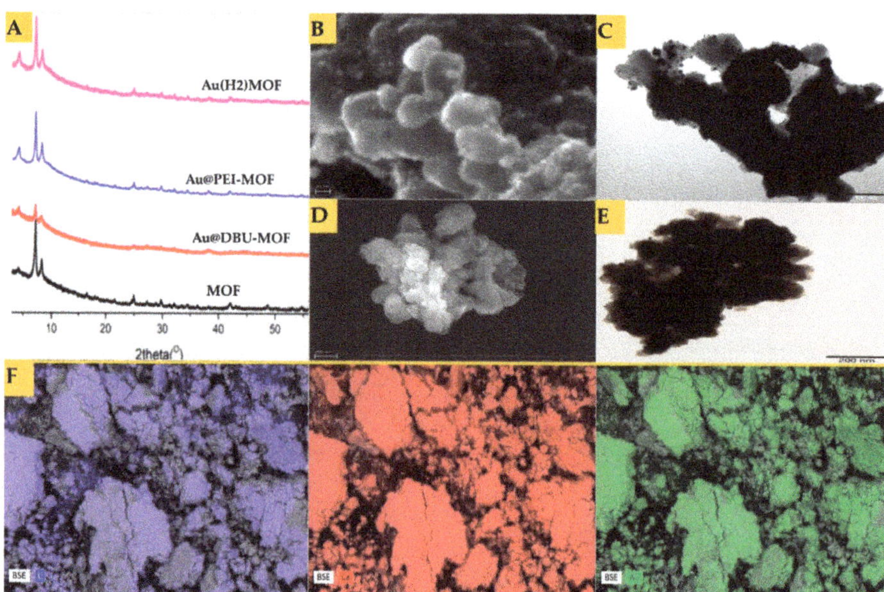

Figure 3. (**A**) XRD powder patterns for the catalysts prepared by method 1 (the lowermost plot is for the pre-synthesized MOF). Panels (**B–E**): SEM and TEM images of 4%Au@DBU-MOF1 (**B** and **C**, respectively) and 4%Au@PEI-MOF1 (**D** and **E**, respectively). Panel (**F**): SEM EDS elemental maps of 4%Au(H_2)-MOF1.

3.3. Synthesis and Characterization of the Catalysts

The aim of all three procedures described in the Section 2—Materials and Methods (Methods 1–3) was to produce catalysts in the form of nanoparticles embedded inside the cavities of MOFs because such a NPs@MOF system ensures the stabilization of the NPs. The catalysts obtained using Method 1 (with pre-synthesized MOFs and abbreviated MOF1) were the most active and selective for the carboxylation reaction, and our discussion will concentrate much on the ones with gold nanoparticles embedded inside cavities: 4%Au@DBU-MOF1, 4%Au@PEI-MOF1, and 4%Au(H_2)@MOF1. The FTIR spectra (shown in Figure S2) indicate strong bands at 1401, 1596 cm^{-1} and 1421, 1576 cm^{-1} in all the samples due to symmetric and antisymmetric carboxylate stretching modes [48,49]. The appearance of weak bands at around 3000 to 3500 cm^{-1} in all spectra is attributed to both O-H (from adsorbed water) and C-H stretching modes or N-H bonds present in 4%Au@DBU-MOF1 and 4%Au@PEI-MOF1. The weak bands at 524, 750, and 110 cm^{-1} are due to the several vibrational modes of the benzene rings from the carboxylate linkers in all of the spectra.

As presented in Figure 3A, the XRD patterns demonstrate similarity regarding their XRD peaks and their 2theta positions to UiO-66(Ce) [45–47], UiO-66(Hf) [50], and MOF-808(Ce) [51] (see Figure S3 in Supplementary Materials), confirming that our catalysts were obtained with good purity. Apart from the XRD pattern of 4%Au@DBU-MOF1 (Figure 1A), the rest of the plot represents sharp peaks and therefore high crystallinity.

As can be seen in Figure 3A, there is no appreciable difference between the XRD results of the MOF and the three catalysts based on this MOF: 4%Au@DBU-MOF1, 4%Au@PEI-MOF1, and 4%Au(H_2)-MOF1. Such retention of the XRD pattern of the MOF after the introduction of Au [50,52] corroborates our expectations that AuNPs are present at a low concentration (confirmed by ICP-MS), uniformly distributed, and held inside the MOF, resulting in no X-ray reflections from Au. The XRD pattern of 4%Au@DBU-MOF1 suggests that the material is amorphous, probably due to the interference of reflection from the much higher number of AuNPs on the surface of the MOF, as is evident in the TEM and SEM images (Figure 3B,C). Large NPs (about 5 to 10 nm) are visible on the catalyst's surface, while small ones are inside the MOF. No small AuNPs are visible on the surface, suggesting they were effectively embedded inside the MOF cavities. Obstruction due to overlap and masking by the host MOF surface could not allow for the concise particle positioning of very small NPs/NCs using TEM imaging [53–55].

Large NPs reside tightly on or close to the surface of the MOF crystals, probably due to an ineffective encapsulation strategy that might have resulted in some Au NPs on the surface of the MOF aggregating during synthesis or TEM imaging [53–55]. In contrast, the XRD of the 4%Au@PEI-MOF1 shows sharp peaks (Figure 3A), pointing to better crystallinity than the DBU catalysts, and is complemented by the TEM images of the PEI catalyst, which shows no particles on the surface. In some images (see Figure 3B–D), we observed the presence of a few rod-like crystals and needle-like forms similar to the BTC-based Ce-MOF reported by other researchers [56], suggesting a presence of Ce^{3+} moieties during synthesis [56]. The 4%Au@PEI-MOF1 and 4%Au(H_2)@MOF1 form more regular crystals, and there is a smaller number of aggregates than in the pristine MOF, suggesting that the presence of the NPs prevents aggregation in the MOFs.

The SEM image of the 4%Au(H_2)@MOF1 synthesized by a modified double solvent method coupled with the hydrogen reduction of Au^{3+} is given in Figure S4 (Supplementary Materials). No AuNPs are present on the surface of the MOF crystals. This shows that this strategy is more efficient in embedding the ultra-small NPs inside the MOF. Moreover, the crystals of the MOF seem to maintain a uniform shape. This probably explains the better performance and reusability of the resultant catalyst, vide infra. The EDS elemental maps shown in Figure 3F support the speculation of the presence of NCs. The EDS elemental maps of the MOF 4%Au(H_2)@MOF1 given in Figure S5 confirm that Ce and Au are uniformly distributed in the catalysts. This suggests that all the Au is well distributed inside the MOF. The representative elemental composition is given in the Supplementary Materials (Figure S6).

Measurements of the MOF's porosity and surface area and a pore analysis were also conducted. As presented in Figure 4A, the N_2 adsorption–desorption graph and the pore size distribution graph provide an estimated total surface area of the MOF of 459 m^2/g, and a pore volume of 0.156 cm^3/g. The pore size distribution shows both micropores and mesopores due to the presence of volume in both regions with more mesopores, as indicated by the distribution plot (approximately 0.5 to 20 nm with two peaks at around 1 nm and 3.5 nm pore size) (Figure 4B). The adsorption plot shows a type 4 isotherm with an H_2a hysteresis loop at a relative pressure of approximately 0.37–0.7, according to the latest IUPAC classification [57]. This type of hysteresis loop is caused by the blocking of the desorption pores and delayed condensation during the physisorption measurement due to connectivity in the pores with small apertures/necks [57].

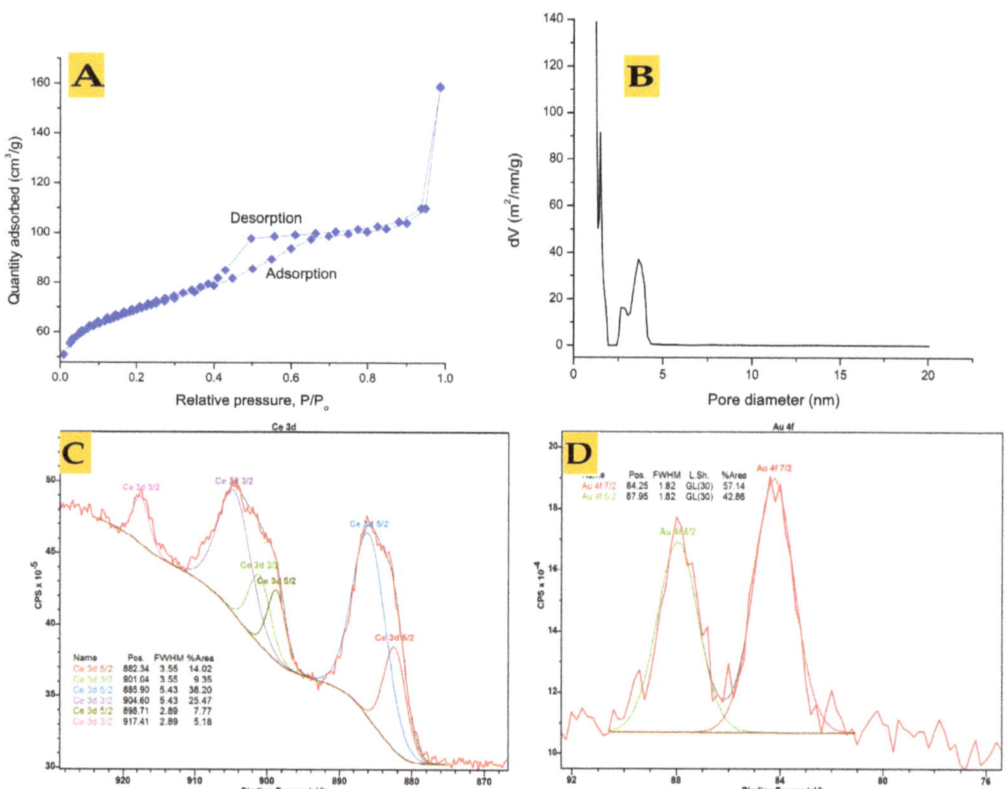

Figure 4. Physisorption analysis of the MOF: (**A**) N_2 adsorption/desorption plot and (**B**) pore size distribution;. Lower panels: XPS spectra of Au(H_2)@MOF1 for: (**C**) Ce and (**D**) Au (with Shirley background subtracted).

The XPS spectra of Au 4f were taken for the most promising system: Au(H_2)@MOF1. NP–support interactions usually play a very important role in supported/composite catalysts [58–61]. Therefore, XPS analysis was performed mainly to determine the chemical states of Au and Ce atoms and the relative percentage composition in the composite catalyst because it is a reliable technique for studying the surface chemistry of the elements. Such information is relevant to checking the presence of the species in reduced or oxidized states and the presence of NP–support interactions. The XPS spectra of Ce and Au are given in Figures 4C and 4D, respectively, while the survey spectra are given in the Supplementary Materials (Figure S8). From the survey spectra of the sample, peaks confirming the presence of the major elements, i.e., for Au 4f, Ce 3d, C 1s, and O 1s, are evident. The relative quantities in at% are C = 66.4, O = 24, Au = 0.15, and Ce = 2.05. The high-resolution XPS spectra in Figure 4D show binding energies of 87.95 eV (Au 4f5/2) and 84.25 (Au 4f7/2) for Au^0. On the other hand, the peak at 885.9 eV in Figure 4C is due to Ce^{3+} [62–64] and contributes 38.2% of the area, which is higher than all other Ce states. Presence of Ce^{3+} in Ce^{4+} MOFs and ceria creates structural defects (oxygen vacancies), resulting in higher catalytic activity. The surface defects in the MOF can strongly interact with Au atoms and modify their electronic structure to stabilize the oxidation states of Au [62–64], which could play a role in the insertion of CO_2 into the alcohol. Therefore, we examined the chemical status of Au in the synthesized catalyst composite. As depicted in Figure 4C, the deconvoluted Au 4f spectra exhibit two main peaks corresponding to metallic Au^0. No dominant peaks for Au^{3+} (normally at around 86.1 and 90.2 eV) are observed, confirming the presence of most of the Au in the zero state [65]. However, there are minor peak branches on the main peaks of the Au binding energy, which are obviously assigned to oxidized Au (Au^{3+} and Au^{+}). Since it is very difficult to have oxidized Au under an excess of H_2 and at 205 °C, the presence of oxidized gold species in the XPS spectra must originate from the strong Au–CeMOF interactions caused by the defects [66–68]. Moreover, the NPs not surrounded by agents containing heteroatoms have better contact with Ce-O moieties in the MOF. Similar effects were observed for Au supported on ceria [68]. Gold existing at different oxidiation states in AuNPs [69] accelerates the charge transfers leading to superior catalytic performance, and were also reported to enhance CO_2 adsorption [63,66]. It is also important to note that the characteristic peaks for C and Ce (Figure S8) of the MOF are not attenuated, which confirms the efficient encapsulation of Au NPs inside the CeMOF, not on the MOF surface.

3.4. *Catalytic Activity of Synthesized X%AuNPs@MOF*

The direct carboxylation of GL with CO_2 is known to form various products depending on the reaction conditions. In this work, effort was put into optimizing the conditions towards selective GLC formation, though the identification, isolation, and quantification of other products is outside the scope of this study, as the by-products of glycerol carbonylation (acetins, dimethyl carbonate, and glyceraldehyde) have been well identified and described in several reviews [4,10,16]. Since the analytical standards of GL and GLC are commercially available, we were able to identify and quantify these compounds accurately by gas chromatography using the internal standard method.

Optimization experiments of reaction conditions for the carbonylation of GL to GLC with carbon dioxide in the presence of X%Au@DBU-MOF1 (X = mol% of Au, see Section 2—Materials and Methods) included the determination of the effect of various factors (dehydrating agent, catalyst composition and amount, CO_2 pressure, temperature, time) on the conversion of GL and the yield of GLC.

3.4.1. Effect of a Dehydrating Agent, Au Content, and Amount of Catalyst

Dehydration is crucial for the carbonylation of GL to GLC by CO_2 [3,70,71] because the evolved water shifts the equilibrium towards reactants. The effect of dehydrating agents ($MgCO_3$, acetonitrile, and adiponitrile) on the conversion of GL and the GLC yield was tested, and the results are presented in Table 1. Since mass transfer limitations are more pronounced in reactions involving a mixture of materials in different physical states, we checked the stirring speed and the optimal mass of $MgCO_3$ to ensure the absence of such limitations. Using 0.5 g of $MgCO_3$ gave a very low conversion, and the highest conversion and uniform stirring was observed when 1 g of $MgCO_3$ was used. Increasing the mass to 1.5 g had no effect on stirring uniformity and no effect on conversion, whereas the presence of 2 g of $MgCO_3$ caused non-uniform stirring, sluggish and depositing lumps on higher parts of the vessel were observed. Thus, 1 g was chosen as an optimal amount of $MgCO_3$ for further experiments.

Table 1. The effect of dehydrating agents on the conversion of glycerol (C_{GL}) and yield of glycerol carbonate (Y_{GLC}) during the carbonylation of GL to GLC [a].

Run	Dehydrating Agent	C_{GL} [%]	Y_{GLC} [%]
1 [b]	$MgCO_3$	49	32
2	Acetonitrile [c]	75	12
3	Adiponitrile [c]	63	30
4	$MgCO_3$	51	39
5	-	trace	-

[a] Reaction conditions: 4%Au@DBU-MOF1, 12.5 mmol GL, 10 mL DMF, 2.5 mL MeOH, 1 g $MgCO_3$, 150 °C, 1.5 MPa CO_2, 13 h. [b] Reaction performed in the absence of MeOH, [c] 2 mL.

The highest yield was obtained for $MgCO_3$ (perhaps due to the stability of $MgCO_3$ in the protic environment [72]), followed by adiponitrile. We observe a much better performance for adiponitrile than for acetonitrile, and such results agree with the data in the literature [73]. Considering the results from Table 1, we decided to study the catalytic activity of other catalysts in the presence of the two most effective dehydrating agents ($MgCO_3$ and adiponitrile) (see runs 1–5 in Table 2). The added advantage of using $MgCO_3$ is that it captures the formed water from the reaction mixture physically without undergoing a reaction to form other products.

Table 2. Conversion of GL (C_{GL}) and yield of GLC (Y_{GLC}) in the carbonylation of GL to GLC in the presence of CO_2, $MgCO_3$, adiponitrile, and various catalysts [a].

Run	Catalyst	Dehydrating Agent	C_{GL} [%]	Y_{GLC} [%]
1	4%Au@DBU-MOF1	$MgCO_3$	51	39
2	4%Au@DBU-MOF1	Adiponitrile [b]	63	30
3	4%Au@PEI-MOF1	$MgCO_3$	65	42
4	4%Au@PEI-MOF1	Adiponitrile [b]	70	37
5	4%Au(H_2)@MOF1	$MgCO_3$	52	44
6	MOF	$MgCO_3$	14 [c]	12
7	4%Cu@PEI-MOF1	$MgCO_3$	25 [d]	17
8	4%Ag@ PEI-MOF1	$MgCO_3$	33 [d]	26
9	4%Au@DBU-MOF2	$MgCO_3$	49 [e]	40
10	4%Au@DBU-MOF3	$MgCO_3$	26	18
11	3%Au@DBU-MOF1	$MgCO_3$	41	28
12	8%Au@DBU-MOF1	$MgCO_3$	68	43
13 [f]	4%Au(H_2)@MOF1	$MgCO_3$	49	7.0
14	-	$MgCO_3$	<1	-

[a] Reaction conditions: 12.5 mmol GL, 10 mL DMF as a solvent, 2.5 mL MeOH, 1 g $MgCO_3$, 150 °C, 1.5 MPa of CO_2, 13 h. [b] 2 mL, [c] reaction time = 25 h, [d] reaction time = 20 h, [e] reaction time = 18 h. [f] Instead of pure glycerol, dry crude glycerol was applied (dried over $MgCO_3$ and molecular sieves).

Three catalysts based on MOF1, i.e., 4%Au(H_2)@MOF1, 4%Au@PEI-MOF1, and 4%Au@DBU-MOF1, exhibit relatively similar activity (with 44, 42, and 39% yield, respectively). Considering the slight increase in yield in the presence of methanol (Table 1, run 1 versus run 4), methanol was applied as a co-solvent since it can be recovered easily if required. There is an additional advantage of using methanol as this solvent might also form dimethyl carbonate (DMC) [74]. As we did not detect DMC in the reaction mixture, we suppose that all DCM (if formed) is subsequently trans-esterified into glycerol carbonate. The high selectivity of 4%Au(H_2)@MOF1 can be attributed to the small and evenly distributed Au particles (clusters or nanoparticles) within the MOF cavities, with relatively uniform particle sizes and shapes. The reaction performed with crude glycerol (as a substrate) gave a relatively good conversion (C_{GL} = 49%) but a very poor yield (Y_{GLC} = 7.0%) of GLC (Table 2, run 13). A representative GC chromatogram showing the peaks of the silylated products is given in the Supplementary Materials (Figure S7).

The TEM images of the 4%Au@DBU-MOF1 and 4%Au@PEI-MOF1 catalysts indicate the presence of NPs within the MOF (Figures 3B and 3D, respectively). However, most large NPs are visible on or close to the surface of the MOF (Figure 3B), and their broad size distribution has been reported to lead to by-products [75,76]. For reduced or doubled Au content (3%Au@DBU-MOF1 or 8%Au@DBU-MOF1, respectively), conversions and yields changed to 41 and 68% and 28 and 43%, respectively (Table 2, runs 11 and 12). The content of Au = 4% (based on the results obtained for 4%Au@DBU-MOF1) was optimal, and therefore, the same content of Au was applied for the synthesis of 4%Au(H_2)@MOF1, 4%Cu@ PEI -MOF1, and 4%Ag@ PEI -MOF1 (runs 5, 7, and 8 respectively).

Table 3 presents the effects of the amount of catalyst 4%Au@DBU-MOF1 on its activity in the carbonylation of GL to GLC. As doubling the catalyst amount from 0.05 to 0.1 causes a small change in conversion and yield for the process carried out for 13 h at 150 °C), the optimal amount of catalyst is 0.05 g.

Table 3. Effect of the amount of catalyst (4%Au@DBU-MOF1) on the conversion of GL (C_{GL}) and the yield of GLC (Y_{GLC}) [a].

Run	Catalyst Amount [g]	C_{GL} [%]	Y_{GLC} [%]
1	0.01	34	22
2	0.05	51	39
3	0.10	58	41
4	0.15	66	44

[a] Reaction conditions: 12.5 mmol GL, 10 mL DMF as solvent, 2.5 mL MeOH, 1 g $MgCO_3$, 150 °C, 1.5 MPa CO_2, 13 h.

3.4.2. Effect of the Reaction Time, Temperature, and Pressure

Based on the results in Table 2, we selected 4%Au@DBU-MOF1 as the most promising catalyst (before 4%Au@PEI-MOF1 and 4%Au(H_2)@MOF1 were synthesized) and checked to what degree the reaction efficiency depends on the reaction time, temperature, and pressure, and the results are presented in Table 4. For processes carried out at 150 °C under 1.5 MPa pressure, an increase in GLC yield from 5 to ca. 40% is observed during the first 13 h, and after that time the further increase is negligible (runs 6–11), probably due to decomposition of GLC or side reactions of GLC, as suggested by gas chromatography analysis (small unidentified peaks appear in the products obtained for a long reaction time). Table 4 also presents the results obtained at different temperatures—from 120 to 170 °C (runs 12 to 16), indicating an increase in conversion and yield with an increase in temperature in the range of 120 to 150 °C (runs 14 to 16). However, above 150 °C, the yield of GLC decreases (runs 12 and 13), perhaps due to the formation of by-products formed during the decomposition of GLC or the reaction of GLC with GL at higher temperatures (the conversion of GC still increases). The effect of the third parameter, the pressure of CO_2 (in the range of 0.5 MPa to 5.0 MPa, see runs 1 to 5 in Table 4), allows one to identify 1.5 MPa as the optimal pressure, as a further increase in CO_2 pressure causes a high increase in conversion but a negligible increase in yield.

Table 4. Effect of reaction time (t), temperature (T), and CO_2 pressure (p) on the conversion of GL (C_{GL}) and yield of GLC (Y_{GLC}) in the presence of 4%Au@DBU-MOF1 [a].

Run	t [h]	T [°C]	p [MPa]	C_{GL} [%]	Y_{GLC} [%]
1	13	150	5.0	86	38
2	13	150	3.5	76	42
3	13	150	2.5	64	33
4	13	150	1.5	51	39
5	25	150	0.5	39	24

Table 4. *Cont.*

Run	t [h]	T [°C]	p [MPa]	C_{GL} [%]	Y_{GLC} [%]
6	18	150	1.5	53	38
7	15	150	1.5	52	40
8	13	150	1.5	51	39
9	9	150	1.5	32	24
10	6	150	1.5	23	16
11	3	150	1.5	13	5
12	13	170	1.5	67	29
13	13	160	1.5	60	34
14	13	150	1.5	51	39
15	13	140	1.5	36	24
16	13	120	1.5	22	7

[a] Reaction conditions: 12.5 mmol GL, 10 mL DMF, 2.5 mL MeOH, 1 g $MgCO_3$.

3.4.3. Regeneration and Efficiency of the Applied Catalysts

In order to examine whether the catalysis was heterogeneous or homogeneous (due to a release of nanocatalysts in the form of nanoclusters from the MOF), the leaching from the 4%Au(H_2)@MOF1 was checked by hot filtration of the reaction mixture at midway through the reaction. To the fresh filtrate (approx. 12 mL), 1.15 g of GL was added and 1 mL of sample was removed for analysis (after homogenization), while the rest was placed in the reactor with fresh $MgCO_3$ (1 g), pressurized with CO_2, and heated to 150 °C over 15 h, and after that time, no conversion of GL was observed, indicating that no appreciable amount of NPs or NCs leached from the MOF to the reaction mixture. This is confirmed by ICP-MS analysis of the fresh and recycled catalysts, which shows negligible difference in the % ratio of Au to Ce after three consecutive cycles. Both experiments indicate also that AuNPs are not released from the MOF, which is an important observation when a decrease in activity is discussed, vide infra.

The separation and recyclability of a catalyst from the products have a crucial implication on the profitability and sustainable application of such a catalyst in industry. To recycle the catalyst, the post-reaction solid residue was washed with dry DMF, dried, cleaned in H_2 as described in the Materials and Methods section, and reused. The activity of 4%Au(H_2)@MOF1 and 4%Au@DBU-MOF1 (expressed as GLC yield) over three consecutive cycles is shown in Figure 5. Both catalysts lose their activity, but the decrease in yield is smaller for 4%Au(H_2)@MOF1 (from 44% to 24% after three cycles) compared to the DBU-based catalyst (from 39 to 9%) after the same number of cycles. The better stability of 4%Au(H_2)@MOF1 is probably attributed to the fine structure of the catalyst, as depicted by the electron microscopy images, which leads to the stability and reusability of the nanocatalysts.

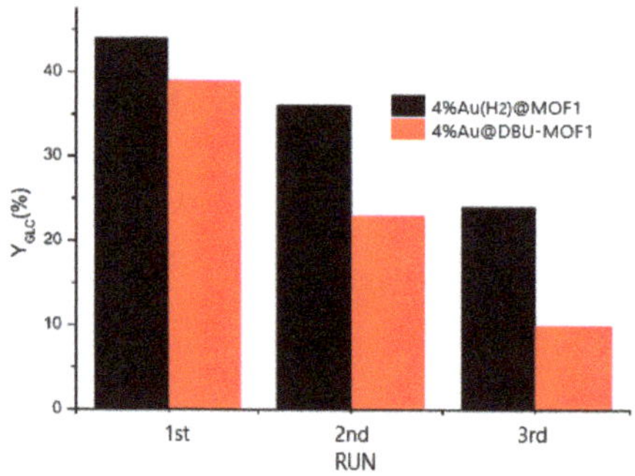

Figure 5. GLC yield over three consecutive cycles in the presence of 4%Au(H_2)@MOF1 (black bars) and 4%Au@DBU-MOF1 (red bars).

However, the loss of activity observed for the DBU-based catalyst is believed to be due to the loss of nanoparticles from the MOF leading to a decrease in AuNP content during subsequent carboxylation cycles [77]. This is in agreement with the SEM and TEM images (Figures 3B and 3C, respectively), where the particles are seen mainly on the surface of the MOF.

During the last three decades, AuNPs have been used as model catalysts in many reactions, and the results and ideas tested on AuNPs have usually been applied to develop new families of catalysts based on less-expensive metals and their bimetallic combinations. Here, we have demonstrated for the first time that AuNPs@MOF can catalyze the reaction of GL with CO_2. The main advantage of the catalytic system designed and prepared by us is that a small amount of Au catalyst is required to obtain a reasonable yield, resulting in a TOF that is higher than other catalytic systems reported previously [16,17] by almost two orders of magnitude (>78 h^{-1}) (see Scheme 1). Another advantage of the AuNPs@MOF composite is that it requires a low overpressure of 1.5 MPa, compared to 4–8 MPa in other systems [4,10,78], which can be attributed to a synergy of the MOF's exceptional ability to adsorb CO_2 and the presence of AuNPs embedded within the MOF. The formation of such unique catalytic centers is fundamentally essential for the future optimization of industrial catalysts. The proposed AuNPs@MOF catalytic system provides a relatively high productivity that is accompanied by a low E-factor, i.e., (mass of waste)/(mass of desired product), because water is the only by-product (the dehydrating agent can be recycled). Taking into account that the other catalytic systems require a long activation time or calcination treatments at elevated temperatures (up to 600 °C), the method of AuNPs @ MOF formation proposed in this work is time and energy saving: preparation in aqueous solution, activation at low temperature, and relatively simple recovery make it a very promising catalytic system that is in line with the assumptions of green chemistry and sustainable processing [79,80]. The catalyst system uses a very dilute amount of Au (0.0064 mmol) and Ce (ca. 0.2 mmol) to yield a high amount of GLC (ca. 50%) based on a calculation made with the use of ICP-MS quantification analysis for 50 mg of catalyst used in the carbonylation of 12.5 mmol of GL.

The catalytic system (Au-based) developed and optimized by us was extended to robust and sustainable systems containing less-expensive metals, i.e., copper and silver (AgNPs and CuNPs). We obtained 25 and 33% GL conversions and 17 and 26% yields for Cu and Ag, respectively. The results are promising given that the reaction system was not optimized for these two metals since the scope of this study was centered on Au.

Higher GLC yields were obtained using other catalytic systems; for example, Lim et al. reported a high GLC yield (86%) using a metal-free catalyst (DBU and CH_2Br_2) [81]. They applied very mild conditions (70 °C, 1 MPa); however, CH_2Br_2 is expensive and brominated by-products are formed. Another catalytic system based on divinylbenzene 3-butyl-1-vinylimidazolium bromide polymer allows for GLC to be obtained at a yield of 81% [82]; however, the relatively high content of bromide in the polymer (more than 17%), the large amounts of chloroform used, and the complicated and time-consuming synthesis make this method uncompetitive in the current shift to green synthesis.

4. Conclusions

In this research, we developed the idea of a catalytic system assembled from two morphologically different solid species, namely metal organic frameworks and nanoparticles, and we tested the applicability of such a system to the carboxylation of glycerol. Several new catalytic systems based on cerium carboxylate MOFs with embedded AuNPs were successfully used to react glycerol with CO_2 to form glycerol carbonate as the main product. The most efficient were two catalysts prepared from pre-synthesized MOF impregnated with Au^{3+} salt and subsequently reduced to AuNPs using H_2 (catalyst 4%Au(H_2)@MOF1) or reduced with $NaBH_4$ (catalyst 4%Au@PEI-MOF1). Although the yield of the process is still not competitive when compared to existing systems, the great and most promising advantage is the highest TOF (>78 h^{-1}) ever reported for this reaction, which demonstrates the much higher performance compared to already known systems. We also tried to test the applicability of the optimal system to crude bioglycerol because it is more environmentally and economically sustainable but hitherto has been ignored. The conversion obtained is promising, but the system needs optimization to obtain significant GLC yields. Other advantages of such hybrid catalysts include the fast and straightforward fabrication of AuNPs@MOF at a much lower temperature than currently used catalysts and mild conditions of operation (they work at the lowest CO_2 overpressure ever reported). The developed catalyst system was finally applied to Ag and Cu with promising results (26 and 17% GLC yields, respectively).

For the thermodynamically demanding reaction of CO_2 with glycerol, the equilibrium has to be shifted toward the products by applying drying agents that drive water out of the reaction mixture, and we found that an inorganic dehydrating agent ($MgCO_3$) is more effective than adiponitrile and acetonitrile. The use of the inorganic drying agent brings an additional advantage because, unlike most other methods, no by-product is obtained. The presented preliminary data are promising, and further efforts are planned to ensure the more effective encapsulation of different metal NPs inside MOFs to enable better yields, longer catalyst life, and more-efficient recovery after the reaction.

Supplementary Materials: The following supporting information can be downloaded at: https://www.mdpi.com/article/10.3390/nano14080650/s1. Description of chemicals; calculation of yield, conversion, TOF, and TON; Table S1: SEM image of 4%Au(H_2)@MOF1; Figure S1: SEM images showing small MOF nanoparticles of less than 100 nm (A) and large octahedral particles surrounded by aggregates (small particles) (B); Figure S2: FTIR of the synthesized MOF and catalysts materials.; Figure S3: Literature XRD patterns of two MOFs and their derivatives; Figure S4: SEM image of 4%Au(H_2)@MOF1; Figure S5: SEM-EDS elemental maps for 4%Au(H_2)@MOF1; Figure S6: EDS elemental composition for 4%Au(H_2)@MOF1; Figure S7: Gas chromatogram of silylated glycerol and glycerol carbonate; Figure S8: Survey XPS spectra of 4%Au(H_2)@MOF1. Refs [83,84] are cited in Supplementary Materials.

Author Contributions: Conceptualization, S.L. and A.K.-S.; methodology, S.L., A.K.-S. and M.W.; validation, S.L. and A.K.-S.; formal analysis, G.L., S.L., A.K.-S. and M.W.; investigation, S.L. and A.K.-S.; synthesis of catalysts, S.L. and A.K.-S.; writing—original draft preparation, G.L., S.L., A.K.-S. and M.W.; writing—review and editing, G.L., S.L., A.K.-S. and M.W.; visualization, S.L.; supervision, G.L. and A.K.-S.; project administration, A.K.-S.; funding acquisition, S.L. and G.L. All authors have read and agreed to the published version of the manuscript.

Funding: Financial support from the University of Warsaw IDUB Excellence Initiative—Research University Microgrant PSP-501-D112-20-0004316 and Ph.D. Scholarship for S.L. are acknowledged.

Data Availability Statement: Data are contained within the article and Supplementary Materials.

Acknowledgments: S.L. expresses his gratitude to Elżbieta Megiel for her help and assistance during the initial stage of the research. The authors thank Michał Chmielewski for performing gas physisorption for surface area and pore measurements free of charge, and thank Laura C. Folkers of STOE & Cie GmbH, Germany, for performing the initial XRD measurements for the MOF catalysts free of charge on their STOE XRD machines.

Conflicts of Interest: The authors declare no conflicts of interest.

References

1. Monteiro, M.R.; Kugelmeier, C.L.; Pinheiro, R.S.; Batalha, M.O.; da Silva César, A. Glycerol from biodiesel production: Technological paths for sustainability. *Renew. Sustain. Energy Rev.* **2018**, *88*, 109–122. [CrossRef]
2. Sonnati, M.O.; Amigoni, S.; de Givenchy, E.P.T.; Darmanin, T.; Choulet, O.; Guittard, F. Glycerol carbonate as a versatile building block for tomorrow: Synthesis, reactivity, properties and applications. *Green Chem.* **2013**, *15*, 283–306. [CrossRef]
3. Christy, S.; Noschese, A.; Lomelí-Rodriguez, M.; Greeves, N.; Lopez-Sanchez, J.A. Recent progress in the synthesis and applications of glycerol carbonate. *Curr. Opin. Green Sustain. Chem.* **2018**, *14*, 99–107. [CrossRef]
4. Lukato, S.; Kasozi, G.N.; Naziriwo, B.; Tebandeke, E. Glycerol carbonylation with CO_2 to form glycerol carbonate: A review of recent developments and challenges. *Curr. Res. Green Sustain. Chem.* **2021**, *4*, 100199. [CrossRef]
5. Tundo, P.; Musolino, M.; Aricò, F. The reactions of dimethyl carbonate and its derivatives. *Green Chem.* **2018**, *20*, 28–85. [CrossRef]

6. Zhou, C.-H.C.; Beltramini, J.N.; Fan, Y.-X.; Lu, G.M. Chemoselective catalytic conversion of glycerol as a biorenewable source to valuable commodity chemicals. *Chem. Soc. Rev.* **2008**, *37*, 527–549. [CrossRef] [PubMed]
7. Ochoa-Gómez, J.R.; Gómez-Jiménez-Aberasturi, O.; Maestro-Madurga, B.; Pesquera-Rodríguez, A.; Ramírez-López, C.; Lorenzo-Ibarreta, L.; Torrecilla-Soria, J.; Villarán-Velasco, M.C. Synthesis of glycerol carbonate from glycerol and dimethyl carbonate by transesterification: Catalyst screening and reaction optimization. *Appl. Catal. A Gen.* **2009**, *366*, 315–324. [CrossRef]
8. Ochoa-Gómez, J.R.; Gómez-Jiménez-Aberasturi, O.; Ramirez-Lopez, C.; Belsué, M. A brief review on industrial alternatives for the manufacturing of glycerol carbonate, a green chemical. *Org. Process Res. Dev.* **2012**, *16*, 389–399. [CrossRef]
9. Su, X.; Lin, W.; Cheng, H.; Zhang, C.; Wang, Y.; Yu, X.; Wu, Z.; Zhao, F. Metal-free catalytic conversion of CO_2 and glycerol to glycerol carbonate. *Green Chem.* **2017**, *19*, 1775–1781. [CrossRef]
10. Rozulan, N.; Halim, S.A.; Razali, N.; Lam, S.S. A review on direct carboxylation of glycerol waste to glycerol carbonate and its applications. *Biomass Convers. Biorefinery* **2022**, *12*, 4665–4682. [CrossRef]
11. Procopio, D.; Di Gioia, M.L. An Overview of the Latest Advances in the Catalytic Synthesis of Glycerol Carbonate. *Catalysts* **2022**, *12*, 50. [CrossRef]
12. Aresta, M.; Dibenedetto, A.; Nocito, F.; Pastore, C. A study on the carboxylation of glycerol to glycerol carbonate with carbon dioxide: The role of the catalyst, solvent and reaction conditions. *J. Mol. Catal. A Chem.* **2006**, *257*, 149–153. [CrossRef]
13. George, J.; Patel, Y.; Pillai, S.M.; Munshi, P. Methanol assisted selective formation of 1, 2-glycerol carbonate from glycerol and carbon dioxide using $^{n}Bu_2SnO$ as a catalyst. *J. Mol. Catal. A Chem.* **2009**, *304*, 1–7. [CrossRef]
14. Ozorio, L.P.; Pianzolli, R.; da Cruz Machado, L.; Miranda, J.L.; Turci, C.C.; Guerra, A.C.; Souza-Aguiar, E.F.; Mota, C.J. Metal-impregnated zeolite Y as efficient catalyst for the direct carbonation of glycerol with CO_2. *Appl. Catal. A Gen.* **2015**, *504*, 187–191. [CrossRef]
15. Ozorio, L.P.; Mota, C.J. Direct carbonation of glycerol with CO_2 catalyzed by metal oxides. *ChemPhysChem* **2017**, *18*, 3260–3265. [CrossRef]
16. Kulal, N.; Vetrivel, R.; Ganesh Krishna, N.S.; Shanbhag, G.V. Zn-Doped CeO_2 Nanorods for Glycerol Carbonylation with CO_2. *ACS Appl. Nano Mater.* **2021**, *4*, 4388–4397. [CrossRef]
17. Liu, J.; Li, Y.; Zhang, J.; He, D. Glycerol carbonylation with CO_2 to glycerol carbonate over CeO_2 catalyst and the influence of CeO_2 preparation methods and reaction parameters. *Appl. Catal. A Gen.* **2016**, *513*, 9–18. [CrossRef]
18. Narayan, N.; Meiyazhagan, A.; Vajtai, R. Metal nanoparticles as green catalysts. *Materials* **2019**, *12*, 3602. [CrossRef]
19. Dhakshinamoorthy, A.; Garcia, H. Catalysis by metal nanoparticles embedded on metal–organic frameworks. *Chem. Soc. Rev.* **2012**, *41*, 5262–5284. [CrossRef] [PubMed]
20. Yang, Q.; Xu, Q.; Jiang, H.-L. Metal–organic frameworks meet metal nanoparticles: Synergistic effect for enhanced catalysis. *Chem. Soc. Rev.* **2017**, *46*, 4774–4808. [CrossRef] [PubMed]
21. Nguyen, H.T.; Tran, Y.; Nguyen, H.N.; Nguyen, T.C.; Gándara, F.; Nguyen, P.T. A series of metal–organic frameworks for selective CO_2 capture and catalytic oxidative carboxylation of olefins. *Inorg. Chem.* **2018**, *57*, 13772–13782. [CrossRef] [PubMed]
22. Li, P.-Z.; Wang, X.-J.; Liu, J.; Lim, J.S.; Zou, R.; Zhao, Y. A Triazole-Containing Metal–Organic Framework as a Highly Effective and Substrate Size-Dependent Catalyst for CO_2 Conversion. *J. Am. Chem. Soc.* **2016**, *138*, 2142–2145. [CrossRef] [PubMed]
23. Mousavi, B.; Chaemchuen, S.; Phatanasri, S.; Chen, C.; Zeng, C.; Ganguly, R.; Zhuiykov, S.; Verpoort, F. Selective cyclodimerization of epichlorohydrin to dioxane derivatives over MOFs. *Arab. J. Chem.* **2020**, *13*, 1088–1093. [CrossRef]
24. Farrusseng, D.; Tuel, A. Perspectives on zeolite-encapsulated metal nanoparticles and their applications in catalysis. *New J. Chem.* **2016**, *40*, 3933–3949. [CrossRef]
25. Ndolomingo, M.J.; Bingwa, N.; Meijboom, R. Review of supported metal nanoparticles: Synthesis methodologies, advantages and application as catalysts. *J. Mater. Sci.* **2020**, *55*, 6195–6241. [CrossRef]
26. Cao, A.; Lu, R.; Veser, G. Stabilizing metal nanoparticles for heterogeneous catalysis. *Phys. Chem. Chem. Phys.* **2010**, *12*, 13499–13510. [CrossRef] [PubMed]
27. Zhu, Q.-L.; Xu, Q. Immobilization of ultrafine metal nanoparticles to high-surface-area materials and their catalytic applications. *Chem* **2016**, *1*, 220–245. [CrossRef]
28. Zanon, A.; Verpoort, F. Metals@ ZIFs: Catalytic applications and size selective catalysis. *Coord. Chem. Rev.* **2017**, *353*, 201–222. [CrossRef]
29. Gao, C.; Lyu, F.; Yin, Y. Encapsulated metal nanoparticles for catalysis. *Chem. Rev.* **2020**, *121*, 834–881. [CrossRef] [PubMed]
30. Lesiak, P.; Bednarska, K.; Lewandowski, W.; Wójcik, M.; Polakiewicz, S.; Bagiński, M.; Osuch, T.; Markowski, K.; Orzechowski, K.; Makowski, M.; et al. Self-Organized, One-Dimensional Periodic Structures in a Gold Nanoparticle-Doped Nematic Liquid Crystal Composite. *ACS Nano* **2019**, *13*, 10154–10160. [CrossRef]
31. Du, Y.; Sheng, H.; Astruc, D.; Zhu, M. Atomically Precise Noble Metal Nanoclusters as Efficient Catalysts: A Bridge between Structure and Properties. *Chem. Rev.* **2020**, *120*, 526–622. [CrossRef] [PubMed]
32. Kehoe, A.B.; Scanlon, D.O.; Watson, G.W. Role of lattice distortions in the oxygen storage capacity of divalently doped CeO_2. *Chem. Mater.* **2011**, *23*, 4464–4468. [CrossRef]
33. Misch, L.M.; Kurzman, J.A.; Derk, A.R.; Kim, Y.-I.; Seshadri, R.; Metiu, H.; McFarland, E.W.; Stucky, G.D. C–H bond activation by Pd-substituted CeO_2: Substituted ions versus reduced species. *Chem. Mater.* **2011**, *23*, 5432–5439. [CrossRef]
34. Tao, Y.; Li, Z.; Ju, E.; Ren, J.; Qu, X. Polycations-functionalized water-soluble gold nanoclusters: A potential platform for simultaneous enhanced gene delivery and cell imaging. *Nanoscale* **2013**, *5*, 6154–6160. [CrossRef] [PubMed]

35. Howarth, A.J.; Peters, A.W.; Vermeulen, N.A.; Wang, T.C.; Hupp, J.T.; Farha, O.K. Best Practices for the Synthesis, Activation, and Characterization of Metal–Organic Frameworks. *Chem. Mater.* **2017**, *29*, 26–39. [CrossRef]
36. Aijaz, A.; Karkamkar, A.; Choi, Y.J.; Tsumori, N.; Rönnebro, E.; Autrey, T.; Shioyama, H.; Xu, Q. Immobilizing highly catalytically active Pt nanoparticles inside the pores of metal–organic framework: A double solvents approach. *J. Am. Chem. Soc.* **2012**, *134*, 13926–13929. [CrossRef] [PubMed]
37. Imperor-Clerc, M.; Bazin, D.; Appay, M.-D.; Beaunier, P.; Davidson, A. Crystallization of β-MnO_2 Nanowires in the Pores of SBA-15 Silicas: In Situ Investigation Using Synchrotron Radiation. *Chem. Mater.* **2004**, *16*, 1813–1821. [CrossRef]
38. Shichibu, Y.; Negishi, Y.; Tsunoyama, H.; Kanehara, M.; Teranishi, T.; Tsukuda, T. Extremely high stability of glutathionate-protected Au25 clusters against core etching. *Small* **2007**, *3*, 835–839. [CrossRef] [PubMed]
39. Kawasaki, H.; Hamaguchi, K.; Osaka, I.; Arakawa, R. ph-Dependent Synthesis of Pepsin-Mediated Gold Nanoclusters with Blue Green and Red Fluorescent Emission. *Adv. Funct. Mater.* **2011**, *21*, 3508–3515. [CrossRef]
40. Li, L.; Liu, W.; Chen, R.; Shang, S.; Zhang, X.; Wang, H.; Zhang, H.; Ye, B.; Xie, Y. Atom-Economical Synthesis of Dimethyl Carbonate from CO_2: Engineering Reactive Frustrated Lewis Pairs on Ceria with Vacancy Clusters. *Angew. Chem. Int. Ed.* **2022**, *61*, e202214490. [CrossRef] [PubMed]
41. Farrag, M. In situ preparation of palladium nanoclusters in cerium metal-organic frameworks Ce-MOF-808, Ce-UiO-66 and Ce-BTC as nanoreactors for room temperature Suzuki cross-coupling reaction. *Microporous Mesoporous Mater.* **2021**, *312*, 110783. [CrossRef]
42. Sun, W.; Li, X.; Sun, C.; Huang, Z.; Xu, H.; Shen, W. Insights into the pyrolysis processes of Ce-MOFs for preparing highly active catalysts of toluene combustion. *Catalysts* **2019**, *9*, 682. [CrossRef]
43. Cheng, X.Q.; Li, S.; Bao, H.; Yang, X.; Li, Z.; Zhang, Y.; Wang, K.; Ma, J.; Ullah, A.; Shao, L. Poly (sodium-p-styrenesulfonate)-grafted UiO-66 composite membranes boosting highly efficient molecular separation for environmental remediation. *Adv. Compos. Hybrid Mater.* **2021**, *4*, 562–573. [CrossRef]
44. Stawowy, M.; Róziewicz, M.; Szczepańska, E.; Silvestre-Albero, J.; Zawadzki, M.; Musioł, M.; Łuzny, R.; Kaczmarczyk, J.; Trawczyński, J.; Łamacz, A. The impact of synthesis method on the properties and CO_2 sorption capacity of UiO-66 (Ce). *Catalysts* **2019**, *9*, 309. [CrossRef]
45. Lu, G.-P.; Li, X.; Zhong, L.; Li, S.; Chen, F. Ru@ UiO-66 (Ce) catalyzed acceptorless dehydrogenation of primary amines to nitriles: The roles of Lewis acid–base pairs in the reaction. *Green Chem.* **2019**, *21*, 5386–5393. [CrossRef]
46. Cavka, J.H.; Jakobsen, S.; Olsbye, U.; Guillou, N.; Lamberti, C.; Bordiga, S.; Lillerud, K.P. A new zirconium inorganic building brick forming metal organic frameworks with exceptional stability. *J. Am. Chem. Soc.* **2008**, *130*, 13850–13851. [CrossRef] [PubMed]
47. Caratelli, C.; Hajek, J.; Cirujano, F.G.; Waroquier, M.; i Xamena, F.X.L.; Van Speybroeck, V. Nature of active sites on UiO-66 and beneficial influence of water in the catalysis of Fischer esterification. *J. Catal.* **2017**, *352*, 401–414. [CrossRef]
48. Atzori, C.; Lomachenko, K.A.; Øien-Ødegaard, S.; Lamberti, C.; Stock, N.; Barolo, C.; Bonino, F. Disclosing the Properties of a New Ce(III)-Based MOF: $Ce_2(NDC)_3(DMF)_2$. *Cryst. Growth Des.* **2019**, *19*, 787–796. [CrossRef]
49. He, J.; Xu, Y.; Wang, W.; Hu, B.; Wang, Z.; Yang, X.; Wang, Y.; Yang, L. Ce(III) nanocomposites by partial thermal decomposition of Ce-MOF for effective phosphate adsorption in a wide pH range. *Chem. Eng. J.* **2020**, *379*, 122431. [CrossRef]
50. Bakuru, V.R.; Velaga, B.; Peela, N.R.; Kalidindi, S.B. Hybridization of Pd Nanoparticles with UiO-66(Hf) Metal-Organic Framework and the Effect of Nanostructure on the Catalytic Properties. *Chem.—A Eur. J.* **2018**, *24*, 15978–15982. [CrossRef]
51. Zhai, Y.; Li, Y.; Hou, Q.; Zhang, Y.; Zhou, E.; Li, H.; Ai, S. Highly sensitive colorimetric detection and effective adsorption of phosphate based on MOF-808 (Zr/Ce). *New J. Chem.* **2022**, *46*, 15405–15413. [CrossRef]
52. Cai, X.; Pan, J.; Tu, G.; Fu, Y.; Zhang, F.; Zhu, W. Pd/UiO-66(Hf): A highly efficient heterogeneous catalyst for the hydrogenation of 2,3,5-trimethylbenzoquinone. *Catal. Commun.* **2018**, *113*, 23–26. [CrossRef]
53. Turner, S.; Lebedev, O.I.; Schröder, F.; Esken, D.; Fischer, R.A.; Tendeloo, G.V. Direct Imaging of Loaded Metal−Organic Framework Materials (Metal@MOF-5). *Chem. Mater.* **2008**, *20*, 5622–5627. [CrossRef]
54. Zhang, J.; Cheng, N.; Ge, B. Characterization of metal-organic frameworks by transmission electron microscopy. *Adv. Phys. X* **2022**, *7*, 2046157. [CrossRef]
55. Liu, L.; Zhang, D.; Zhu, Y.; Han, Y. Bulk and local structures of metal–organic frameworks unravelled by high-resolution electron microscopy. *Commun. Chem.* **2020**, *3*, 99. [CrossRef] [PubMed]
56. Zhang, X.; Hou, F.; Yang, Y.; Wang, Y.; Liu, N.; Chen, D.; Yang, Y. A facile synthesis for cauliflower like CeO_2 catalysts from Ce-BTC precursor and their catalytic performance for CO oxidation. *Appl. Surf. Sci.* **2017**, *423*, 771–779. [CrossRef]
57. Thommes, M.; Kaneko, K.; Neimark, A.V.; Olivier, J.P.; Rodriguez-Reinoso, F.; Rouquerol, J.; Sing, K.S.W. Physisorption of gases, with special reference to the evaluation of surface area and pore size distribution (IUPAC Technical Report). *Pure Appl. Chem.* **2015**, *87*, 1051–1069. [CrossRef]
58. Lee, Y.; He, G.; Akey, A.J.; Si, R.; Flytzani-Stephanopoulos, M.; Herman, I.P. Raman analysis of mode softening in nanoparticle $CeO_{2-\delta}$ and Au-$CeO_{2-\delta}$ during CO oxidation. *J. Am. Chem. Soc.* **2011**, *133*, 12952–12955. [CrossRef] [PubMed]
59. Huang, P.; Wu, F.; Zhu, B.; Gao, X.; Zhu, H.; Yan, T.; Huang, W.; Wu, S.; Song, D. CeO_2 nanorods and gold nanocrystals supported on CeO_2 nanorods as catalyst. *J. Phys. Chem. B* **2005**, *109*, 19169–19174. [CrossRef] [PubMed]
60. Xu, Q.; Kharas, K.C.; Datye, A. The preparation of highly dispersed Au/Al_2O_3 by aqueous impregnation. *Catal. Lett.* **2003**, *85*, 229–235. [CrossRef]

61. Lopez-Sanchez, J.A.; Dimitratos, N.; Hammond, C.; Brett, G.L.; Kesavan, L.; White, S.; Miedziak, P.; Tiruvalam, R.; Jenkins, R.L.; Carley, A.F. Facile removal of stabilizer-ligands from supported gold nanoparticles. *Nat. Chem.* **2011**, *3*, 551–556. [CrossRef] [PubMed]
62. Zhou, Z.; Kooi, S.; Flytzani-Stephanopoulos, M.; Saltsburg, H. The role of the interface in CO oxidation on Au/CeO2 multilayer nanotowers. *Adv. Funct. Mater.* **2008**, *18*, 2801–2807. [CrossRef]
63. Camellone, M.F.; Fabris, S. Reaction mechanisms for the CO oxidation on Au/CeO_2 catalysts: Activity of substitutional Au^{3+}/Au^{+} cations and deactivation of supported Au^{+} adatoms. *J. Am. Chem. Soc.* **2009**, *131*, 10473–10483. [CrossRef] [PubMed]
64. Abad, A.; Concepción, P.; Corma, A.; García, H. A Collaborative Effect between Gold and a Support Induces the Selective Oxidation of Alcohols. *Angew. Chem. Int. Ed.* **2005**, *44*, 4066–4069. [CrossRef]
65. Crist, V.B. *Handbooks of Monochromatic XPS Spectra: The Elements and Native Oxides*; John Wiley & Sons: New York, NY, USA, 2000.
66. Si, R.; Flytzani-Stephanopoulos, M. Shape and crystal-plane effects of nanoscale ceria on the activity of Au-CeO_2 catalysts for the water–gas shift reaction. *Angew. Chem.* **2008**, *120*, 2926–2929. [CrossRef]
67. Zhou, Z.; Flytzani-Stephanopoulos, M.; Saltsburg, H. Decoration with ceria nanoparticles activates inert gold island/film surfaces for the CO oxidation reaction. *J. Catal.* **2011**, *280*, 255–263. [CrossRef]
68. Horváth, A.; Beck, A.; Stefler, G.r.; Benkó, T.; Sáfrán, G.R.; Varga, Z.; Gubicza, J.; Guczi, L. Silica-supported Au nanoparticles decorated by CeO_2: Formation, morphology, and CO oxidation activity. *J. Phys. Chem. C* **2011**, *115*, 20388–20398. [CrossRef]
69. Casaletto, M.P.; Longo, A.; Martorana, A.; Prestianni, A.; Venezia, A.M. XPS study of supported gold catalysts: The role of Au^{0} and $Au^{+\delta}$ species as active sites. *Surf. Interface Anal.* **2006**, *38*, 215–218. [CrossRef]
70. Li, H.; Jiao, X.; Li, L.; Zhao, N.; Xiao, F.; Wei, W.; Sun, Y.; Zhang, B. Synthesis of glycerol carbonate by direct carbonylation of glycerol with CO_2 over solid catalysts derived from Zn/Al/La and Zn/Al/La/M (M = Li, Mg and Zr) hydrotalcites. *Catal. Sci. Technol.* **2015**, *5*, 989. [CrossRef]
71. Shen, Q.; Yan, H.; Yuan, X.; Li, R.; Kong, D.; Zhang, W.; Zhang, H.; Liu, Y.; Chen, X.; Feng, X. Tailoring morphology of MgO catalyst for the enhanced coupling reaction of CO_2 and glycerol to glycerol carbonate. *Fuel* **2023**, *335*, 126972. [CrossRef]
72. Takeuchi, K.; Matsumoto, K.; Fukaya, N.; Sato, K.; Choi, J.-C. Synthesis of Glycerol Carbonate from Glycerol and CO_2 Using CaO as a Dehydrating Agent. *Asian J. Org. Chem.* **2022**, *11*, E202200212. [CrossRef]
73. Razali, N.; McGregor, J. Improving product yield in the direct carboxylation of glycerol with CO_2 through the tailored selection of dehydrating agents. *Catalysts* **2021**, *11*, 138. [CrossRef]
74. Tomishige, K.; Gu, Y.; Nakagawa, Y.; Tamura, M. Reaction of CO_2 With Alcohols to Linear-, Cyclic-, and Poly-Carbonates Using CeO_2-Based Catalysts. *Front. Energy Res.* **2020**, *8*, 117. [CrossRef]
75. Somorjai, G.A.; Park, J.Y. Molecular Factors of Catalytic Selectivity. *Angew. Chem. Int. Ed.* **2008**, *47*, 9212–9228. [CrossRef] [PubMed]
76. Zhang, M.; Liu, Q.; Long, H.; Sun, L.; Murayama, T.; Qi, C. Insights into Au nanoparticle size and chemical state of Au/ZSM-5 catalyst for catalytic cracking of n-octane to increase propylene production. *J. Phys. Chem. C* **2021**, *125*, 16013–16023. [CrossRef]
77. Wang, S.; Wang, J.; Sun, P.; Xu, L.; Okoye, P.U.; Li, S.; Zhang, L.; Guo, A.; Zhang, J.; Zhang, A. Disposable baby diapers waste derived catalyst for synthesizing glycerol carbonate by the transesterification of glycerol with dimethyl carbonate. *J. Clean. Prod.* **2019**, *211*, 330–341. [CrossRef]
78. Koranian, P.; Dalai, A.K.; Sammynaiken, R. Production of glycerol carbonate from glycerol and carbon dioxide using metal oxide catalysts. *Chem. Eng. Sci.* **2024**, *286*, 119687. [CrossRef]
79. Sheldon, R.A. The E factor 25 years on: The rise of green chemistry and sustainability. *Green Chem.* **2017**, *19*, 18–43. [CrossRef]
80. Constable, D.J.; Curzons, A.D.; Cunningham, V.L. Metrics to 'green'chemistry—Which are the best? *Green Chem.* **2002**, *4*, 521–527. [CrossRef]
81. Lim, Y.N.; Lee, C.; Jang, H.-Y. Metal-Free Synthesis of Cyclic and Acyclic Carbonates from CO_2 and Alcohols. *Eur. J. Org. Chem.* **2014**, *2014*, 1823–1826. [CrossRef]
82. Song, X.; Wu, Y.; Pan, D.; Zhang, J.; Xu, S.; Gao, L.; Wei, R.; Xiao, G. Functionalized DVB-based polymer catalysts for glycerol and CO_2 catalytic conversion. *J. CO2 Util.* **2018**, *28*, 326–334. [CrossRef]
83. Liu, H.; Li, Q.-Q.; Zhou, L.; Deng, B.; Pan, P.-H.; Zhao, S.-Y.; Liu, P.; Wang, Y.-Y.; Li, J.-L. Confinement of Organic Dyes in UiO-66-Type Metal–Organic Frameworks for the Enhanced Synthesis of [1,2,5]Thiadiazole[3,4-g]benzoimidazoles. *J. Am. Chem. Soc.* **2023**, *145*, 17588–17596. [CrossRef] [PubMed]
84. Melillo, A.; Cabrero-Antonino, M.; Ferrer, B.; Dhakshinamoorthy, A.; Baldoví, H.G.; Navalón, S. MOF-on-MOF Composites with UiO-66-Based Materials as Photocatalysts for the Overall Water Splitting under Sunlight Irradiation. *Energy Fuels* **2023**, *37*, 5457–5468. [CrossRef]

nanomaterials

MDPI

Communication

Stability of Zr-Based UiO-66 Metal–Organic Frameworks in Basic Solutions

Jun Yeong Kim [1,†], Jiwon Kang [1,†], Seungheon Cha [1], Haein Kim [1], Dopil Kim [1], Houng Kang [2], Isaac Choi [1,*] and Min Kim [1,*]

[1] Department of Chemistry, Chungbuk National University, Cheongju 28644, Republic of Korea; junyeongkim1111@gmail.com (J.Y.K.); kangjiwon1001@gmail.com (J.K.); chaseungheon045@gmail.com (S.C.); haeinkim9951@gmail.com (H.K.); ehvlfdl0218@gmail.com (D.K.)

[2] Department of Chemistry Education, Chungbuk National University, Cheongju 28644, Republic of Korea; hkang@chungbuk.ac.kr

* Correspondence: isaac.choi@chungbuk.ac.kr (I.C.); minkim@chungbuk.ac.kr (M.K.)

† These authors contributed equally to this work.

Abstract: Although Zr-based metal–organic frameworks (MOFs) exhibit robust chemical and physical stability in the presence of moisture and acidic conditions, their susceptibility to nucleophilic attacks from bases poses a critical challenge to their overall stability. Herein, we systematically investigate the stability of Zr-based UiO-66 (UiO = University of Oslo) MOFs in basic solutions. The impact of 11 standard bases, including inorganic salts and organic bases, on the stability of these MOFs is examined. The destruction of the framework is confirmed through powder X-ray diffraction (PXRD) patterns, and the monitored dissolution of ligands from the framework is assessed using nuclear magnetic resonance (NMR) spectroscopy. Our key findings reveal a direct correlation between the strength and concentration of the base and the destruction of the MOFs. The summarized data provide valuable insights that can guide the practical application of Zr-based UiO-66 MOFs under basic conditions, offering essential information for their optimal utilization in various settings.

Keywords: metal–organic frameworks; zirconium; UiO-66; basic solution; stability

1. Introduction

Metal–organic frameworks (MOFs) represent a class of materials based on metal–ligand coordination, characterized by a three-dimensional structure and inherent porosity. The versatility of MOFs is underscored by the utilization of various metal salts and clusters coupled with a diverse array of coordinating ligands (e.g., carboxylate or pyridyl groups) employed in their synthesis. This diversity allows for a broad spectrum of MOF compositions, imparting distinctive chemical and physical properties to these materials [1–6].

One defining feature of MOFs is their predictable reticular chemistry, which enables the exploration of similar structures with varying ligand lengths and identical frameworks originating from identical coordinating sites but featuring different functional groups [7–10]. This aspect has been extensively investigated in the field of MOF chemistry. The porosity of MOFs has been exploited for efficient gas storage and molecular separation, leveraging sieving effects, and their unique structural characteristics have found applications in diverse fields [11–16]. The tunable pores of MOFs hold promise for selective molecular conveyance, such as drug delivery applications [17–19]. Furthermore, the recurring combinations of metal–ligand coordination lead to a distinctive catalytic efficacy in reactions, contribute to optical properties and have the potential for applications in energy-related processes [4,20,21]. However, the practical applications of MOFs have been hindered by their relatively low stability under moisture or acidic/basic conditions. Addressing these stability challenges is crucial for unlocking the full potential of these sophisticated materials for real-world applications in various industries as well as in daily life [22–27]. The stability of MOFs has been investigated through theoretical and experimental approaches [28–36], and the effects of buffers, amino acids, and cell media on MOFs have been extensively examined [37–39].

For this reason, diverse strategies for the increase in their stability have been employed during the preparation steps [23,40,41]. Notably, the nature of the metal and ligand, as elucidated by the hard–soft acid–base (HSAB) theory, is pivotal [42,43]. Additionally, introducing hydrophobic characteristics and safeguarding frameworks have been the subject of extensive investigation [44–46]. For instance, the choice of metal ions and ligands significantly influences stability. Applying the HSAB theory, MOFs featuring hard carboxylate ligands, particularly those anchored to zirconium (Zr, a hard metal ion), exhibit superior stability against water compared to MOFs that incorporate softer metal ions such as zinc [47,48]. Consequently, Zr-based MOFs have emerged as a focal point in a plethora of research, demonstrating enhanced stability and versatility for various applications, including industrial uses [49–54].

The pioneering work on Zr-based MOFs is exemplified by the UiO series, employing benzene-1,4-dicarboxylate (BDC) ligands, as reported by Lillerud et al. [55]. Subsequent studies have explored manipulating the connectivity of secondary building units and the coordination of ligands, transitioning from dicarboxylic acids to tricarboxylic and tetracarboxylic acids. The exceptional stability of Zr-based MOFs in water and acidic conditions, coupled with their diverse structural possibilities and relative ease of accessibility, position them as representative MOFs across various applications spanning diverse research fields [49–54].

Despite the numerous advantages of Zr-based MOFs, the persistent challenge of low stability under basic conditions remains unresolved. Recurrent nucleophilic attacks by hydroxide ions (OH^-) on Zr–carboxylate coordination bonds lead to the dissolution of Zr-based MOFs [56]. This instability is intricately linked to structural issues and the properties of the pore environments. Notably, Zr-based UiO-66 MOFs are commonly reported to be unstable under basic conditions [56], with a limitation level at pH 12 [41]. This inherent instability poses a significant obstacle, especially in the realm of catalytic applications in organic reactions, where bases are often used as essential additives. For instance, new bond-forming cross-coupling reactions, such as the Suzuki–Miyaura reaction and Buchwald–Hartwig amination, require a wide range of bases [57], translating the fact that the low stability of Zr-MOFs under basic conditions restricts their widespread use in applications in organic syntheses [58–62].

To address these concerns, our work focuses on the systematic study of Zr-MOFs in basic solutions to determine their stability. The UiO-66 system, which is based on the simplest BDC ligand, was chosen for investigation because of its ease of accessibility and high chemical stability. Our investigations encompass various inorganic and organic bases, and the insights gained from this study offer valuable information for the practical application of Zr-MOFs, particularly in scenarios involving basic conditions and catalytic applications.

2. Results and Discussion

2.1. Preparation of Zr-Based UiO-66 MOFs

UiO-66, the primary Zr-based MOF investigated in this study, was synthesized following Farha's solvothermal protocol [63]. The synthesis involved $ZrCl_4$ and H_2BDC (benzene-1,4-dicarboxylic acid or terephthalic acid) with formic acid modulators. Detail procedures are provided in the Supplementary Materials (p. S2). Acid modulators are commonly employed in the solvothermal synthesis of MOFs to regulate crystallization speed and enhance reproducibility. However, using acid modulators introduced structural defects into the pristine UiO-66 MOF, a characteristic arising from the consistent protocol employed [64–66].

The crystallinity and stability of the synthesized MOFs were initially validated using powder X-ray diffraction (PXRD) pattern analysis. The PXRD patterns of the as-synthesized MOFs were compared with a simulated pattern derived from a reported structure (Figure 1).

Any deviations, such as peak broadening in UiO-66, indicated framework destruction. The dissolution of ligands in the MOFs under basic conditions was monitored through physical mass measurements after basic treatment to further assess the stability. Additionally, ^{1}H nuclear magnetic resonance (NMR) spectroscopy data acquired after acid digestion provided insight into ligand dissolution. An internal standard method was employed to quantify the BDC ligands within the framework, utilizing 1,3,5-trimethoxy benzene as the NMR internal standard. Detail protocols for PXRD and NMR are provided in the Supplementary Materials (p. S2). Before subjecting UiO-66 to the basic treatment, the pristine MOFs exhibited 91% of the expected BDC ligands compared to the ideal structure (Figure 1). This 9% discrepancy could be attributed to structural defects resulting from the use of formic acid modulators. Further comparisons of the remaining BDC ligand amounts (by ^{1}H NMR) post-basic treatments shed light on the variations in stability. Notably, the introduction of additional functional groups to the ligand was not considered in this analysis. This omission was due to the potential impact of functional groups (e.g., nitro or tetramethyl groups) on MOF stability [41,67], which could introduce confounding variables into the study.

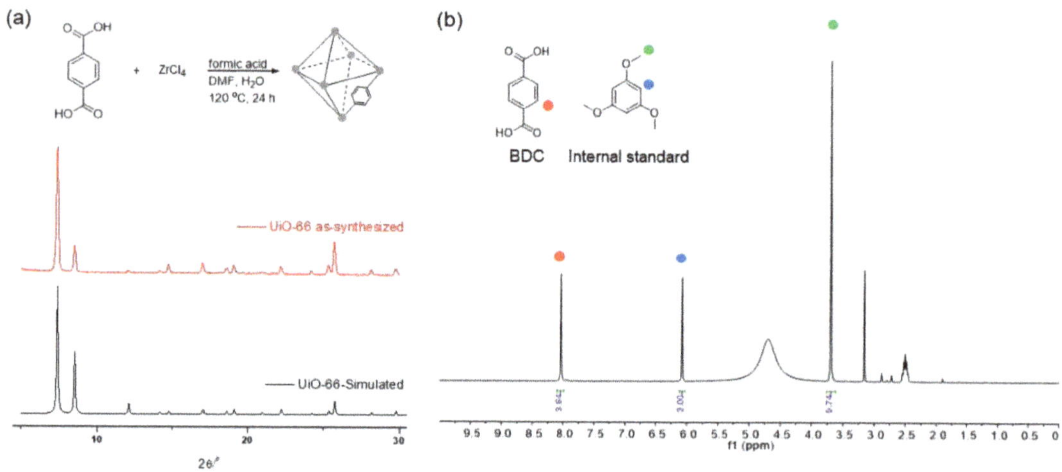

Figure 1. (**a**) Preparation of UiO-66 MOF and PXRD and (**b**) NMR data of pristine UiO-66 MOF (the digested sample) before base treatments.

2.2. *Effects of Inorganic Base Solutions on UiO-66 MOF*

To assess the stability of the UiO-66 MOF under basic conditions, we selected representative potassium salts, including KOH, K_3PO_4, K_2CO_3, $KHCO_3$, KOAc, and K_2SO_4. These inorganic bases were chosen to ensure a comprehensive evaluation while excluding cationic effects on MOF stability. The pK_b values of these bases ranged from −1.7 to 12 (Table 1) [68–70]. Initial stability tests involved incubating 30 mg of UiO-66 MOF in each 0.1 M aqueous basic solution for 1 h, followed by recovery through centrifugation and PXRD analysis after washing with pure methanol to remove trapped water molecules.

Table 1. pK_b value of tested inorganic bases toward UiO-66 MOF.

Base	KOH	K_3PO_4	K_2CO_3	$KHCO_3$	KOAc	K_2SO_4
pK_b	−1.7	1.6	4.1	7.2	9.2	12

As illustrated in Figure 2, the strongest base, KOH, led to the total destruction of the UiO-66 structure within 1 h. Conversely, weak bases, such as $KHCO_3$, KOAc, and K_2SO_4, generally retained their PXRD patterns after 0.1 M of the solution treatment. Therefore, detailed analyses were performed with K_3PO_4, $KHCO_3$, and KOAc by varying the concentration and treatment time.

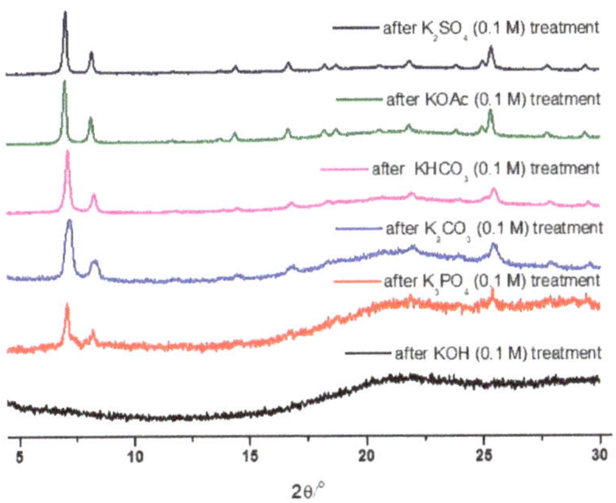

Figure 2. PXRD patterns of the recovered UiO-66 samples after the 0.1 M basic solution treatment.

The weak base, KOAc, demonstrated good tolerance for the UiO-66 MOF, with no changes in the PXRD patterns observed until treatment with a 1.0 M KOAc solution for 1 d. Structural decomposition commenced at a 2.0 M solution, and substantial destruction occurred with a 3.0 M KOAc solution for 1 h, leaving only 31% remaining BDC ligands in the solid-state MOF (Figure 3) compared to the ideal UiO-66 MOF. Since the starting MOF had 91% of BDC ligands in their structure (and 9% of the defect), 60% of the ligand was removed from the framework during base treatments.

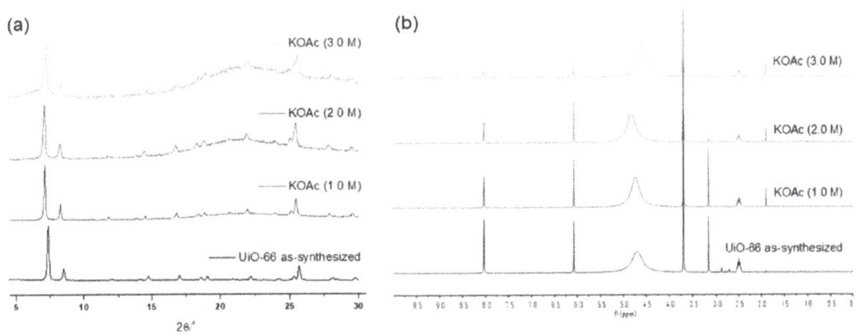

Figure 3. (**a**) PXRD patterns and (**b**) NMR spectral changes after treatment with the KOAc solution for 1 h.

Specific concentration effects were examined using medium-strength bases, $KHCO_3$ and K_3PO_4. $KHCO_3$ exhibited a PXRD peak that broadened at 0.2 M of the solution treatment for 1 h, with complete framework destruction confirmed at 0.3 M of solution incubation for 1 h, leaving less than 10% of the remaining ligands in the solid state (which meant an 81% loss from pristine MOF, Figure 4). Between K_3PO_4 and $KHCO_3$, the stronger base, K_3PO_4, annihilated the UiO-66 structure with a 0.05 M solution within 1 h, with almost no remaining BDC ligand, as confirmed via the ^1H NMR analysis after acid digestion (Figure 4).

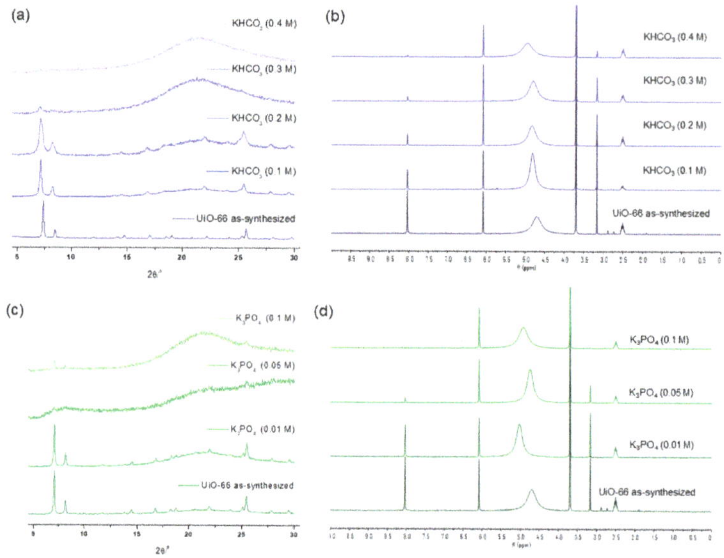

Figure 4. (**a**) PXRD patterns and (**b**) NMR spectral changes after $KHCO_3$ solution treatment for 1 h. (**c**) PXRD patterns and (**d**) NMR spectral changes after K_3PO_4 solution treatment for 1 h.

The concentration limits and remaining ligands are summarized in Figure 5. Notably, KOH exterminated all MOF solid materials with a 0.1 M solution for 1 h, while KOAc exhibited tolerance even with a 1.0 M solution for 1 d. K_3PO_4 led to over half the destruction with a 0.05 M solution for 1 h, and $KHCO_3$ resulted in similar framework destruction with a 0.2 M solution for 1 h of incubation. Consequently, it was confirmed that the acetate series is a promising inorganic base additive for using Zr-MOFs under basic conditions.

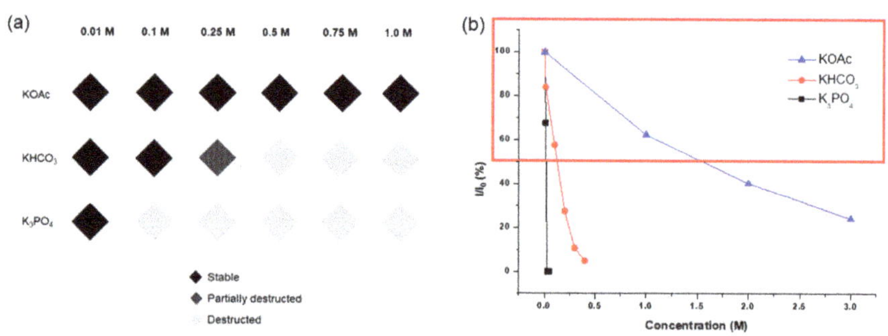

Figure 5. (**a**) Concentration limits for inorganic base treatments of UiO-66 and (**b**) the remaining BDC ligands in the recovered UiO-66 after each base treatment.

2.3. *Effects of Organic Base Solutions on UiO-66 MOF*

To assess the stability of UiO-66 MOFs under basic conditions, standard organic bases were examined, including DBU (1,8-diazabicyclo[5.4.0]undec-7-ene), Et_3N (triethylamine), $BnNH_2$ (benzylamine), DABCO (1,4-diazabicyclo[2.2.2]octane), and pyridine. Initial stability tests were conducted using 0.5 M of aqueous solutions, and the pK_b values of these organic bases ranged from 0.5 to 8.77 (Table 2) [71]. The strong bases, DBU and Et_3N, resulted in the complete destruction of the UiO-66 frameworks within 1 h of the 0.5 M concentration test (Figure 6). Furthermore, the concentration limits were explored using DBU, DABCO, and pyridine.

Table 2. pK_b values of the tested organic bases toward the UiO-66 MOF.

Base	DBU	Et_3N	$BnNH_2$	DABCO	Pyridine
pK_b	0.5	3.3	4.6	5.2	8.8

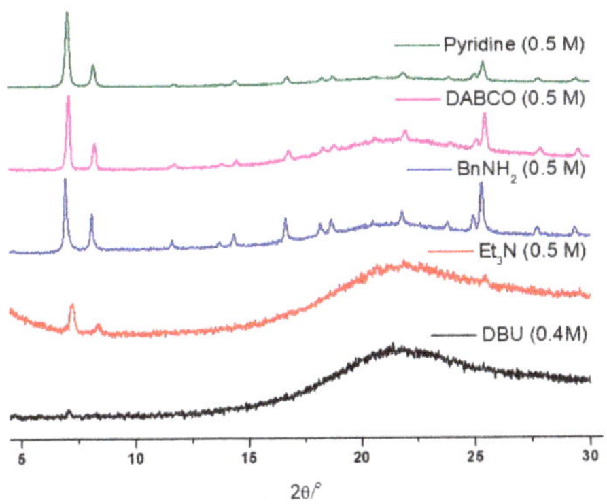

Figure 6. PXRD patterns of the recovered UiO-66 samples after the organic basic solution treatment.

The concentration variation revealed that 0.3 M was the limit for a one-hour treatment of UiO-66 MOFs with the DBU solution. Although all the primary peaks in the PXRD of UiO-66 were retained with a 0.2 M solution treatment, 0.3 M of the DBU solution removed all the primary peaks from the PXRD patterns (Figure 7). Meanwhile, the portion of the remaining BDC ligand in the solid state was drastically decreased in the 0.3 M DBU solution treatment (Figure 7). In contrast, DABCO, which is often used as a nitrogen donor ligand for MOF synthesis, showed good compatibility with the Zr-MOFs. Although some peak broadenings were observed, all primary peaks of the UiO-66 frameworks were completely retained. In addition, ligand dissolution was much lower than that in the DBU treatment (Figure 7). Notably, higher concentration tests (>0.5 M) were unsuccessful because of the low solubility of DBU and DABCO in water. Finally, in the case of the weakest pyridine, the PXRD pattern of the UiO-66 MOF retained its sharpness until the 1.0 M treatment, and the BDC ligands remained in their solid state (Figure 7).

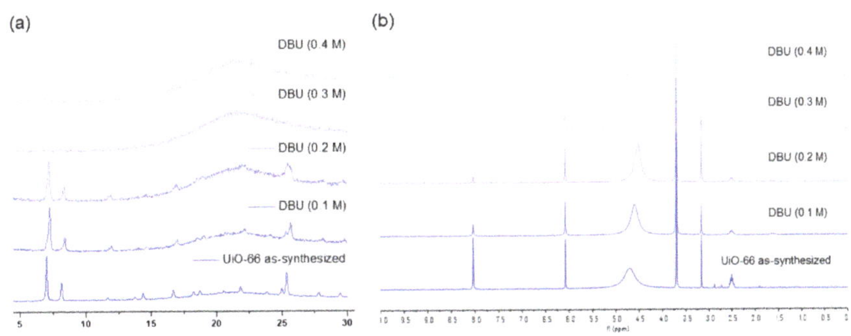

Figure 7. *Cont.*

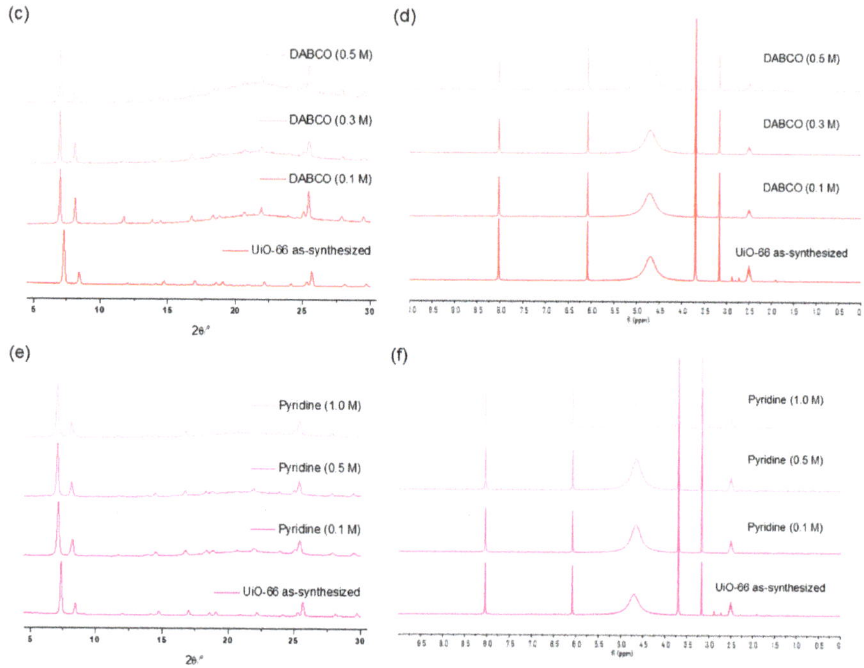

Figure 7. (**a**) PXRD patterns and (**b**) NMR spectral changes after DBU solution treatment for 1 h. (**c**) PXRD patterns and (**d**) NMR spectral changes after DABCO solution treatment for 1 h. (**e**) PXRD patterns and (**f**) NMR spectral changes after pyridine solution treatment for 1 h.

Six samples, each treated with 0.1 M of K_2CO_3, $KHCO_3$, and K_2SO_4 as inorganic bases and 0.5 M of $BnNH_2$, DABCO, and pyridine as organic bases, were carefully chosen for in-depth analysis to evaluate their morphology and porosity. The selection was based on the successful recovery of measurable and analyzable samples following the respective base treatments. Notably, the morphology of the recovered samples exhibited no discernible damage, as validated by the scanning electron microscope (SEM) images presented in Figure 8. Both the SEM images and the bulk crystallinity, as confirmed via the PXRD analysis depicted in Figure 7, indicated an overall influence on the entire MOF particles rather than the partial dissolution of specific MOF components.

Although the morphology of the MOF particles remained intact, the porosity experienced alterations due to the basic treatment. Gas adsorption experiments unveiled reductions in the N_2 uptake following base treatments, with stronger bases demonstrating a more pronounced impact on porosity than their weaker counterparts (Figure 9). Notably, the N_2 uptake underwent an approximate 50% decrease under the $BnNH_2$ treatment among the organic bases. A subsequent Brunauer–Emmett–Teller (BET) surface area analysis was performed. While the pristine UiO-66 MOF boasted a BET surface area of 1777 m^2/g, the recovered sample generally displayed a diminished surface area. Specifically, K_2CO_3 yielded the lowest surface area (524 m^2/g), and weak bases like $KHCO_3$, K_2SO_4, and pyridine had a minimal impact on the surface area (see Table 3).

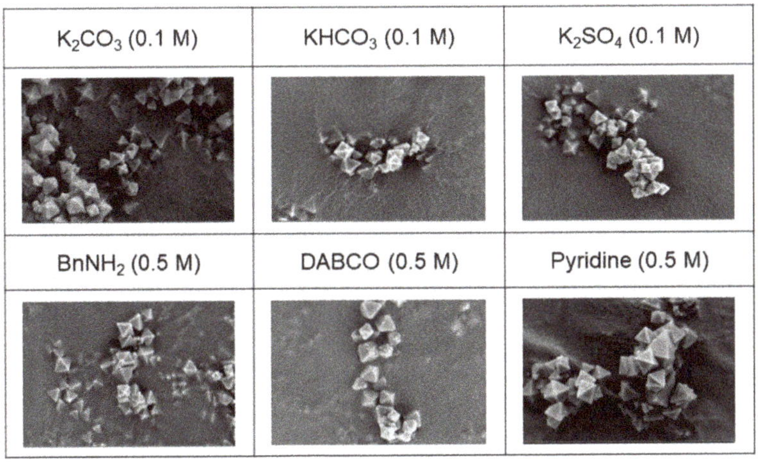

Figure 8. SEM images (×20,000) of recovered UiO-66 after base treatments.

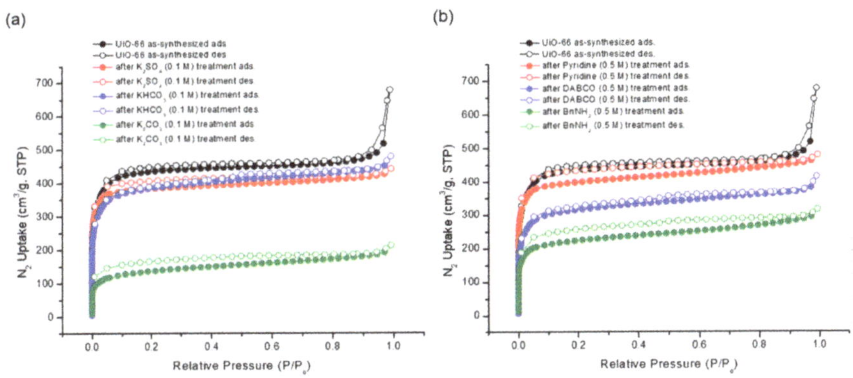

Figure 9. N_2 adsorption (at 77 K) experiments of (**a**) recovered UiO-66 after inorganic base treatments and (**b**) recovered UiO-66 after organic base treatments.

Table 3. BET surface area of recovered UiO-66 after base treatments.

Base	- (Pristine)	K_2CO_3 (0.1 M)	$KHCO_3$ (0.1 M)	K_2SO_4 (0.1 M)	$BnNH_2$ (0.5 M)	DABCO (0.5 M)	Pyridine (0.5 M)
BET surface area (m^2/g)	1777	524	1540	1566	881	1060	1622

Finally, inductively coupled plasma–optical emission spectrometry (ICP-OES) analysis was conducted to examine the metal-to-ligand ratio of the MOF following the base treatment. Given that the general formula of UiO-66 MOF is $Zr_6O_4(OH)_4$(BDC-ligand)$_6$, the dissolution of the ligand in a basic solution could lead to an increase in the metal ratio [72–74]. While UiO-66 exhibits a theoretical zirconium content of 32.9 wt%, the pristine UiO-66 (as-synthesized form) contains 34.0 wt% due to structural defects. Following the base treatment, strong bases, such as K_2CO_3 (inorganic) and $BnNH_2$ (organic), exhibited significant increases in the Zr ratio according to ICP-OES analysis (refer to Table 4).

Table 4. The ICP-OES of recovered UiO-66 after base treatments.

Base	- (Pristine)	K_2CO_3 (0.1 M)	$KHCO_3$ (0.1 M)	K_2SO_4 (0.1 M)	$BnNH_2$ (0.5 M)	DABCO (0.5 M)	Pyridine (0.5 M)
Zr wt%	34.0	45.2	25.7	30.0	49.4	31.0	35.6

2.4. Comprehensive Concentration Limits and pK_b Values for UiO-66 MOF

A comprehensive investigation demonstrated that KOAc and pyridine were the bases with the highest concentration limits for Zr-based UiO-66 MOFs. As an organic base, pyridine demonstrated remarkable ligand preservation, with more than 80% remaining intact within the solid-state frameworks. Similarly, the KOAc treatment retained over 60% of the ligands, as shown in Figure 10. During the DABCO base test, approximately 70% of the BDC ligands were dissolved from the solid framework into solution. Despite ligand dissolution, the PXRD patterns were adequately retained, and the morphology of UiO-66 remained intact.

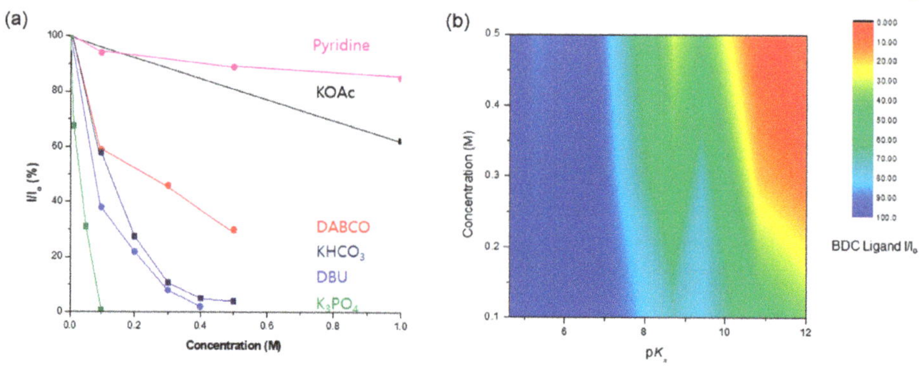

Figure 10. (**a**) Concentration limits of UiO-66 MOFs against the basic solution. (**b**) The correlation of pK_a and concentration for the conservation of the MOF structure.

The relationship between the strength of the base (expressed by the pK_b value) and concentration was directly linked to the dissolution of the MOF frameworks. High basicity (i.e., low pK_b and high pK_a) proved critical for MOF destruction owing to a strong nucleophilic attack, while high concentrations also played a crucial role in MOF dissolution. Figure 10b illustrates the correlation between pK_a and the concentration of the basic solution. The red color signifies the MOF framework dissolution, whereas the blue portion denotes the conserved structure of the MOFs. This visual representation underscores the interplay between the base strength and concentration in determining the fate of MOF stability under basic conditions.

3. Conclusions

Zr-based MOFs have gained widespread attention and applications owing to their excellent chemical and physical stability, ease of accessibility, and tunability. However, the practical deployment of these MOFs in industrial and daily-life scenarios is constrained by their inherent susceptibility to basic conditions. This limitation is indeed critical in a plethora of organic syntheses that use basic additives.

This study systematically probed the stability of Zr-based UiO-66 MOFs in various basic solutions, in which inorganic and organic bases commonly found in chemistry laboratories were investigated. The findings revealed that highly basic solutions (pK_b < 3.3) led to the complete destruction of UiO-66 MOFs within 1 h at a 0.1 M concentration. Subsequent examinations focused on bases that have a higher pK_b than 3.3. Among these experiments, UiO-66 MOFs exhibited overall stability in both the KOAc and pyridine solutions, while

PXRD and NMR analyses detected some framework destruction and ligand dissolution during the $KHCO_3$ and DABCO solution treatments. Notably, K_3PO_4 and DBU exhibited low concentration limits for the degradation of the UiO-66 MOF.

In conclusion, when designing experiments involving Zr-MOFs under basic conditions, considering the concentration and pK_b values of the chosen base is essential. In this context, we believe that Figure 10 can serve as a key guide to practitioners in selecting bases for UiO-66 MOFs under aqueous conditions. Notably, MOF stability may also be influenced by metal salts, ligand functionalization, solvents, etc. Therefore, additional systematic consideration, both from the literature and empirical perspectives, is imperative for advancing the practical applications of MOFs.

Supplementary Materials: The following supporting information can be downloaded at: https://www.mdpi.com/article/10.3390/nano14010110/s1, General methods, preparation of UiO-66, procedure of base treatment, PXRD measurement, acid digestion of MOFs for NMR measurement, N_2 isotherm for UiO-66, and appendix-PXRD and ^1H NMR data of UiO-66. Ref. [63] is cited in the supplementary materials.

Author Contributions: J.Y.K., J.K., S.C., H.K. (Haein Kim) and D.K. equally contributed to this work. J.Y.K. and J.K. performed the stability studies with organic bases. S.C. and H.K. (Haein Kim) performed the stability test with inorganic bases. D.K. focused on the properties of MOFs. H.K. (Houng Kang) and I.C. summarized overall basicity and stability. H.K. (Houng Kang), I.C. and M.K. prepared the manuscript. All authors have read and agreed to the published version of the manuscript.

Funding: This work is supported by the National Research Foundation of Korea (NRF), funded by the Ministry of Science and ICT (2022R1A2C1009706). In addition, Dopil Kim (a PhD candidate) was also financially supported by the NRF funded by the Ministry of Education (RS-2023-00274850).

Data Availability Statement: Data will be available upon request.

Conflicts of Interest: The authors declare no conflicts of interest.

References

1. Zhou, H.-C.; Long, J.R.; Yaghi, O.M. Introduction to Metal–Organic Frameworks. *Chem. Rev.* **2012**, *112*, 673–674. [CrossRef] [PubMed]
2. Lu, W.; Wei, Z.; Gu, Z.-Y.; Liu, T.-F.; Park, J.; Park, J.; Tian, J.; Zhang, M.; Zhang, Q.; Gentle Iii, T.; et al. Tuning the structure and function of metal–organic frameworks via linker design. *Chem. Soc. Rev.* **2014**, *43*, 5561–5593. [CrossRef] [PubMed]
3. Zhou, H.-C.J.; Kitagawa, S. Metal–Organic Frameworks (MOFs). *Chem. Soc. Rev.* **2014**, *43*, 5415–5418. [CrossRef] [PubMed]
4. Bavykina, A.; Kolobov, N.; Khan, I.S.; Bau, J.A.; Ramirez, A.; Gascon, J. Metal–Organic Frameworks in Heterogeneous Catalysis: Recent Progress, New Trends, and Future Perspectives. *Chem. Rev.* **2020**, *120*, 8468–8535. [CrossRef] [PubMed]
5. Jiang, J.; Yaghi, O.M. Brønsted Acidity in Metal–Organic Frameworks. *Chem. Rev.* **2015**, *115*, 6966–6997. [CrossRef] [PubMed]
6. Zhu, L.; Liu, X.-Q.; Jiang, H.-L.; Sun, L.-B. Metal–Organic Frameworks for Heterogeneous Basic Catalysis. *Chem. Rev.* **2017**, *117*, 8129–8176. [CrossRef] [PubMed]
7. O'Keeffe, M. Design of MOFs and intellectual content in reticular chemistry: A personal view. *Chem. Soc. Rev.* **2009**, *38*, 1215–1217. [CrossRef]
8. Chen, Z.; Hanna, S.L.; Redfern, L.R.; Alezi, D.; Islamoglu, T.; Farha, O.K. Reticular chemistry in the rational synthesis of functional zirconium cluster-based MOFs. *Coord. Chem. Rev.* **2019**, *386*, 32–49. [CrossRef]
9. Jiang, H.; Alezi, D.; Eddaoudi, M. A reticular chemistry guide for the design of periodic solids. *Nat. Rev. Mater.* **2021**, *6*, 466–487. [CrossRef]
10. Freund, R.; Canossa, S.; Cohen, S.M.; Yan, W.; Deng, H.; Guillerm, V.; Eddaoudi, M.; Madden, D.G.; Fairen-Jimenez, D.; Lyu, H.; et al. 25 Years of Reticular Chemistry. *Angew. Chem. Int. Ed.* **2021**, *60*, 23946–23974. [CrossRef]
11. Zhao, X.; Wang, Y.; Li, D.-S.; Bu, X.; Feng, P. Metal–Organic Frameworks for Separation. *Adv. Mater.* **2018**, *30*, 1705189. [CrossRef] [PubMed]
12. Lim, D.-W.; Ha, J.; Oruganti, Y.; Moon, H.R. Hydrogen separation and purification with MOF-based materials. *Mater. Chem. Front.* **2021**, *5*, 4022–4041. [CrossRef]
13. Wang, L.; Huang, H.; Zhang, X.; Zhao, H.; Li, F.; Gu, Y. Designed metal-organic frameworks with potential for multi-component hydrocarbon separation. *Coord. Chem. Rev.* **2023**, *484*, 215111. [CrossRef]
14. Bae, C.; Gu, M.; Jeon, Y.; Kim, D.; Kim, J. Metal–organic frameworks for NH_3 adsorption by different NH_3 operating pressures. *Bull. Korean Chem. Soc.* **2023**, *44*, 112–124. [CrossRef]

15. Mariella Babu, A.; Varghese, A. Electrochemical deposition for metal organic Frameworks: Advanced Energy, Catalysis, sensing and separation applications. *J. Electroanal. Chem.* **2023**, *937*, 117417. [CrossRef]
16. Kim, J.; Na, C.; Son, Y.; Prabu, M.; Yoon, M. Stilbene ligand-based metal–organic frameworks for efficient dye adsorption and nitrobenzene detection. *Bull. Korean Chem. Soc.* **2023**, *44*, 507–515. [CrossRef]
17. Sun, Z.; Li, T.; Mei, T.; Liu, Y.; Wu, K.; Le, W.; Hu, Y. Nanoscale MOFs in nanomedicine applications: From drug delivery to therapeutic agents. *J. Mater. Chem. B* **2023**, *11*, 3273–3294. [CrossRef]
18. Gatou, M.A.; Vagena, I.A.; Lagopati, N.; Pippa, N.; Gazouli, M.; Pavlatou, E.A. Functional MOF-Based Materials for Environmental and Biomedical Applications: A Critical Review. *Nanomaterials* **2023**, *13*, 2224. [CrossRef]
19. Pourmadadi, M.; Omrani, Z.; Forootan, Z.; Ebadi, M.S.; Yazdian, F. UiO-66 nanoparticles as a drug delivery system: A comprehensive review. *J. Drug Deliv. Sci. Technol.* **2023**, *86*, 104690. [CrossRef]
20. Dhakshinamoorthy, A.; Li, Z.; Garcia, H. Catalysis and photocatalysis by metal organic frameworks. *Chem. Soc. Rev.* **2018**, *47*, 8134–8172. [CrossRef]
21. Liu, J.; Goetjen, T.A.; Wang, Q.; Knapp, J.G.; Wasson, M.C.; Yang, Y.; Syed, Z.H.; Delferro, M.; Notestein, J.M.; Farha, O.K.; et al. MOF-enabled confinement and related effects for chemical catalyst presentation and utilization. *Chem. Soc. Rev.* **2022**, *51*, 1045–1097. [CrossRef] [PubMed]
22. Kirchon, A.; Feng, L.; Drake, H.F.; Joseph, E.A.; Zhou, H.-C. From fundamentals to applications: A toolbox for robust and multifunctional MOF materials. *Chem. Soc. Rev.* **2018**, *47*, 8611–8638. [CrossRef] [PubMed]
23. Ding, M.; Cai, X.; Jiang, H.-L. Improving MOF stability: Approaches and applications. *Chem. Sci.* **2019**, *10*, 10209–10230. [CrossRef] [PubMed]
24. Cheng, P.; Wang, C.; Kaneti, Y.V.; Eguchi, M.; Lin, J.; Yamauchi, Y.; Na, J. Practical MOF Nanoarchitectonics: New Strategies for Enhancing the Processability of MOFs for Practical Applications. *Langmuir* **2020**, *36*, 4231–4249. [CrossRef] [PubMed]
25. Freund, R.; Zaremba, O.; Arnauts, G.; Ameloot, R.; Skorupskii, G.; Dincă, M.; Bavykina, A.; Gascon, J.; Ejsmont, A.; Goscianska, J.; et al. The Current Status of MOF and COF Applications. *Angew. Chem. Int. Ed.* **2021**, *60*, 23975–24001. [CrossRef]
26. Ntouros, V.; Kousis, I.; Pisello, A.L.; Assimakopoulos, M.N. Binding Materials for MOF Monolith Shaping Processes: A Review towards Real Life Application. *Energies* **2022**, *15*, 1489. [CrossRef]
27. Lee, J.; Lee, J.; Kim, J.Y.; Kim, M. Covalent connections between metal–organic frameworks and polymers including covalent organic frameworks. *Chem. Soc. Rev.* **2023**, *52*, 6379–6416. [CrossRef]
28. Acuna-Yeomans, E.; Gutiérrez-Sevillano, J.J.; Dubbeldam, D.; Calero, S. A simulation study of linker vacancy distribution and its effect on UiO-66 stability. *Microporous Mesoporous Mater.* **2024**, *366*, 112922. [CrossRef]
29. An, Y.; Lv, X.; Jiang, W.; Wang, L.; Shi, Y.; Hang, X.; Pang, H. The stability of MOFs in aqueous solutions—Research progress and prospects. *Green Chem. Eng.* **2023**, in press. [CrossRef]
30. Pramanik, B.; Sahoo, R.; Das, M.C. pH-stable MOFs: Design principles and applications. *Coord. Chem. Rev.* **2023**, *493*, 215301. [CrossRef]
31. Semivrazhskaya, O.O.; Salionov, D.; Clark, A.H.; Casati, N.P.M.; Nachtegaal, M.; Ranocchiari, M.; Bjelić, S.; Verel, R.; van Bokhoven, J.A.; Sushkevich, V.L. Deciphering the Mechanism of Crystallization of UiO-66 Metal-Organic Framework. *Small* **2023**, *19*, 2305771. [CrossRef]
32. Ahmadijokani, F.; Molavi, H.; Rezakazemi, M.; Tajahmadi, S.; Bahi, A.; Ko, F.; Aminabhavi, T.M.; Li, J.-R.; Arjmand, M. UiO-66 metal–organic frameworks in water treatment: A critical review. *Prog. Mater. Sci.* **2022**, *125*, 100904. [CrossRef]
33. Bůžek, D.; Demel, J.; Lang, K. Zirconium Metal–Organic Framework UiO-66: Stability in an Aqueous Environment and Its Relevance for Organophosphate Degradation. *Inorg. Chem.* **2018**, *57*, 14290–14297. [CrossRef] [PubMed]
34. He, S.; Li, L.-X.; Zhang, L.-T.; Zeng, S.; Feng, C.; Chen, X.-X.; Zhou, H.-L.; Huang, X.-C. Elucidating influences of defects and thermal treatments on CO2 capture of a Zr-based metal–organic framework. *Chem. Eng. J.* **2024**, *479*, 147605. [CrossRef]
35. Li, L.-X.; He, S.; Zeng, S.; Chen, W.-T.; Ye, J.-W.; Zhou, H.-L.; Huang, X.-C. Equipping carbon dots in a defect-containing MOF via self-carbonization for explosive sensing. *J. Mater. Chem. C* **2023**, *11*, 321–328. [CrossRef]
36. Safy, M.E.A.; Amin, M.; Haikal, R.R.; Elshazly, B.; Wang, J.; Wang, Y.; Wöll, C.; Alkordi, M.H. Probing the Water Stability Limits and Degradation Pathways of Metal–Organic Frameworks. *Chem.—A Eur. J.* **2020**, *26*, 7109–7117. [CrossRef]
37. Ahmad, R.; Rizaldo, S.; Gohari, M.; Shanahan, J.; Shaner, S.E.; Stone, K.L.; Kissel, D.S. Buffer Effects in Zirconium-Based UiO Metal–Organic Frameworks (MOFs) That Influence Enzyme Immobilization and Catalytic Activity in Enzyme/MOF Biocatalysts. *ACS Omega* **2023**, *8*, 22545–22555. [CrossRef]
38. Bůžek, D.; Adamec, S.; Lang, K.; Demel, J. Metal–organic frameworks vs. buffers: Case study of UiO-66 stability. *Inorg. Chem. Front.* **2021**, *8*, 720–734. [CrossRef]
39. Wang, Y.; Yin, J.; Cao, W.; Fu, Y.; Kong, X. The instability of a stable metal-organic framework in amino acid solutions. *Nano Res.* **2022**, *15*, 6607–6612. [CrossRef]
40. Burtch, N.C.; Jasuja, H.; Walton, K.S. Water Stability and Adsorption in Metal–Organic Frameworks. *Chem. Rev.* **2014**, *114*, 10575–10612. [CrossRef]
41. Wang, K.; Li, Y.; Xie, L.-H.; Li, X.; Li, J.-R. Construction and application of base-stable MOFs: A critical review. *Chem. Soc. Rev.* **2022**, *51*, 6417–6441. [CrossRef]
42. Hamisu, A.M.; Ariffin, A.; Wibowo, A.C. Cation exchange in metal-organic frameworks (MOFs): The hard-soft acid-base (HSAB) principle appraisal. *Inorganica Chim. Acta* **2020**, *511*, 119801. [CrossRef]

43. Wang, L.; Li, J.; Cheng, L.; Song, Y.; Zeng, P.; Wen, X. Application of hard and soft acid base theory to uncover the destructiveness of Lewis bases to UiO-66 type metal organic frameworks in aqueous solutions. *J. Mater. Chem. A* **2021**, *9*, 14868–14876. [CrossRef]
44. Fernandez, C.A.; Nune, S.K.; Annapureddy, H.V.; Dang, L.X.; McGrail, B.P.; Zheng, F.; Polikarpov, E.; King, D.L.; Freeman, C.; Brooks, K.P. Hydrophobic and moisture-stable metal–organic frameworks. *Dalton Trans.* **2015**, *44*, 13490–13497. [CrossRef] [PubMed]
45. Jayaramulu, K.; Geyer, F.; Schneemann, A.; Kment, Š.; Otyepka, M.; Zboril, R.; Vollmer, D.; Fischer, R.A. Hydrophobic Metal–Organic Frameworks. *Adv. Mater.* **2019**, *31*, 1900820. [CrossRef] [PubMed]
46. Ding, M.; Jiang, H.-L. Improving Water Stability of Metal–Organic Frameworks by a General Surface Hydrophobic Polymerization. *CCS Chem.* **2020**, *3*, 2740–2748. [CrossRef]
47. Kim, M.; Cohen, S.M. Discovery, development, and functionalization of Zr(iv)-based metal–organic frameworks. *CrystEngComm* **2012**, *14*, 4096–4104. [CrossRef]
48. Kim, M.; Cahill, J.F.; Fei, H.; Prather, K.A.; Cohen, S.M. Postsynthetic Ligand and Cation Exchange in Robust Metal–Organic Frameworks. *J. Am. Chem. Soc.* **2012**, *134*, 18082–18088. [CrossRef]
49. Bai, Y.; Dou, Y.; Xie, L.-H.; Rutledge, W.; Li, J.-R.; Zhou, H.-C. Zr-based metal–organic frameworks: Design, synthesis, structure, and applications. *Chem. Soc. Rev.* **2016**, *45*, 2327–2367. [CrossRef]
50. Chen, Y.; Zhang, X.; Mian, M.R.; Son, F.A.; Zhang, K.; Cao, R.; Chen, Z.; Lee, S.-J.; Idrees, K.B.; Goetjen, T.A.; et al. Structural Diversity of Zirconium Metal–Organic Frameworks and Effect on Adsorption of Toxic Chemicals. *J. Am. Chem. Soc.* **2020**, *142*, 21428–21438. [CrossRef]
51. Feng, L.; Day, G.S.; Wang, K.-Y.; Yuan, S.; Zhou, H.-C. Strategies for Pore Engineering in Zirconium Metal-Organic Frameworks. *Chem* **2020**, *6*, 2902–2923. [CrossRef]
52. Hou, J.; Wang, H.; Zhang, H. Zirconium Metal–Organic Framework Materials for Efficient Ion Adsorption and Sieving. *Ind. Eng. Chem. Res.* **2020**, *59*, 12907–12923. [CrossRef]
53. Fu, J.; Wu, Y.-n. A Showcase of Green Chemistry: Sustainable Synthetic Approach of Zirconium-Based MOF Materials. *Chem.—A Eur. J.* **2021**, *27*, 9967–9987. [CrossRef] [PubMed]
54. Zhang, X.; Tong, S.; Huang, D.; Liu, Z.; Shao, B.; Liang, Q.; Wu, T.; Pan, Y.; Huang, J.; Liu, Y.; et al. Recent advances of Zr based metal organic frameworks photocatalysis: Energy production and environmental remediation. *Coord. Chem. Rev.* **2021**, *448*, 214177. [CrossRef]
55. Cavka, J.H.; Jakobsen, S.; Olsbye, U.; Guillou, N.; Lamberti, C.; Bordiga, S.; Lillerud, K.P. A New Zirconium Inorganic Building Brick Forming Metal Organic Frameworks with Exceptional Stability. *J. Am. Chem. Soc.* **2008**, *130*, 13850–13851. [CrossRef] [PubMed]
56. Howarth, A.J.; Liu, Y.; Li, P.; Li, Z.; Wang, T.C.; Hupp, J.T.; Farha, O.K. Chemical, thermal and mechanical stabilities of metal–organic frameworks. *Nat. Rev. Mater.* **2016**, *1*, 15018. [CrossRef]
57. Biffis, A.; Centomo, P.; Del Zotto, A.; Zecca, M. Pd Metal Catalysts for Cross-Couplings and Related Reactions in the 21st Century: A Critical Review. *Chem. Rev.* **2018**, *118*, 2249–2295. [CrossRef] [PubMed]
58. Dhakshinamoorthy, A.; Asiri, A.M.; Garcia, H. Formation of C–C and C–Heteroatom Bonds by C–H Activation by Metal Organic Frameworks as Catalysts or Supports. *ACS Catal.* **2019**, *9*, 1081–1102. [CrossRef]
59. Luo, S.; Zeng, Z.; Zeng, G.; Liu, Z.; Xiao, R.; Chen, M.; Tang, L.; Tang, W.; Lai, C.; Cheng, M.; et al. Metal Organic Frameworks as Robust Host of Palladium Nanoparticles in Heterogeneous Catalysis: Synthesis, Application, and Prospect. *ACS Appl. Mater. Interfaces* **2019**, *11*, 32579–32598. [CrossRef]
60. Khoury, C.; Gadipelly, C.; Pappuru, S.; Shpasser, D.; Gazit, O.M. Progress in the Design of Cooperative Heterogeneous Catalytic Materials for C–C Bond Formation. *Adv. Funct. Mater.* **2020**, *30*, 1901385. [CrossRef]
61. Wang, Q.; Astruc, D. State of the Art and Prospects in Metal–Organic Framework (MOF)-Based and MOF-Derived Nanocatalysis. *Chem. Rev.* **2020**, *120*, 1438–1511. [CrossRef]
62. Singh, S.; Kumar, P.; Soni, A.; Nemiwal, M. An Exploration on Copper-Based Metal-Organic Frameworks as Propitious Heterogeneous Catalyst for Coupling Reactions. *ChemistrySelect* **2023**, *8*, e202204279. [CrossRef]
63. Katz, M.J.; Brown, Z.J.; Colón, Y.J.; Siu, P.W.; Scheidt, K.A.; Snurr, R.Q.; Hupp, J.T.; Farha, O.K. A facile synthesis of UiO-66, UiO-67 and their derivatives. *Chem. Commun.* **2013**, *49*, 9449–9451. [CrossRef] [PubMed]
64. Shearer, G.C.; Vitillo, J.G.; Bordiga, S.; Svelle, S.; Olsbye, U.; Lillerud, K.P. Functionalizing the Defects: Postsynthetic Ligand Exchange in the Metal Organic Framework UiO-66. *Chem. Mater.* **2016**, *28*, 7190–7193. [CrossRef]
65. Cai, G.; Jiang, H.-L. A Modulator-Induced Defect-Formation Strategy to Hierarchically Porous Metal–Organic Frameworks with High Stability. *Angew. Chem. Int. Ed.* **2017**, *56*, 563–567. [CrossRef] [PubMed]
66. Park, H.; Kim, S.; Jung, B.; Park, M.H.; Kim, Y.; Kim, M. Defect Engineering into Metal–Organic Frameworks for the Rapid and Sequential Installation of Functionalities. *Inorg. Chem.* **2018**, *57*, 1040–1047. [CrossRef] [PubMed]
67. Kim, D.; Kang, M.; Ha, H.; Hong, C.S.; Kim, M. Multiple functional groups in metal–organic frameworks and their positional regioisomerism. *Coord. Chem. Rev.* **2021**, *438*, 213892. [CrossRef]
68. Olmstead, W.N.; Margolin, Z.; Bordwell, F.G. Acidities of water and simple alcohols in dimethyl sulfoxide solution. *J. Org. Chem.* **1980**, *45*, 3295–3299. [CrossRef]
69. Bickmore, B.R.; Tadanier, C.J.; Rosso, K.M.; Monn, W.D.; Eggett, D.L. Bond-valence methods for pKa prediction: Critical reanalysis and a new approach1 1Associate editor: C. M. Eggleston. *Geochim. Et Cosmochim. Acta* **2004**, *68*, 2025–2042. [CrossRef]

70. Pines, D.; Ditkovich, J.; Mukra, T.; Miller, Y.; Kiefer, P.M.; Daschakraborty, S.; Hynes, J.T.; Pines, E. How Acidic Is Carbonic Acid? *J. Phys. Chem. B* **2016**, *120*, 2440–2451. [CrossRef]
71. Tshepelevitsh, S.; Kütt, A.; Lõkov, M.; Kaljurand, I.; Saame, J.; Heering, A.; Plieger, P.G.; Vianello, R.; Leito, I. On the Basicity of Organic Bases in Different Media. *Eur. J. Org. Chem.* **2019**, *2019*, 6735–6748. [CrossRef]
72. Bodylska, W.; Fandzloch, M.; Szukiewicz, R.; Lukowiak, A. Cation-Exchange in Metal-Organic Framework as a Strategy to Obtain New Material for Ascorbic Acid Detection. *Nanomaterials* **2022**, *12*, 4480. [CrossRef] [PubMed]
73. Ye, G.; Wan, L.; Zhang, Q.; Liu, H.; Zhou, J.; Wu, L.; Zeng, X.; Wang, H.; Chen, X.; Wang, J. Boosting Catalytic Performance of MOF-808(Zr) by Direct Generation of Rich Defective Zr Nodes via a Solvent-Free Approach. *Inorg. Chem.* **2023**, *62*, 4248–4259. [CrossRef] [PubMed]
74. Luzuriaga, M.A.; Benjamin, C.E.; Gaertner, M.W.; Lee, H.; Herbert, F.C.; Mallick, S.; Gassensmith, J.J. ZIF-8 degrades in cell media, serum, and some—But not all—Common laboratory buffers. *Supramol. Chem.* **2019**, *31*, 485–490. [CrossRef] [PubMed]

nanomaterials

MDPI

Article

Bismuth-Based Metal–Organic Framework as a Chemiresistive Sensor for Acetone Gas Detection

Ashraf Ali [1], Yaser E. Greish [2], Reem H. Alzard [2], Lamia A. Siddig [2], Ahmed Alzamly [2], Naser Qamhieh [1] and Saleh T. Mahmoud [1,*]

[1] Department of Physics, United Arab Emirates University, Al-Ain 15551, United Arab Emirates
[2] Department of Chemistry, United Arab Emirates University, Al-Ain 15551, United Arab Emirates
* Correspondence: saleh.thaker@uaeu.ac.ae

Abstract: Analyzing acetone in the exhaled breath as a biomarker has proved to be a non-invasive method to detect diabetes in humans with good accuracy. In this work, a Bi-gallate MOF doped into a chitosan (CS) matrix containing an ionic liquid (IL) was fabricated to detect acetone gas with a low detection limit of 10 ppm at an operating temperature of 60 °C and 5 V operating bias. The sensor recorded the highest response to acetone in comparison to other test gases, proving its high selectivity along with long-term stability and repeatability. The sensor also exhibited ultra-fast response and recovery times of 15 ± 0.25 s and 3 ± 0.1 s, respectively. Moreover, the sensor membrane also exhibited flexibility and ease of fabrication, making it ideal to be employed as a real-time breath analyzer.

Keywords: acetone sensor; breath analyzer; Bi-gallate MOF; chitosan

1. Introduction

As the world advances in technology and industries, this has also caused pollution of the environment to accelerate. The polluted environment harbors diseases that affect the pace of human existence. Focusing on one form of pollution, in line with the advancement in technology, there has been an exponential decline in the quality of the air we breathe. Low quality of the air has adverse effects on our health and lifestyle. In response to this scenario, the diagnosis of diseases triggered by this poor air quality at early stages greatly increases the chances of early treatment and the betterment of the individual's health.

Amazingly, the human breath incorporates a lot of gases that can divulge information on the health state, but only if we have a way to decipher the data. Hence, sensors that can detect and distinguish these gases in real-time offer an inexpensive, non-invasive approach to gaining diagnostic information on the diseases people may have developed from inhaling polluted air. The gases that are exhaled by humans mostly contain a mixture of CO_2, N_2, H_2O, O_2, and trace levels of other volatile organic compounds (VOCs) such as ethane or acetone, to name a few. These gases assist as biomarkers in the detection of many diseases such as diabetes, lung cancer, etc. [1–5]. Acetone is a product of a biological process between the human and the invading micro-organisms. It then infuses into the bloodstream and is transported to the lungs. From here, it becomes a key component in the exhaled breath [6], enabling us to detect its concentration, based on which we can non-invasively distinguish numerous diseases. According to the World Health Organization (WHO), the level of acetone in a healthy human is supposed to be 0.2–1.8 ppm, in contrast to a diabetic patient, in whom it is between 1.25 and 2.5 ppm [7]. Determining the concentrations in terms of parts per billion (ppb) has proved to be quite a challenge, but there have been a few materials [8,9] that have been up for the task.

Researchers have not only been making groundbreaking advances in materials with the potential for acetone detection, but they have also been using acetone as a cleaning agent and solvent for their experimentations [10,11]. However, it has been demonstrated that prolonged exposure to acetone vapors exceeding 173 ppm can cause long-term health issues such as central nervous system anesthesia, skin and eye irritation, narcosis, nausea, headaches, and dizziness [12–16]. It is also flammable and explosive, with lower and upper explosive limits of 2.6% and 12.8%, respectively [17].

Metal–organic frameworks (MOFs) have been demonstrated to be a key component in developing devices that can detect hazardous gases evolving due to various processes. MOFs consist of metal cations lined by organic ligand molecules [18,19]. With properties like high porosity and tunable surface area diversity in structures, they have various applications such as gas separation and storage [20], catalysis [21], and energy applications and as sensing materials [22–25]. However, there is one obstacle for some of them, which is related to their high electrical resistance. As standalone materials, they would not serve the purpose of their designs, but when they are combined into an organic–inorganic matrix with ionic liquids (ILs), they can exhibit changes in their properties that can be recorded and evaluated.

Sensors based on these materials tend to operate under various principles such as impedance sensing [26], chemicapacitive sensing [27], chemiresistive sensing, Kelvin probe [28], capacitive sensing [29], field effect transistors [30,31], optical sensing, fluorescence sensing for the detection of various materials such as volatile gases, hazardous gases, and explosive compounds [32–34], ion sensing [35], biosensing [36,37], humidity sensing [38–40], pH sensing [41,42], and temperature sensing [43]. There are rapid photophysical, electrical, or mechanical changes in the properties of the material, with various circumstances that influence the behavior depending on the concentration of the analyte, active materials, ability to bond, electron accepting–donating ability, and hydrogen bonding, to name a few [44]. From the aforementioned principles, one of the simplest is chemiresistive sensing, which measures the change in resistance of the sensing material when exposed to the target gas. The mechanism of sensing in these materials is attributed to the transfer of electrons or holes, which results in an interaction between the surface of the sensing material and the target gas molecules via adsorption or surface reactions. Other advantages that these chemiresistive sensors have to offer are low cost of fabrication, easy integration with other electronic components for commercial devices, low operational costs, and ease of miniaturization [23,45,46]. In addition, the ability to detect trace-level analytes with efficiency and accuracy enables them to be successfully commercialized.

The change in resistance in these sensing materials is dependent on the type of material. With the diversity in the MOF and MOF-based materials, the sensing mechanism can also be altered. The interaction of the test gas on the surface can either donate an electron to the material or deprive it of one, thereby causing a change in resistance [47]. The linkers or the active functional organic groups that are used in the synthesis of these materials serve as effective adsorption sites that facilitate the transfer of charges within the system. Depending on the test gas, the reduction or oxidization reaction [48], due to its interaction with the material, can also be the cause for the change in the resistance of the material. Another parameter is the change in volume of the MOF on interaction with the gas modulating the number of electrons that are transferred during the interaction [49]. Depending on the concentration of the gas, the response of the sensor is recorded. As the synthesis methods of the materials advance, the MOFs show enhanced conductivities and porosities that increase the sensitivities multifold.

Bismuth-based materials have been traditionally used for cosmetics and drug delivery. Wang et al. [50] reported an elaborate study deciphering the structure of the Bi-MOF. Furthermore, Z. Wang et al. [51] reported a detailed review of the Bi-based MOFs and their derivatives, outlining that they have been traditionally used in catalysis applications [21,52,53], electrocatalysis, sensors [54], CO_2 capture [51], electrochemical energy storage [20,55], biomedical imaging [56], drug delivery, fluorescence sensing [51], absorption, and separation [52,57]. Mirica et al. [58], meanwhile, reported the synthesis and characterization of Bi-based MOFs for the detection of VOC compounds such as acetone, MeOH, and EtOH. They demonstrated that the material can also detect NO and NH_3 at room temperature. Some of the reported sensors from the literature have been consolidated in Table 1 for comparison with our work.

Table 1. Sensor performance comparison with literature-reported values.

Sensor/Material	Target Gas	Optimum Operating Temperature (°C)	Detection Limit (ppm)	Ref.
Bi-gallate MOF/CS/IL membrane	Acetone	60	10	This Work
MOF-5/CS/IL membrane	H_2S	RT	1	[59]
ZIF-67	Acetone	220 250	100 50	[60] [61]
ZIF-67/ZIF-8	Acetone	275	1	[62]
ZnO/ZIF-CoZn	Acetone	250	10	[63]
Hierarchical MOF derived ZnO-Co_3O_4	Acetone	450	5	[64]
Bi(HHTP)	NH_3 NO Acetone MeOH EtOH	RT RT RT RT RT	0.29 0.15 41.2 278 185	[58]

Our group has demonstrated that chitosan (CS) polymer incorporation with glycerol, as an IL [65], can detect H_2S gas at 15 ppm operating at 80 °C [24]. Conventionally, it has also been used in combination with MOFs and ZIFs [25,59] to enhance the detection of H_2S at room temperature. To the best of the authors' knowledge, the Bi-gallate MOF has not been used as an acetone sensor in combination with an organic matrix. Hence, in this work, we present the possibility of employing the Bi-gallate MOF in combination with CS/IL matrix as an acetone sensor with a breath analyzer application for the detection of diabetic patients.

2. Materials and Methods

2.1. Materials

Bismuth (III) nitrate pentahydrate ($Bi(NO_3)_3{\cdot}5H_2O$), gallic acid (3,4,5-trihydroxybenzoic acid), and anhydrous dimethylformamide (DMF) were bought from Sigma-Aldrich, St. Louis, MO, USA. Chitosan (MW = 50,000–190,000 Da) ($\geq$75%) and acetic acid were purchased from Polysciences, Warrington, PA, USA. Glycerol, as an ionic liquid (IL) (99.5%), was purchased from Quarek Corp company, Denver, CO, USA. All chemicals were used without further purification.

2.2. Synthesis of Bi-Gallate MOF

Bi-gallate MOF was synthesized following a slightly modified version of a previously published procedure [50]. In a scintillation vial, gallic acid (17 mg, 2 mmol) was dissolved in 3 mL of deionized water. Then, drops of 4 M NH_4OH were added to the solution until the pH reached 8.5. In a separate scintillation vial, $Bi(NO_3)_3{\cdot}5H_2O$ (24 mg, 1 mmol) was dissolved in 1 mL of DMF and then gradually added with continuous stirring to the gallic acid solution. The vial was sealed and placed in a preheated oven set at 85 °C for a duration of 24 h. After completion of the reaction, the resulting yellow powder was filtered and successively washed with distilled water and ethanol. Then, the Bi-gallate compound was subjected to activation in a vacuum oven at 90 °C for 24 h to remove any trapped solvent molecules.

2.3. Fabrication of the Bi-Gallate MOF/CS/IL Membrane

The Bi-gallate MOF/CS/IL membrane was fabricated by dissolving 2 wt% (0.4 g) of the Bi-gallate MOF with 0.4 g of CS and 5 vol% of IL into 20 mL of 3% acetic acid solution.

The solution was then stirred for 24 h at room temperature and 1400 RPM on a magnetic stirrer. The prepared solution was then cast on a Petri dish and dried at 70 °C for 18 h in a drying oven. A flexible and uniform membrane formed, as shown in Figure 1. The thickness of the membrane was determined to be 0.20 mm using a screw gauge.

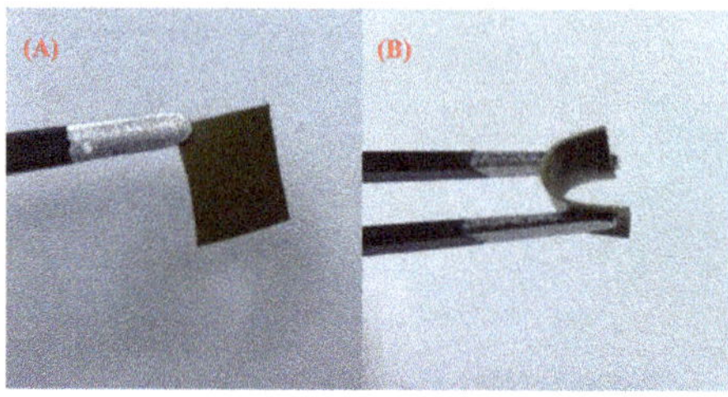

Figure 1. (**A**) A 1 cm × 1 cm portion of the membrane. (**B**) Demonstration of high flexibility.

2.4. Characterization

Powder X-ray diffraction (PXRD) was conducted using a Rigaku MiniFlex X-ray diffractometer (Rigaku, Tokyo, Japan) with a CuKα radiation tube (wavelength = 1.542 Å) operated at 40 kV. PXRD measurements were taken within a 3° to 50° 2θ range at a rate of 2° min^{-1}. Scanning electron microscopy (SEM) was performed using JSM-6010LA, JEOL, Tokyo, Japan, with an operating voltage of 5 keV. The secondary electron imaging configuration was used to procure the images with a working distance of 14 mm. The samples were made conductive with a gold sputtering system. The surface area and porosity were examined using nitrogen sorption analysis at 77 K. The N_2 adsorption–desorption isotherm indicated gas adsorption (cm^3 g^{-1}) in relation to relative pressure (P/P_0). Here, P represents N_2 equilibrium pressure, and P_0 represents saturated vapor pressure at 77 K. Prior to surface measurements, powder samples were placed in a glass tube and subjected to vacuum at 353 K for 3 h. Thermogravimetric analysis (TGA) was performed using a TGA-50 Shimadzu analyzer (Shimadzu, Kyoto, Japan) with an aluminum pan sample holder. FTIR analysis was carried out using a Shimadzu IRAffinity-1S with a scan range from 400 to 4000 cm^{-1}.

2.5. Sensor Fabrication and Gas Testing

The sensor prototype was fabricated by sandwiching the active layer between a Cu plate with dimensions 1.5 cm × 1.5 cm as the bottom electrode and a stainless steel mesh with a grid size of 250 μm × 250 μm serving as the top electrode [59,66]. The layers were confined using temperature-resistive Kapton tape. The device was connected to the sensing system with electrical probes which were housed in a Teflon chamber connected to Bronkhorst mass flow controllers (MFCs). The setup was sealed to avoid leaking of the test gas and placed inside the fume hood for safety. The humidity of the chamber was maintained close to 0% throughout the testing sequences. The gas testing programs were sequenced to expose the sample to test gas in between cycles of synthetic air to exfoliate any residual test gas molecules.

3. Results and Discussion

3.1. Structural and Morphological Characterization of Bi-Gallate MOF and Bi-MOF/CS/IL Membrane

The structural analysis of the Bi-gallate MOF and Bi-gallate MOF/CS/IL membranes was carried out using the XRD patterns obtained. The XRD pattern in Figure 2 confirmed the structure and phase purity of the synthesized Bi-gallate MOF. By comparing the resulting diffraction pattern with the simulated pattern derived from single-crystal data as previously reported [50], it was confirmed that the Bi-gallate we prepared was successfully and purely synthesized. The Bi-gallate material exhibited an orthorhombic lattice and unit cell dimensions of a = 8.80 Å, b = 4.66 Å, c = 24.09 Å. The XRD pattern of the Bi-gallate MOF/CS/IL with a broad hump confirmed the incorporation of the Bi-gallate MOF into the CS matrix, as can be seen in the top pattern in Figure 2. The characteristic peaks representing the (110), (310), (321), and (421) planes were observed.

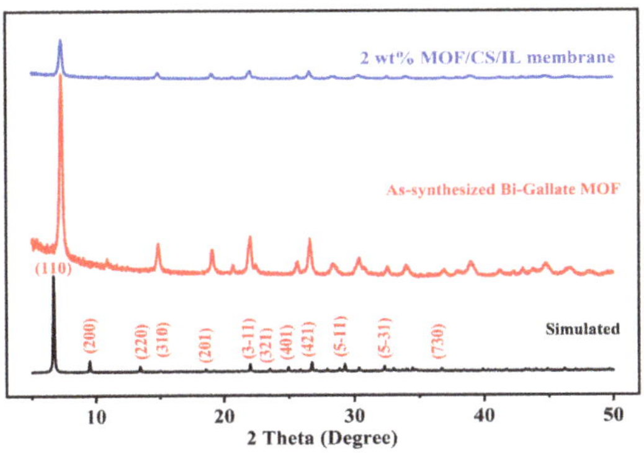

Figure 2. PXRD pattern of the Bi-gallate MOF and the composite membrane.

The FTIR analysis of Bi-gallate MOF and Bi-gallate MOF/CS/IL membrane confirmed the interaction within the framework of Bi-gallate and confirmed its presence within the composite, as shown in Figure 3. The disappearance of the hydroxyl group (-OH) band confirmed the chelation of bismuth metal with the gallic acid linker. Furthermore, a band was observed around 1671 cm^{-1}, which is attributed to the presence of (C=O) of a carboxylate (COO^-) group. Analyzing the spectra of the Bi-gallate MOF/CS/IL membrane, a broad band at 3400 cm^{-1} was observed in the FTIR spectrum, and is attributed to the NH and OH stretching vibration, as well as the intermolecular hydrogen bonds of chitosan. Additionally, the bands at 2900 and 3000 cm^{-1} belong to the symmetric and asymmetric stretching of the C-H bond of chitosan, respectively. Meanwhile, the C-O bending and C-O stretching vibrations of chitosan were determined at 1100 and 980 cm^{-1}, respectively. The presence of Bi-gallate MOF within the chitosan membrane was further confirmed through the sharp bands at 1700 and 1209 cm^{-1}, which, respectively, belong to C=O and C-O stretching of the gallate linker. As can be observed in Figure 3, the band frequencies that are attributed to the carboxylate group (at 1671 cm^{-1}) and the Bi-O group (at 439 cm^{-1}) remained intact, which also confirms the stability of the MOF framework within the chitosan membrane.

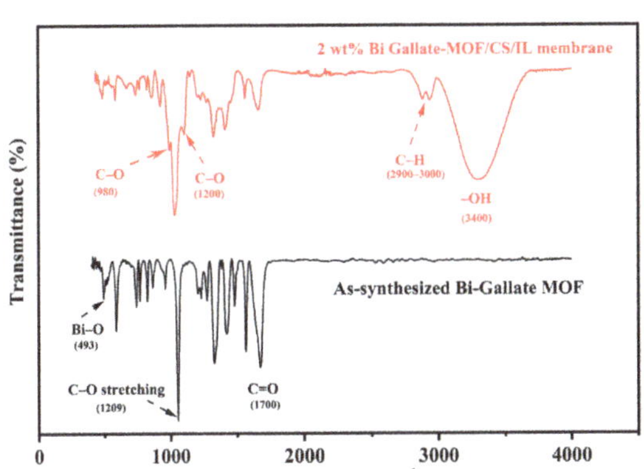

Figure 3. FTIR curves of the as-synthesized Bi-gallate MOF powder and the composite membrane.

The thermal analysis of the as-synthesized Bi-gallate MOF was performed using TGA under a nitrogen atmosphere (Figure 4). In the temperature range of room temperature to 150 °C, the sample experienced a weight reduction of approximately 10%, corresponding to the loss of water molecules. This finding aligned with previously reported results [50]. Decomposition of the Bi-gallate commenced in the range of 260–350 °C, and is attributed to the breakdown of the gallate linker of the MOF structure. From the TGA curves of the Bi-gallate MOF/CS/IL membrane, we can surmise that the first weight loss observed would be due to the loss of water molecules from the pores of the framework. The drastic loss in weight observed between 150 °C and 225 °C indicates the onset of the decomposition of the gallate linker from the matrix, followed by a gradual loss in weight due to the continued decomposition of the remaining linkers.

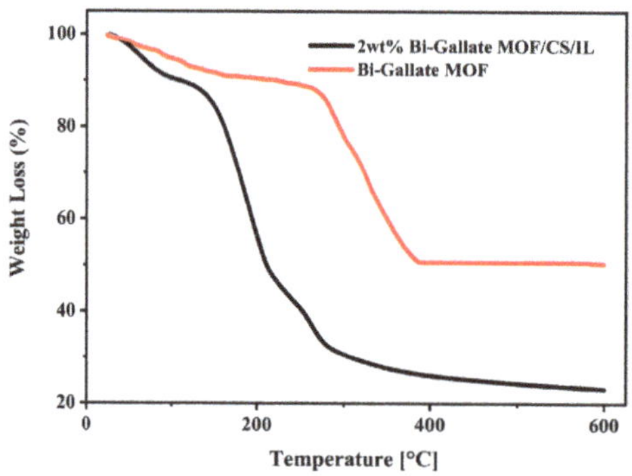

Figure 4. TGA curves of the Bi-gallate MOF powder and the composite membrane.

The SEM analysis of the Bi-gallate MOF powder, shown in Figure 5A, shows that the morphology of the as-synthesized powder is comparable to the reported morphologies of the same MOF [50]. The SEM micrographs of the Bi-gallate MOF/CS/IL membranes shown in Figure 5B displayed the incorporation of the MOFs into the matrix. The cross-section of the membrane showed numerous pores, as can be seen in Figure 5C. The EDX spectra we recorded of the as-synthesized MOF (Figure 5D) and the composite membrane (Figure 5E) showed the homogeneous distribution of the Bi-gallate MOF particles within the chitosan matrix.

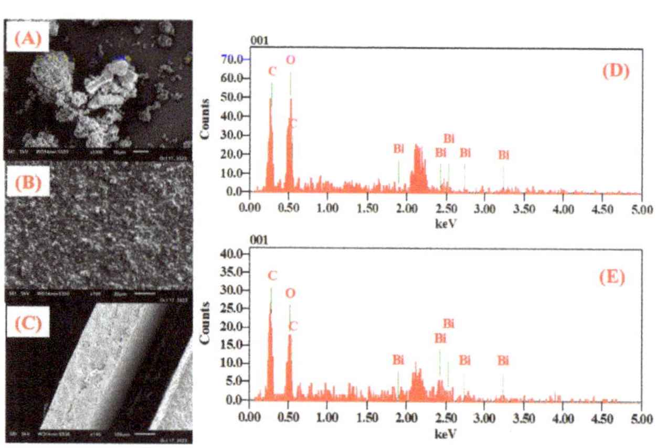

Figure 5. (**A**,**D**) SEM image and EDX spectra of the as-synthesized Bi-gallate MOF. (**B**,**E**) SEM image and EDX spectra of the Bi-gallate MOF/CS/IL membrane. (**C**) Cross-section of the membrane.

The as-synthesized Bi-gallate MOF powder was also subjected to N_2 adsorption measurements to evaluate the micro- and macroporosity in the MOF. The recorded loop shown in Figure 6 follows a type I isotherm with a calculated Brunauer–Emmett–Teller (BET) surface area of 31.58 m^2/g, and shows a maximum pore volume of 0.0135 cm^3/g.

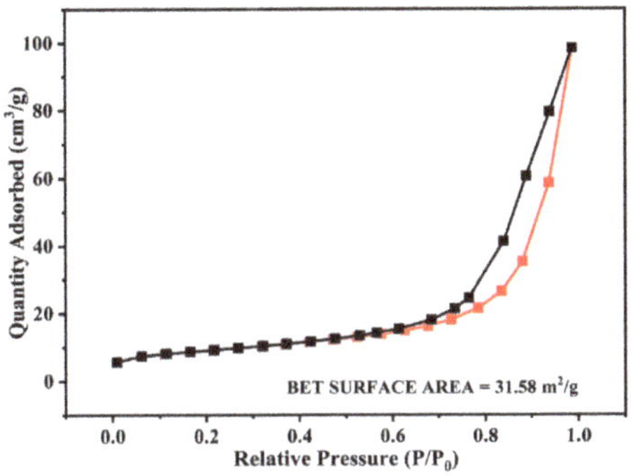

Figure 6. N_2-sorption isotherms of as-synthesized Bi-gallate MOF.

3.2. Gas Sensing Performance

The sensor prototype was set up as detailed in our previous reports [59,66]. The CS/IL matrix was doped with different concentrations of Bi-gallate MOF. The membranes were subjected to different test gases at 100 ppm to evaluate their response. It was evaluated that 2 wt% doping of the Bi-gallate MOF into the CS/IL matrix was most sensitive toward acetone gas at an operating temperature of 60 °C and a bias voltage of 5 V. The response of the sensor was evaluated using Equation (1):

$$S\ (\%) = \frac{R_g - R_a}{R_a} \times 100 = \frac{\Delta R}{R_a} \times 100 \quad (1)$$

where R_a is the resistance of the sensor in synthetic air and R_g is the resistance in the presence of the test gas.

The sensor was evaluated in terms of sensitivity, which showed a response to 10 ppm of acetone gas at 60 °C, as plotted in Figure 7. The inset graph shows the sensor's response toward different concentrations of acetone gas. The sensor was further analyzed toward other test gases at 100 ppm and an operating temperature of 60 °C.

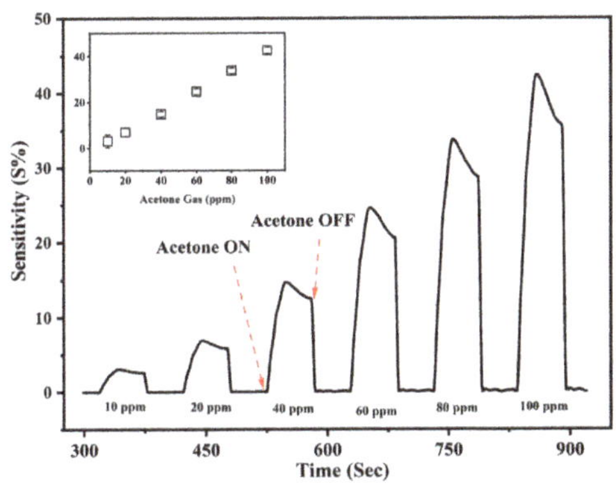

Figure 7. Sensitivity of the Bi-gallate MOF/CS/IL membrane as a function of time and acetone concentration measured at 60 °C. Inset: Sensitivity for corresponding gas concentration.

The other aspects of evaluation were repeatability and stability. The tests were performed with exposure to 100 ppm of acetone, with synthetic air flushing in between each cycle to exfoliate any residual molecules from the previous cycle. The stability response was calculated as 42.41 ± 1.8%, whereas the repeatability of the sensor was calculated as 39.95 ± 1.4% (Figure 8).

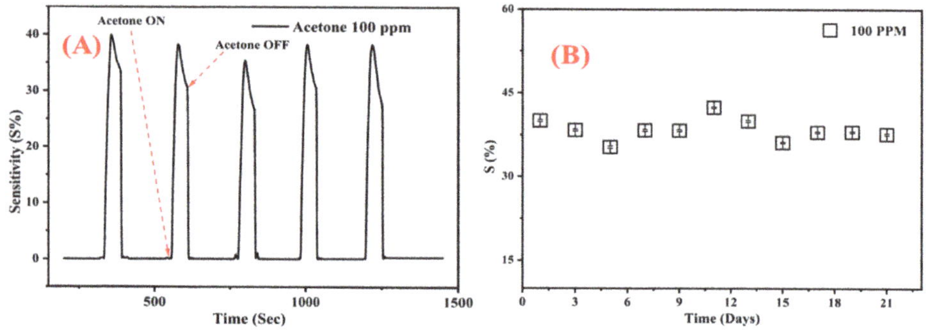

Figure 8. (**A**) Repeatability of the sensor. (**B**) Long-term stability of the Bi-gallate MOF/CS/IL membrane to 100 ppm of acetone gas at 60 °C.

Another aspect of evaluating the sensor's performance is the response and recovery times, which can be defined as the time taken from gas exposure for the sensor to reach 90% of its recorded response, and the time taken for the sensor to recover to 10% of its initial resistance from the shutdown of the gas, respectively. From the results shown in Figure 9A, the responses recorded were calculated as 15 ± 0.25 s and 3 ± 0.1 s, respectively. Yet another vital parameter is the selectivity among other test gases. It was recorded that the sensitivity toward acetone was the highest among other test gases such as H_2, H_2S, C_2H_4, CO, and CO_2, as shown in Figure 9B.

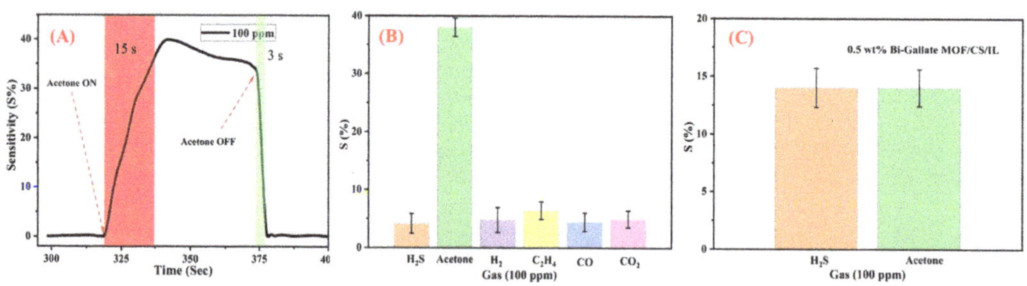

Figure 9. (**A**) The calculated response and recovery time of the sensor. (**B**) Selectivity of the sensor containing 2 wt% Bi-gallate MOF in comparison to other test gases. (**C**) Selectivity of the sensor containing 0.5 wt% Bi-gallate MOF to H_2S and acetone gases.

3.3. Gas Sensing Mechanism

The mechanism of the standalone CS–IL membrane was outlined by Hani et al. [24]. The functional groups along the chitosan polymeric matrix (-OH and -NH_2) interact with the highly hydroxylated IL molecules through the formation of a network of H bonding. In the presence of hydroxy- and carboxy-functionalized Bi-gallate MOF particles in the matrix, an extended network of H bonding is produced, as shown in Figure 10. Compared with slightly polar or non-polar gas molecules, H_2S and acetone showed adsorption to this H-bonded network by taking part in the network through H bonding, as well. This was observed when a low proportion of the Bi-gallate MOF (0.5 wt%) was added, as shown in

Figure 9C. Upon increasing the proportion of Bi-gallate MOF to 2 wt%, a higher tendency of acetone to adsorb onto the extensively formed H-bonded network was observed, as shown in Figure 9B. Compared with H_2S molecules with a dipole moment of 0.95 D, the preferential adsorption of acetone molecules is attributed to its higher polarity (with a dipole moment of up to 4.19 D). It should be mentioned that the highly polar acetone molecules interact with the H-bonded network of chitosan containing IL and Bi-gallate through the attraction of its highly polar C=O group to the H atoms along the other components of the sensor. Accordingly, an increase in the sensitivity of the composite sensor membrane was observed when increasing the Bi-gallate component of the sensor.

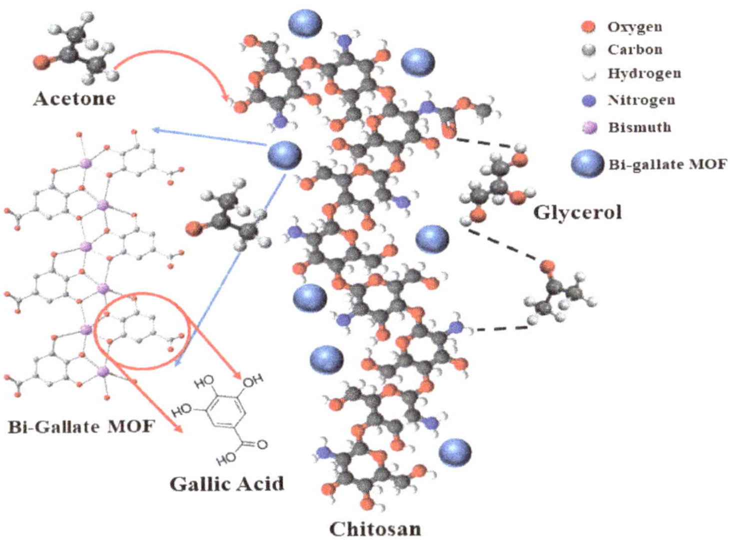

Figure 10. Sensing mechanism of the Bi-gallate-doped CS-IL membrane.

4. Conclusions

This study demonstrates the potential of fabricating a fast response and ultra-fast recovery sensor based on Bi-gallate MOF doped into a chitosan matrix containing IL. The fabricated membrane was investigated for its sensing performance. Previous studies conducted by our group demonstrated that CS/IL membranes showed potential for sensing toward H_2S gas, and doping the matrix with 2 wt% Bi-gallate MOF showed sensitivity toward acetone vapor alone with a 5 V bias. The detection limit of the prototype is 10 ppm of acetone at 60 °C, with ultra-fast response and recovery times of 15 ± 0.25 s and 3 ± 0.1 s, respectively. The components of the membrane do not cause any harm to the environment, hence making the prototype highly eco-friendly. The proposed membrane can be used as an acetone sensor with ultra-fast response and recovery times, which can be used as a real-time breath analyzer.

Author Contributions: Conceptualization, Y.E.G. and S.T.M.; data curation, A.A. (Ashraf Ali), R.H.A. and N.Q.; formal analysis, A.A. (Ashraf Ali); funding acquisition, S.T.M.; investigation, A.A. (Ashraf Ali), R.H.A. and L.A.S.; methodology, A.A. (Ashraf Ali), R.H.A. and L.A.S.; project administration, S.T.M.; supervision, S.T.M.; validation, A.A. (Ashraf Ali), Y.E.G., A.A. (Ahmed Alzamly) and S.T.M.; writing—original draft, A.A. (Ashraf Ali), Y.E.G. and S.T.M.; writing—review and editing, A.A. (Ashraf Ali), Y.E.G., R.H.A., L.A.S. and S.T.M. All authors have read and agreed to the published version of the manuscript.

Funding: This work was financially supported by United Arab Emirates University (grants code G00004169, 12R165-Research Center-ZCHS-2-2023, and grants code USRP-G00003232 with fund code 31R238-M4).

Data Availability Statement: The data that support the findings of this study are available from the corresponding authors upon request.

Acknowledgments: The authors are thankful for the financial support and Facilities provided by United Arab Emirates University.

Conflicts of Interest: The authors declare no conflict of interest.

References

1. Güntner, A.T.; Pineau, N.J.; Chie, D.; Krumeich, F.; Pratsinis, S.E. Selective sensing of isoprene by Ti-doped ZnO for breath diagnostics. *J. Mater. Chem. B* **2016**, *4*, 5358–5366. [CrossRef] [PubMed]
2. Chen, H.; Bo, R.; Shrestha, A.; Xin, B.; Nasiri, N.; Zhou, J.; Di Bernardo, I.; Dodd, A.; Saunders, M.; Lipton-Duffin, J. NiO–ZnO nanoheterojunction networks for room-temperature volatile organic compounds sensing. *Adv. Opt. Mater.* **2018**, *6*, 1800677. [CrossRef]
3. de Lacy Costello, B.; Amann, A.; Al-Kateb, H.; Flynn, C.; Filipiak, W.; Khalid, T.; Osborne, D.; Ratcliffe, N.M. A review of the volatiles from the healthy human body. *J. Breath Res.* **2014**, *8*, 014001. [CrossRef] [PubMed]
4. Rudnicka, J.; Kowalkowski, T.; Buszewski, B. Searching for selected VOCs in human breath samples as potential markers of lung cancer. *Lung Cancer* **2019**, *135*, 123–129. [CrossRef] [PubMed]
5. Guntner, A.T.; Abegg, S.; Konigstein, K.; Gerber, P.A.; Schmidt-Trucksass, A.; Pratsinis, S.E. Breath sensors for health monitoring. *ACS Sens.* **2019**, *4*, 268–280. [CrossRef] [PubMed]
6. Schnabel, R.; Fijten, R.; Smolinska, A.; Dallinga, J.; Boumans, M.-L.; Stobberingh, E.; Boots, A.; Roekaerts, P.; Bergmans, D.; van Schooten, F.J. Analysis of volatile organic compounds in exhaled breath to diagnose ventilator-associated pneumonia. *Sci. Rep.* **2015**, *5*, 17179. [CrossRef] [PubMed]
7. Masikini, M.; Chowdhury, M.; Nemraoui, O. Metal oxides: Application in exhaled breath acetone chemiresistive sensors. *J. Electrochem. Soc.* **2020**, *167*, 037537. [CrossRef]
8. Majhi, S.M.; Ali, A.; Greish, Y.E.; El-Maghraby, H.F.; Qamhieh, N.N.; Hajamohideen, A.R.; Mahmoud, S.T. Accordion-like-Ti_3C_2 MXene-based gas sensors with sub-ppm level detection of acetone at room temperature. *ACS Appl. Electron. Mater.* **2022**, *4*, 4094–4103. [CrossRef]
9. Majhi, S.M.; Ali, A.; Greish, Y.E.; El-Maghraby, H.F.; Mahmoud, S.T. V_2CT_X MXene-based hybrid sensor with high selectivity and ppb-level detection for acetone at room temperature. *Sci. Rep.* **2023**, *13*, 3114. [CrossRef]
10. Xu, X.; Chen, Y.; Ma, S.; Li, W.; Mao, Y. Excellent acetone sensor of La-doped ZnO nanofibers with unique bead-like structures. *Sens. Actuators B* **2015**, *213*, 222–233. [CrossRef]
11. Xiao, T.; Wang, X.-Y.; Zhao, Z.-H.; Li, L.; Zhang, L.; Yao, H.-C.; Wang, J.-S.; Li, Z.-J. Highly sensitive and selective acetone sensor based on C-doped WO_3 for potential diagnosis of diabetes mellitus. *Sens. Actuators B* **2014**, *199*, 210–219. [CrossRef]
12. Toda, K.; Furue, R.; Hayami, S. Recent progress in applications of graphene oxide for gas sensing: A review. *Anal. Chim. Acta* **2015**, *878*, 43–53. [CrossRef] [PubMed]
13. Du, Y.; Xue, Q.; Zhang, Z.; Xia, F. Great enhancement in H_2 response using graphene-based Schottky junction. *Mater. Lett.* **2014**, *135*, 151–153. [CrossRef]
14. Xu, K.; Fu, C.; Gao, Z.; Wei, F.; Ying, Y.; Xu, C.; Fu, G. Nanomaterial-based gas sensors: A review. *Instrum. Sci. Technol.* **2018**, *46*, 115–145. [CrossRef]
15. Perreault, F.; De Faria, A.F.; Elimelech, M. Environmental applications of graphene-based nanomaterials. *Chem. Soc. Rev.* **2015**, *44*, 5861–5896. [CrossRef]
16. Amiri, V.; Roshan, H.; Mirzaei, A.; Neri, G.; Ayesh, A.I. Nanostructured metal oxide-based acetone gas sensors: A review. *Sensors* **2020**, *20*, 3096. [CrossRef]
17. Drmosh, Q.A.; Olanrewaju Alade, I.; Qamar, M.; Akbar, S. Zinc Oxide-Based Acetone Gas Sensors for Breath Analysis: A Review. *Chem. Asian J.* **2021**, *16*, 1519–1538. [CrossRef]
18. Li, H.; Eddaoudi, M.; O'Keeffe, M.; Yaghi, O.M. Design and synthesis of an exceptionally stable and highly porous metal-organic framework. *Nature* **1999**, *402*, 276–279. [CrossRef]
19. Tranchemontagne, D.J.; Hunt, J.R.; Yaghi, O.M. Room temperature synthesis of metal-organic frameworks: MOF-5, MOF-74, MOF-177, MOF-199, and IRMOF-0. *Tetrahedron* **2008**, *64*, 8553–8557. [CrossRef]
20. Murray, L.J.; Dincă, M.; Long, J.R. Hydrogen storage in metal–organic frameworks. *Chem. Soc. Rev.* **2009**, *38*, 1294–1314. [CrossRef]
21. Dhakshinamoorthy, A.; Li, Z.; Garcia, H. Catalysis and photocatalysis by metal organic frameworks. *Chem. Soc. Rev.* **2018**, *47*, 8134–8172. [CrossRef] [PubMed]
22. Campbell, M.G.; Dincă, M. Metal–organic frameworks as active materials in electronic sensor devices. *Sensors* **2017**, *17*, 1108. [CrossRef] [PubMed]

23. Yi, F.Y.; Chen, D.; Wu, M.K.; Han, L.; Jiang, H.L. Chemical sensors based on metal–organic frameworks. *ChemPlusChem* **2016**, *81*, 675–690. [CrossRef] [PubMed]
24. Abu-Hani, A.F.; Greish, Y.E.; Mahmoud, S.T.; Awwad, F.; Ayesh, A.I. Low-temperature and fast response H_2S gas sensor using semiconducting chitosan film. *Sens. Actuators B* **2017**, *253*, 677–684. [CrossRef]
25. Ali, A.; Alzamly, A.; Greish, Y.E.; Alzard, R.H.; El-Maghraby, H.F.; Qamhieh, N.; Mahmoud, S.T. Enhancing Hydrogen Sulfide Detection at Room Temperature Using ZIF-67-Chitosan Membrane. *Membranes* **2023**, *13*, 333. [CrossRef] [PubMed]
26. Achmann, S.; Hagen, G.; Kita, J.; Malkowsky, I.M.; Kiener, C.; Moos, R. Metal-organic frameworks for sensing applications in the gas phase. *Sensors* **2009**, *9*, 1574–1589. [CrossRef] [PubMed]
27. Hosseini, M.; Zeinali, S.; Sheikhi, M. Fabrication of capacitive sensor based on Cu-BTC (MOF-199) nanoporous film for detection of ethanol and methanol vapors. *Sens. Actuators B* **2016**, *230*, 9–16. [CrossRef]
28. Stassen, I.; Bueken, B.; Reinsch, H.; Oudenhoven, J.; Wouters, D.; Hajek, J.; Van Speybroeck, V.; Stock, N.; Vereecken, P.; Van Schaijk, R. Towards metal–organic framework based field effect chemical sensors: UiO-66-NH_2 for nerve agent detection. *Chem. Sci.* **2016**, *7*, 5827–5832. [CrossRef]
29. Yassine, O.; Shekhah, O.; Assen, A.H.; Belmabkhout, Y.; Salama, K.N.; Eddaoudi, M. H2S sensors: Fumarate-based fcu-MOF thin film grown on a capacitive interdigitated electrode. *Angew. Chem.* **2016**, *128*, 16111–16115. [CrossRef]
30. Panda, T.; Banerjee, R. High Charge Carrier Mobility in two dimensional indium(III) isophthalic acid based frameworks. *Proc. Natl. Acad. Sci. India Sect. A Phys. Sci.* **2014**, *84*, 331–336. [CrossRef]
31. Wu, G.; Huang, J.; Zang, Y.; He, J.; Xu, G. Porous field-effect transistors based on a semiconductive metal–organic framework. *J. Am. Chem. Soc.* **2017**, *139*, 1360–1363. [CrossRef] [PubMed]
32. Hu, Z.; Deibert, B.J.; Li, J. Luminescent metal–organic frameworks for chemical sensing and explosive detection. *Chem. Soc. Rev.* **2014**, *43*, 5815–5840. [CrossRef] [PubMed]
33. Hu, Z.; Pramanik, S.; Tan, K.; Zheng, C.; Liu, W.; Zhang, X.; Chabal, Y.J.; Li, J. Selective, sensitive, and reversible detection of vapor-phase high explosives via two-dimensional mapping: A new strategy for MOF-based sensors. *Cryst. Growth Des.* **2013**, *13*, 4204–4207. [CrossRef]
34. Pramanik, S.; Hu, Z.; Zhang, X.; Zheng, C.; Kelly, S.; Li, J. A Systematic Study of Fluorescence-Based Detection of Nitroexplosives and Other Aromatics in the Vapor Phase by Microporous Metal–Organic Frameworks. *Chem. Eur. J.* **2013**, *19*, 15964–15971. [CrossRef] [PubMed]
35. Wang, Y.; Wu, Y.; Xie, J.; Ge, H.; Hu, X. Multi-walled carbon nanotubes and metal–organic framework nanocomposites as novel hybrid electrode materials for the determination of nano-molar levels of lead in a lab-on-valve format. *Analyst* **2013**, *138*, 5113–5120. [CrossRef] [PubMed]
36. Ma, W.; Jiang, Q.; Yu, P.; Yang, L.; Mao, L. Zeolitic imidazolate framework-based electrochemical biosensor for in vivo electrochemical measurements. *Anal. Chem.* **2013**, *85*, 7550–7557. [CrossRef] [PubMed]
37. Shen, W.-J.; Zhuo, Y.; Chai, Y.-Q.; Yuan, R. Ce-based metal-organic frameworks and DNAzyme-assisted recycling as dual signal amplifiers for sensitive electrochemical detection of lipopolysaccharide. *Biosens. Bioelectron.* **2016**, *83*, 287–292. [CrossRef]
38. Zhang, Y.; Chen, Y.; Zhang, Y.; Cong, H.; Fu, B.; Wen, S.; Ruan, S. A novel humidity sensor based on NH 2-MIL-125(Ti) metal organic framework with high responsiveness. *J. Nanopart. Res.* **2013**, *15*, 2014. [CrossRef]
39. Yu, Y.; Zhang, X.-M.; Ma, J.-P.; Liu, Q.-K.; Wang, P.; Dong, Y.-B. Cu(I)-MOF: Naked-eye colorimetric sensor for humidity and formaldehyde in single-crystal-to-single-crystal fashion. *Chem. Commun.* **2014**, *50*, 1444–1446. [CrossRef]
40. Yu, Y.; Ma, J.-P.; Dong, Y.-B. Luminescent humidity sensors based on porous Ln^{3+}-MOFs. *CrystEngComm* **2012**, *14*, 7157–7160. [CrossRef]
41. Harbuzaru, B.V.; Corma, A.; Rey, F.; Jordá, J.L.; Ananias, D.; Carlos, L.D.; Rocha, J. A miniaturized linear pH sensor based on a highly photoluminescent self-assembled europium(III) metal–organic framework. *Angew. Chem. Int. Ed.* **2009**, *48*, 6476–6479. [CrossRef] [PubMed]
42. Li, H.-Y.; Wei, Y.-L.; Dong, X.-Y.; Zang, S.-Q.; Mak, T.C. Novel Tb-MOF embedded with viologen species for multi-photofunctionality: Photochromism, photomodulated fluorescence, and luminescent pH sensing. *Chem. Mater.* **2015**, *27*, 1327–1331. [CrossRef]
43. Cui, Y.; Zhu, F.; Chen, B.; Qian, G. Metal–organic frameworks for luminescence thermometry. *Chem. Commun.* **2015**, *51*, 7420–7431. [CrossRef] [PubMed]
44. Kreno, L.E.; Leong, K.; Farha, O.K.; Allendorf, M.; Van Duyne, R.P.; Hupp, J.T. Metal–organic framework materials as chemical sensors. *Chem. Rev.* **2012**, *112*, 1105–1125. [CrossRef] [PubMed]
45. Li, H.-Y.; Zhao, S.-N.; Zang, S.-Q.; Li, J. Functional metal–organic frameworks as effective sensors of gases and volatile compounds. *Chem. Soc. Rev.* **2020**, *49*, 6364–6401. [CrossRef]
46. Koo, W.-T.; Jang, J.-S.; Kim, I.-D. Metal-organic frameworks for chemiresistive sensors. *Chem* **2019**, *5*, 1938–1963. [CrossRef]
47. Campbell, M.G.; Sheberla, D.; Liu, S.F.; Swager, T.M.; Dincă, M. Cu_3(hexaiminotriphenylene)$_2$: An electrically conductive 2D metal–organic framework for chemiresistive sensing. *Angew. Chem. Int. Ed.* **2015**, *54*, 4349–4352. [CrossRef]
48. Aubrey, M.L.; Wiers, B.M.; Andrews, S.C.; Sakurai, T.; Reyes-Lillo, S.E.; Hamed, S.M.; Yu, C.-J.; Darago, L.E.; Mason, J.A.; Baeg, J.-O. Electron delocalization and charge mobility as a function of reduction in a metal–organic framework. *Nat. Mater.* **2018**, *17*, 625–632. [CrossRef]

49. Rubio-Giménez, V.; Almora-Barrios, N.; Escorcia-Ariza, G.; Galbiati, M.; Sessolo, M.; Tatay, S.; Martí-Gastaldo, C. Origin of the chemiresistive response of ultrathin films of conductive metal–organic frameworks. *Angew. Chem.* **2018**, *130*, 15306–15310. [CrossRef]
50. Wang, Y.; Takki, S.; Cheung, O.; Xu, H.; Wan, W.; Öhrström, L.; Inge, A.K. Elucidation of the elusive structure and formula of the active pharmaceutical ingredient bismuth subgallate by continuous rotation electron diffraction. *Chem. Commun.* **2017**, *53*, 7018–7021. [CrossRef]
51. Wang, Z.; Zeng, Z.; Wang, H.; Zeng, G.; Xu, P.; Xiao, R.; Huang, D.; Chen, S.; He, Y.; Zhou, C. Bismuth-based metal–organic frameworks and their derivatives: Opportunities and challenges. *Coord. Chem. Rev.* **2021**, *439*, 213902. [CrossRef]
52. Kalmutzki, M.J.; Diercks, C.S.; Yaghi, O.M. Metal–organic frameworks for water harvesting from air. *Adv. Mater.* **2018**, *30*, 1704304. [CrossRef] [PubMed]
53. Ye, S.; Yan, M.; Tan, X.; Liang, J.; Zeng, G.; Wu, H.; Song, B.; Zhou, C.; Yang, Y.; Wang, H. Facile assembled biochar-based nanocomposite with improved graphitization for efficient photocatalytic activity driven by visible light. *Appl. Catal. B Environ.* **2019**, *250*, 78–88. [CrossRef]
54. Wang, H.; Zeng, Z.; Xu, P.; Li, L.; Zeng, G.; Xiao, R.; Tang, Z.; Huang, D.; Tang, L.; Lai, C. Recent progress in covalent organic framework thin films: Fabrications, applications and perspectives. *Chem. Soc. Rev.* **2019**, *48*, 488–516. [CrossRef] [PubMed]
55. Yi, H.; Yan, M.; Huang, D.; Zeng, G.; Lai, C.; Li, M.; Huo, X.; Qin, L.; Liu, S.; Liu, X. Synergistic effect of artificial enzyme and 2D nano-structured Bi_2WO_6 for eco-friendly and efficient biomimetic photocatalysis. *Appl. Catal. B Environ.* **2019**, *250*, 52–62. [CrossRef]
56. Lu, K.; Aung, T.; Guo, N.; Weichselbaum, R.; Lin, W. Nanoscale metal–organic frameworks for therapeutic, imaging, and sensing applications. *Adv. Mater.* **2018**, *30*, 1707634. [CrossRef]
57. Liu, S.; Lu, X.F.; Xiao, J.; Wang, X.; Lou, X.W. Bi_2O_3 nanosheets grown on multi-channel carbon matrix to catalyze efficient CO_2 electroreduction to HCOOH. *Angew. Chem.* **2019**, *131*, 13966–13971. [CrossRef]
58. Aykanat, A.; Jones, C.G.; Cline, E.; Stolz, R.M.; Meng, Z.; Nelson, H.M.; Mirica, K.A. Conductive Stimuli-Responsive Coordination Network Linked with Bismuth for Chemiresistive Gas Sensing. *ACS Appl. Mater. Interfaces* **2021**, *13*, 60306–60318. [CrossRef]
59. Ali, A.; Alzamly, A.; Greish, Y.E.; Bakiro, M.; Nguyen, H.L.; Mahmoud, S.T. A Highly Sensitive and Flexible Metal–Organic Framework Polymer-Based H_2S Gas Sensor. *ACS Omega* **2021**, *6*, 17690–17697. [CrossRef]
60. Ding, D.; Xue, Q.; Lu, W.; Xiong, Y.; Zhang, J.; Pan, X.; Tao, B. Chemically functionalized 3D reticular graphene oxide frameworks decorated with MOF-derived Co_3O_4: Towards highly sensitive and selective detection to acetone. *Sens. Actuators B* **2018**, *259*, 289–298. [CrossRef]
61. Cui, W.; Kang, X.; Zhang, X.; Zheng, Z.; Cui, X. Facile synthesis of porous cubic microstructure of Co_3O_4 from ZIF-67 pyrolysis and its Au doped structure for enhanced acetone gas-sensing. *Phys. E Low-Dimens. Syst. Nanostruct.* **2019**, *113*, 165–171. [CrossRef]
62. Xiao, J.; Diao, K.; Zheng, Z.; Cui, X. MOF-derived porous ZnO/Co_3O_4 nanocomposites for high performance acetone gas sensing. *J. Mater. Sci. Mater. Electron.* **2018**, *29*, 8535–8546. [CrossRef]
63. Yao, M.S.; Tang, W.X.; Wang, G.E.; Nath, B.; Xu, G. MOF thin film-coated metal oxide nanowire array: Significantly improved chemiresistor sensor performance. *Adv. Mater.* **2016**, *28*, 5229–5234. [CrossRef] [PubMed]
64. Jang, J.-S.; Koo, W.-T.; Kim, D.-H.; Kim, I.-D. In situ coupling of multidimensional MOFs for heterogeneous metal-oxide architectures: Toward sensitive chemiresistors. *ACS Cent. Sci.* **2018**, *4*, 929–937. [CrossRef]
65. Chiappe, C.; Signori, F.; Valentini, G.; Marchetti, L.; Pomelli, C.S.; Bellina, F. Novel (glycerol) borate-based ionic liquids: An experimental and theoretical study. *J. Phys. Chem. B* **2010**, *114*, 5082–5088. [CrossRef]
66. Ali, A.; AlTakroori, H.H.; Greish, Y.E.; Alzamly, A.; Siddig, L.A.; Qamhieh, N.; Mahmoud, S.T. Flexible $Cu_3(HHTP)_2$ MOF Membranes for Gas Sensing Application at Room Temperature. *Nanomaterials* **2022**, *12*, 913. [CrossRef]

nanomaterials

MDPI

Article

Tuning Electromagnetic Parameters Induced by Synergistic Dual-Polarization Enhancement Mechanisms with Amorphous Cobalt Phosphide with Phosphorus Vacancies for Excellent Electromagnetic Wave Dissipation Performance

Bo Wen [1,2,3], Yunzi Miao [1,2,3], Zhijie Zhang [4], Na Li [1,2,3], Jiyuan Xiao [1,2,3], Yushuo Li [1,2,3], Jiangtao Feng [4], Shujiang Ding [1] and Guorui Yang [1,2,3,*]

1 School of Chemistry, Xi'an Jiaotong University, Xi'an 710049, China; wenbom@stu.xjtu.edu.cn (B.W.); myznofwplz@163.com (Y.M.); lina0831@xjtu.edu.cn (N.L.); xjy@nwafu.edu.cn (J.X.); lys11544@163.com (Y.L.); dingsj@mail.xjtu.edu.cn (S.D.)

2 Engineering Research Center of Energy Storage Materials and Devices, Ministry of Education, Four Joint Subjects One Union, Xi'an Jiaotong University, Xi'an 710049, China

3 School-Enterprise Joint Research Center for Power Battery Recycling & Circulation Utilization Technology, Xi'an Jiaotong University, Xi'an 710049, China

4 Department of Environmental Science and Engineering, Xi'an Jiaotong University, Xi'an 710049, China; zzhijie@stu.xjtu.edu.cn (Z.Z.)

* Correspondence: yangguorui@xjtu.edu.cn

Abstract: The understanding of amorphous and heterojunction materials has been widely used in the field of electromagnetic wave absorption due to their unique physical and chemical properties. However, the effectiveness of individual strategies currently used is still limited. Well-designed compositions and amorphous structures simplify the effect of different polarization mechanisms on the absorption of electromagnetic waves. In this work, through the carbonization and controlled phosphating of one-dimensional Co Metal–Organic Framework (Co-MOF) nanorods, the synthesis of complex components and amorphous CoP_x with phosphorus vacancies is successfully accomplished, thus adjusting the optimization of electromagnetic parameters. Phosphorus-vacancy-induced defective polarization loss and enhanced-electronegativity-differences-induced dipole polarization loss synergistically as a dual-polarization strategy significantly improved the electromagnetic parameters and impedance matching. In conclusion, the electromagnetic parameters of the Co@CoP_x@C composites are indeed significantly regulated, with reflection losses of −55 dB and a bandwidth of up to 5.5 GHz. These innovative research ideas provide instructive thinking for the development of amorphous absorbers with vacancies.

Keywords: Co metal–organic framework; amorphous CoP_x/carbon composites; phosphorus vacancy; dipole polarization; defect polarization; electromagnetic wave absorption

1. Introduction

Electromagnetic (EM) wave absorbents possess great application potential in preventing EM interference and human health damage due to a reliable EM wave attenuation ability, resulting in the emergence of a large number of new materials and revealing avant-garde EM wave attenuation mechanisms in recent years [1–5]. Commonly, the attenuation mechanisms include conduction loss, magnetic loss, and dielectric loss, as well as the multiple reflection and scattering of EM waves caused by microstructures [6–8]. Dielectric losses generally rely on various polarization losses to provide strong attenuation EM wave characteristics, while magnetic losses rely on resonance and eddy current to provide smaller levels of EM wave attenuation capabilities [9,10]. The magnetic loss capacity is generally much lower than the dielectric loss, and the significance of it existing is that it can be used to improve impedance matching and promote EM wave absorption [11]. Conduction loss, multiple reflections, and scattering generally only serve to promote the EM wave absorption [12]. Therefore, the hot research in the field of EM wave absorption focuses on the regulation of multi-component and complex structures, leading to the facilitation of the attenuation characteristics of EM waves by using complex and changeable loss mechanisms [13,14].

To enrich the types of loss mechanisms in absorbents, the composition and structure regulation of absorbents are commonly used strategies. Naturally, absorbents with superior electrical conductivity, including graphene, carbon nanotubes, and conductive polymers, can rely on excellent electrical conductivity and functional groups to contribute to favorable conductive loss and polarization loss [15–17]. Magnetic materials include magnetic metal elements, alloys, and ferrite materials, in which magnetic metal elements and alloys generally only provide magnetic loss, while ferrite materials also provide additional dielectric loss [18–20]. A large number of studies have proved that in magnetic materials/graphene, magnetic materials/carbon nanotubes, and magnetic materials/conductive polymers, the abundant heterojunctions promote enhanced interfacial polarization loss [21]. Of course, in addition to the above composite materials, oxides/carbon sulfide/carbon, selenide/carbon, and other composite materials have also become a class of materials that contribute to EM wave attenuation [22–24]. Therefore, combined with the above analysis, the combination of different types of materials can achieve the enhancement of EM wave absorption ability. However, the above-mentioned magnetic composite material has poor corrosion resistance and a single heterogeneous structure, which leads to a large number of limitations in the actual use process.

At present, due to their high intrinsic magnetism, relatively higher electrical conductivity, corrosion resistance, and preferable stability, the transition metal phosphide heterojunctions are widely used in catalysis, batteries, sensors, and other fields, but there is still less research in the field of EM wave absorption [25–27]. The excellent properties of heterojunctions are attributed to the following explanations. First, charge transfer at the complex heterojunction interfaces results in a reallocation of electrons, which improves the interfacial polarization ability. Second, the complex heterogeneous structure can realize the regulation of the defect structure and help to improve the polarization of the defects. Third, the abundant edge activity is also conducive to electron transfer, which enhances the interfacial polarization loss. For example, Fan et al. [28] prepared CoP/RGO composites with a reflection loss (RL) value of −52.60 dB at 2.05 mm through the synergistic enhancement of conductive loss and polarization loss. Furthermore, constructing a synergistic combination of multiple heterojunctions is another strategy to improve the attenuation of EM waves, as they tend to have better synergistic attenuation. Wang et al. [29] synthesized 3D flower-like CoNi-P/C composites that exhibited an optimal RL value of −65.5 dB depending on the excellent dielectric loss such as conductive loss and polarization loss. However, most of the above heterojunctions are crystal–crystalline heterojunctions, and crystal–amorphous heterojunctions are less studied. Moreover, the further formation of an amorphous structure is conducive to further increasing the defect loss and internal electron transfer.

Herein, we successfully synthesized amorphous Co@CoP_x@C composites with phosphorus vacancies with the controlled phosphating of MOF-derived metal/carbon composites. After partial phosphating, the impact of cobalt metal radioactivity on living organisms and the impact on crop growth can be reduced. Depending on the charge transfer at the interface of the multiple heterojunctions, the interfacial polarization loss is enhanced. The constructed phosphorus vacancies also promote the formation of more defect polarization. Additionally, phosphorus-doped porous carbon is also proved to be an important cause of enhanced polarization loss. Benefitting from the above improvement in polarization loss, the prepared Co@CoP_x@C composites realize the effective attenuation of EM waves with lower RL characteristics (−55 dB) and a wider absorption bandwidth (5.5 GHz). This work on the construction of amorphous heterojunctions with abundant phosphorus vacancies reveals the mechanism of polarization and defect effects on EM wave absorption characteristics and provides a new strategy for further designing EM wave absorbents with excellent loss capacity by depending on their enhanced polarization and defect loss capabilities.

2. Experimental Section

2.1. Materials

Cobalt acetate tetrahydrate ($Co(Ac)_2{\cdot}4H_2O$, AR, 98.5%), benzene-1, 3, 5-tricarboxylic acid (H_3BTC, 98%), and Selenium sulfide (SeS_2, 99%) were purchased from Sinopharm Chemical Reagent Co., Ltd., Shanghai, China.

2.2. Preparation of Co-MOF Nanorods

The Co-MOF nanorods were synthesized with the following methods. First, 4 mmol $Co(Ac)_2{\cdot}4H_2O$ was dissolved in 40 mL deionized water to form a clarified solution. Then, 4 mmol H_3BTC was added into the above solution, and the solution was stirred for 12 h. Finally, the product was washed with ethanol and deionized water before being dried at 60 °C overnight.

2.3. Preparation of Co@C and Co@CoP$_x$@C Composites

The synthesized Co-MOF nanorods were put in a quartz boat and annealed at 700 °C for 2 h with a heat rate of 2.0 °C min^{-1} under an argon atmosphere; then, Co@C composite material was obtained. Further, the Co@CoP$_x$@C composites were obtained by placing different molar quantities of $NaH_2PO_2{\cdot}H_2O$ (1, 5, 10) and Co@C at the left and right ends of a porcelain boat and calcining at 300 °C under the protection of argon for 2 h.

2.4. Characterization

The crystal structures were measured using X-ray diffraction (XRD, D8 Advance, Germany). A field-emission scanning electron microscope (FE-SEM, GeminiSEM 500, Oberkochen, Germany) and transmission electron microscope (TEM, ThermoFisher, Talos-F200X JEM-F200, Waltham, MA, USA) were used to confirm the morphology of the composites. The surface composition of composites was analyzed with X-ray photoelectron spectra (XPS, KRATOS, AXIS ULtrabld). The Raman (Raman, Renishaw, inVia Qontor) characterization test demonstrates the graphitization degree of the products. The specific surface area was obtained based on the Brunauer–Emmett–Teller (BET) theory (ASAP 2020 device).

3. Results and Discussion

Figure 1a demonstrates the schematic diagram of the generation of Co@CoP$_x$@C composite material after a multi-step reaction. Co-MOF nanorods were prepared with the stirring method at room temperature, as described in the experimental section. Co@CoP$_x$@C composites can be obtained by first carbonizing the Co-MOF nanorods and then placing $NaH_2PO_2{\cdot}H_2O$ and Co@C in the upstream and downstream of the porcelain boat with different contents of $NaH_2PO_2{\cdot}H_2O$ for low-temperature calcination, respectively.

To prove that Co-MOF nanorods were successfully synthesized, the X-ray diffraction (XRD) pattern is depicted in Figure 1b, demonstrating that the prepared precursors were Co-MOF nanorods with good crystallinity. Co-MOF nanorods with smooth surfaces were detected with SEM images, as shown in Figure 1c. The SEM images in Figure 1d verify that the Co@C composite can maintain the original nanorods' morphology after calcination with an apparently coarse surface. On top of that, the phosphating samples remained rod-like with rougher surfaces (Figure 1e–g).

Further analysis of the crystal structure of Co@C composites using XRD patterns testified to the presence of three diffraction peaks at 44.4°, 51.7°, and 76.0°, corresponding to the (111), (200), and (220) crystal planes of Co, respectively (JCPDS No. 15-0806) (Figure 2a) [30]. The intensity of the diffraction peaks on the (111) crystal plane of Co decreased with the increase in the degree of phosphatization, and the two crystal planes (200) and (220) almost disappeared. Herein, attributed to the possibility of the existence of amorphous CoP$_x$, the diffraction peaks related to CoP_x were not clearly found on the diffraction peaks of the phosphated samples. P doping is an effective method for introducing defects in carbon substrates. As displayed in Figure 2b, I_D/I_G values reveal that two peaks located at about 1340 and 1590 cm^{-1} are associated with the D-band and G-band of the carbon, respectively [31,32]. The relative intensity values of I_D/I_G in different Co@ CoP_x @C composites are slightly higher than those of Co@C composites, facilitating the provision of reliable evidence of increased defects in the carbon.

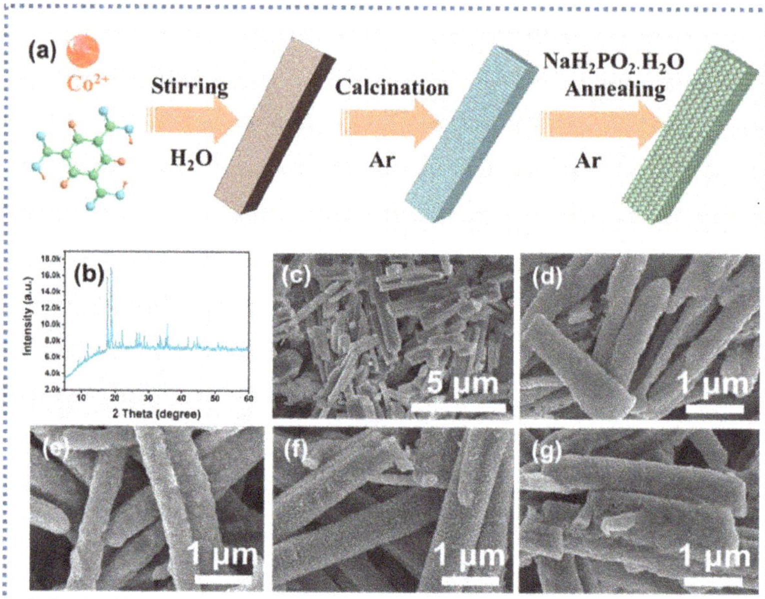

Figure 1. (**a**) Schematic of the synthesis of Co@CoP_x@C composites; (**b**) XRD spectra of Co-MOF nanorods; SEM images of (**c**) Co-MOF, (**d**) Co@C, (**e**) Co@CoP_x@C-1, (**f**) Co@CoP_x@C-5, and (**g**) Co@CoP_x@C-10 nanorods.

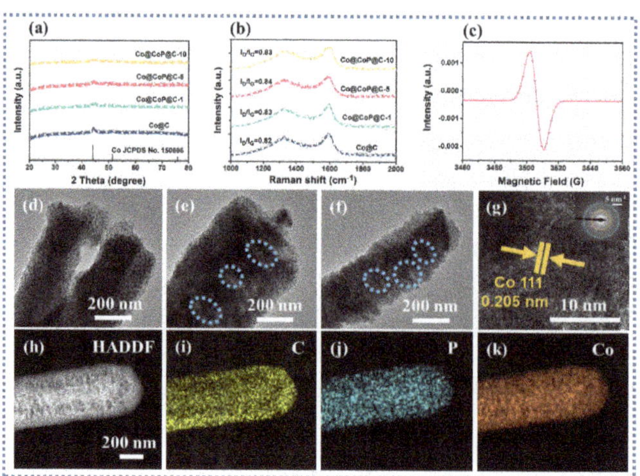

Figure 2. (**a**) XRD spectra, (**b**) Raman spectra, (**c**) EPR, (**d–f**) TEM images of the Co@CoP_x@C-1, Co@CoP_x@C-5, and Co@CoP_x@C-10 composites; (**g**) HRTEM image of the Co@CoP_x@C-10 composites, (h) HAADF, and (**i–k**) the EDS mapping images of the Co@CoP_x@C-10 composites.

Corroborating the existence of phosphorus vacancies in Co@CoP$_x$@C composites, electron paramagnetic resonance (EPR) was reliably used as a direct and effective technique for recognizing phosphorus vacancies [33]. As detected in the analysis results, the presence of phosphorous vacancies implies that the vacancy-induced defect polarization loss will appear (Figure 2c). Transmission electron microscopy (TEM) images of Co@CoP$_x$@C composites ascertain that with the increasing degree of phosphatization, the appearance of visible large pore structures on the surface of the nanorods may be caused by the loss of carbon during the phosphatization process (Figure 2d–f). The HRTEM image flags up that the interplanar distance of 0.205 corresponds to the Co nanoparticles (111) plane, proving the presence of Co particles in Co@CoP$_x$@C-5 composites (Figure 2g) [34]. The high-angle annular dark-field (HAADF) images and the corresponding energy spectrometer (EDS) mapping images confirm the presence of the elements C, P, and Co (Figure 2h–k). In contrast, the particle sizes of different samples can be statistically derived from Figure 2d–f and Figure S1. The average particle sizes of 5.13, 10.45, 11.77, and 11.9 nm in Co@C, Co@CoP$_x$@C-1, Co@CoP$_x$@C-5, and Co@CoP$_x$@C-10 composites testify to the markedly increased particle sizes of the phosphated samples (Figure S2). The pore characteristics of Co@C and Co@CoP$_x$@C composites were analyzed using nitrogen adsorption–desorption isotherms.

Specific surface area measurement is necessary to determine the structure of the samples after phosphating as both the particle size and carbon content of the samples are significantly altered. The adsorption–desorption curves in Figure S3a show that the Co@C, Co@CoP$_x$@C-1, Co@CoP$_x$@C-5, and Co@CoP$_x$@C-10 composites can be found to have the same adsorption–desorption curves, all of which are of type IV isotherms showing numerous mesoporous structures. Based on the calculations, the specific surface areas of Co@C, Co@CoP$_x$@C-1, Co@CoP$_x$@C-5, and Co@CoP$_x$@C-10 composites are 71.856, 37.693, 54.933, and 131.632 m^2/g, with the average pore size distributions of 1.90, 2.02, 2.32, and 6.06 nm (calculated with the BJH method) (Figure S3b). It is deduced that the specific surface areas of Co@CoP$_x$@C-1 and Co@CoP$_x$@C-5 composites are significantly lower than that of Co@C composites due to the increase in particle size after partial phosphatization. The specific surface area of Co@CoP$_x$@C-10 composites is larger than that of Co@C composites, which comes from the fact that a greater degree of phosphatization can result in the volatilization of the carbon component.

The transmutation of the element valence and composition for the products were revealed by X-ray photoelectron spectroscopy (XPS) [35]. As shown in Figure 3(a1–d1), the three distinctive elements C, Co, and accessional P can be detected in the survey spectrum of the samples. The comparable C1s of Co@C, Co@CoP$_x$@C-1, Co@CoP$_x$@C-5, and Co@CoP$_x$@C-10 composites speculate the inability of the phosphorylation process to significantly modify the structure of the carbon, and and C-C/C=C C-O, C=O (284.8, 285.6, 289.0 eV) peaks exist in the carbon structure for all samples exist in the carbon structure for all samples (Figure 3(a2–d2)) [36]. The typical XPS peaks of the metallic Co fraction also decreased significantly with the intensification of the phosphatization process and are divided into several peaks (Co^0 778.0 eV and 794.2 eV; Co^{2+} 781.7 eV and 797.5 eV; satellite peaks 802.5) (Figure 3(a3–d3)), indicating the gradual conversion of metallic Co to CoP [37]. As for the spectrum of P 2p, the peak at a binding energy of 130.2 eV stems from the 2P electron of phosphorus in CoP. For Co@CoP$_x$@C-1 composites, the absence of P-O is attributed to the fact that the less phosphated sample is protected by a carbon layer (Figure S4a) [38]. As the Co@CoP$_x$@C-5 and Co@CoP$_x$@C-10 composites are exposed to air, an oxidation reaction occurs on the surface, producing another peak attributed to P-O at 134.1 eV (Figure S4b,c).

Commonly, the complex permittivity ($\varepsilon_r = \varepsilon' - j\varepsilon''$) and permeability ($\mu_r = \mu' - j\mu''$) are closely correlated with the mechanism affecting the attenuation of EM waves [39–41]. Analysis of the differences in the EM parameters of the samples is authoritative in supervising the excogitation of absorbents with preeminent electromagnetic wave absorption properties. Considering the presence of dispersion behavior, the incremental frequency will induce a decrease in the ε' values [42]. Consequently, the curve trend analysis based on Figure 4a deduces that the ε' values of all samples are expected to drop from 9.57, 14.95, 12.73, and 9.95 to 4.22, 5.83, 6.22, and 4.78 with increasing frequency, where the ε' values of Co@CoP$_x$@C-1 and Co@CoP$_x$@C-5 composites are relatively larger, connoting an eminent charge storage capacity. The cause for the decrease in the ε' values of the Co@CoP$_x$@C-1, Co@CoP$_x$@C-5, and Co@CoP$_x$@C-10 composites can be attributed to the decrease in carbon content and the increase in specific surface area. According to the free electron theory, the variation in ε'' value is closely related to the electrical conductivity. As presented in Figure 4b, the difference in the presented curves infers that the ε'' value gradually decreases with increasing phosphorylation, which may be due to the decrease in conductivity. The conductivity test can also prove that the declining conductivity of the phosphated products is accompanied by the growth of the phosphating degree [43]. Typically, the Debye polarization relaxation model is relied upon to identify polarization loss and conduction loss. The ε'-ε'' can be expressed as the following equation [44–46]:

$$\left(\varepsilon' - \frac{\varepsilon_s + \varepsilon_\infty}{2}\right)^2 + (\varepsilon'')^2 = \left(\frac{\varepsilon_s - \varepsilon_\infty}{2}\right)^2 \tag{1}$$

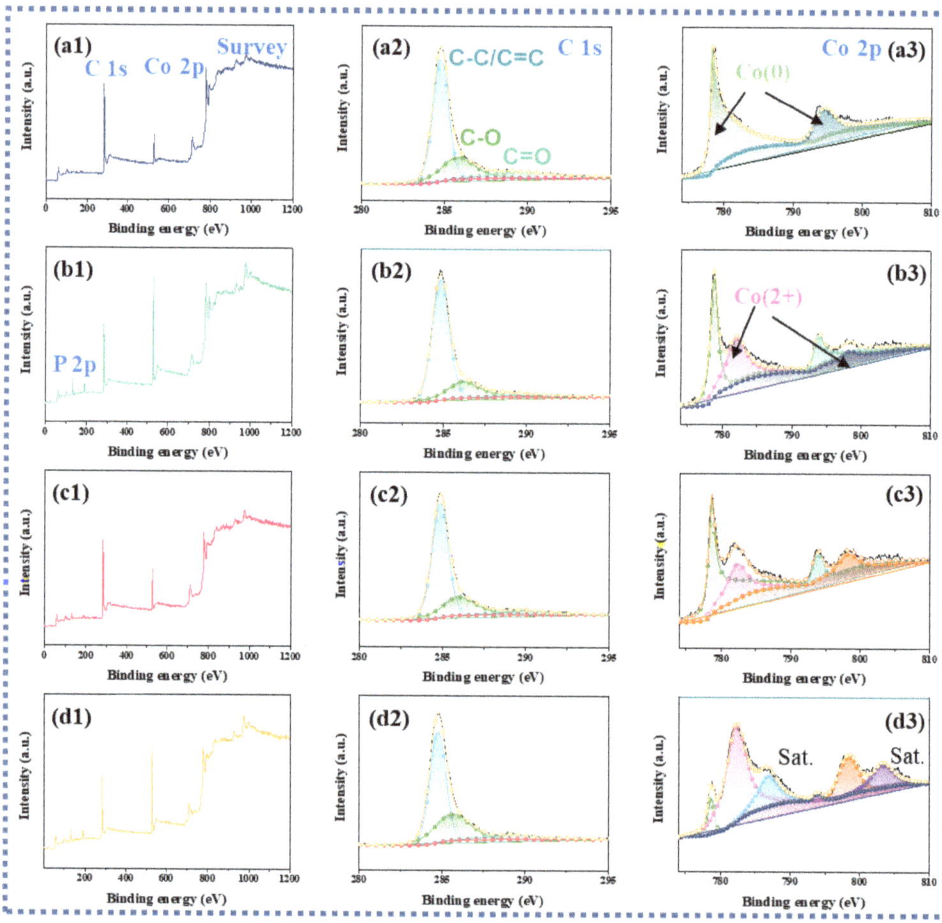

Figure 3. (**a1**–**d1**) XPS survey spectrum, (**a2**–**d2**) high-resolution C 1s, (**a3**–**d3**) high-resolution Co 2p for Co@C, Co@CoP$_x$@C-1, Co@CoP$_x$@C-5, and Co@CoP$_x$@C-10 composites.

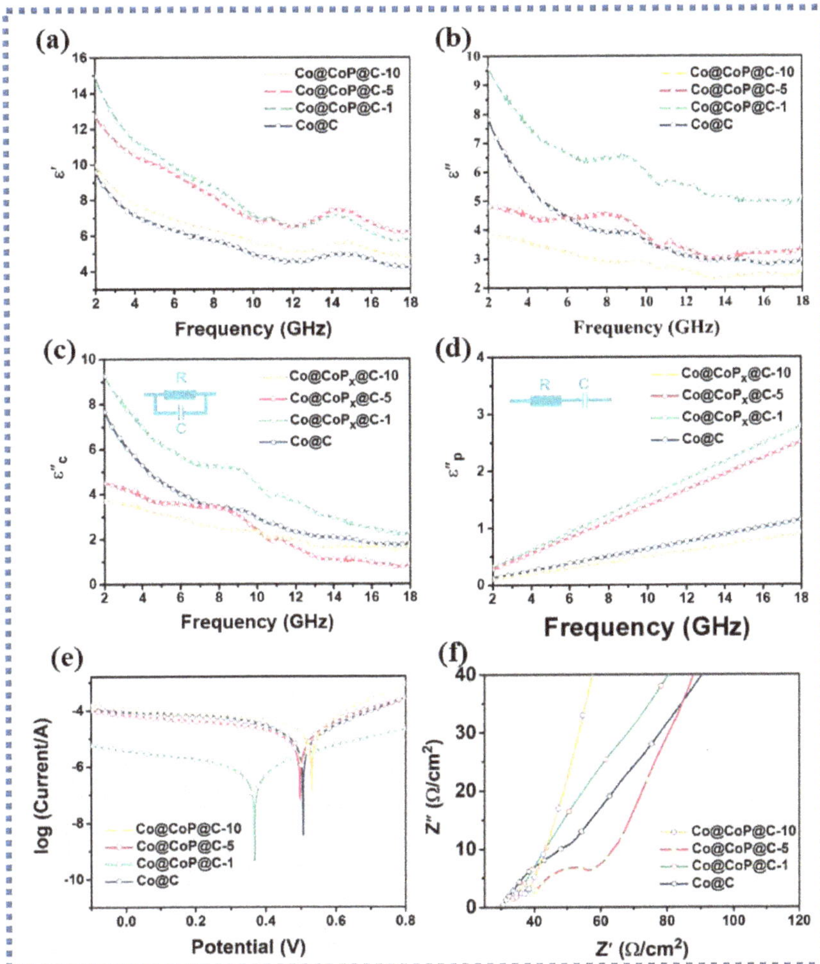

Figure 4. (**a**) Real part ε′, (**b**) imaginary part ε″, (**c**) conduction loss and (**d**) polarization loss, (**e**) Tafel curves, (**f**) Nyquist plots of Co@C, Co@CoP$_x$@C-1, Co@CoP$_x$@C-5, and Co@CoP$_x$@C-10 composites.

The Cole–Cole curve obtained by analyzing the relationship between ε′ and ε′ is plotted in Figure S5, where each Debye relaxation process comes from a semicircular. It is deduced that the presence of multiple semicircular portions in the low-frequency region and the apparent presence of straight portions in the high-frequency region in Co@C and Co@CoP$_x$@C-1 composites prove the presence of conduction loss and multiple polarization loss mechanisms. The presence of only multiple semicirculars in Co@CoP$_x$@C-5 and Co@CoP$_x$@C-10 composites suggests the predominance of polarization loss.

To further visualize the attenuation degree of conductivity loss and polarization loss on EM wave absorption, the dissipation capacity generated by polarization loss and conductivity loss can be calculated separately based on the following formula [47]:

$$\varepsilon''(\omega) = \varepsilon''_p + \varepsilon''_c = (\varepsilon''_s - \varepsilon''_\infty)\frac{\omega\tau}{1+\omega^2\tau^2} + \frac{\sigma}{\varepsilon_0\omega} \quad (2)$$

The corresponding electromagnetic fitting parameters of the samples are shown in Table S1. As presented in Figure 4c, the results confirm that the Co@CoP$_x$@C-1 composite has the highest conduction loss, while both Co@CoP_x@C-5 and Co@CoP_x@C-10 composites have a lower conduction loss than the Co@C composite. Additionally, the polarization loss capability analysis calculated for the Co@CoP_x@C-1 and Co@CoP_x@C-5 composites showed a stronger polarization loss capability (Figure 4d). The Tafel curve in Figure 4e shows that the overpotential of the Co@CoP_x@C-1 composite is much smaller than that of other composites, which means a superior interfacial electron migration ability and the strongest interfacial polarization loss ability. Correspondingly, the interfacial polarization loss capacity relationship of Co@C, Co@CoP_x@C-5, and Co@CoP_x@C-10 composites is as follows: Co@CoP_x@C-5 > Co@C > Co@CoP_x@C-10, and there is little difference in the interface polarization loss capacity. The Electrochemical Impedance Spectroscopy (EIS) plot in Figure 4f further demonstrates that the impedance of the phosphorylated composite increases, which is not favorable for electron transfer. In a comprehensive analysis, the interfacial polarization loss and conduction loss are not the main polarization loss type of the Co@CoP_x@C-5 composite. Based on the enhancement of the P-0 dipole and the presence of vacancies, the main polarization loss mechanism originates from the dipole-evoked dipole polarization loss and vacancy-evoked defect polarization loss.

Given that the cobalt metal content will certainly decrease after phosphorylation, which will inevitably make for a decrease in the saturation magnetization of the composites, the magnetic properties of all samples were tested using VSM. As exhibited in Figure 5a, it is evident that the Co@C, Co@CoP_x@C-1, Co@CoP_x@C-5, and Co@CoP_x@C-10 composites have saturated magnetization (Ms) values of 105.5, 58.8, 46.1, and 42.1 emu g^{-1}, respectively, inferring that the magnetic storage capacity gradually decreases. For all samples, magnetic loss plays an extremely feeble role in the EM wave absorption process. Modulated by the small real part (μ') and imaginary part (μ'') values, the magnetic loss is weak to the dissipation of the incident EM waves (Figure S6a,b). In view of the association between the $\tan\delta_\varepsilon$ and the $\tan\delta_\mu$, a smaller difference will promote the improvement in impedance matching. Analysis of the data in Figure S6c,d yields a significant decrease in the $\tan\delta_\mu$ values for the Co@CoP_x@C-5 and Co@CoP_x@C-10 composites, whereas the change in the $\tan\delta_\mu$ values is insignificant, which indicates that the dielectric loss not only constitutes the main mechanism for the dissipation of the EM waves but also promotes the improvement of the impedance matching.

The magnetic loss can be further distinguished from the magnetic loss mechanism by using C_0, which is calculated as follows [48,49]:

$$C_0 = \mu''(\mu')^{-2}f^{-1} \tag{3}$$

The C_0 curve in Figure 5b can be used to analyze the specific loss type of magnetic loss. The C_0 curve of all composites fluctuates in the low-frequency range, indicating that natural resonance is dominant in the low-frequency range. In the middle- and high-frequency range, the C_0 curve remains almost unchanged, demonstrating the dominance of eddy current loss [50]. On account of the research results on the internal mechanisms, the collaborative efficiency of various loss mechanisms in the absorbents was further analyzed, consisting of the attenuation constant (α) representing the attenuation capacity and the impedance matching standing for the reflection capacity of EM waves on the surface of the absorbents [51]. An explanation of the curve in Figure 5c is that higher attenuation constants of the Co@CoP_x@C-1 and Co@CoP_x@C-5 composites demonstrate the synergistic enhancement effect of various loss mechanisms, such as interfacial polarization loss, defect polarization loss, dipole loss, and conduction loss.

Nevertheless, considering that improper impedance matching can cause the reflection of EM waves on the surface of the material, superior attenuation ability does not necessarily precipitate excellent EM wave absorption performance. Commonly, the improvement in impedance matching allows EM waves to enter the material as much as possible, rather than reflecting on the material surface, thereby promoting more EM waves to be attenuated. The associated impedance matching plot (Figure 5d) represents four different $|Z_{in}/Z_0|$ curves. A critical analysis shows that the Z_{in}/Z_0 values of the Co@CoP_x@C-1 composite are below 1 throughout the entire frequency range, implying that the EM waves are mainly reflected [52]. Furthermore, by analyzing the Z_{in}/Z_0 values for the Co@C, Co@CoP$_x$@C-5, and Co@CoP$_x$@C-10 composites, it can be concluded that the absorption frequency of the phosphorylated samples significantly shifts the to the left when Z_{in}/Z_0 is equal to 1, which helps to achieve EM wave absorption in low-frequency region.

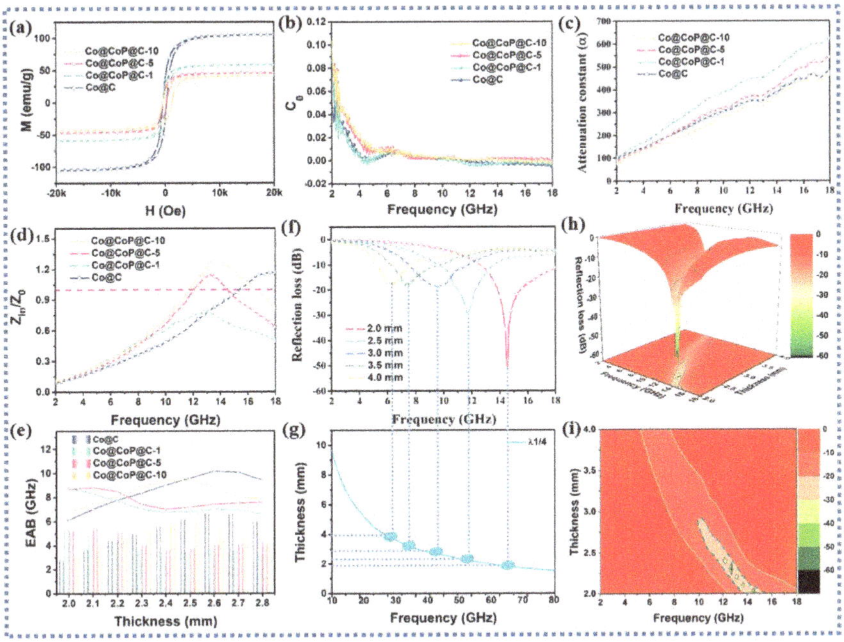

Figure 5. (**a**) Hysteresis loops, (**b**) C_0 curves, (**c**) attenuation constant (α), (**d**) impedance matching $|Z_{in}/Z_0|$ curves, (**e**) EAB comparison plots, (**f**) 2D RL curves, (**g**) $\lambda 1/4$ curves, (**h**) and 3D plots and (**i**) 2D surface plots of Co@C, Co@CoP$_x$@C-1, Co@CoP$_x$@C-5, and Co@CoP$_x$@C-10 composites.

Theoretically, the RL values of the absorbents depend on the permittivity and permeability; the calculated formula is as follows [53,54]:

$$Z_{in} = Z_0\sqrt{\frac{\mu_r}{\varepsilon_r}}tanh\left[j\frac{2\pi}{c}\sqrt{\mu_r\varepsilon_r}fd\right] \tag{4}$$

$$R_L(\text{dB}) = 20log\left|\frac{Z_{in} - Z_0}{Z_{in} + Z_0}\right| \tag{5}$$

where ε_r and μ_r represent complex permittivity and complex permeability, Z represents impedance matching, and RL represents reflection loss. To more visually characterize the EM wave absorption properties of the samples, the 2D plot, $\lambda 1/4$ curve, 3D plot, EAB curve, and 2D plane show that the prepared Co@CoP@C-5 composites accomplish effective attenuation of EM waves with a minimum reflection loss (RL$_{min}$) of −55 dB and an effective absorption bandwidth (EAB) of 5.5 GHz at a thickness of 2.0 mm, and RL curves also indulge the $\lambda 1/4$ curve. Data for the comparison samples calculated in Figure S7 manifest that the RL$_{min}$ of Co@C, Co@CoP$_x$@C-1, and Co@CoP$_x$@C-10 are −29.86 dB, −14.66 dB, and −45.38 dB at the thicknesses of 2.5 mm, 2.0 mm, and 3.0 mm, respectively, and the EAB values are up to 6.16, 5.36, and 4.56 GHz, respectively. It can be straightforwardly inferred that the Co@CoP$_x$@C-5 composite not only materializes extremely excellent EM wave absorption performance at a thin thickness but also allows for EM wave attenuation over a wide range of frequencies.

The controllable synthesis of anion-rich-vacancy Co@CoP$_x$@C composites due to different contents of phosphide reveals the dual-polarization loss mechanisms of action with defective polarization loss and dipole polarization loss dominating the polarization loss capability. The detailed EM wave loss mechanisms are shown in Figure 6. Firstly, the one-dimensional porous nanorod structure has a large specific surface area and porosity, which provides more electron transport channels and facilitates rapid electron hopping and migration, thus increasing the conduction loss. Secondly, three substances with different electron hopping behaviors form an abundant heterogeneous interface containing Co/C, CoP$_x$/C, and Co/CoP$_x$, which is conducive to achieving enhanced interfacial electron transfer to promote interfacial polarization loss. Thirdly, the phosphorus vacancy defects in CoP$_x$ nanorods lead to enhanced defect polarization, and the presence of P-O bonds also promotes the enhancement of defect polarization. In addition, the dipoles in carbon, including C-O/C=O, are the main factors that promote the loss of dipole polarization. Fifthly, the magnetic properties can also further contribute to magnetic loss.

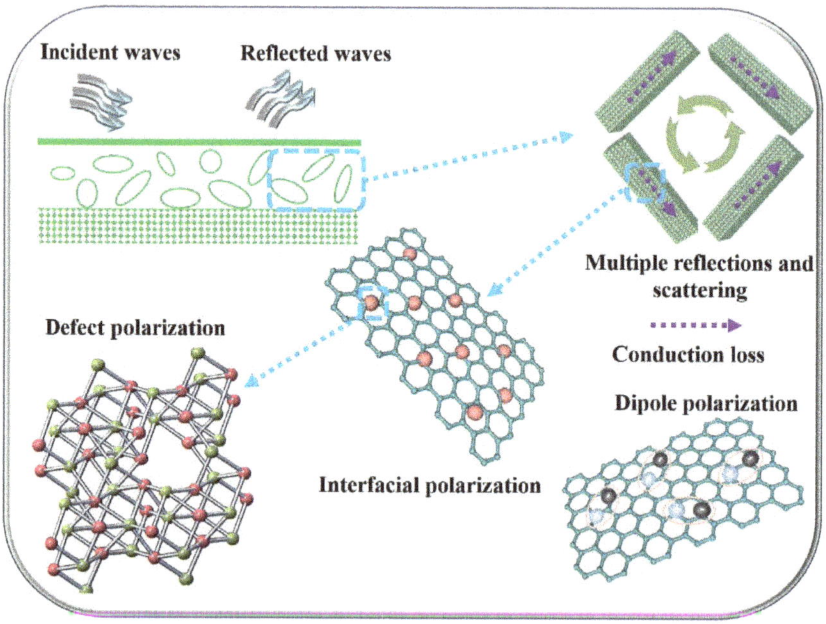

Figure 6. Schematic illustration of the mechanisms of EM wave absorption for amorphous Co@CoP$_x$@C composites.

4. Conclusions

The controllable regulation of polarization loss in dielectric loss is achieved based on suitable EM parameters. It is demonstrated, relying on extensive reliable characterization, that the optimization of EM parameters to promote a dual-polarization loss strategy is conducive to the enhancement of EM wave absorption performance. Of course, the main influencing factor can be attributed to the fact that CoP$_x$ with defects on the uniformly loaded porous carbon structure modulates the interfacial and defective polarization, resulting in a controllable regulation of polarization loss. As a result, the RL_{min} of the Co@CoP$_x$@C composites reaches −55 dB at only 2.0 mm, and the EAB is 5.5 GHz. Herein, with an eye to the amorphous CoP$_x$ with phosphorus vacancies, we reveal that the defect and interface polarization strategy is beneficial for the regulation of the dielectric constant, thus expanding the research ideas for the application of EM parameter modulation in the field of EM wave absorption.

Supplementary Materials: The following supporting information can be downloaded at: https://www.mdpi.com/article/10.3390/nano13233025/s1, Figure S1. TEM images of the Co@C composites; Figure S2. The average particle sizes of Co@C, Co@CoP$_x$@C-1, Co@CoP$_x$@C-5 and Co@CoP$_x$@C-10 composites; Figure S3. (a) Nitrogen adsorption/desorption isotherms and (b) the corresponding pore size distribution of Co@C, Co@CoP$_x$@C-1, Co@CoP$_x$@C-5 and Co@CoP$_x$@C-10 composites; Figure S4. high-resolution P 2p of (a) Co@CoP$_x$@C-1, (b) Co@CoP$_x$@C-5 and (c) Co@CoP$_x$@C-10 composites; Figure S5. Cole-Cole curves for the (a) Co@ Co@C, (b) Co@CoP$_x$@C-1, (c) Co@CoP$_x$@C-5 and (d) Co@CoP$_x$@C-5 composites; Table S1. Relevant electromagnetic fitting parameters of the samples; Figure S6. (a) real parts μ', (b) imaginary parts of complex permeability, (c) dielectric loss tangent ($\tan\delta_\varepsilon$) and (d) magnetic loss tangent ($\tan\delta_\mu$) of Co@C, Co@CoP$_x$@C-1, Co@CoP$_x$@C-5 and Co@CoP$_x$@C-10 composites; Figure S7. 2D RL curves and 3D plots of (a,b) Co@C, (c,d) Co@CoP$_x$@C-1 and (e,f) Co@CoP$_x$@C-10 composites.

Author Contributions: B.W.: design of the experiment and writing—original draft preparation; Y.M., Z.Z. and N.L.: writing—review and editing; J.X. and Y.L.: visualization; J.F.: conceptualization; G.Y.: funding acquisition; S.D.: supervision. All authors have read and agreed to the published version of the manuscript.

Funding: This work was supported by the National Natural Science Foundation of China (51908458), Natural Science Foundation of Shannxi Province, China (2020JQ-036), Key projects of Shannxi Province, China (2021GXLH-Z-070), and Fundamental Research Funds for the Central Universities of China (xzy022022004). We would like to thank the Instrument Analysis Center of Xi'an Jiaotong University for their assistance in instrument testing.

Data Availability Statement: The data are not publicly available due to privacy or ethical restrictions.

Conflicts of Interest: The authors declare no conflict of interest.

References

1. Vanskevice, I.; Kazakova, M.A.; Macutkevic, J.; Semikolenova, N.V.; Banys, J. Dielectric Properties of Hybrid Polyethylene Composites Containing Cobalt Nanoparticles and Carbon Nanotubes. *Materials* **2022**, *15*, 1876. [CrossRef] [PubMed]
2. Bleija, M.; Platnieks, O.; Macutkevic, J.; Starkova, O.; Gaidukovs, S. Comparison of Carbon-Nanoparticle-Filled Poly(Butylene Succinate-co-Adipate) Nanocomposites for Electromagnetic Applications. *Nanomaterials* **2022**, *12*, 3671. [CrossRef] [PubMed]
3. Bleija, M.; Platnieks, O.; Macutkevic, J.; Banys, J.; Starkova, O.; Grase, L.; Gaidukovs, S. Poly(Butylene Succinate) Hybrid Multi-Walled Carbon Nanotube/Iron Oxide Nanocomposites: Electromagnetic Shielding and Thermal Properties. *Polymers* **2023**, *15*, 515. [CrossRef]
4. Li, Q.; Zhang, Z.; Qi, L.P.; Liao, Q.L.; Kang, Z.; Zhang, Y. Toward the Application of High Frequency Electromagnetic Wave Absorption by Carbon Nanostructures. *Adv. Sci.* **2019**, *6*, 1801057. [CrossRef] [PubMed]
5. Wen, B.; Yang, H.B.; Lin, Y.; Ma, L.; Qiu, Y.; Hu, F.F.; Zheng, Y.N. Synthesis of Core-Shell Co@S-doped Carbon@ Mesoporous N-Doped Carbon Nanosheets with a Hierarchically Porous Structure for Strong Electromagnetic Wave Absorption. *J. Mater. Chem. A* **2021**, *9*, 3567–3575. [CrossRef]
6. Nan, H.Y.; Luo, F.; Jia, H.Y.; Deng, H.W.; Qing, Y.C.; Huang, Z.B.; Wang, C.H.; Chen, Q. Balancing between Polarization and Conduction Loss toward Strong Electromagnetic Wave Absorption of Hard Carbon Particles with Morphology Heterogeneity. *ACS Appl. Mater. Interfaces* **2022**, *14*, 19836–19846. [CrossRef] [PubMed]
7. Yang, H.B.; Wen, B.; Wang, L. Carbon Nanotubes Modified CoZn/C Composites with Rambutan-like Applied to Electromagnetic Wave Absorption. *Appl. Surf. Sci.* **2020**, *509*, 145336. [CrossRef]
8. Gao, Z.G.; Ma, Z.H.; Lan, D.; Zhao, Z.H.; Zhang, L.M.; Wu, H.J.; Hou, Y.L. Synergistic Polarization Loss of MoS_2-Based Multiphase Solid Solution for Electromagnetic Wave Absorption. *Adv. Funct. Mater.* **2022**, *32*, 2112294. [CrossRef]
9. Li, C.; Qi, X.S.; Gong, X.; Peng, Q.; Chen, Y.L.; Xie, R.; Zhong, W. Magnetic-Dielectric Synergy and Interfacial Engineering to Design Yolk-Shell Structured CoNi@void@C and CoNi@void@C@MoS_2 Nanocomposites with Tunable and Strong Wideband Microwave Absorption. *Nano Res.* **2022**, *15*, 6761–6771. [CrossRef]
10. Wen, C.Y.; Li, X.; Zhang, R.X.; Xu, C.Y.; You, W.B.; Liu, Z.W.; Zhao, B.; Che, R.C. High-Density Anisotropy Magnetism Enhanced Microwave Absorption Performance in $Ti_3C_2T_x$ MXene@Ni Microspheres. *ACS Nano* **2022**, *16*, 1150–1159. [CrossRef]
11. Yin, Y.C.; Liu, X.F.; Wei, X.J.; Yu, R.H.; Shui, J.L. Porous CNTs/Co Composite Derived from Zeolitic Imidazolate Framework: A Lightweight, Ultrathin, and Highly Efficient Electromagnetic Wave Absorber. *ACS Appl. Mater. Interfaces* **2016**, *8*, 34686–34698. [CrossRef] [PubMed]
12. Cheng, Y.; Zhao, H.Q.; Lv, H.L.; Shi, T.F.; Ji, G.B.; Hou, Y.L. Lightweight and Flexible Cotton Aerogel Composites for Electromagnetic Absorption and Shielding Applications. *Adv. Funct. Mater.* **2019**, *6*, 1900796. [CrossRef]
13. Wang, S.Y.; Peng, S.S.; Zhong, S.T.; Jiang, W. Construction of SnO_2/Co_3Sn_2@C and SnO_2/Co_3Sn_2@Air@C Hierarchical Heterostructures for Efficient Electromagnetic Wave Absorption. *J. Mater. Chem. C* **2018**, *6*, 9465–9474. [CrossRef]

14. Zhang, H.X.; Jia, Z.R.; Wang, B.B.; Wu, X.M.; Sun, T.; Liu, X.H.; Bi, L.; Wu, G.L. Construction of Remarkable Electromagnetic Wave Absorber from Heterogeneous Structure of Co-$CoFe_2O_4$@Mesoporous Hollow Carbon Spheres. *Chem. Eng. J.* **2021**, *421*, 129960. [CrossRef]
15. Wen, G.S.; Zhao, X.C.; Liu, Y.; Zhang, H.; Wang, C. Facile Synthesis of RGO/Co@Fe@Cu Hollow Nanospheres with Efficient Broadband Electromagnetic Wave Absorption. *Chem. Eng. J.* **2019**, *372*, 1–11. [CrossRef]
16. Xue, W.; Yang, G.; Bi, S.; Zhang, J.Y.; Hou, Z.L. Construction of Caterpillar-Like Hierarchically Structured Co/MnO/CNTs Derived from MnO_2/ZIF-8@ZIF-67 for Electromagnetic Wave Absorption. *Carbon* **2021**, *173*, 521–527. [CrossRef]
17. Zhang, M.; Zhao, L.B.; Zhao, W.X.; Wang, T.; Yuan, L.Y.; Guo, Y.Y.; Xie, Y.X.; Cheng, T.T.; Meng, A.L.; Li, Z.J. Boosted Electromagnetic Wave Absorption Performance from Synergistic Induced Polarization of SiCNWs@MnO_2@PPy Heterostructures. *Nano Res.* **2023**, *16*, 3558–3569. [CrossRef]
18. Liu, J.R.; Itoh, M.; Terada, M.; Horikawa, T.; Machida, K.I. Enhanced Electromagnetic Wave Absorption Properties of Fe Nanowires in Gigaherz Range. *Appl. Phys. Lett.* **2007**, *91*, 093101. [CrossRef]
19. Yang, L.Q.; Wang, Y.; Lu, Z.; Cheng, R.R.; Wang, N.; Li, Y.F. Construction of Multi-Dimensional NiCo/C/CNT/rGO Aerogel by MOF Derivative for Efficient Microwave Absorption. *Carbon* **2023**, *205*, 411–421. [CrossRef]
20. Xie, X.B.; Wang, B.L.; Wang, Y.K.; Ni, C.; Sun, X.Q.; Du, W. Spinel Structured $MFe_{(2)}O_4$ (M = Fe, Co, Ni, Mn, Zn) and Their Composites for Microwave Absorption: A review. *Chem. Eng. J.* **2021**, *428*, 131160. [CrossRef]
21. Wu, Z.C.; Cheng, H.W.; Jin, C.; Yang, B.T.; Xu, C.Y.; Pei, K.; Zhang, H.B.; Yang, Z.Q.; Che, R.C. Dimensional Design and Core-shell Engineering of Nanomaterials for Electromagnetic Wave Absorption. *Adv. Mater.* **2022**, *34*, 2107538. [CrossRef] [PubMed]
22. Zhao, G.L.; Lv, H.P.; Zhou, Y.; Zheng, X.T.; Wu, C.; Xu, C. Self-Assembled Sandwich-like MXene-Derived Nanocomposites for Enhanced Electromagnetic Wave Absorption. *ACS Appl. Mater. Interfaces* **2019**, *10*, 42925–42932. [CrossRef]
23. Li, Y.Y.; Gai, L.X.; Song, G.L.; An, Q.D.; Xiao, Z.Y.; Zhai, S.R. Enhanced Properties of CoS_2/Cu_2S Embedded N/S Co-Doped Mesh-Like Carbonaceous Composites for Electromagnetic Wave Absorption. *Carbon* **2021**, *186*, 238–252. [CrossRef]
24. Hou, T.Q.; Jia, Z.R.; Wang, B.B.; Li, H.B.; Liu, X.H.; Chi, Q.G.; Wu, G.L. Metal-Organic Framework-Derived $NiSe_2$-$CoSe_2$@C/$Ti_3C_2T_x$ Composites as Electromagnetic Wave Absorbers. *Chem. Eng. J.* **2021**, *422*, 130079. [CrossRef]
25. Sun, R.; Qu, M.X.; Peng, L.; Yang, W.W.; Wang, Z.H.; Bai, Y.; Sun, K.N. Regulating Electrochemical Kinetics of CoP by Incorporating Oxygen on Surface for High-Performance Li-S Batteries. *Small* **2023**, *19*, e2302092. [CrossRef]
26. Ha, D.H.; Han, B.H.; Risch, M.; Giordano, L.; Yao, K.P.C.; Karayaylali, P.; Shao-Horn, Y. Activity and Stability of Cobalt Phosphides for Hydrogen Evolution upon Water Splitting. *Nano Energy* **2016**, *29*, 37–45. [CrossRef]
27. Pradhan, B.; Kumar, G.S.; Dalui, A.; Khan, A.H.; Satpati, B.; Ji, Q.M.; Shrestha, L.K.; Ariga, K.; Acharya, S. Shape-Controlled Cobalt Phosphide Nanoparticles as Volatile Organic Solvent Sensor. *J. Mater. Chem. C* **2016**, *4*, 4967–4977. [CrossRef]
28. Fan, W.; Qu, H.J.; Zhao, Q.L.; Sun, X.; Wang, T.; Yu, X.Y.; Pan, J.J.; He, J.P. Cobalt Phosphide Decorated on Reduced Graphene Oxide with Enhanced Microwave Absorption Performance. *J. Alloys Compd.* **2022**, *925*, 166636. [CrossRef]
29. Wang, R.L.; He, M.; Zhou, Y.M.; Nie, S.X.; Wang, Y.J.; Liu, W.Q.; He, Q.; Wu, W.T.; Bu, X.H.; Yang, X.M. Self-Assembled 3D Flower-like Composites of Heterobimetallic Phosphides and Carbon for Temperature-Tailored Electromagnetic Wave Absorption. *ACS Appl. Mater. Interfaces* **2019**, *11*, 38361–38371. [CrossRef]
30. Ma, L.; Wang, R.; Li, Y.H.; Liu, X.F.; Zhang, Q.Q.; Dong, X.Y.; Zang, S.Q. Apically Co-Nanoparticles-Wrapped Nitrogen-Doped Carbon Nanotubes from a Single-Source MOF for Efficient Oxygen Reduction. *J. Mater. Chem. A* **2018**, *6*, 24071–24077. [CrossRef]
31. Zhang, D.Q.; Wang, H.H.; Cheng, J.Y.; Han, C.Y.; Yang, X.Y.; Xu, J.Y.; Shan, G.C.; Zheng, G.P.; Cao, M.S. Conductive WS_2-NS/CNTs Hybrids Based 3D Ultra-Thin Mesh Electromagnetic Wave Absorbers with Excellent Absorption Performance. *Appl. Surf. Sci.* **2020**, *528*, 147052. [CrossRef]
32. Duan, Y.L.; Jiang, B.; Ma, C.; Wang, X.; Wang, Y.B.; Li, R.; Yang, W.; Li, Y.F. Construction of MnO-Skeleton Cross-Linked by Carbon Nanotubes Networks for Efficient Microwave Absorption. *J. Colloid Interface Sci.* **2021**, *602*, 778–788. [CrossRef]
33. Zhuang, G.X.; Chen, Y.W.; Zhuang, Z.Y.; Yu, Y.; Yu, J.G. Oxygen Vacancies in Metal Oxides: Recent Progress towards Advanced Catalyst Design. *Sci. China Mater.* **2020**, *63*, 2089–2118. [CrossRef]
34. Li, Z.N.; Han, X.J.; Ma, Y.; Liu, D.W.; Wang, Y.H.; Xu, P.; Li, C.L.; Du, Y. MOFs-Derived Hollow Co/C Microspheres with Enhanced Microwave Absorption Performance. *ACS Sustain. Chem. Eng.* **2018**, *6*, 8904–8913. [CrossRef]
35. Li, Y.; Liu, X.F.; Nie, X.Y.; Yang, W.W.; Wang, Y.D.; Yu, R.H.; Shui, J.L. Multifunctional Organic-Inorganic Hybrid Aerogel for Self-Cleaning, Heat-Insulating, and Highly Efficient Microwave Absorbing Material. *Adv. Funct. Mater.* **2019**, *29*, 1807624. [CrossRef]
36. Yan, H.Y.; Guo, Y.; Bai, X.Z.; Qi, J.W.; Lu, H.P. Facile Constructing Ti3C2Tx/TiO2@C Heterostructures for Excellent Microwave Absorption Properties. *J. Colloid Interface Sci.* **2023**, *654*, 1483–1491. [CrossRef]
37. Liu, T.; Li, P.; Yao, N.; Cheng, G.Z.; Chen, S.L.; Luo, W.; Yin, Y.D. CoP-Doped MOF-Based Electrocatalyst for pH-Universal Hydrogen Evolution Reaction. *Angew. Chem. Int. Ed.* **2019**, *58*, 4679–4684. [CrossRef]
38. Guan, C.; Xiao, W.; Wu, H.J.; Liu, X.M.; Zang, W.J.; Zhang, H.; Ding, J.; Feng, Y.P.; Pennycook, S.J.; Wang, J. Hollow Mo-doped CoP Nanoarrays for Efficient overall Water Splitting. *Nano Energy* **2018**, *48*, 73–80. [CrossRef]
39. Wu, C.; Chen, Z.F.; Wang, M.M.; Cao, X.; Zhang, Y.; Song, P.; Zhang, T.Y.; Ye, X.L.; Yang, Y.; Gu, W.H.; et al. Confining Tiny MoO_2 Clusters into Reduced Graphene Oxide for Highly Efficient Low Frequency Microwave Absorption. *Small* **2020**, *16*, 2001686. [CrossRef]
40. Cui, L.R.; Tian, C.H.; Tang, L.L.; Han, X.J.; Wang, Y.H.; Liu, D.W.; Xu, P.; Li, C.L.; Du, Y.C. Space-Confined Synthesis of Core-shell $BaTiO_3$@Carbon Microspheres as a High-Performance Binary Dielectric System for Microwave Absorption. *ACS Appl. Mater. Interfaces* **2019**, *11*, 31182–31190. [CrossRef]

41. Sun, G.B.; Dong, B.X.; Cao, M.H.; Wei, B.Q.; Hu, C.W. Hierarchical Dendrite-Like Magnetic Materials of Fe_3O_4, Gamma-Fe_2O_3, and Fe with High Performance of Microwave Absorption. *Chem. Mater.* **2011**, *23*, 1587–1593. [CrossRef]
42. Lyu, L.F.; Wang, F.L.; Zhang, X.; Qiao, J.; Liu, C.; Liu, J.R. CuNi Alloy/Carbon Foam Nanohybrids as High-Performance Electromagnetic Wave Absorbers. *Carbon* **2021**, *172*, 488–496. [CrossRef]
43. Lv, H.L.; Ji, G.B.; Liang, X.H.; Zhang, H.Q.; Du, Y.W. A Novel Rod-Like MnO_2@Fe Loading on Graphene Giving Excellent Electromagnetic Absorption Properties. *J. Mater. Chem. C* **2015**, *3*, 5056–5064. [CrossRef]
44. Qiu, Y.; Yang, H.B.; Cheng, Y.; Bai, X.Y.; Wen, B.; Lin, Y. Constructing a Nitrogen-Doped Carbon and Nickel Composite Derived from a Mixed Ligand Nickel-Based a Metal-Organic Framework toward Adjustable Microwave Absorption. *Nanoscale* **2021**, *13*, 9204–9216. [CrossRef] [PubMed]
45. Wen, B.; Yang, H.B.; Lin, Y.; Qiu, Y.; Cheng, Y.; Jin, L.X. Novel Bimetallic MOF Derived Hierarchical Co@C Composites Modified with Carbon Nanotubes and Its Excellent Electromagnetic Wave Absorption Properties. *J. Colloid Interface Sci.* **2023**, *605*, 657–666. [CrossRef] [PubMed]
46. Wen, B.; Zhang, J.T.; Yang, G.R.; Jing, D.W.; Yin, X.C.; Fan, L.; Nasir, M.S.; Ding, S.J. Optimal Electrical Conductivity and Interfacial Polarization Induced by Loaded Nanoparticles on Carbon Nanotubes for Excellent Electromagnetic Wave Absorption Performance. *J. Colloid Interface Sci.* **2022**, *626*, 759–767. [CrossRef] [PubMed]
47. Liu, J.; Tao, J.Q.; Gao, L.L.; He, X.X.; Wei, B.; Gu, Y.S.; Yao, Z.J.; Zhou, J.T. Morphology-Size Synergy Strategy of SiC@C Nanoparticles towards Lightweight and Efficient Microwave Absorption. *Chem. Eng. J.* **2022**, *433*, 134484. [CrossRef]
48. Qiu, Y.; Lin, Y.; Yang, H.B.; Wang, L.; Wang, M.Q.; Wen, B. Hollow Ni/C Microspheres Derived from Ni-Metal Organic Framework for Electromagnetic Wave Absorption. *Chem. Eng. J.* **2020**, *383*, 123207. [CrossRef]
49. Wen, B.; Yang, H.B.; Lin, Y.; Ma, L.; Qiu, Y.; Hu, F.F. Controlling the Heterogeneous Interfaces of S, Co Co-Doped porous Carbon Nanosheets for Enhancing the Electromagnetic Wave Absorption. *J. Colloid Interface Sci.* **2021**, *586*, 208–218. [CrossRef]
50. Cai, L.; Pan, F.; Zhu, X.J.; Dong, Y.Y.; Shi, Y.Y.; Xiang, Z.; Cheng, J.; Jiang, H.J.; Shi, Z.; Lu, W. Etching Engineering and Electrostatic Self-Assembly of N-Doped MXene/Hollow Co-ZIF Hybrids for High-Performance Microwave Absorbers. *Chem. Eng. J.* **2022**, *434*, 133865. [CrossRef]
51. Zhang, X.C.; Shi, Y.A.; Xu, J.; Ouyang, Q.Y.; Zhang, X.; Zhu, C.L.; Zhang, X.L.; Chen, Y.J. Identification of the Intrinsic Dielectric Properties of Metal Single Atoms for Electromagnetic Wave Absorption. *Nano-Micro Lett.* **2022**, *14*, 27. [CrossRef] [PubMed]
52. Wen, B.; Yang, G.R.; Zhou, X.Y.; Ding, S.J. Intelligent Diffusion Regulation Induced In-Situ Growth of Cobalt Nanoclusters on Carbon Nanotubes for Excellent Electromagnetic Wave Absorption. *J. Colloid Interface Sci.* **2023**, *634*, 74–85. [CrossRef] [PubMed]
53. He, Y.Y.; Xie, P.Y.; Li, S.; Wang, Y.N.; Liao, D.G.; Liu, H.X.; Zhou, L.; Chen, Y.H. Multifunctional Carbon Foam with Hollow Microspheres and a Concave-Convex Microstructure for Adjustable Electromagnetic Wave Absorption and Wearable Applications. *J. Mater. Chem. A* **2021**, *9*, 25982–25998. [CrossRef]
54. Liang, J.; Chen, J.; Shen, H.Q.; Hu, K.T.; Zhao, B.N.; Kong, J. Hollow Porous Bowl-like Nitrogen-Doped Cobalt/Carbon Nanocomposites with Enhanced Electromagnetic Wave Absorption. *Chem. Mater.* **2021**, *33*, 1789–1798. [CrossRef]

nanomaterials

MDPI

Article

Post-Synthetic Modification of an Amino-Functionalized Metal–Organic Framework for Highly In Situ Luminescent Detection of Mercury (II)

Chen Ji †, Li Pei †, Junyi Qin, Pengyan Wu *, Nuo Su, Ting Zhang, Yexin Zhang and Jian Wang *

Jiangsu Key Laboratory of Green Synthetic Chemistry for Functional Materials, School of Chemistry and Materials Science, Jiangsu Normal University, Xuzhou 221116, China

* Correspondence: wpyan@jsnu.edu.cn (P.W.); wjian@jsnu.edu.cn (J.W.)

† These authors contributed equally to this work.

Abstract: A sulfur-containing metal–organic framework, donated as UiO-66-NSMe, was prepared by the post-synthetic modification (PSM) of UiO-66-NH_2 with 2-(Methylthio)benzaldehyde, and the successful synthesis of PSM was confirmed by X-ray photoelectron spectroscopy (XPS), FT-IR and ^1H NMR studies. According to the characteristics of mercury thiophilic, UiO-66-NSMe could be used as a luminescent sensor for Hg^{2+} detection with a high selectivity and sensitivity ($K_{sv} = 2.5 \times 10^4$ M^{-1}; LOD = 20 nM), which could be attributed to the coordination between sulfur sites and Hg^{2+} based on XPS results. In practical applications, UiO-66-NSMe yielded satisfactory recovery rates (ranging from 96.1% to 99.5%) when it was employed for detecting Hg^{2+} in spiked environmental samples. Furthermore, UiO-66-NSMe was successfully employed to detect mercury (II) residues with the in situ rapid nondestructive imaging in simulated fresh agricultural products. Thus, this PSM strategy could provide good guidance for environmental protection methodologies in the future.

Keywords: metal–organic frameworks (MOFs); fluorescent sensor; post-synthetic modification (PSM); sulfur-containing; mercury (II)

1. Introduction

Rapid industrialization has released significant amounts of pollutants, including heavy metal ions, into the environment. Among the most toxic and hazardous contaminants, mercury is extensively used in batteries, barometers, float valves, thermometers, and other devices [1]. However, when introduced into natural water environments through microbial processes, ionic mercury readily converts into neurotoxic methylmercury. Even trace levels of mercury bioaccumulation can lead to organ and health issues, such as digestive, kidney, and nervous system disorders [2]. Therefore, the detection of mercury ions in aqueous solutions is of paramount importance. Numerous techniques have been employed to detect mercury, including atomic absorption spectrometry (AAS), cold vapor atomic fluorescence spectroscopy (AFS), mass spectrometry (MS), and liquid chromatography (LC) [3,4]. However, these methods encounter challenges related to intricate sample preparation and costly equipment, which restrict their extensive in-field application. Consequently, achieving prompt response and the highly selective and sensitive detection of Hg^{2+} in water remains a formidable challenge.

Metal–organic frameworks (MOFs) [5–7] materials have garnered significant attention in the last two decades due to their high surface area, exceptional stability, adjustable porosity, customizable chemical functionality, and potential applications, particularly in gas storage and separation [8,9], heterogeneous catalysis [10,11], chemical sensing [12–15], proton conduction [16], and drug delivery [17]. Specifically, as fluorescent sensors, MOFs offer the advantage of precise structural design and control by selecting different metal ions and organic ligands to meet the needs of various analytes. Simultaneously, they can selectively adsorb molecules with specific sizes or properties through the sieving action of their pores, thereby concentrating the analytes within MOFs. This can enhance the sensitivity of detecting target molecules. Some fluorescent sensors based on MOFs for detecting Hg^{2+} have been documented [18–20]. For example, Wen et al. developed two amino-decorated MOFs for the selective and quantitative detection of Hg^{2+}, attributed to the chelating interaction between Hg^{2+} and the pendant amino motif of the ligand [21]. Our group reported the first example of novel dual-emission MOF-implicated ratiometric sensor for mercury (II) in pure water with an LOD of about 2 nM, attributed to Hg^{2+}-induced structural collapse [22]. More recently, Wang et al. designed and synthesized several FRET-based MOFs with different donor/acceptor ratios. The MOF with an optimized donor/acceptor ratio of 7.0 was employed as an efficient fluorescence turn-on sensor for Hg (II) ions, demonstrating good sensitivity and selectivity [23].

However, certain challenges, such as the compatibility of the functional groups, limited linker solubility, and inadequate chemical, or thermal stability in some cases, can hinder direct synthesis. Moreover, there are several factors that restrict their widespread use, including their inability to function in pure water, susceptibility to interference from other metal ions, and the inability to achieve precise metal ion detection specificity. Cohen and Burrows, among others, have established that the post-synthetic modification (PSM) of MOFs can yield new active sites as an effective and flexible strategy [24,25]. This approach enhances selectivity and augments detection sensitivity, offering a promising solution to the aforementioned problems. As an example, Lang and his colleagues utilized PSM in a two-dimensional metal–organic framework through photodimerization. The resulting PSM product displayed the capability to detect Al^{3+} ions through a luminescent quenching mechanism, offering significantly enhanced selectivity and sensitivity [26].

In this study, we focus on the exceptional water stability of UiO-series MOFs, attributed to the robust interaction between zirconium (IV) (Zr^{4+}) clusters and carboxylic ligands [27]. To this end, we synthesized a conventional UiO-66-type MOF (UiO-66-NH_2) utilizing a microwave-assisted method, which yields primary amine groups suitable for subsequent post-synthetic functionalization. Through the post-synthetic modification of UiO-66-NH_2 with 2-(Methylthio)benzaldehyde, a novel sulfur-containing MOF (UiO-66-NSMe) was created. Notably, when tested in the HEPES buffer, UiO-66-NSMe exhibited a significantly improved ability to selectively detect Hg^{2+} compared to the original compound. Furthermore, we delve into the mechanistic understanding behind this enhanced selectivity. The in situ imaging of UiO-66-NSMe for mercury (II) ion detection has also been developed.

2. Materials and Methods

2.1. Materials

All chemicals used in this study were of reagent-grade quality and were obtained from commercial sources without the need for additional purification. Specifically, 2-aminoterephthalic acid (99%) and acetic acid (99.5%) were purchased from the Shanghai Aladdin Biochemical Technology Co., Ltd. (Shanghai, China). N,N-Dimethylformamide (DMF, 99.5%), methanol (99.5%), Zirconium (IV) chloride ($ZrCl_4$, 98%), 2-(Methylthio)benzaldehyde (97%), and HEPES (99%) were purchased from Beijing Innochem Science & Technology Co., Ltd. (Beijing, China). Mercury (II) nitrate hydrate (98%) and the other metal salts were purchased from Shanghai Fourth Chemical Reagent Company (Shanghai, China). Stock solution (2×10^{-2} M) of the aqueous nitrate salts of K^+, Na^+, Li^+, Cd^{2+}, Zn^{2+}, Mn^{2+}, Fe^{2+}, Cu^{2+}, Ba^{2+}, Sr^{2+}, Ca^{2+}, Mg^{2+}, Ni^{2+}, Co^{2+}, Pb^{2+}, Ag^+, Fe^{3+}, Cr^{3+}, Al^{3+}, and Hg^{2+} were prepared for further experiments. Deionized water generated from the purification chain was used for all experiments. Lettuce was purchased at the Zhaishan Farmers' Market (No. 8 Chengnan Avenue, Quanshan District, Xuzhou, China).

2.2. Characterization Methods

The X-ray powder diffraction (XRD) patterns of UiO-66-NH_2, UiO-66-NSMe, and UiO-66-NSMe after Hg^{2+} detection were recorded on a Bruker D8 Advance X-ray powder diffractometer using Cu–Kα (λ = 1.5405 Å) radiation. FT-IR spectra were recorded as KBr pellets on Bruker Optics TENSOR 27 FT-IR spectrophotometer (Bruker, Billerica, MA, USA). X-ray photoelectron spectroscopy (XPS) experiments were performed using a PHI QUANTUM2000 surface analysis instrument. The morphologies of the prepared samples were recorded by a Field Emission Scanning Electron Microscopy (SEM) by Hitachi (SU8010) (Tokyo, Japan). Samples were treated via Pt sputtering for 90 s before observation. Nitrogen sorption isotherms were recorded on an Autosorb-IQ2 instrument at 77 K. ^{1}H NMR spectra were recorded using Me_4Si as an internal standard on a Bruker-400 spectrometer. The Zr^{4+} contents before and after Hg^{2+} detection were measured by Inductively Coupled Plasma Spectrometer (Perkin Elmer, Waltham, MA, USA). The solution fluorescence spectra were measured on JASCO FP6500. Both excitation and emission slit widths were 5 nm, and fluorescence measurements were carried out in a 1 cm quartz cuvette while stirring the suspension of UiO-66-NH_2 or UiO-66-NSMe.

2.3. Synthesis of UiO-66-NH_2

Following the methodology outlined in the references [28], the synthesis of UiO-66-NH_2 was achieved through a microwave-assisted procedure. Initially, a combination of 2-aminoterephthalic acid (0.75 mmol, 0.135 g) and $ZrCl_4$ (0.75 mmol, 0.175 g) was introduced into a solution composed of DMF (20 mL) and acetic acid (2.4 mL). This mixture was then subjected to microwave reactor and heating at 120 °C for a duration of 1 h. After cooling to room temperature, the solution underwent centrifugation, resulting in a precipitate that underwent a thorough washing process using DMF and MeOH, which was repeated three times, to effectively remove any unreacted ligand/metal. The resulting gray-colored powder was collected and subsequently air-dried.

2.4. Synthesis of UiO-66-NSMe

A quantity of UiO-66-NH_2 weighing 100 mg was dispersed within a solution of 2-(Methylthio)benzaldehyde (1.2 mmol, 0.178 g) in MeOH (50 mL), along with two drops of acetic acid. The resulting mixture was subjected to stirring at 75 °C for a duration of 24 h. Following the completion of the reaction, the resultant product, UiO-66-NSMe, was obtained in a colorless form. The obtained products were collected through a filtration process, subsequently washed with MeOH, and finally air-dried.

2.5. Digestion and ^{1}H NMR on UiO-66-NSMe

Approximately 10 mg of UiO-66-NSMe materials was desiccated under vacuum at 100 °C for 8 h. Subsequently, it were subjected to digestion by d^6-DMSO (500 μL) and 48% HF (30 μL) via sonication for 1 h until a clear solution was achieved. The resulting solution was then subjected to ^{1}H NMR analysis.

2.6. Detection of Hg^{2+}

In the typical procedure, 1 mg of finely ground UiO-66-NSMe sample was carefully dispersed in 3 mL of HEPES buffer with a pH of 7.4. Following a 5 min sonication period to ensure thorough mixing, the initial fluorescence spectra of UiO-66-NSMe were recorded. Subsequently, the fluorescence spectra of UiO-66-NSMe were measured upon the addition of Hg^{2+} ions, using an excitation wavelength of 360 nm. To assess the selectivity of UiO-66-NSMe for Hg^{2+}, similar experiments were conducted with various other metal ions, each undergoing a comparable sensing procedure in the solution.

In the analysis of Hg^{2+} in spiked samples, we employed lake water, tap water, and rainwater devoid of Hg^{2+} as representative sample matrices. Spiked lake water, tap water, and rainwater were subjected to direct detection immediately following spiking. To simulate the direct detection of Hg^{2+} residue on the surfaces of garden produce, we sprayed lettuce samples with solutions containing varying concentrations of Hg^{2+}. After a 2 min interval, a UiO-66-NSMe water suspension was applied to the smeared surfaces, followed by another 2 min wait period. Subsequently, we observed the samples under UV light. To prepare the fluorescent test paper, a standard light-yellow wood-colored filter paper was cut into a square measuring 2 cm in both length and width. Subsequently, the filter paper was immersed in a UiO-66-NSMe solution and allowed to air-dry naturally.

2.7. Typical Procedure for Cyanosilylation of Aldehydes

Into a 3 mL CH_2Cl_2 mixture containing 1.2 mmol of Trimethylsilyl cyanide ($(CH_3)_3SiCN$, TMSCN) and 0.5 mmol of aromatic aldehyde, UiO-66-NSMe or UiO-66-NSMe$\supset$$Hg^{2+}$ (5 mg) was introduced. The resulting mixture was stirred at room temperature for a duration of 3 h. The progression of the reaction was monitored via thin-layer chromatography (TLC), and conversions were quantified through ^{1}H-NMR analysis.

3. Results

3.1. Characterization of PSM Product UiO-66-NSMe

A reaction involving $ZrCl_4$ and 2-aminoterephthalic acid in DMF utilizing microwave synthesis yielded UiO-66-NH_2. The resulting material exhibited powder X-ray diffraction (PXRD) patterns that were notably congruent with those previously reported for the UiO-66 topology (Figure 1a). Additionally, scanning electron microscopy (SEM) observations displayed a consistently uniform octahedral morphology (Figure 1b), aligning with the findings in the existing literature [29,30], all confirming the successful fabrication of UiO-66-NH_2. Subsequently, sulfur-containing functional groups were covalently incorporated into the framework by exploiting Schiff base reactions between amino groups and aldehyde groups within the framework structure.

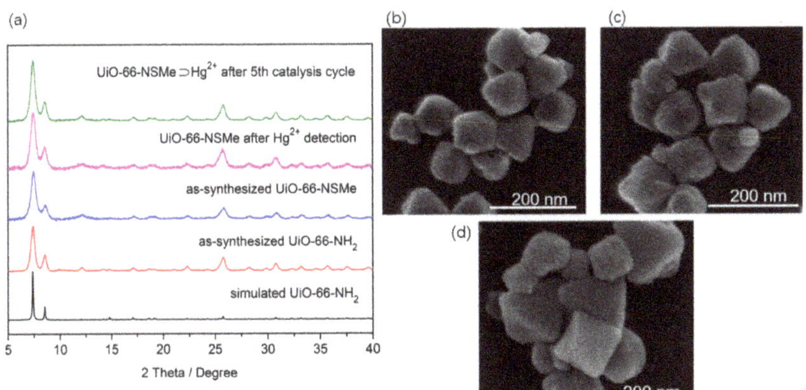

Figure 1. (**a**) Powder X-ray diffraction (XRD) profiles for the simulated (black) and as-synthesized UiO-66-NH_2 (red), as-synthesized UiO-66-NSMe (blue), UiO-66-NSMe after Hg^{2+} detection (magenta), and UiO-66-NSMe$\supset$$Hg^{2+}$ after the fifth catalysis cycle (olive). (**b**) SEM imaging of UiO-66-NH_2. (**c**) SEM imaging of UiO-66-NSMe. (**d**) SEM imaging of UiO-66-NSMe after Hg^{2+} detection.

By conducting a reflux reaction involving UiO-66-NH_2 and 2-(Methylthio)benzaldehyde with glacial acetic acid as the catalyst, UiO-66-NSMe was obtained. The synthesized material underwent filtration, washing, and drying steps. The PXRD patterns demonstrated that the structural integrity of UiO-66-NMSe remained unchanged during the process of post-synthetic modification (Figure 1a). Simultaneously, the morphology of the samples was examined using SEM. As depicted in Figure 1c, the images of UiO-66-NSMe exhibited a uniformly dispersed octahedral structure. These observations are in line with the PXRD analysis, indicating that the PSM process applied to UiO-66-NH_2 did not exert a discernible influence on its microstructure.

The successful synthesis of UiO-66-NSMe was subjected to a comprehensive analysis. A comparison of the FT-IR spectra of UiO-66-NSMe with UiO-66-NH_2, depicted in Figure 2a, revealed new peaks at approximately 2972–2924 cm^{-1}. These peaks correspond to the characteristic C-H stretching vibrations of the methyl group in UiO-66-NSMe. Importantly, a new peak appeared at 1618 cm^{-1}, which is indicative of the characteristic C=N stretching vibrations specific to UiO-66-NSMe. Simultaneously, the peak at 1334 cm^{-1}, which was attributed to the C-N stretching vibration of UiO-66-NH_2, nearly disappeared. The alterations in the FT-IR spectra collectively demonstrated that the structure of UiO-66-NSMe is characterized by the absence of free amino groups, replaced by the creation of new C=N groups via the Schiff base reaction. In addition, methyl groups associated with -SMe are unambiguously identified within the structure of UiO-66-NSMe.

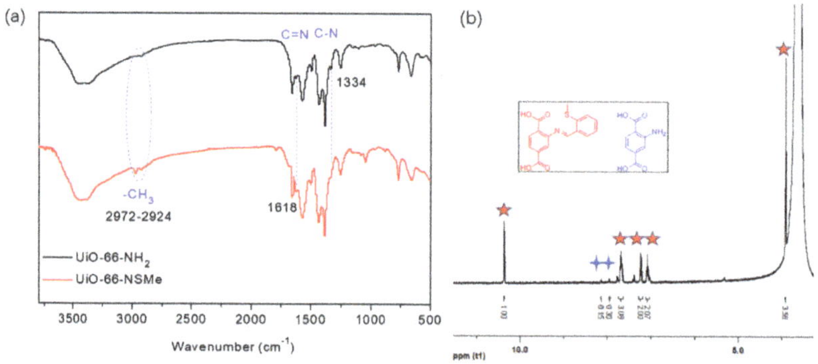

Figure 2. (**a**) FT-IR spectra of UiO-66-NH_2 and UiO-66-NSMe. (**b**) ^1H NMR spectra of UiO-66-NSMe digested in DMSO-d_6/HF.

Another solid evidence supporting the formation of Schiff base reaction products is derived from ^1H NMR studies. In DMSO-d_6/HF, UiO-66-NSMe displayed a resonance signal at 10.37 ppm, corresponding to the -CH=N- proton, together with a methyl proton signal at 3.92 ppm. The integration area between these signals is consistent with the anticipated product, affirming the success of the PSM process (Figure 2b). A modification rate of up to 87% was determined by calculating the ratio of aromatic region at PSM products (Figure 2b, red star) to the proton signal of 2-aminoterephthalic acid (Figure 2b, blue cross) in ^1H NMR after the digestion of UiO-66-NSMe. Furthermore, the XPS full spectra of UiO-66-NSMe (Figure 4) unveiled a new peak at 163.07 eV in the S 2p region, providing further confirmation of successful PSM and the resultant formation of UiO-66-NSMe. At the same time, the elemental composition obtained by XPS for UiO-66-NSMe indicated that PSM extent is calculated as 84%, which is quite close to the ^1H NMR result (Table S1).

N_2 sorption isotherms were subsequently measured at 77 K on the as-synthesized UiO-66-NH_2 and UiO-66-NSMe samples. It was observed that the surface area had decreased from 963 m^2 g^{-1} to 425 m^2 g^{-1}, further confirming the successful formation of PSM products (Figure 3). Nonetheless, UiO-66-NSMe still maintains its pores, ensuring the ingress and egress of guest molecules or ions. All these data collectively highlight the successful covalent post-synthetic modification of UiO-66-NH_2, in which a Schiff base reaction was effectively established between the amino group within the MOF and the aldehyde group, all while maintaining the structural integrity.

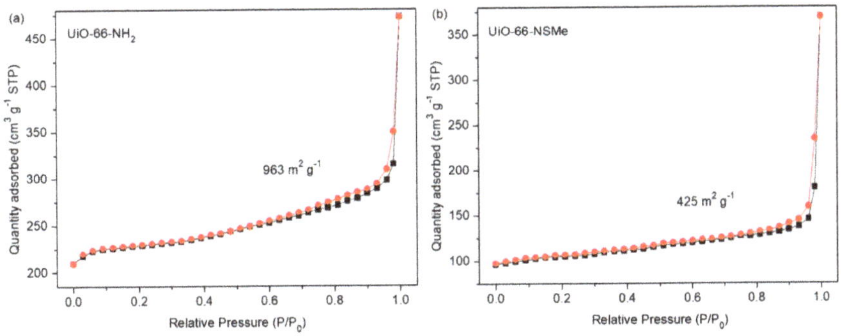

Figure 3. N_2 sorption isotherms for compounds UiO-66-NH_2 (**a**) and UiO-66-NSMe (**b**) at 77 K. The adsorption and desorption branches are shown with black square and red circle, respectively.

3.2. Luminescent Detection of Metal Ions in HEPES Buffer Based on UiO-66-NSMe

Considering the presence of sulfur atoms within the UiO-66-NSMe framework, we have delved into its metal-ion-sensing capabilities, particularly its exploitation of the thiophilic nature of mercury ions [31–33]. Under excitation at 360 nm, the UiO-66-NSMe suspension in HEPES buffer exhibited a pronounced fluorescence emission peak at approximately 450 nm (Figure 4a). Luminescence titrations involving Hg^{2+} were then conducted with the UiO-66-NSMe suspension to assess its capacity for detecting mercury (II) ions. Notably, rapid and substantial fluorescence quenching was observed upon the addition of Hg^{2+} solution to the UiO-66-NSMe suspension, reaching concentrations up to 0.56 mM (Figure 4a). This quenching effect can be quantified using the Stern–Volmer equation, $I_0/I = 1 + K_{sv}[M]$, where I denotes the luminescence intensity after Hg^{2+} addition, I_0 is the initial luminescence intensity of UiO-66-NSMe, [M] represents the molar concentration of the introduced Hg^{2+}, and K_{sv} signifies the Stern–Volmer constant. Through calculation based on the experimental data in Figure 3a, a K_{sv} value of 2.5×10^4 M^{-1} was determined (Figure 4b). Following the 3δ IUPAC criteria, the limit of detection for Hg^{2+} is approximately 20 nM. Therefore, UiO-66-NSMe proves to be a proficient material for detecting mercury (II) ions, with its heightened sensitivity for Hg^{2+} attributed to the affinity-driven interaction between Hg^{2+} and the sulfur atoms present.

To assess the selectivity of UiO-66-NSMe towards Hg^{2+}, its luminescence responses were examined upon the introduction of various other metal ions, including Cr^{3+}, Al^{3+}, K^+, Co^{2+}, Sr^{2+}, Cd^{2+}, Na^+, Mg^{2+}, Ba^{2+}, Pb^{2+}, Ni^{2+}, Cu^{2+}, Zn^{2+}, Ag^+, Ca^{2+}, Fe^{3+}, Li^+, and Mn^{2+} (Figures 4c and S1). The fluorescence spectra of UiO-66-NSMe exhibited minimal changes in the presence of K^+, Co^{2+}, Mg^{2+}, Ba^{2+}, Pb^{2+}, Ni^{2+}, Al^{3+}, Li^+, Cd^{2+}, and Zn^{2+}. Slight fluorescence enhancement was observed with Sr^{2+}, Mn^{2+}, and Na^+. In UiO-66-NSMe, the fluorescence quenching induced by Cr^{3+}, Ca^{2+}, Ag^{2+}, Cu^{2+}, and Fe^{3+} was only 40% of their highest quenching efficiency. In sharp contrast, the addition of Hg^{2+} ions (0.56 mM) led to a fluorescence quenching exceeding 93%. Comparatively, the introduction of Hg^{2+} ions into the original UiO-66-NH_2 suspension resulted in only a 33% fluorescence quench. Meanwhile, the anti-interference capability of UiO-66-NSMe for Hg^{2+} detection was further explored by competing experiments (Figure 4c, green bars). The experimental result reveals that the coexistence of interferers had no obvious interference on the Hg^{2+} detection. This distinction highlights the remarkable selectivity of UiO-66-NSMe for Hg^{2+}. The framework modification through PSM, leading to the introduction of sulfur (S) moieties, emerges as an effective approach to enhance the detection of mercury (II) ions.

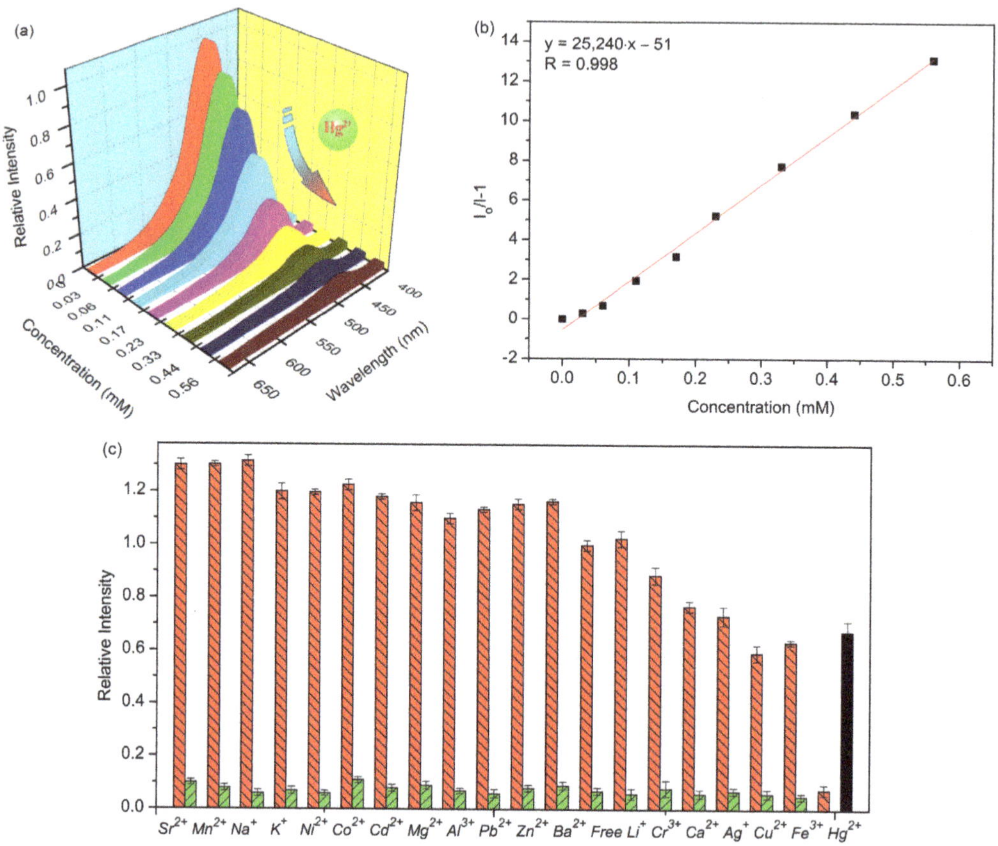

Figure 4. (**a**) Luminescence spectra of UiO-66-NSMe upon incremental addition of Hg^{2+} up to 0.56 mM in HEPES buffer, with excitation at 360 nm. (**b**) Stern–Volmer plot of UiO-66-NSMe quenched by Hg^{2+}. (**c**) The red bars stand for the luminescence intensities of UiO-66-NSMe when the selected metal ions (1.12 mM) are present. The green ones stand for the change of the emission when continuing to add 0.56 mM Hg^{2+} (The black bar represents the fluorescence intensity of UiO-66-NH_2 upon adding 0.56 mM of Hg^{2+}).

To further validate the detection mechanism of Hg^{2+} on UiO-66-NSMe, the PXRD patterns of the UiO-66-NSMe samples following Hg^{2+} detection matched well with the original pattern (Figure 1a). This alignment demonstrates that the primary framework structure remained unchanged. Moreover, inductively coupled plasma (ICP) measurements revealed minimal Zr^{4+} content present in the filtrate (Table S2), effectively excluding the possibilities of compound dissolution and the exchange of Hg^{2+} and Zr^{4+} within the lattice, which could lead to the drastic luminescence quenching observed. These findings are additionally supported by the sustained SEM morphology even after detecting Hg^{2+} ions (Figure 1d). Collectively, these results reinforce the notion that the significant luminescence quenching in response to Hg^{2+} is not attributed to structural changes or dissolution but rather is due to specific interaction between Hg^{2+} and UiO-66-NSMe.

Furthermore, the X-ray photoelectron spectroscopy (XPS) spectra of UiO-66-NSMe were measured both before and after Hg^{2+} detection. After Hg^{2+} detection, a new peak corresponding to Hg 4f at a binding energy of approximately 100.91 eV emerged in the full spectrum of UiO-66-NSMe. This unequivocally confirms the adsorption of Hg^{2+} within the channels of UiO-66-NSMe (Figure 5). Intriguingly, a shift from 163.07 eV to 164.04 eV in the binding energy of the S 2p peak was observed before and after Hg^{2+} detection. Such a shift implies an augmented electron density around the sulfur (S) atoms, likely stemming from the coordination interaction between Hg^{2+} and S atoms within UiO-66-NSMe. Thus, combined with the structure of UiO-66-NSMe, it is possible that Hg^{2+} ions are chelating coordinated by the S and N sites in UiO-66-NSMe.

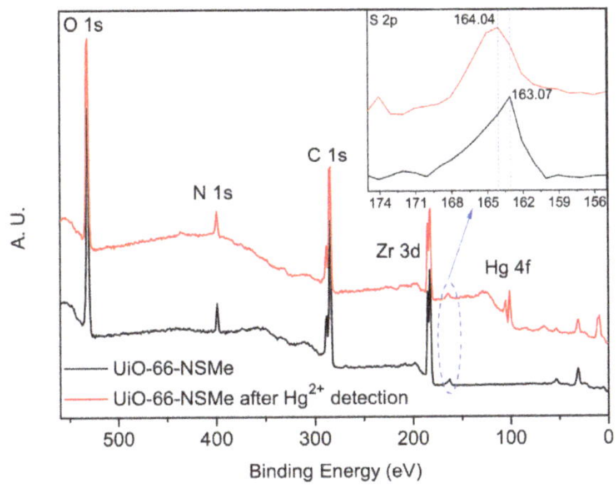

Figure 5. XPS spectra of UiO-66-NSMe before and after Hg^{2+} detection.

As the UiO-66-NSMe detection of Hg^{2+} relies on the coordination between mercury and sulfur sites, the material can be recycled after detecting Hg^{2+}. This newly coordinated Hg^{2+} material, donated as UiO-66-NSMe⊃Hg^{2+}, possesses additional metal centers compared to the original material, thereby enhancing its Lewis acidic properties. In its role as a catalyst, UiO-66-NSMe⊃Hg^{2+} demonstrates improved catalytic activity for Lewis acid-catalyzed cyanosilane reactions. To illustrate this enhancement, we conducted a model reaction using benzaldehyde and $(CH_3)_3SiCN$ at room temperature in CH_2Cl_2 for 3 h. With UiO-66-NSMe⊃Hg^{2+} (5 mg) as catalysts, the expected product, 2-phenyl-2-((trimethylsilyl)oxy)acetonitrile, was obtained in >99% yield. In contrast, when UiO-66-NSMe was used as the catalyst, it yielded a 58% 2-phenyl-2-((trimethylsilyl)oxy)acetonitrile with no observed byproducts. This substantial difference can be attributed to the enhanced Lewis acid properties conferred by UiO-66-NSMe⊃Hg^{2+}.

Importantly, UiO-66-NSMe⊃Hg^{2+} could be readily recovered from the catalytic system through centrifugation and subsequently reused in fresh conditions. Remarkably, this catalyst exhibited no significant decrease in efficiency; the yield for 2-phenyl-2-((trimethylsilyl)oxy)acetonitrile was 97% for the second run, 98% for the third run, 96% for the fourth run, and 97% for the fifth run. The stability of UiO-66-NSMe⊃Hg^{2+} was confirmed by XRD after the fifth catalysis cycle (Figure 1a). This exceptional catalytic cycling performance further underscores the robust interaction between UiO-66-NSMe and mercury (II) ions, reinforcing the previous findings of coordination between mercury (II) ions and sulfur sites.

With these impressive catalytic results in hand, we extended our investigation to explore the generality and scope of the current protocol across a variety of aromatic aldehydes in conjunction with $(CH_3)_3SiCN$. As presented in Table 1 (Figures S2–S7), when utilizing UiO-66-NSMe⊃Hg^{2+} as the catalyst, the yields for 4-methoxybenzaldehyde, 4-nitrobenzaldehyde, and 4-chlorobenzaldehyde reached 96%, 95%, and 93%, respectively. These yields were almost twice as high as when UiO-66-NSMe was used as the catalyst. However, as the size of the substrates increased, catalytic efficiency exhibited a substantial decrease. For instance, 1-naphthaldehyde yielded only 51%, while the larger 3,5-di-tert-butylbenzaldehyde resulted in a mere 15% yield when UiO-66-NSMe⊃Hg^{2+} served as the catalyst. These outcomes underscore that Hg^{2+}-modified UiO-66-NSMe performs well not only in the cyanosilylation of aromatic aldehydes but also displays promising size-selective behavior. This presents a bright prospect for the design of more active catalyst materials and offers an effective means for the reutilization of test materials.

Table 1. Results of the cyanosilylation reaction catalyzed by UiO-66-NSMe and UiO-66-NSMe⊃Hg^{2+}.

Entry	Substrates	Yields (%)	
		UiO-66-NSMe⊃Hg^{2+}	**UiO-66-NSMe**
1	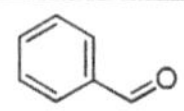	>99	58
2	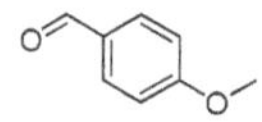	96	55
3	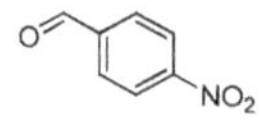	95	52
4	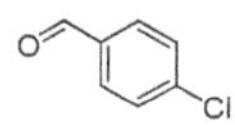	93	54
5	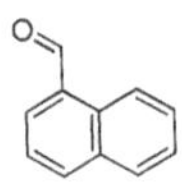	51	33
6	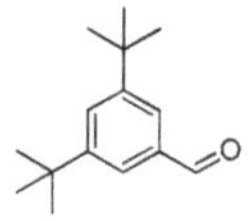	15	6

Reaction conditions: aromatic aldehydes (0.50 mmol), catalyst (5 mg), $(CH_3)_3SiCN$ (1.2 mmol), CH_2Cl_2 (3 mL) are added to the cuvette. The cuvette is stirred at room temperature for 3 h.

3.3. *In Situ Imaging Detection of Hg^{2+} in Real Water, Vegetables, and Test Paper*

To evaluate the reliability for practical Hg^{2+} detection by using the prepared material in real water samples, we employed a "turn-off" fluorescent sensor based on UiO-66-NSMe. Given that Hg^{2+} ions were not naturally present in rainwater, lake water (Yunlong Lake, Yunlong district, Xuzhou, China), or tap water (No. 1 Beijing South Road, Tongshan district, Xuzhou, China), we conducted recovery experiments using spiked samples. The added amounts of Hg^{2+} were set at 10 μM, 20 μM, and 30 μM, respectively. The fluorescence-based method for detecting Hg^{2+}, relying on UiO-66-NSMe in real water samples, as indicated in Table 2 (Figures S8 and S9), yielded recovery rates ranging from 96.1% to 99.5%. These outstanding detection results demonstrate that our designed UiO-66-NSMe possesses the capability of reliable detection for Hg^{2+} in real-world samples.

Table 2. Spike and recovery test of Hg^{2+} by using UiO-66-NSMe in real water samples.

Samples	Spiked (μM)	Found (μM)	Recovery (%)
Rainwater	10	9.65	96.5
	20	19.52	97.6
	30	29.34	97.8
Lake water	10	9.95	99.5
	20	19.78	98.9
	30	28.83	96.1
Tap water	10	9.84	98.4
	20	19.64	98.2
	30	29.12	97.1

Given UiO-66-NSMe's remarkable ability to rapidly and efficiently detect Hg^{2+}, we conducted further investigations to assess its potential for on-site detection. UiO-66-

NSMe was employed to simulate the detection of mercury (II) residue on the surface of garden produce through direct imaging technology. Because vegetables are sensitive agricultural products that require a lot of water to irrigate, exogenous mercury (II) ions can enter them. Since Hg^{2+} ions were not naturally present on the lettuce we obtained, we simulated mercury (II) contamination by applying varying concentrations of Hg^{2+} solution to their surfaces. Subsequently, luminescence was observed under UV light at 365 nm following the addition of the UiO-66-NSMe suspension. As illustrated in Figure 5, a significant reduction in UiO-66-NSMe luminescence on the surface of lettuce was observed, progressively diminishing until complete quenching as the Hg^{2+} concentration increased from 0.04 mM to 0.56 mM (Figure 6). Collectively, these results strongly indicate that UiO-66-NSMe holds great promise for the rapid and on-site detection of mercury (II) residues in real samples through direct imaging techniques.

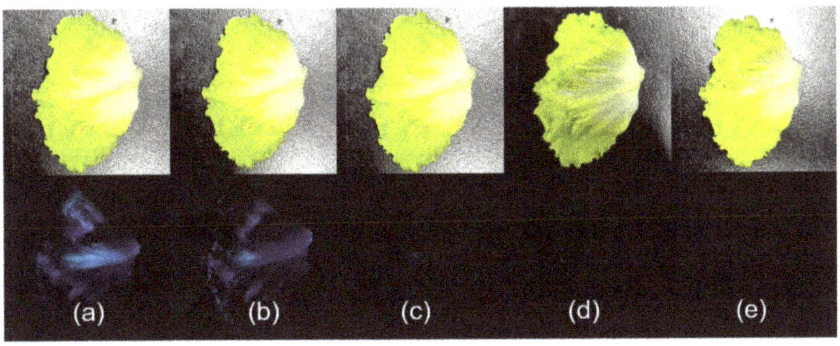

Figure 6. Noticeable differences in fluorescence became evident when UiO-66-NSMe suspension (0.33 g/L) was introduced to the surface of the lettuce under ultraviolet light, and the change in fluorescence was observed within 30 s. Lettuce was subjected to spraying with solutions of varying concentrations: 0 (**a**), 0.04 (**b**), 0.18 (**c**), 0.35 (**d**), and 0.56 (**e**) mM Hg^{2+}, simulating mercury (II) residue.

To facilitate simple and portable detection, we developed a fluorescence test paper for the rapid assessment of Hg^{2+} in aqueous solutions. This test paper was created by immersing a light-yellow wood-colored Whatman filter paper measuring 2.0 × 2.0 cm^2 into a dispersion of ground UiO-66-NSMe in water and subsequently allowing it to dry at room temperature. For the detection of minuscule quantities of Hg^{2+} ions, 1 μL of aqueous solutions containing varying concentrations of Hg^{2+} ions was placed onto the UiO-66-NSMe test papers. As depicted in Figure 7, when exposed to UV light at 365 nm, the test papers exhibited varying colors of different intensities that were discernible to the naked eye, corresponding to the concentration levels of Hg^{2+} ions. The minimum detectable amount of Hg^{2+} ions was calculated to be at the remarkably low level of 13.7 ng. All these practice experimental findings underscore the advantages of UiO-66-NSMe as an efficient, sensitive, and selective sensor for Hg^{2+} ions.

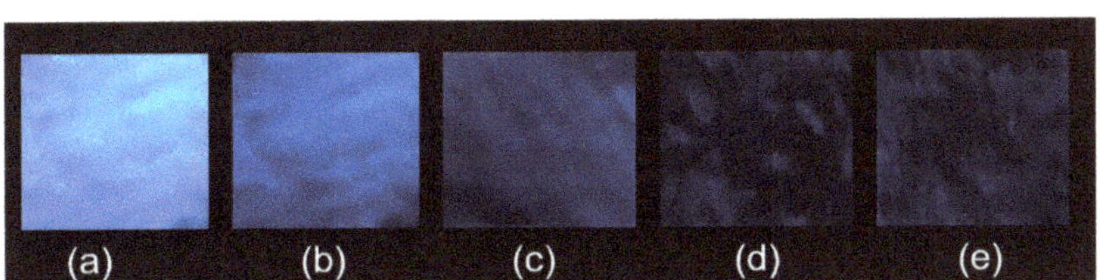

Figure 7. Photographs depicting the fluorescence quenching of UiO-66-NSMe on test strips for the visual detection of small amounts of Hg^{2+} under 365 nm UV light: (**a**) test strip; Hg^{2+} of different concentrations (**b**) 0.04 mM, (**c**) 0.18 mM, (**d**) 0.35 mM, (**e**) 0.56 mM.

4. Conclusions

In conclusion, we have successfully synthesized a novel framework with the -SCH_3 functional group, designated as UiO-66-NSMe, through post-synthetic modification (PSM). The effective conversion was verified by FT-IR, XPS, and ^{1}H NMR analyses, yielding a PSM rate of up to 87%. Luminescence-based studies of metal ion detection demonstrated that UiO-66-NSMe displays exceptional selectivity and sensitivity towards Hg^{2+} in HEPES buffer. Additionally, the comprehensive detection mechanism was attributed to the presence of a coordination interaction between Hg^{2+} and S atoms within UiO-66-NSMe. The MOF sensor was also utilized for the detection of Hg^{2+} in tap water, rainwater, and river water spiked with the contaminant. Additionally, it was employed to detect mercury (II) residues with in situ rapid nondestructive imaging in simulated fresh agricultural products. Moreover, a straightforward and portable test paper enabled the visual detection of Hg^{2+} ions at the nanogram level by the naked eye. We believe that this study holds promise for advancing environmental detection methodology in the future.

Supplementary Materials: The following supporting information can be downloaded at: https://www.mdpi.com/article/10.3390/nano13202784/s1, Table S1: XPS elemental compositions (atomic %); Figure S1: The fluorescence spectra of UiO-66-NSMe in HEPES buffer upon the addition of 0.56 mM of various metal ions; Figures S2 to S7: ^{1}H NMR characterization of catalytic products by as-synthesized UiO-66-NSMe⊃Hg^{2+}; Table S2: The ICP result on the filtrate of UiO-66-NSMe after Hg^{2+} detection; Figure S8: The relative fluorescence intensity of UiO-66-NSMe with varying Hg^{2+} concentrations; Figure S9: Spike test of Hg^{2+} by using UiO-66-NSMe in real water samples.

Author Contributions: Conceptualization, P.W. and J.W.; methodology, C.J. and L.P.; software, J.Q. and Y.Z.; validation, C.J., L.P., N.S. and T.Z.; formal analysis, C.J. and L.P.; investigation, J.Q.; data curation, C.J. and L.P.; writing—original draft preparation, J.W.; writing—review and editing, P.W. and J.W.; visualization, N.S.; supervision, P.W. and J.W.; project administration, J.W. All authors have read and agreed to the published version of the manuscript.

Funding: This research was funded by the Major Basic Research Project of Natural Science Foundation of the Jiangsu Higher Education Institutions (21KJA150001), the Natural Science Foundation of Jiangsu Province for Outstanding Youth (No. BK20180105), the Top-notch Academic Programs Project of Jiangsu Higher Education Institutions (TAPP), and the Jiangsu Innovation and Entrepreneurship Training Program for College Students (No. 202210320142Y).

Data Availability Statement: All data are contained within the article.

Conflicts of Interest: The authors declare no conflict of interest.

References

1. Kim, H.N.; Ren, W.X.; Kim, J.S.; Yoon, J. Fluorescent and colorimetric sensors for detection of lead, cadmium, and mercury ions. *Chem. Soc. Rev.* **2012**, *41*, 3210–3244. [CrossRef] [PubMed]
2. Driscoll, C.T.; Mason, R.P.; Chan, H.M.; Jacob, D.J.; Pirrone, N. Mercury as a Global Pollutant: Sources, Pathways, and Effects. *Environ. Sci. Technol.* **2013**, *47*, 4967–4983. [CrossRef] [PubMed]
3. Nolan, E.M.; Lippard, S.J. Tools and Tactics for the Optical Detection of Mercuric Ion. *Chem. Rev.* **2008**, *108*, 3443–3480. [CrossRef]
4. Liu, J.L.; Han, Q.; Hu, E.Y.; Yang, C.; Yin, M.M. Determination of Trace Mercury in Water Samples by Cloud Point Extraction Coupled with Atomic Fluorescence Spectrometry. *J. Anal. Chem.* **2023**, *78*, 303–309. [CrossRef]
5. Yaghi, O.M.; Li, G.; Li, H. Selective binding and removal of guests in a microporous metal–organic framework. *Nature* **1995**, *378*, 703–706. [CrossRef]
6. Kitagawa, S.; Kitaura, R.; Noro, S. Functional porous coordination polymers. *Angew. Chem. Int. Ed.* **2004**, *43*, 2334–2375. [CrossRef]
7. Férey, G. Hybrid porous solids: Past, present, future. *Chem. Soc. Rev.* **2008**, *37*, 191–214. [CrossRef]
8. Li, J.R.; Kuppler, R.J.; Zhou, H.C. Selective gas adsorption and separation in metal–organic frameworks. *Chem. Soc. Rev.* **2009**, *38*, 1477–1504. [CrossRef]
9. Li, B.; Wen, H.M.; Zhou, W.; Chen, B. Porous Metal–Organic Frameworks for Gas Storage and Separation: What, How, and Why? *J. Phys. Chem. Lett.* **2014**, *5*, 3468–3479. [CrossRef]
10. Huang, Y.B.; Liang, J.; Wang, X.S.; Cao, R. Multifunctional metal–organic framework catalysts: Synergistic catalysis and tandem reactions. *Chem. Soc. Rev.* **2017**, *46*, 126–157. [CrossRef]

11. Zhu, L.; Liu, X.Q.; Jiang, H.L.; Sun, L.B. Metal–organic frameworks for heterogeneous basic catalysis. *Chem. Rev.* **2017**, *117*, 8129–8176. [CrossRef] [PubMed]
12. Cui, Y.; Yue, Y.; Qian, G.; Chen, B. Luminescent functional metal–organic frameworks. *Chem. Rev.* **2012**, *112*, 1126–1162. [CrossRef] [PubMed]
13. Lustig, W.P.; Mukherjee, S.; Rudd, N.D.; Desai, A.V.; Li, J.; Ghosh, S.K. Metal–organic frameworks: Functional luminescent and photonic materials for sensing applications. *Chem. Soc. Rev.* **2017**, *46*, 3242–3285. [CrossRef] [PubMed]
14. Yan, B. Luminescence response mode and chemical sensing mechanism for lanthanide-functionalized metal–organic framework hybrids. *Inorg. Chem. Front.* **2021**, *8*, 201–233. [CrossRef]
15. Huangfu, M.; Wang, M.; Lin, C.; Wang, J.; Wu, P. Luminescent metal–organic frameworks as chemical sensors based on "mechanism–response": A review. *Dalton Trans.* **2021**, *50*, 3429–3449. [CrossRef]
16. Wong, N.E.; Ramaswamy, P.; Lee, A.S.; Gelfand, B.S.; Bladek, K.J.; Taylor, J.M.; Spasyuk, D.M.; Shimizu, G.K.H. Tuning Intrinsic and Extrinsic Proton Conduction in Metal–Organic Frameworks by the Lanthanide Contraction. *J. Am. Chem. Soc.* **2017**, *139*, 14676–14683. [CrossRef]
17. Rocca, J.D.; Liu, D.; Lin, W. Nanoscale Metal–Organic Frameworks for Biomedical Imaging and Drug Delivery. *Acc. Chem. Res.* **2011**, *44*, 957–968. [CrossRef]
18. Yang, L.; Li, X.; Qin, C.; Shao, K.Z.; Su, Z.M. A fluorescent sensor for highly selective sensing of nitro explosives and Hg(ii) ions based on a 3D porous layer metal–organic framework. *CrystEngComm* **2016**, *18*, 4765–4771. [CrossRef]
19. Wang, M.; Guan, J.; Liu, S.; Chen, K.; Gao, Z.; Liu, Q.; Chen, X. Dual-ligand lanthanide metal–organic framework probe for ratiometric fluorescence detection of mercury ions in wastewater. *Microchim. Acta* **2023**, *190*, 359. [CrossRef]
20. El Taher, B.J.; Sabouni, R.; Ghommem, M. Luminescent metal organic framework for selective detection of mercury in aqueous media: Microwave-based synthesis and evaluation. *Colloid. Surface. A* **2020**, *607*, 125477. [CrossRef]
21. Wen, L.; Zheng, X.; Lv, K.; Wang, C.; Xu, X. Two Amino-Decorated Metal–Organic Frameworks for Highly Selective and Quantitatively Sensing of Hg^{II} and Cr^{VI} in Aqueous Solution. *Inorg. Chem.* **2015**, *54*, 7133–7135. [CrossRef] [PubMed]
22. Wu, P.; Liu, Y.; Liu, Y.; Wang, J.; Li, Y.; Liu, W.; Wang, J. Cadmium-Based Metal–Organic Framework as a Highly Selective and Sensitive Ratiometric Luminescent Sensor for Mercury(II). *Inorg. Chem.* **2015**, *54*, 11046–11048. [CrossRef] [PubMed]
23. Wang, X.; Mal, A.; Gui, B.; Wang, C. Tuning Energy Transfer in Metal-Organic Frameworks for Fluorescence Turn-on Sensing of Hg(II) Ions. *Chin. J. Chem.* **2023**, *41*, 1051–1056. [CrossRef]
24. Cohen, S.M. Postsynthetic Methods for the Functionalization of Metal–Organic Frameworks. *Chem. Rev.* **2012**, *112*, 970–1000. [CrossRef] [PubMed]
25. Burrows, A.D.; Frost, C.G.; Mahon, M.F.; Richardson, C. Post-Synthetic Modification of Tagged Metal–Organic Frameworks. *Angew. Chem. Int. Ed.* **2008**, *47*, 8482–8486. [CrossRef]
26. Li, W.X.; Gu, J.H.; Li, H.X.; Dai, M.; Young, D.J.; Li, H.Y.; Lang, J.P. Post-synthetic Modification of a Two-Dimensional Metal–Organic Framework via Photodimerization Enables Highly Selective Luminescent Sensing of Aluminum(III). *Inorg. Chem.* **2018**, *57*, 13453–13460. [CrossRef]
27. Cavka, J.H.; Jakobsen, S.; Olsbye, U.; Guillou, N.; Lamberti, C.; Bordiga, S.; Lillerud, K.P. A New Zirconium Inorganic Building Brick Forming Metal Organic Frameworks with Exceptional Stability. *J. Am. Chem. Soc.* **2008**, *130*, 13850–13851. [CrossRef]
28. Ge, J.; Liu, L.; Shen, Y. Facile synthesis of amine-functionalized UiO-66 by microwave method and application for methylene blue adsorption. *J. Porous Mater.* **2017**, *24*, 647–655. [CrossRef]
29. Huang, A.; Wan, L.; Caro, J. Microwave-assisted synthesis of well-shaped UiO-66-NH_2 with high CO_2 adsorption capacity. *Mater. Res. Bull.* **2018**, *98*, 308–313. [CrossRef]
30. Taddei, M.; Dau, P.V.; Cohen, S.M.; Ranocchiari, M.; van Bokhoven, J.A.; Costantino, F.; Sabatini, S.; Vivani, R. Efficient microwave assisted synthesis of metal–organic framework UiO-66: Optimization and scale up. *Dalton Trans.* **2015**, *44*, 14019–14026. [CrossRef]
31. Sapianik, A.A.; Barsukova, M.O.; Kovalenko, K.A.; Samsonenko, D.G.; Fedin, V.P. Heterometallic MOFs constructed from thiophene and furandicarboxylate ligands for heavy metal luminescence sensing. *Dalton Trans.* **2021**, *50*, 2807–2814. [CrossRef] [PubMed]
32. Fu, L.; Xie, K.; Wang, A.; Lyu, F.; Ge, J.; Zhang, L.; Zhang, H.; Su, W.; Hou, Y.L.; Zhou, C.; et al. High selective detection of mercury (II) ions by thioether side groups on metal-organic frameworks. *Anal. Chim. Acta* **2019**, *1081*, 51–58. [CrossRef] [PubMed]
33. Volynkin, S.S.; Demakov, P.A.; Shuvaeva, O.V.; Kovalenko, K.A. Metal-organic framework application for mercury speciation using solid phase extraction followed by direct thermal release–electrothermal atomization atomic absorption spectrophotometric detection (ETA AAS). *Anal. Chim. Acta* **2021**, *1177*, 338795. [CrossRef] [PubMed]

nanomaterials

MDPI

Article

Specific Recognition and Adsorption of Volatile Organic Compounds by Using MIL-125-Based Porous Fluorescence Probe Material

Qiuyu Wu †, Feiyang Tian †, Wenqian Chen *, Jianying Wang * and Bo Lei *

Key Laboratory of Organic Compound Pollution Control Engineering, Ministry of Education, School of Environmental and Chemical Engineering, Shanghai University, Shanghai 200444, China; wuqiuyuhua@126.com (Q.W.); feiyangtian3638@163.com (F.T.)

* Correspondence: wenqianchen@shu.edu.cn (W.C.); wjytech@163.com (J.W.); lb870213@shu.edu.cn (B.L.)

† These authors contributed equally to this work.

Abstract: The severity of the volatile organic compounds (VOCs) issue calls for effective detection and management of VOC materials. Metal-organic frameworks (MOFs) are organic-inorganic hybrid crystals with promising prospects in luminescent sensing for VOC detection and identification. However, MOFs have limitations, including weak response signals and poor sensitivity towards VOCs, limiting their application to specific types of VOC gases. To address the issue of limited recognition and single luminosity for specific VOCs, we have introduced fluorescent guest molecules into MOFs as reference emission centers to enhance sensitivity. This composite material combines the gas adsorption ability of MOFs to effectively adsorb VOCs. We utilized (MIL-125/NH_2-MIL-125) as the parent material for adsorbing fluorescent molecules and selected suitable solid fluorescent probes (FGFL-B_1) through fluorescence enhancement using thioflavin T and MIL-125. FGFL-B_1 exhibited a heightened fluorescence response to various VOCs through charge transfer between fluorescent guest molecules and ligands. The fluorescence enhancement effect of FGFL-B_1 on tetrahydrofuran (THF) was particularly pronounced, accompanied by a color change from yellow to yellowish green in the presence of CCl_4. FGFL-B_1 demonstrated excellent adsorption properties for THF and CCl_4, with saturated adsorption capacities of 655.4 mg g^{-1} and 811.2 mg g^{-1}, respectively. Furthermore, FGFL-B_1 displayed strong luminescence stability and reusability, making it an excellent sensing candidate. This study addresses the limitations of MOFs in VOC detection, opening avenues for industrial and environmental applications.

Keywords: metal organic frameworks; porous fluorescent probe; VOCs

1. Introduction

Environmental pollution has become a pressing global issue with the release of hazardous chemical pollutants into the atmosphere through industrial emissions, steam leakage, and fossil fuel combustion [1,2]. Among these pollutants, volatile organic compounds (VOCs) pose a significant threat to human health [3]. There is an urgent need to develop VOC detection sensors with high sensitivity capable of detecting a variety of gases. The selection of sensor materials plays a crucial role in achieving efficient VOC detection [4]. Metal-organic frameworks (MOFs) are highly effective porous materials for the removal of harmful gases [5–7]. They offer numerous advantages over traditional materials such as zeolite-type materials [8], mesoporous silica [9], resins [10], and activated carbons with additives [11,12]. Unlike zeolite-related inorganic hybrid materials that require a template for formation, MOFs rely primarily on a solvent as the main templating molecule [13,14]. To ensure efficient capture of harmful gases through interaction with MOFs, it is crucial to employ MOFs with appropriate pore size and shape [15]. MOFs are functional materials known for their tailorable porosity, making them attractive candidates for VOC capture and sensing [16].

MOFs are a class of crystalline micro-mesoporous hybrid materials with diverse potential applications [17–20]. They possess designable structures, adjustable chemical functions, a low-density skeleton, an ultra-high specific surface area, and functional permanent pore space [21–23]. Due to their porous structure and broad luminescence behavior, MOFs are commonly used as gas adsorbents and luminescent sensors [24–27]. These MOF sensors exhibit rapid response, high sensitivity, and non-invasive operation [28–30]. Luminescent MOF sensors often utilize lanthanide elements as metal centers or specific organic ligands [31–34]. The selection of materials for such sensors still has limitations [35]. The disadvantage of MOFs is that excitation and emission energy transfer primarily occur within a single component, making fluorescence advantageous only for specific volatile organic compounds (VOCs) that closely interact with the luminescent groups [35–37].

Introducing luminescent groups in MOFs as reference emission centers is a simple and effective strategy to enhance the luminescence sensitivity of MOF sensors, allowing for widespread applications. Yao et al. used a porous supramolecular framework to adsorb guest dye molecules, creating solid materials with high fluorescence. This provides a simple, low-cost, and efficient way to create luminescent solid materials [38]. Similarly, Tian created solid luminous materials by adsorbing guest dye molecules into a symmetrical urea-based polypore supramolecular assembly of tetramethyl calabinus [6] for the detection of VOCs [39]. By adopting this method, luminescent groups can be introduced into MOFs as reference emission centers, effectively addressing the issue of limited luminescence and specific VOC recognition in traditional MOFs. Furthermore, taking advantage of the excellent gas adsorption performance of MOFs, there is potential to utilize them as parent materials for simultaneous VOC detection and adsorption. MIL-125, a popular Ti-based MOF, stands out due to its high thermal stability, solid framework, large specific surface area, and two types of pore cages (effective reachable diameters of 12.55 Å and 6.13 Å, respectively, are octahedral and tetrahedral cages, connected by triangular windows of 5–7 Å). These properties make MIL-125 an exceptional material for gas adsorption [40]. In 2014, Kim et al. reported the adsorption and catalytic properties of MIL-125 and NH_2-MIL-125 [41]. Recently, Kim et al. found that the formaldehyde adsorption ability of NH_2-MIL-125 was significantly better than that of other MOF materials [42]. However, MIL-125 and NH_2-MIL-125 have not been widely utilized for VOC detection. Our objective is to employ MIL-125 and NH_2-MIL-125 as parent materials for developing fluorescent materials capable of detecting and adsorbing VOCs.

The fluorescence probe FGFL-B_1 (composed of MIL-125 and thioflavin T dye molecules) showed the strongest fluorescence enhancement effect on tetrahydrofuran (THF), and it exhibited distinct fluorescence discoloration when specifically recognizing CCl_4. This method not only has a simple material fabrication step and strong operability but also expands the application range of MIL-125 and other MOF materials as fluorescent probes. The study aimed to address practical challenges in VOC detection and treatment, offering potential applications in industrial and living environments.

2. Experimental

2.1. Materials

All reagent-grade chemicals were used without further purification: NH_2-terephthalate acid (H_2ATA, RG, Adamas-beta, Titan, Shanghai, China), terephthalate acid (H_2BDC, RG, Adamas-beta, Titan, Shanghai, China), tetrabutyl titanate (TBT, AR, Greagent, Titan, Shanghai, China), anhydrous methanol (AR, Greagent, Titan, Shanghai, China), N, N-dimethylformamide (DMF, AR, Greagent, Titan, Shanghai, China), dansyl chloride (Leyan, Shanghai, China), pyren-1-amine (bidepharm, Shanghai, China), 8-hydroxyquinoline (Adamas-beta, Titan, Shanghai, China), umbelliferone (Adamas-beta, Titan, Shanghai, China), thioflavin T (ThT, bidepharm, Shanghai, China) benzene (C_6H_6, AR, General-Reagent, Titan, Shanghai, China), toluene ($C_6H_5CH_3$, AR, General-Reagent, Titan, Shanghai, China), dichloromethane (CH_2Cl_2, AR, General-Reagent, Titan, Shanghai, China), carbon tetrachloride (CCl_4, AR, General-Reagent, Titan, Shanghai, China), formaldehyde

(HCHO, AR, General-Reagent, Titan, Shanghai, China), and tetrahydrofuran (C_4H_8O, THF, Safe Dry, Adamas-beta, Titan, Shanghai, China).

2.2. Methods

2.2.1. Precursor Synthesis

The MIL-125 and NH_2-MIL-125 were prepared by a solvothermal method, which was modified from a previous report [43].

Terephthalate acid (H_2BDC, 3 mmol) was completely dissolved in N, N-dimethylformamide (DMF, 9 mL), and anhydrous methanol (3 mL). Then, 0.75 mmol of tetrabutyl titanate (TBT) was added to the solution under an N_2 atmosphere to form a suspension. After vigorous stirring for 30 min, the suspension was reacted in a 50 mL Teflon-sealed autoclave at 423 K for 24 h. Finally, the material was cooled down to room temperature. The white MIL-125 powders were obtained after being washed with DMF and anhydrous methanol, dried at 333 K, and then vacuum dried at 423 K for 3 h (In order to remove water and DMF). Yellow NH_2-MIL-125 powders were prepared in the same way, except that H_2BDC was replaced with NH_2-terephthalate acid (H_2ATA).

2.2.2. Porous Fluorescence Probe Synthesis

15 mg of each of the five dyes (thioflavin T (ThT), 8-hydroxyquinoline, dansyl chloride, pyren-1-amine, umbelliferone) were placed in five small bottles containing 20 mL acetonitrile. Ultrasonic treatment was carried out for 15 min to form a uniformly dispersed dye solution. Then 50 mg of NH_2-MIL-125/MIL-125 was added to this uniformly dispersed dye solution and let stand for 3.5 h. After centrifugation, the lower layer was precipitated and dried at 333 K for 12 h to obtain (FGFL-$A_{1–5}$ and FGFL-$B_{1–5}$) materials. In this study, MIL-125/NH_2-MIL-125 was used as the parent material to adsorb five fluorescent dye molecules (thioflavin T, 8-hydroxyquinoline, dansyl chloride, pyrene-1 amine, and cymflorone), labeled as 1, 2, 3, 4, and 5 in the paper. Different composites were prepared, with the complex formed by MIL-125 named FGFL-B_{1-5} and the composites formed by NH_2-MIL-125 named FGFL-$A_{1–5}$. By calculating the mass of dye before and after adsorption, the amounts of fluorophores 1–5 in 50 mg MIL-125 were 6.7 mg, 4.8 mg, 1.4 mg, 1.1 mg, and 5.4 mg, respectively, and 5.3 mg, 2.9 mg, 1 mg, 0.9 mg, and 2.3 mg in 50 mg of NH_2-MIL-125, respectively.

2.3. Adsorption and Fluorescence Response of FGFL-$A_{1–5}$/FGFL-$B_{1–5}$ for VOCs

A glass bottle (10 mL) containing 50 mg of FGFL-$A_{1–5}$/FGFL-$B_{1–5}$ was placed into the sealable glass container (100 mL) and vacuumed with a vacuum pump until the weight of the FGFL-$A_{1–5}$/FGFL-$B_{1–5}$ remained the same. Then, another glass bottle (10 mL) containing a small amount of volatile organic compounds (C_6H_6, $C_6H_5CH_3$, THF, CCl_4, C_2Cl_2, HCHO) was placed in the vacuum-sealed glass container and sealed for 3.5 h. After the adsorption of different VOCs, the fluorescence of FGFL-$A_{1–5}$/FGFL-$B_{1–5}$ was detected successively. The weight variation was measured, and corresponding solid-state fluorescence spectra were determined at intervals of approximately 20–240 min over several hours in order to obtain the vapor adsorption profile.

2.4. Measurement of Fluorescence Spectra of Solid FGFL-$A_{1–5}$/FGFL-$B_{1–5}$

Fluorescence spectra of solid FGFL-$A_{1–5}$/FGFL-$B_{1–5}$ before and after adsorbing VOCs were recorded at room temperature by using a fluorescence spectrometer (F-320 PL, Guangdong Technology, Guangzhou, China), respectively.

2.5. Characterization

A powder X-ray diffractometer (PXRD, Bruker D8 Advance) was used to characterize the crystal structure of the synthesized catalysts by Cu Kα radiation operated at 40 kV and 40 mA. Patterns were collected using a scan speed of 2 s/step and a step size of 0.02° at a 2θ range from 5° to 50°. The morphologies and microstructures of the prepared materials were observed using a scanning electron microscope (SEM, Gemini SEM300, ZEISS, Germany).

The Brunauer–Emmett–Teller (BET) surface areas of the samples were measured using a nitrogen adsorption instrument (Micromeritics ASAP 2460 analyzer, Beishide, Beijing, China) at liquid-nitrogen temperature. The IR experiments were carried out on a Nicolet 380 FT-IR spectrometer. The IR spectra of pure samples were collected without diluting with KBr. Photoluminescence (PL) spectra were recorded using a Guangdong Technology F-320 PL spectrophotometer with an excitation wavelength of 385 nm. All the liquid-state ^{1}H NMR spectra in this work were obtained on a Bruker 400 MHz spectrometer in dimethyl sulfoxide solution. Ultraviolet-visible (UV-vis) diffuse reflectance spectra (DRS) of the prepared samples were recorded using a UV-vis spectrometer (UV-2600, Shimadzu, Kyoto, Japan) with a background of $BaSO_4$.

3. Results and Discussion

3.1. Structure and Property of FGFL-$A_{1–5}$/FGFL-$B_{1–5}$

We successfully synthesized MIL-125 and NH_2-MIL-125 by the solvothermal method [40]. The powder X-ray diffraction (PXRD) patterns are shown in Figure 1. The main diffraction peaks of NH_2-MIL-125 (Figure 1a) and MIL-125 (Figure 1b) were consistent with the simulated spectra, indicating that both MIL-125 and NH_2-MIL-125 have been successfully prepared [40]. Different fluorescent probe materials were prepared by introducing dye molecules into MIL-125/NH_2-MIL-125 by the impregnation method. The PXRD pattern of NH_2-MIL-125/MIL-125 (FGFL-$A_{1–5}$/FGFL-$B_{1–5}$) remained unchanged after the adsorption of fluorescent dye molecules, indicating that the crystal structure of NH_2-MIL-125/MIL-125 was not affected by the adsorption process.

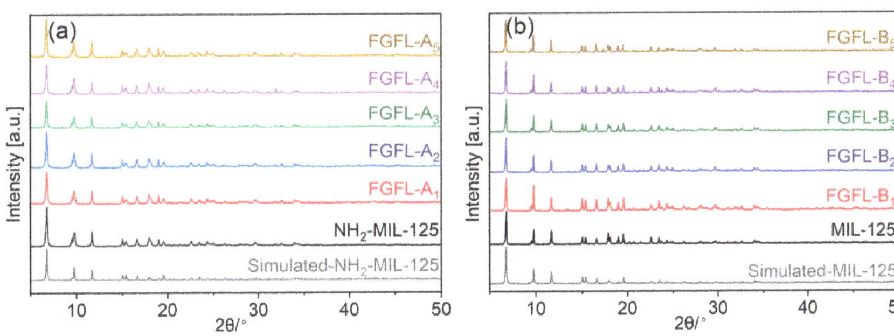

Figure 1. PXRD patterns of (**a**) NH_2-MIL-125, FGFL-$A_{1–5}$, and (**b**) MIL-125, FGFL-$B_{1–5}$ (Thioflavin T, 8-hydroxyquinoline, dansyl chloride, pyrene-1 amine, and umbelliferone were referred to as 1, 2, 3, 4 and 5).

To investigate the impact of dye molecules on the morphology of NH_2-MIL-125/MIL-125, we examined both the pristine materials and composites using a Scanning Electron Microscope (SEM). Figure 2 shows that MIL-125 and FGFL-$B_{1–5}$ exhibit similar disc-like shapes with a diameter of approximately 300 nm and a thickness of around 100 nm. Similarly, the morphology and size of NH_2-MIL-125 and FGFL-$A_{1–5}$, as shown in Figure S1, remained unchanged. The surface morphology and structure, as observed from the PXRD and SEM patterns, also did not show significant alterations, and no dye was detected.

Nitrogen sorption experiments demonstrated the highly porous nature of NH_2-MIL-125 and MIL-125, characterized by a type I isotherm typical of micro-porous solids (Figure 3), which exhibited a BET surface area of 1031.60 m^2 $g^{™1}$ and 1252.45 m^2 g^{-1}, and a micro-pore volume of 0.4806 cm^3 g^{-1} and 0.6015 cm^3 g^{-1}, respectively (Figure S2, Table S1). Compared to NH_2-MIL-125/MIL-125, the specific BET surface area and micro-pore volume of FGFL-$A_{1–5}$/FGFL-$B_{1–5}$ were significantly reduced, indicating successful adsorption of fluorescent dye molecules onto NH_2-MIL-125/MIL-125 (Table S1). Notably, the thioflavin T (ThT) composite exhibited the minimum BET surface area and micro-pore volume. In the case of FGFL-A_1, formed using NH_2-MIL-125 as a template, the BET surface area decreased from 1031.60 m^2 g^{-1} to 271.45 m^2 g^{-1}, and the micropore volume decreased from 0.4806 cm^3 g^{-1} to 0.1000 cm^3 g^{-1}. The BET surface area of MIL-125 composite (FGFL-B_1) decreased from 1252.25 m^2 g^{-1} to 356.05 m^2 g^{-1}, and the micro-pore volume decreased from 0.6015 cm^3 g^{-1} to 0.1615 cm^3 g^{-1}. Therefore, NH_2-MIL-125/MIL-125 exhibited the highest ThT adsorption capacity.

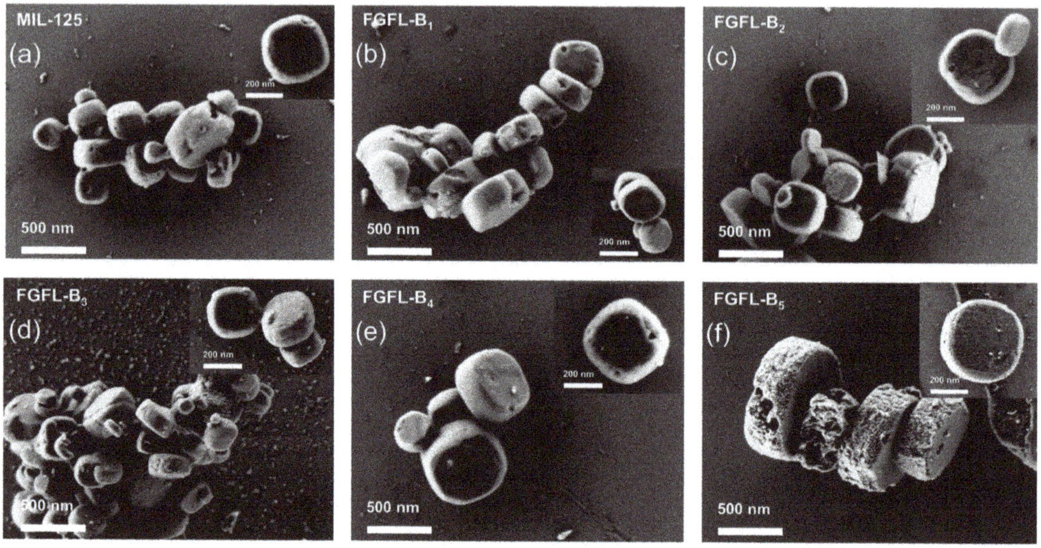

Figure 2. (**a**–**f**) SEM images of MIL-125, FGFL-B_1, FGFL-B_2, FGFL-B_3, FGFL-B_4, and FGFL-B_5, respectively.

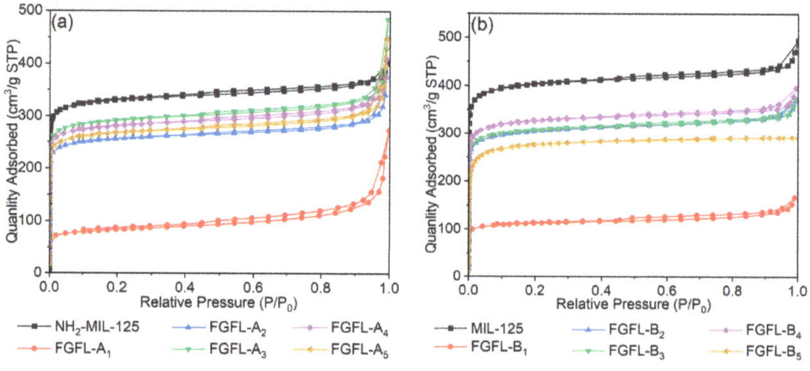

Figure 3. N_2 adsorption–desorption isotherms of (**a**) NH_2-MIL-125, FGFL-$A_{1–5}$, (**b**) MIL-125, FGFL-$B_{1–5}$.

3.2. Fluorescence of FGFL-$A_{1–5}$/FGFL-$B_{1–5}$

The fluorescence spectra of FGFL-$A_{1–5}$/FGFL-$B_{1–5}$ were tested to identify the optimal fluorescent probe materials. The results revealed that ThT exhibited a strong fluorescence enhancement effect when adsorbed onto NH_2-MIL-125/MIL-125 (Figures 4, S3 and S10), while other dye molecules showed minimal fluorescence response. Therefore, ThT was deemed the most suitable fluorescent guest molecule for NH_2-MIL-125/MIL-125. Additionally, FGFL-B_1 demonstrated significantly better fluorescence enhancement than FGFL-A_1 (Figure 4b). The electrophotographs captured under sunlight and UV (365 nm) illumination (Figure S4) demonstrated that FGFL-B_1, a fluorescent material formed after ThT adsorption by MIL-125, emitted vibrant yellow light under UV (365 nm) illumination.

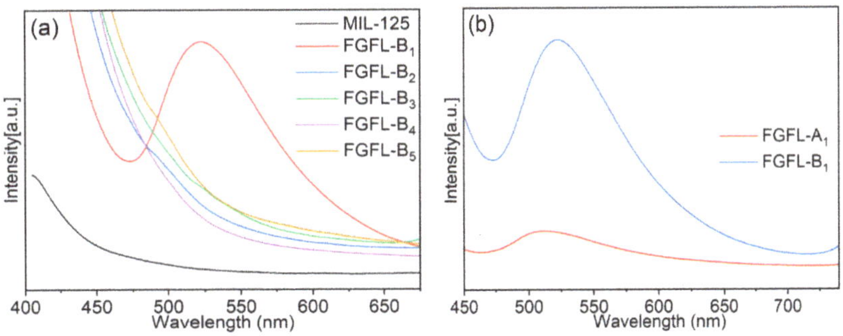

Figure 4. (**a**) Fluorescence spectra of MIL-125 and FGFL-$B_{1–5}$, (**b**) Fluorescence spectra of FGFL-A_1 and FGFL-B_1.

To analyze whether ThT undergoes simple physical adsorption or chemical coupling with MIL-125/NH_2-MIL-125, we conducted IR spectroscopy on ThT, NH_2-MIL-125/MIL-125, and FGFL-$A_{1–5}$/FGFL-$B_{1–5}$. As shown in Figure 5, the carboxyl group characteristic peak of NH_2-MIL-125/MIL-125 showed at 3.140–3.204 (log1380–log1600 cm^{-1}) [40]. An evident larger wavenumber shift at 3.23 (log1700 cm^{-1}) was observed in the infrared spectra of FGFL-$A_{1–5}$/FGFL-$B_{1–5}$, indicating the adsorption and coupling of ThT molecules with NH_2-MIL-125/MIL-125. The larger wavenumber shift was more pronounced in the FGFL-B_1 infrared spectra. Figure S5 displays the Raman spectra of NH_2-MIL-125/MIL-125, ThT, and FGFL-A_1/FGFL-B_1, with FGFL-A_1/FGFL-B_1 revealing peaks from both NH_2-MIL-125/MIL-125 and the fluorescent dye ThT. Additionally, Figure S6a,b illustrates the liquid nuclear magnetic hydrogen spectra of ThT, MIL-125, and FGFL-B_1. The peak at 8.14 ppm of ThT belongs to the H peak on the benzene ring on the benzylamine ring; the H peak (8.06 ppm) on the benzene ring of the organic ligand terephthalic acid of MIL-125 was significantly offset in FGFL-B_1, indicating that ThT forms H···O conjugations with MIL-125 [44]. The ThT molecule consists of a benzylamine ring and a phenyl sulfide ring. When the benzylamine ring and phenyl sulfide ring rotate freely around the C-C bond in the natural state, the ThT fluorescence signal is weak. Once this rotation is limited by some structures, the ThT fluorescence is enhanced [45–47]. When ThT fluorescent guest molecules were introduced into NH_2-MIL-125/MIL-125 as light sources and coupled with them via H···O interaction, the restriction of intramolecular rotation (RIR) of ThT fluorescence was caused to some extent. Charge transfer from fluorescent guest molecules to organic ligands was formed, resulting in a strong fluorescence response [48–50]. In order to confirm this phenomenon, we tested the solid UV-vis absorption spectra of ThT, MIL-125, and FGFL-B_1, respectively. As can be seen from Figure S7a, the absorption peak of FGFL-B_1 formed by the combination of ThT and MIL-125 had a significant red shift compared with MIL-125. As a chromophore, ThT is prone to electron transition. The FGFL-B_1 material formed by the combination of MIL-125 and ThT had a significant red shift relative to the absorption wavelength of MIL-125 and a significant blue shift relative to the absorption wavelength of ThT, which corresponded to the above phenomenon [51,52]. The band gaps of ThT, MIL-125, and FGFL-B_1 can be calculated from the solid UV-vis absorption spectra using the Kubelka–Munk function [53]. It can be seen from Figure S7b that by combining with ThT, the band gap of MIL-125 was reduced from 3.4 eV to 2.4 eV.

3.3. Fluorescence Sensors and Storage Performance of VOCs over FGFL-B_1

Environmental pollution is a pressing global issue caused by the discharge of harmful chemical pollutants into the atmosphere. Volatile organic compounds (benzene, toluene, tetrahydrofuran, carbon tetrachloride, dichloromethane, methylene chloride, formaldehyde) are the main cause of environmental air pollution and pose risks to human health [54–56]. For instance, THF exhibits stimulatory and anesthetic effects. Inhalation causes upper respiratory irritation, nausea, dizziness, headache, and central nervous system depression [57,58]. CCl_4

and its decomposition products can be absorbed through the respiratory tract, and skin contact can result in rapid absorption. CCl_4 is particularly damaging to peripheral nerves, especially the liver [59]. Therefore, the development of fluorescent probes and porous materials for the detection and adsorption of VOC gases is crucial to addressing this issue.

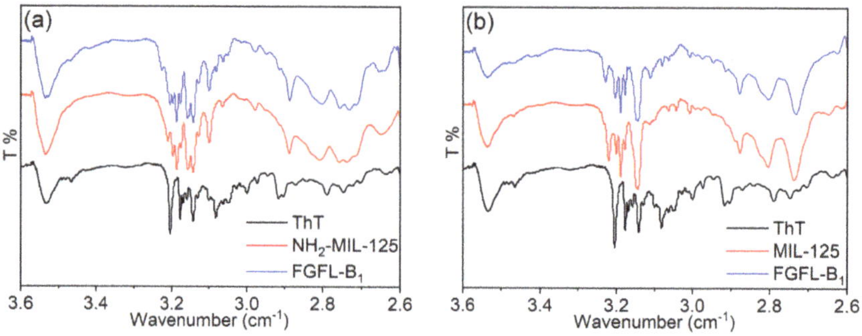

Figure 5. (**a**) IR spectra of ThT, NH_2-MIL-125, and FGFL-A_1, (**b**) IR spectra of ThT, MIL-125, and FGFL-B_1.

In this paper, fluorescent molecules were selected through fluorescence spectra testing to determine the most suitable candidates. FGFL-A_1 and FGFL-B_1 were chosen as fluorescent probes for investigating the detection and adsorption properties of six common volatile organic compounds (VOCs): dichloromethane (CH_2Cl_2), carbon tetrachloride (CCl_4), tetrahydrofuran (THF), formaldehyde (HCHO), benzene (C_6H_6), and toluene ($C_6H_5CH_3$). As shown in Figure 6a,b, FGFL-B_1, and FGFL-A_1 demonstrated varying degrees of fluorescence response towards the VOCs. However, the fluorescence enhancement effect of FGFL-A_1 was very low, which was negligible compared with that of FGFL-B_1 (Figure 6c,d). Therefore, FGFL-B_1 is considered the most suitable fluorescent probe material. After FGFL-B_1 and FGFL-B_1 adsorbed with different VOCs were formed into a suspension in acetonitrile solvent, different fluorescence responses could be clearly observed under UV light (Figure 6e). Notably, FGFL-B_1 showed the most significant fluorescence enhancement effect towards THF gas, with fluorescence intensity 36 times higher than FGFL-B_1 (Figure 6b). Additionally, the adsorption of CCl_4 gas induced a noticeable red shift in FGFL-B_1 (Figure 6b). This red shift resulted in a distinct transformation in fluorescence color from yellow to yellowish green (After the adsorption of CCl_4, the maximum emission peak of FGFL-B_1 undergoes a red shift from 530 nm to 560 nm, while it remained unchanged at 530 nm after the adsorption of other VOCs). To demonstrate the color transformation more effectively, test strips of FGFL-B_1 were fabricated, which visibly turned yellowish green upon adsorption of CCl_4 (Figure 6f). This test strip method offers the potential for faster and easier detection of simulated factory or indoor gas leaks.

FGFL-B_1 was formed by the recombination of ThT (4 Å) within the MIL-125 channel [60,61]. Upon the entry of VOC gas into the FGFL-B_1 channel, collisions with ThT molecules generate an effect [62]. The interaction between ThT and MIL-125 restricts the rotation of the benzene ring, thereby limiting non-radiative energy consumption pathways and reinforcing radiative transitions. As a result, fluorescence is substantially enhanced [63–65]. To illustrate this effect, we present the example of THF adsorption (Figure S8). Following the adsorption of THF, the fluorescence intensity of FGFL-B_1 was significantly enhanced, accompanied by a change in the full width at half maximum (FWHM) from 73.63 nm to 85.04 nm [66]. Electronegative molecules induce a pronounced red shift in the spectra of the material. CCl_4, known for its strong electronegativity, triggers intramolecular charge transfer of ThT, resulting in a red shift effect and fluorescence color transformation [67,68]. Therefore, FGFL-B_1 can selectively recognize CCl_4. These findings demonstrate that the incorporation of fluorescent guest molecules (ThT) as luminescent clusters significantly improves the fluorescence detection performance of materials and overcomes the limitations of conventional MOFs, which only exhibit a fluorescence response to specific gases (Figure 7).

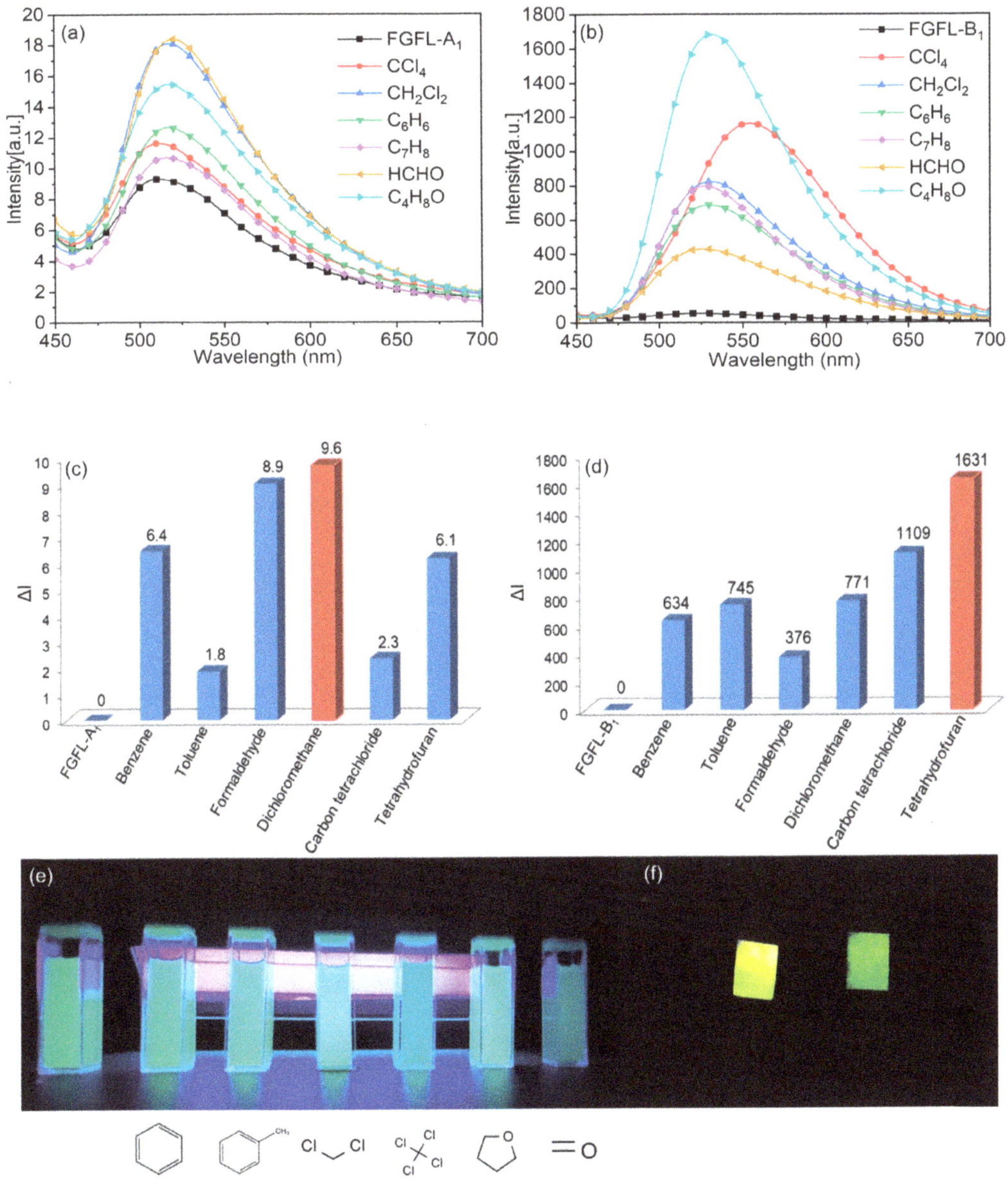

Figure 6. A general survey of the fluorescence spectra of (**a**) FGFL-A_1 and (**b**) FGFL-B_1 loaded with the six selected VOCs. Relative fluorescence intensities of (**c**) FGFL-A_1 and (**d**) FGFL-B_1 in response to the six selected VOCs. (**e**) Photos under UV (365 nm) of FGFL-B_1 and FGFL-B_1 loaded with six selected VOCs. (**f**) Photos under UV (365 nm) of FGFL-B_1 (**left**) and FGFL-B_1 loaded with CCl_4 (**right**).

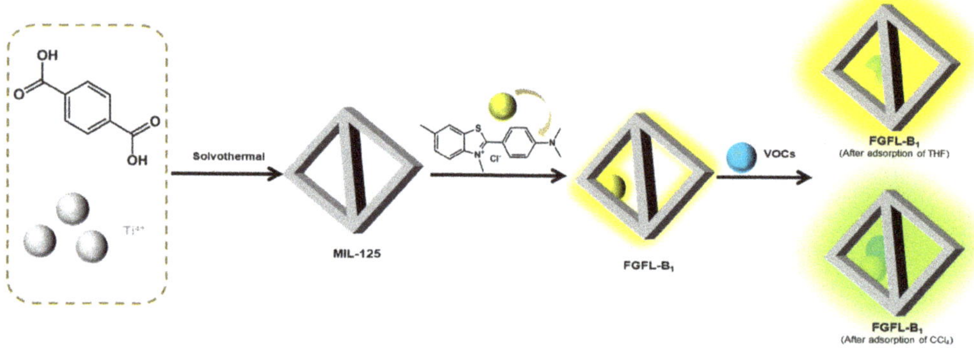

Figure 7. Schematic diagram of FGFL-B_1 adsorption detection of VOCs.

The titrated fluorescence spectra of FGFL-B_1 were analyzed in detail to monitor the real-time adsorption of THF and CCl_4 (Figure 8a,b). The fluorescence intensity of solid FGFL-B_1 at 530 nm and 560 nm increased with the increase in the adsorption time of THF and CCl_4. As shown in Figure 8e,f, it took approximately 2 h for FGFL-B_1 to reach its maximum fluorescence intensity with THF and CCl_4, while only about 20 min were needed to shift the peak position to 560 nm. Although the fluorescence intensity and peak position of FGFL-B_1 remained unchanged, the absorption rates of THF and CCl_4 continued to increase within 3 h (Figure 8c,d). Comparing Figure 8c–f, it can be observed that the adsorption amount and time of THF and CCl_4 required to achieve the maximum equilibrium value of FGFL-B_1 fluorescence intensity are much lower than those required by saturated adsorption. The size of ThT (4 Å) is large, and the barrier entering porous MIL-125 is much larger than that of VOCs molecules. Therefore, the amount of ThT that can interact with MIL-125 is significantly lower than that of VOC molecules. Therefore, the time required for fluorescence to reach maximum equilibrium and the amount of gas is far less than the time of saturation adsorption and gas volume. By comparing the mass of FGFL-B_1 before and after the saturated adsorption of THF and CCl_4, we calculated that the adsorption properties of FGFL-B_1 for THF and CCl_4 were 655 mg g^{-1} and 812 mg g^{-1}, respectively (Table S2). The adsorption performance of other VOC gases is shown in Table S1. Additionally, the ΔI curves representing the adsorption capacity of THF and CCl_4 on FGFL-B_1 (Figure 8g,h) were plotted using the adsorption-time curve data (Figure 8c,d) and intensity (ΔI) versus time curve data (Figure 8e,f) obtained from the interpolation method. The detection limits (DL) [39] of FGFL-B_1 for THF and CCl_4 were determined to be 1.41×10^{-4} mol g^{-1} and 4.66×10^{-5} mol g^{-1}, respectively.

The detection of VOCs by FGFL-B_1 is based on a fluorescence reaction triggered by the interaction of gases after entering the material. VOCs can be detached from FGFL-B_1, allowing the material to be recycled [39]. To validate this, we conducted adsorption and desorption experiments using THF and CCl_4 as examples. Lifetime tests demonstrated that the adsorption and desorption of THF and CCl_4 on FGFL-B_1 were not only reversible but also highly stable. Numerous adsorption and desorption experiments consistently indicated that the detection properties of FGFL-B_1 for THF and CCl_4, such as fluorescence intensity, remained unchanged (Figure S9c,d). Furthermore, a comparison of the PXRD patterns of FGFL-B_1 before and after the adsorption and desorption of THF and CCl_4 revealed no significant changes, suggesting that the adsorption or desorption of THF and CCl_4 had minimal impact on FGFL-B_1 (Figure S9a,b). These results provided strong evidence for the excellent stability of FGFL-B_1.

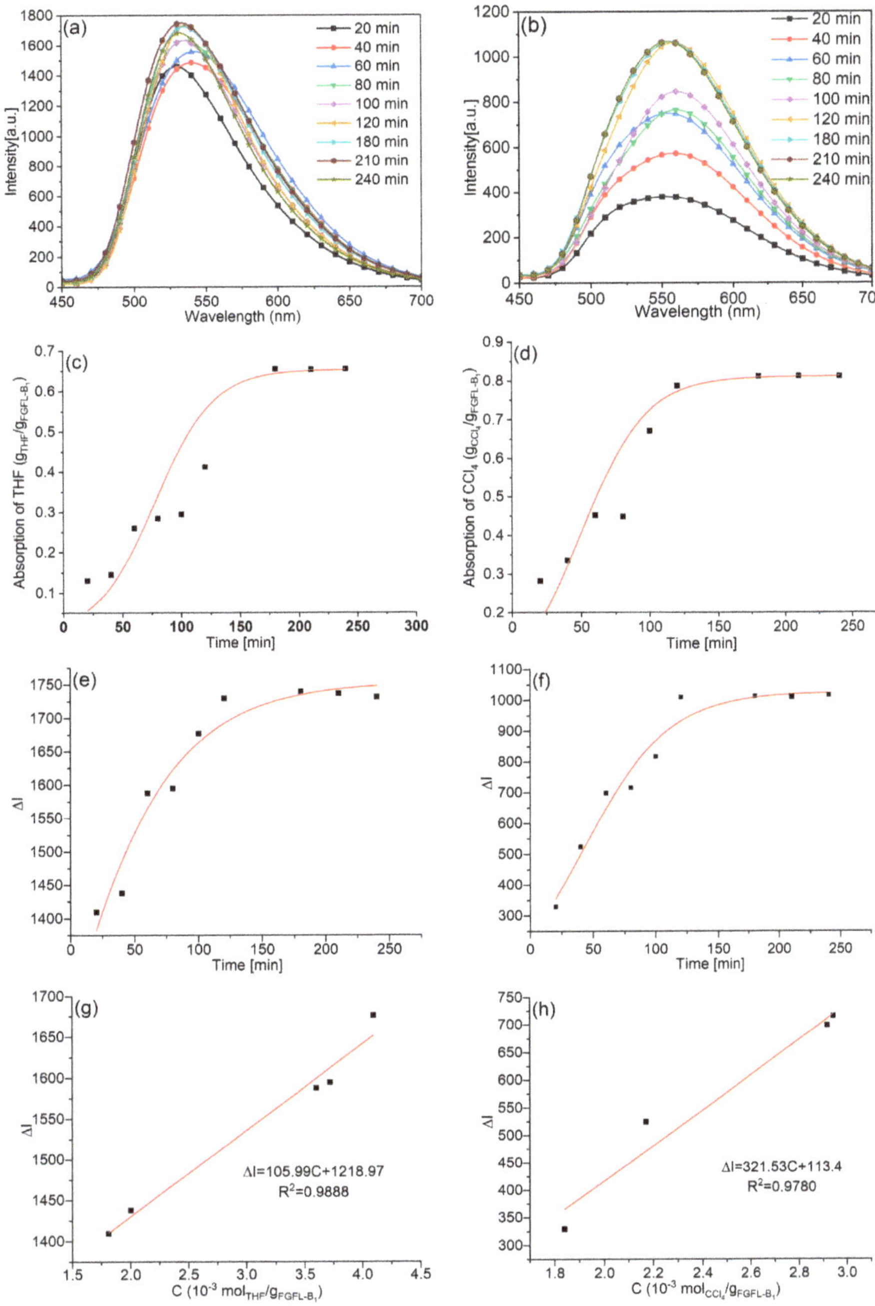

Figure 8. The titrated fluorescence spectra of FGFL-B_1 loaded (**a**) THF and (**b**) CCl_4. Adsorption profile of the loading of (**c**) THF and (**d**) CCl_4 in FGFL-B_1 upon increasing the adsorption time. Change in the fluorescence intensity of FGFL-B_1 by loading (**e**) THF and (**f**) CCl_4 upon increasing the adsorption time. Plot of ΔI versus the amount of (**g**) THF and (**h**) CCl_4 adsorbed by solid FGFL-B_1.

4. Conclusions

We investigated the fluorescence enhancement effect of FGFL-B_1, a composite of the ThT dye molecules and MIL-125 H···O conjugation, on VOCs by introducing various fluorescent dye molecules into MIL-125 and NH_2-MIL-125 synthesized via the solvothermal method. The presence of steric hindrance in ThT molecules both hampers and enhances their intramolecular rotation, resulting in a remarkable fluorescence enhancement effect on THF. The fluorescence intensity increases by a factor of 36 upon THF adsorption. Moreover, due to the strong electronegativity of CCl_4, FGFL-B_1 exhibits selective recognition of CCl_4, leading to a distinct yellow-to-yellowish-green fluorescence color change. Remarkably, FGFL-B_1 demonstrates excellent adsorption capacities for THF and CCl_4, with values of 655.4 mg g^{-1} and 811.2 mg g^{-1}, respectively. As a result, the prepared porous fluorescence probe material, FGFL-B_1, holds great potential for detecting and adsorbing mixed gases in both industrial and domestic environments. This method opens up new possibilities for fluorescence detection of adsorption properties in Ti-based MOF materials and other MOF materials.

Supplementary Materials: The following supporting information can be downloaded at: https://www.mdpi.com/article/10.3390/nano13192732/s1, Figure S1. (a–f) SEM images of NH_2-MIL-125, FGFL-A_1, FGFL-A_2, FGFL-A_3, FGFL-A_4 and FGFL-A_5, respectively; Figure S2. The corresponding BJH pore size distribution of the as-prepared samples of (a) NH_2-MIL-125, FGFL-A_1, FGFL-A_2, FGFL-A_3, FGFL-A_4 and FGFL-A_5 (b) MIL-125, FGFL-B_1, FGFL-B_2, FGFL-B_3, FGFL-B_4 and FGFL-B_5. Table S1: BET surface area and T-Plot micro-pore volume of NH_2-MIL-125, FGFL-A_{1-5}, MIL-125, and FGFL-B_{1-5}; Figure S3. Fluorescence spectra of (a) FGFL-A_1, (b) FGFL-A_2, (c) FGFL-A_3, (d) FGFL-A_4 and (e) FGFL-A_5; Figure S4. Electronic photos of ThT, MIL-125, and FGFL-B_1 (a–c) under sunlight and (d–f) under UV (365 nm); Figure S5. Raman spectra of (a) ThT, NH_2-MIL-125, and FGFL-A_1, (b) ThT, MIL-125, and FGFL-B_1; Figure S6. (a) The liquid-state ^{1}H NMR spectra in dimethyl sulfoxide of ThT, FGFL-B_1, and MIL-125; (b) The liquid-state ^{1}H NMR spectra in dimethyl sulfoxide of ThT, FGFL-B_1, MIL-125 (7.87–8.25 ppm); (c) The solid-state ^{1}H NMR spectra of MIL-125, FGFL-B_1 and ThT; Figure S7. (a) The UV-vis absorption spectra of ThT, MIL-125, and FGFL-B_1; (b) The plot of $(ahv)^{1/2}$ versus hv of ThT, MIL-125, and FGFL-B_1; Figure S8. The fluorescence spectra of FGFL-B_1 and FGFL-B_1 loaded with THF. Table S2: Adsorption properties of FGFL-B_1 for the six selected VOCs; Figure S9. After FGFL-B_1 adsorbed VOCs, the VOCs were removed by vacuum drying at 333 K. PXRD patterns and fluorescence intensity were measured before and after the cycle stability experiment. The numbers 1, 2, 3, 4, 5, and 6 represent the number of adsorption and desorption cycles. A general survey of the PXRD patterns of (a) FGFL-B_1 before and after loading with CCl_4, and (b) FGFL-B_1 before and after loading with THF. A general survey of the fluorescence spectra of (c) FGFL-B_1 before and after loading with CCl_4, (d) FGFL-B_1 before and after loading with THF; Figure S10. Fluorescence spectra of (a) FGFL-B_1, (b) FGFL-B_2, (c) FGFL-B_3, (d) FGFL-B_4, and (e) FGFL-B_5.

Author Contributions: Conceptualization, W.C.; methodology, Q.W.; writing—original draft preparation, Q.W. and F.T.; writing—review and editing, W.C., J.W. and B.L.; supervision, W.C. All authors have read and agreed to the published version of the manuscript.

Funding: This research was funded by the National Natural Science Foundation of China (No. 12104286), Program for Professor of Special Appointment in Shanghai (QD2019005).

Data Availability Statement: The data presented in this study are available on request from the corresponding author.

Conflicts of Interest: The authors declare no conflict of interest.

References

1. Pawelec, B.; Navarro, R.M.; Campos-Martin, J.M.; Fierro, J.L.G. Retracted Article: Towards near Zero-Sulfur Liquid Fuels: A Perspective Review. *Catal. Sci. Technol.* **2011**, *1*, 23–42. [CrossRef]
2. Colvile, R.N.; Hutchinson, E.J.; Mindell, J.S.; Warren, R.F. The Transport Sector as a Source of Air Pollution. *Atmos. Environ.* **2001**, *35*, 1537–1565. [CrossRef]
3. Guo, Y.; Wen, M.; Li, G.; An, T. Recent Advances in VOC Elimination by Catalytic Oxidation Technology onto Various Nanoparticles Catalysts: A Critical Review. *Appl. Catal. B* **2021**, *281*, 119447. [CrossRef]

4. Zheng, J.-P.; Ou, S.; Zhao, M.; Wu, C.-D. A Highly Sensitive Luminescent Dye@MOF Composite for Probing Different Volatile Organic Compounds. *ChemPlusChem* **2016**, *81*, 758–763. [CrossRef]
5. Hamon, L.; Serre, C.; Devic, T.; Loiseau, T.; Millange, F.; Férey, G.; Weireld, G.D. Comparative Study of Hydrogen Sulfide Adsorption in the MIL-53(Al, Cr, Fe), MIL-47(V), MIL-100(Cr), and MIL-101(Cr) Metal−Organic Frameworks at Room Temperature. *J. Am. Chem. Soc.* **2009**, *131*, 8775–8777. [CrossRef] [PubMed]
6. Shi, J.; Zhao, Z.; Xia, Q.; Li, Y.; Li, Z. Adsorption and Diffusion of Ethyl Acetate on the Chromium-Based Metal–Organic Framework MIL-101. *J. Chem. Eng. Data* **2011**, *56*, 3419–3425. [CrossRef]
7. Zhao, Z.; Li, X.; Li, Z. Adsorption Equilibrium and Kinetics of P-Xylene on Chromium-Based Metal Organic Framework MIL-101. *Chem. Eng. J.* **2011**, *173*, 150–157. [CrossRef]
8. Pires, J.; Carvalho, A.; de Carvalho, M.B. Adsorption of Volatile Organic Compounds in Y Zeolites and Pillared Clays. *Microporous Mesoporous Mater.* **2001**, *43*, 277–287. [CrossRef]
9. Chu, Y.-H.; Kim, H.-J.; Song, K.-Y.; Shul, Y.-G.; Jung, K.-T.; Lee, K.; Han, M.-H. Preparation of Mesoporous Silica Fiber Matrix for VOC Removal. *Catal. Today* **2002**, *74*, 249–256. [CrossRef]
10. Liu, P.; Long, C.; Li, Q.; Qian, H.; Li, A.; Zhang, Q. Adsorption of Trichloroethylene and Benzene Vapors onto Hypercrosslinked Polymeric Resin. *J. Hazard. Mater.* **2009**, *166*, 46–51. [CrossRef]
11. Ma, X.; Yang, L.; Wu, H. Removal of Volatile Organic Compounds from the Coal-Fired Flue Gas by Adsorption on Activated Carbon. *J. Clean. Prod.* **2021**, *302*, 126925. [CrossRef]
12. Lee, K.J.; Miyawaki, J.; Shiratori, N.; Yoon, S.-H.; Jang, J. Toward an Effective Adsorbent for Polar Pollutants: Formaldehyde Adsorption by Activated Carbon. *J. Hazard. Mater.* **2013**, *260*, 82–88. [CrossRef] [PubMed]
13. Lin, Y.; Massa, W.; Dehnen, S. "Zeoball" $[Sn_{36}Ge_{24}Se_{132}]^{24-}$: A Molecular Anion with Zeolite-Related Composition and Spherical Shape. *J. Am. Chem. Soc.* **2012**, *134*, 4497–4500. [CrossRef]
14. Férey, G. Hybrid Porous Solids: Past, Present, Future. *Chem. Soc. Rev.* **2008**, *37*, 191–214. [CrossRef] [PubMed]
15. Xiao, B.; Wheatley, P.S.; Zhao, X.; Fletcher, A.J.; Fox, S.; Rossi, A.G.; Megson, I.L.; Bordiga, S.; Regli, L.; Thomas, K.M.; et al. High-Capacity Hydrogen and Nitric Oxide Adsorption and Storage in a Metal−Organic Framework. *J. Am. Chem. Soc.* **2007**, *129*, 1203–1209. [CrossRef]
16. Shen, Y.; Tissot, A.; Serre, C. Recent Progress on MOF-Based Optical Sensors for VOC Sensing. *Chem. Sci.* **2022**, *13*, 13978–14007. [CrossRef]
17. Zhou, H.-C.; Kitagawa, S. (Eds.) Metal-Organic Frameworks (MOFs). *Chem. Soc. Rev.* **2014**, *43*, 5415–5418. [CrossRef]
18. Long, J.R.; Yaghi, O.M. The Pervasive Chemistry of Metal-Organic Frameworks. *Chem. Soc. Rev.* **2009**, *38*, 1213–1214. [CrossRef]
19. Zhou, H.-C.; Long, J.R.; Yaghi, O.M. Introduction to Metal-Organic Frameworks. *Chem. Rev.* **2012**, *112*, 673–674. [CrossRef]
20. Lee, J.; Farha, O.K.; Roberts, J.; Scheidt, K.A.; Nguyen, S.T.; Hupp, J.T. Metal-Organic Framework Materials as Catalysts. *Chem. Soc. Rev.* **2009**, *38*, 1450–1459. [CrossRef]
21. Tannert, N.; Sun, Y.; Hastürk, E.; Nießing, S.; Janiak, C. A Series of New Urea-MOFs Obtained via Post-Synthetic Modification of NH_2-MIL-101(Cr): SO_2, CO_2 and H_2O Sorption. *Z. Für Anorg. Und Allg. Chem.* **2021**, *647*, 1124–1130. [CrossRef]
22. Suh, M.P.; Park, H.J.; Prasad, T.K.; Lim, D.-W. Hydrogen Storage in Metal-Organic Frameworks. *Chem. Rev.* **2012**, *112*, 782–835. [CrossRef]
23. Ingleson, M.J.; Barrio, J.P.; Bacsa, J.; Dickinson, C.; Park, H.; Rosseinsky, M.J. Generation of a Solid Brønsted Acid Site in a Chiral Framework. *Chem. Commun.* **2008**, *11*, 1287–1289. [CrossRef] [PubMed]
24. Kitagawa, S.; Kitaura, R.; Noro, S. Functional Porous Coordination Polymers. *Angew. Chem. Int. Ed. Engl.* **2004**, *43*, 2334–2375. [CrossRef] [PubMed]
25. Mueller, U.; Schubert, M.; Teich, F.; Puetter, H.; Schierle-Arndt, K.; Pastré, J. Metal-Organic Frameworks-Prospective Industrial Applications. *J. Mater. Chem.* **2006**, *16*, 626–636. [CrossRef]
26. Wang, M.; Guo, L.; Cao, D. Amino-Functionalized Luminescent Metal-Organic Framework Test Paper for Rapid and Selective Sensing of SO_2 Gas and Its Derivatives by Luminescence Turn-On Effect. *Anal. Chem.* **2018**, *90*, 3608–3614. [CrossRef]
27. Gan, L.; Andres-Garcia, E.; Mínguez Espallargas, G.; Planas, J.G. Adsorptive Separation of CO_2 by a Hydrophobic Carborane-Based Metal-Organic Framework under Humid Conditions. *ACS Appl. Mater. Interfaces* **2023**, *15*, 5309–5316. [CrossRef]
28. Li, Y.-A.; Zhao, C.-W.; Zhu, N.-X.; Liu, Q.-K.; Chen, G.-J.; Liu, J.-B.; Zhao, X.-D.; Ma, J.-P.; Zhang, S.; Dong, Y.-B. Nanoscale UiO-MOF-Based Luminescent Sensors for Highly Selective Detection of Cysteine and Glutathione and Their Application in Bioimaging. *Chem. Commun.* **2015**, *51*, 17672–17675. [CrossRef]
29. Ke, F.; Yuan, Y.-P.; Qiu, L.-G.; Shen, Y.-H.; Xie, A.-J.; Zhu, J.-F.; Tian, X.-Y.; Zhang, L.-D. Facile Fabrication of Magnetic Metal-Organic Framework Nanocomposites for Potential Targeted Drug Delivery. *J. Mater. Chem.* **2011**, *21*, 3843–3848. [CrossRef]
30. Hu, Z.; Deibert, B.J.; Li, J. Luminescent Metal-Organic Frameworks for Chemical Sensing and Explosive Detection. *Chem. Soc. Rev.* **2014**, *43*, 5815–5840. [CrossRef]
31. Song, L.; Du, S.-W.; Lin, J.-D.; Zhou, H.; Li, T. A 3D Metal-Organic Framework with Rare 3-Fold Interpenetrating Dia-g Nets Based on Silver(I) and Novel Tetradentate Imidazolate Ligand: Synthesis, Structure, and Possible Ferroelectric Property. *Cryst. Growth Des.* **2007**, *7*, 2268–2271. [CrossRef]
32. Du, J.-L.; Hu, T.-L.; Li, J.-R.; Zhang, S.-M.; Bu, X.-H. Metal Coordination Architectures of 2,3-Bis(Triazol-1-Ylmethyl)Quinoxaline: Effect of Metal Ion and Counterion on Complex Structures. *Eur. J. Inorg. Chem.* **2008**, *2008*, 1059–1066. [CrossRef]

33. Wu, M.-M.; Wang, J.-Y.; Sun, R.; Zhao, C.; Zhao, J.-P.; Che, G.-B.; Liu, F.-C. The Design of Dual-Emissive Composite Material $[Zn_2(HL)_3]^+$@MOF-5 as Self-Calibrating Luminescent Sensors of Al^{3+} Ions and Monoethanolamine. *Inorg. Chem.* **2017**, *56*, 9555–9562. [CrossRef]
34. Geranmayeh, S.; Mohammadnejad, M.; Abbasi, A. Ln Based Metal-Organic Framework for Fluorescence "Turn Off-On" Sensing of Hg^{2+}. *J. Fluoresc.* **2023**, *33*, 1017–1026. [CrossRef] [PubMed]
35. Allendorf, M.D.; Bauer, C.A.; Bhakta, R.K.; Houk, R.J.T. Luminescent Metal-Organic Frameworks. *Chem. Soc. Rev.* **2009**, *38*, 1330–1352. [CrossRef] [PubMed]
36. Dong, M.-J.; Zhao, M.; Ou, S.; Zou, C.; Wu, C.-D. A Luminescent Dye@MOF Platform: Emission Fingerprint Relationships of Volatile Organic Molecules. *Angew. Chem. Int. Ed.* **2014**, *53*, 1575–1579. [CrossRef]
37. An, J.; Shade, C.M.; Chengelis-Czegan, D.A.; Petoud, S.; Rosi, N.L. Zinc-Adeninate Metal-Organic Framework for Aqueous Encapsulation and Sensitization of Near-Infrared and Visible Emitting Lanthanide Cations. *J. Am. Chem. Soc.* **2011**, *133*, 1220–1223. [CrossRef]
38. Yao, Y.-Q.; Zhang, Y.-J.; Zhang, Y.-Q.; Tao, Z.; Ni, X.-L.; Wei, G. Multiple Efficient Fluorescence Emission from Cucurbit[10]Uril-$[Cd_4Cl_{16}]^{8-}$-Based Pillared Diamond Porous Supramolecular Frameworks. *ACS Appl. Mater. Interfaces* **2017**, *9*, 40760–40765. [CrossRef]
39. Tian, F.Y.; Cheng, R.X.; Zhang, Y.Q.; Tao, Z.; Zhu, Q.J. Specific Recognition of Methanol Using a Symmetric Tetramethylcucurbit[6]Uril-Based Porous Supramolecular Assembly Incorporating Adsorbed Dyes. *Aust. J. Chem.* **2020**, *73*, 1065. [CrossRef]
40. Dan-Hardi, M.; Serre, C.; Frot, T.; Rozes, L.; Maurin, G.; Sanchez, C.; Férey, G. A New Photoactive Crystalline Highly Porous Titanium(IV) Dicarboxylate. *J. Am. Chem. Soc.* **2009**, *131*, 10857–10859. [CrossRef]
41. Kim, S.-N.; Kim, J.; Kim, H.-Y.; Cho, H.-Y.; Ahn, W.-S. Adsorption/Catalytic Properties of MIL-125 and NH_2-MIL-125. *Catal. Today* **2013**, *204*, 85–93. [CrossRef]
42. Kim, B.; Lee, Y.-R.; Kim, H.-Y.; Ahn, W.-S. Adsorption of Volatile Organic Compounds over MIL-125-NH_2. *Polyhedron* **2018**, *154*, 343–349. [CrossRef]
43. Fu, Y.; Sun, D.; Chen, Y.; Huang, R.; Ding, Z.; Fu, X.; Li, Z. An Amine-Functionalized Titanium Meta-Organic Framework Photocatalyst with Visible-Light-Induced Activity for CO_2 Reduction. *Angew. Chem. Int. Ed.* **2012**, *51*, 3364–3367. [CrossRef] [PubMed]
44. Cui, Y.; Yue, Y.; Qian, G.; Chen, B. Luminescent Functional Metal-Organic Frameworks. *Chem. Rev.* **2012**, *112*, 1126–1162. [CrossRef]
45. Sjekloća, L.; Ferré-D'Amaré, A.R. Binding between G Quadruplexes at the Homodimer Interface of the Corn RNA Aptamer Strongly Activates Thioflavin T Fluorescence. *Cell Chem. Biol.* **2019**, *26*, 1159–1168.e4. [CrossRef]
46. Mohanty, J.; Choudhury, S.D.; Pal, H.; Bhasikuttan, A.C. Early Detection of Insulin Fibrillation: A Fluorescence Lifetime Assay to Probe the Pre-Fibrillar Regime. *Chem. Commun.* **2012**, *48*, 2403–2405. [CrossRef]
47. Nielsen, L.; Khurana, R.; Coats, A.; Frokjaer, S.; Brange, J.; Vyas, S.; Uversky, V.N.; Fink, A.L. Effect of Environmental Factors on the Kinetics of Insulin Fibril Formation: Elucidation of the Molecular Mechanism. *Biochemistry* **2001**, *40*, 6036–6046. [CrossRef]
48. Mei, J.; Leung, N.L.C.; Kwok, R.T.K.; Lam, J.W.Y.; Tang, B.Z. Aggregation-Induced Emission: Together We Shine, United We Soar! *Chem. Rev.* **2015**, *115*, 11718–11940. [CrossRef]
49. Liu, J.; Meng, Q.; Zhang, X.; Lu, X.; He, P.; Jiang, L.; Dong, H.; Hu, W. Aggregation-Induced Emission Enhancement Based on 11,11,12,12,-Tetracyano-9,10-Anthraquinodimethane. *Chem. Commun.* **2013**, *49*, 1199–1201. [CrossRef]
50. He, J.; Xu, B.; Chen, F.; Xia, H.; Li, K.; Ye, L.; Tian, W. Aggregation-Induced Emission in the Crystals of 9,10-Distyrylanthracene Derivatives: The Essential Role of Restricted Intramolecular Torsion. *J. Phys. Chem. C* **2009**, *113*, 9892–9899. [CrossRef]
51. Krebs, F.C.; Spanggaard, H. An Exceptional Red Shift of Emission Maxima upon Fluorine Substitution. *J. Org. Chem.* **2002**, *67*, 7185–7192. [CrossRef] [PubMed]
52. Choudhury, S.D.; Mohanty, J.; Pal, H.; Bhasikuttan, A.C. Cooperative Metal Ion Binding to a Cucurbit[7]Uril−Thioflavin T Complex: Demonstration of a Stimulus-Responsive Fluorescent Supramolecular Capsule. *J. Am. Chem. Soc.* **2010**, *132*, 1395–1401. [CrossRef] [PubMed]
53. Solís, R.R.; Gómez-Avilés, A.; Belver, C.; Rodriguez, J.J.; Bedia, J. Microwave-Assisted Synthesis of NH_2-MIL-125(Ti) for the Solar Photocatalytic Degradation of Aqueous Emerging Pollutants in Batch and Continuous Tests. *J. Environ. Chem. Eng.* **2021**, *9*, 106230. [CrossRef]
54. Vandenbroucke, A.M.; Morent, R.; De Geyter, N.; Leys, C. Non-Thermal Plasmas for Non-Catalytic and Catalytic VOC Abatement. *J. Hazard. Mater.* **2011**, *195*, 30–54. [CrossRef]
55. Guieysse, B.; Hort, C.; Platel, V.; Munoz, R.; Ondarts, M.; Revah, S. Biological Treatment of Indoor Air for VOC Removal: Potential and Challenges. *Biotechnol. Adv.* **2008**, *26*, 398–410. [CrossRef] [PubMed]
56. Manos, M.J.; Malliakas, C.D.; Kanatzidis, M.G. Heavy-Metal-Ion Capture, Ion-Exchange, and Exceptional Acid Stability of the Open-Framework Chalcogenide $(NH_4)_4In_{12}Se_{20}$. *Chem. Eur. J.* **2007**, *13*, 51–58. [CrossRef]
57. Moody, D.E. The Effect of Tetrahydrofuran on Biological Systems: Does a Hepatotoxic Potential Exist. *Drug Chem. Toxicol.* **1991**, *14*, 319–342. [CrossRef]
58. Draper, A.J.; Madan, A.; Parkinson, A. Inhibition of Coumarin 7-Hydroxylase Activity in Human Liver Microsomes. *Arch. Biochem. Biophys.* **1997**, *341*, 47–61. [CrossRef]

59. Fan, H.; Tu, T.; Zhang, X.; Yang, Q.; Liu, G.; Zhang, T.; Bao, Y.; Lu, Y.; Dong, Z.; Dong, J.; et al. Sinomenine Attenuates Alcohol-Induced Acute Liver Injury via Inhibiting Oxidative Stress, Inflammation and Apoptosis in Mice. *Food Chem. Toxicol.* **2022**, *159*, 112759. [CrossRef]
60. Biancalana, M.; Koide, S. Molecular Mechanism of Thioflavin-T Binding to Amyloid Fibrils. *Biochim. Et Biophys. Acta (BBA) Proteins Proteom.* **2010**, *1804*, 1405–1412. [CrossRef]
61. Harel, M.; Sonoda, L.K.; Silman, I.; Sussman, J.L.; Rosenberry, T.L. Crystal Structure of Thioflavin T Bound to the Peripheral Site of Torpedo Californica Acetylcholinesterase Reveals How Thioflavin T Acts as a Sensitive Fluorescent Reporter of Ligand Binding to the Acylation Site. *J. Am. Chem. Soc.* **2008**, *130*, 7856–7861. [CrossRef] [PubMed]
62. Takashima, Y.; Martínez, V.M.; Furukawa, S.; Kondo, M.; Shimomura, S.; Uehara, H.; Nakahama, M.; Sugimoto, K.; Kitagawa, S. Molecular Decoding Using Luminescence from an Entangled Porous Framework. *Nat. Commun.* **2011**, *2*, 168. [CrossRef]
63. Fang, M.-M.; Yang, J.; Li, Z. Recent Advances in Purely Organic Room Temperature Phosphorescence Polymer. *Chin. J. Polym. Sci.* **2019**, *37*, 383–393. [CrossRef]
64. Wang, Y.; Yang, J.; Tian, Y.; Fang, M.; Liao, Q.; Wang, L.; Hu, W.; Tang, B.Z.; Li, Z. Persistent Organic Room Temperature Phosphorescence: What Is the Role of Molecular Dimers? *Chem. Sci.* **2020**, *11*, 833–838. [CrossRef]
65. Xie, Y.; Li, Z. The Development of Mechanoluminescence from Organic Compounds: Breakthrough and Deep Insight. *Mater. Chem. Front.* **2020**, *4*, 317–331. [CrossRef]
66. Liu, Y.; Xiao, X.; Ran, Y.; Bin, Z.; You, J. Molecular Design of Thermally Activated Delayed Fluorescent Emitters for Narrowband Orange–Red OLEDs Boosted by a Cyano-Functionalization Strategy. *Chem. Sci.* **2021**, *12*, 9408–9412. [CrossRef] [PubMed]
67. Chakraborty, S.; Bhattacharjee, D.; Hussain, S.A. Formation of Nanoscale Aggregates of a Coumarin Derivative in Langmuir-Blodgett Film. *Appl. Phys. A* **2013**, *111*, 1037–1043. [CrossRef]
68. Yang, J.; Huang, J.; Li, Q.; Li, Z. Blue AIEgens: Approaches to Control the Intramolecular Conjugation and the Optimized Performance of OLED Devices. *J. Mater. Chem. C* **2016**, *4*, 2663–2684. [CrossRef]

nanomaterials

MDPI

Article

Synthesis of Aluminum-Based Metal–Organic Framework (MOF)-Derived Carbon Nanomaterials and Their Water Adsorption Isotherm

Dasom Jeong [1,2], Seong Cheon Kim [1,3], Taeseop An [1], Dongho Lee [4], Haejin Hwang [2], Siyoung Q. Choi [3,5] and Jeasung Park [1,*]

1 Green and Sustainable Materials R&D Department, Korea Institute of Industrial Technology (KITECH), Cheonan 31056, Republic of Korea; jdasom0130@kitech.re.kr (D.J.); grey@kitech.re.kr (S.C.K.); ats2660@kitech.re.kr (T.A.)

2 Department of Materials Science and Engineering, INHA University, Incheon 22212, Republic of Korea; hjhwang@inha.ac.kr

3 Department of Chemical and Biomolecular Engineering, Korea Advanced Institute of Science and Technology (KAIST), Daejeon 34141, Republic of Korea; sqchoi@kaist.ac.kr

4 Process R&D Center, Hanwha Solutions R&D Institute, Daejeon 34128, Republic of Korea; dhlee@hanwha.com

5 KAIST Institute for the Nanocentury, Korea Advanced Institute of Science and Technology (KAIST), Daejeon 34141, Republic of Korea

* Correspondence: jpark@kitech.re.kr

Abstract: The characteristics of water vapor adsorption depend on the structure, porosity, and functional groups of the material. Metal–organic framework (MOF)-derived carbon (MDC) is a novel material that exhibits a high specific area and tunable pore sizes by exploiting the stable structure and porosity of pure MOF materials. Herein, two types of aluminum-based MOFs were used as precursors to synthesize hydrophobic microporous C-MDC and micro-mesoporous A-MDC via carbonization and activation depending on the type of ligands in the precursors. C-MDC and A-MDC have different pore characteristics and exhibit distinct water adsorption properties. C-MDC with hydrophobic properties and micropores exhibited negligible water adsorption (108.54 mgg^{-1}) at relatively low pressures (P/P_0~0.3) but showed a rapid increase in water adsorption ability (475.7 mgg^{-1}) at relative pressures of about 0.6. A comparison with the isotherm model indicated that the results were consistent with the theories, which include site filling at low relative pressure and pore filling at high relative pressure. In particular, the Do–Do model specialized for type 5 showed excellent agreement.

Keywords: metal–organic framework (MOF); MOF-derived carbon (MDC); CAU-10-H; aluminum-fumarate; water adsorption; isotherm model

1. Introduction

Porous materials have received considerable attention in various fields, including in the environmental and energy sectors, owing to their potential for addressing a range of environmental issues [1]. In particular, porous materials with excellent physical water adsorption capabilities and high chemical stability have attracted attention for applications in adsorption cooling, desalination, and desiccant air-cooling systems [2–4]. The performance of an adsorbent is determined by its ability to adsorb water at different pressure ranges, depending on the specific application requirements [5].

Metal–organic frameworks (MOFs) are a class of potential materials used as adsorbents. They are porous coordination polymers composed of inorganic metal ions or metal clusters and organic ligands linked together to form three-dimensional crystalline frameworks. MOFs exhibit high chemical stability owing to high surface area and well-ordered structure and have advantageous properties, such as tunable porosity and modified functional groups [6–8]. Materials used as adsorbents must have an appropriate pressure range to suit the intended application. Various studies have reported the modification of functional groups in MOFs for adjusting the relative pressure range by exploiting their hydrophobic or hydrophilic properties [9,10]. However, controlling the relative pressure range has limitations because the range is not very wide, and it can adversely affect the water adsorption capacity [11].

MOFs can be transformed into more stable nanomaterials such as carbon materials, metal oxides, and hybrid materials, as required. MOF-derived materials can be manufactured without additional templates. Moreover, the size and volume of their pores can be freely controlled, and hence, they can be used for doping and hybridization [12–14]. Owing to these characteristics, MOF-derived materials have received considerable attention and have been widely studied in the fields of electrochemical energy storage and conversion devices, including supercapacitors, batteries, and fuel cells [15,16]. Among them, MOF-derived carbon (MDC) is a novel carbon material that utilizes the stable structure and porosity of pure MOFs and exhibits a high specific surface area and tunable porosity [17]. The properties of the MDC depend on the structure of the precursor material, the MOF. In addition, various structures can be formed using different heat treatments and activation conditions, although the underlying reasons for this structure-forming phenomenon are not yet fully understood. Hence, MDCs have garnered considerable attention in the fields of electrochemistry and energy, where excellent electrical conductivity and porous carbon are necessary [15,18,19]. While various studies have investigated the water adsorption properties of biomass-derived activated carbon, there is a lack of research on the water adsorption properties of MDCs [20–23]. The properties and composition of biomass-derived activated carbon can vary due to natural environmental factors, making it difficult to control the pore size. However, MDCs can be manufactured with consistent properties and compositions, and thus have a competitive advantage for commercial use.

According to the International Union of Pure and Applied Chemistry (IUPAC), there are seven types of isotherm water vapor adsorption curves [24]. Types 1, 2, 4, and 6 represent the water adsorption properties of hydrophilic materials that exhibit adsorption at relatively low pressures. In contrast, types 3, 5, and 7 represent the water adsorption properties of hydrophobic materials that do not exhibit water adsorption at low relative pressures and show adsorption characteristics as the relative pressure increases [25]. The adsorption mechanism also varies depending on pore size. Carbon materials with micropores exhibit a sharp increase in water adsorption in the relative pressure range of 0.6, whereas carbon materials with mesopores demonstrate high water adsorption characteristics at a relative pressure of ~0.8, and materials with macropores show high water adsorption properties at relative pressures above 0.9 [26,27].

CAU-10-H is a MOF composed of aluminum clusters and isophthalic acid organic ligands. CAU-10-H exhibits excellent chemical and water stability. Because of the low cost of aluminum salts and isophthalate ligands, it is an excellent material for commercial use. CAU-10-H comprises helical chains of *-cis* corner-sharing AlO_6 octahedra interconnected by isophthalate ions, forming a three-dimensional structure with one-dimensional channels of spherical shape measuring 4.0 × 7.0 Å [9,28–30].

Aluminum-fumarate (Al-fu) is a MOF composed of aluminum clusters and fumaric acid organic ligands. The manufacturing costs of Al-fu are low because aluminum is used as the metal cluster, similar to that in CAU-10-H. Al-fu exhibits high chemical and water stability and excellent water adsorption capabilities. Al-fu comprises helical chains of *-trans* corner sharing AlO_6 octahedra interconnected by fumarate ions, forming a three-dimensional structure with one-dimensional channels of spherical shape measuring 5.7 × 6.0 Å [31,32].

Large-scale synthesis of the two aluminum-based MOFs (CAU-10-H and Al-fu) is feasible owing to the low cost of ligands and metal salts, along with an eco-friendly synthesis process using water as the solvent. Hence, they have a high potential for commercial development [S.1, S.2].

In this study, we synthesized MDCs via carbonization and activation using two aluminum-based MOFs as pure materials. The synthesized MDCs exhibited different structures and pore characteristics owing to the differences in the structure and elements of the precursor MOFs. Differences in pore distribution led to distinct water adsorption properties governed by the water adsorption mechanism. We analyze the pore structure, physical and chemical properties, and their correlation with the water adsorption mechanism and demonstrate the potential of MDCs as water adsorbents.

2. Materials and Methods

2.1. Materials

The reagents used for the synthesis were aluminum sulfate 14–18 H_2O (51.0–57.5%, SAMCHUN, Pyeongtaek-si, Republic of Korea), sodium aluminate (38.0–42.0% Na_2O, 51.0–55.0% Al_2O_3, JUNSEI, Tokyo, Japan), isophthalic acid (99.0%, SAMCHUN, Seoul, Republic of Korea), sodium hydroxide (98.0%, SAMCHUN, Republic of Korea), fumaric acid (99.0%, SAMCHUN, Republic of Korea), urea (99.0%, SAMCHUN, Republic of Korea), ethanol (94.0%, DUKSAN, Seoul, Republic of Korea), and deionized water.

2.2. Synthesis of CAU-10-H

CAU-10-H was synthesized via a procedure described in the literature with water as the solvent [33]. The CAU-10-H was formed via the coordination bonding of aluminum clusters and organic ligands, with sodium hydroxide acting as a modulator.

Sol. 1: 3.6 g (0.090 mol) of sodium hydroxide was added to 90 mL of deionized water in a 200 mL beaker and dissolved for 1 h. Subsequently, 7.476 g (0.045 mol) of isophthalic acid was added to the solution and dissolved at 35 °C for 3 h. Sol. 2: in another 100 mL beaker, 11.263 g (0.017 mol) of aluminum sulfate was dissolved in 33.75 mL of deionized water. Sol. 3: in another 100 mL beaker, 0.922 g (0.011 mol) of sodium aluminate was dissolved in 22.5 mL of deionized water. Solution 1 and 7.5 mL ethanol were mixed in a three-neck flask and stirred for 10 min. Thereafter, solutions 2 and 3 were added simultaneously, and the mixture was refluxed in an oil bath of 160 °C for 20 h. The synthesized solution was washed twice with deionized water and ethanol, filtered using a paper filter, and dried in a vacuum oven at 80 °C overnight to obtain CAU-10-H.

2.3. Synthesis of Aluminum Fumarate MOF

Al-fu was synthesized via a reflux reaction. Al-fu was formed via the coordination bonding of aluminum clusters and the organic ligand, fumaric acid, with urea acting as a modulator.

Next, 78.64 g (0.125 mol) of aluminum sulfate and 22.48 g (0.375 mol) of urea were added to 400 mL of deionized water in a 600 mL of beaker and stirred until fully dissolved. Next, 28.96 g (0.250 mol) of fumaric acid was added to the solution and stirred for 30 min at 25 °C. The mixture was then refluxed in an oil bath of 80 °C for 24 h, followed by refluxing at 110 °C for another 24 h. The resulting solution was washed three times with deionized water, filtered using a paper filter, and dried in a vacuum oven at 80 °C overnight to obtain Al-fu.

2.4. Synthesis of MDCs

The CAU-10-H and Al-fu synthesized herein were used as precursors for synthesizing MDCs. C-14PF-50 (HANTECH, Ulsan, Republic of Korea) semi-tube furnace with a tube size of 50 × 46 × 250 mm was used for carbonization, and nitrogen gas (99.999%) was supplied at a flow rate of 200 cc/min. Sodium hydroxide (98.0%, SAMCHUN, Republic of Korea) was used as the wet etching solution for the carbonized CAU-10-H and carbonized Al-fu.

To synthesize CAU-10-H MDC, 1 g of CAU-10-H in a ceramic boat was placed in the tube furnace. The furnace is heated at a rate of 5 °C per minute until it reaches 900 °C, and then the sample was carbonized at 900 °C for 3 h under a nitrogen atmosphere. The carbonized sample was cooled to room temperature and was poured into 100 mL of 8 M sodium hydroxide solution and activated by stirring for 1 h at 95 °C. The activated sample was washed thoroughly with deionized water, filtered using a paper filter, and dried overnight in a vacuum oven at 80 °C to obtain CAU-10-H MDC.

Al-fu MDC was synthesized via carbonization of Al-fu at 800 °C for 3 h and activation by stirring for 1 h at 90 °C in 10 M sodium hydroxide solution. The other synthesis conditions were identical to those used for the synthesis of CAU-10-H MDC.

Hereafter, the CAU-10-H MDC is referred to as C-MDC, and the Al-fu MDC is referred to as A-MDC.

2.5. Material Characterization

The crystal structures of the MOFs and MDCs were examined via X-ray diffraction (XRD) analysis with Cu-Kα radiation, current of 30 mA, and voltage of 40 kV using a LabX XRD-6100 (Shimadzu, Kyoto, Japan); the scan mode was continuous with 2θ in the range 3–80° and the measurement speed was 10°/min. The elemental composition and chemical bonding states of the sample surfaces were investigated via X-ray photoelectron spectroscopy (XPS) using a K-Alpha instrument (Thermo Fisher Scientific, Waltham, MA, USA) with an Al X-ray source (1486.7 eV). The microstructure and distribution of components in the MOFs and MDCs were analyzed via FE-SEM and EDS using JSM-6701F (JEOL, Tokyo, Japan). The carbon, nitrogen, hydrogen, and oxygen contents of the MOFs and MDCs were examined via elemental analysis (EA) using a Flash*Smart*™ Elemental Analyzer (Thermo Fisher Sciencific, Waltham, MA, USA). Each sample was measured three times and the average value was used for quantitative calculations. The presence of functional groups on the MDC sample surfaces was investigated by measuring the infrared absorption via FTIR analysis. A Nicolet 6700 (Thermo Fisher Scientific, Waltham, MA, USA) was used for the FTIR analysis, and measurements were performed in the attenuated total reflectance measurement mode in the range of 4000–660 cm^{-1}. The pore size, volume, and distribution of the MOFs and MDCs were determined via nitrogen gas adsorption (77 K) analysis using a BELSORP MAX (MicrotracBEL, Osaka, Japan). Before the analysis, all samples were pretreated under vacuum conditions at 180 °C for 8 h. TEM and EDS analyses were conducted using a Tecnai G2 G30 S-Twin (FEI, Naples, Italy). Raman spectroscopy was conducted at a wavelength of 532 nm using a LabRAM Revolution (HORIBA, Kyoto, Japan).

2.6. Water Vapor Adsorption Evaluation and Isotherm Model

The water adsorption capacities of the MOFs and MDCs were investigated via water adsorption analyses using MicrotracBEL BELSORP MAX (MicrotracBEL, Osaka, Japan) under isothermal conditions at 298 K and 308 K. The adsorption and desorption ranges were set at relative pressures (P/P_0) of 0–0.990 and 0.300–0.990, respectively. For pressure measurement, 5 sensors with a range of 133 kPa (1000 Torr) and 2 sensors with a range of 1.33 kPa (10 Torr) were used, each having an error range of 0.15% and 0.5%, respectively.

Two theoretical models were used to validate the water sorption properties of the MDCs obtained from the water adsorption analyses and further expand their practical application range. The first model used was the Henry–Sips model, expressed as follows: [34]

$$q = \beta K_H\left(\frac{p}{p_0}\right) + (1-\beta)\frac{q_m(K_s p/p_0)^{1/n}}{1+(K_s p/p_0)^{1/n}} \quad (1)$$

$$K = K_0 exp\left(\frac{-\Delta H}{RT}\right) \quad (2)$$

$$\beta = exp\left(-\alpha\frac{p}{p_0}\right),\ n = A + \frac{B}{T} \quad (3)$$

where q is the equilibrium adsorption capacity, q_m is maximum adsorption capacity, p is the equilibrium pressure (in kPa), p_0 is the saturation pressure (in kPa), K_H is the Henry constant, K_s is the Sips constant, and n is the degree of surface nonhomogeneity.

The second model is the Do–Do model, expressed by the following equation [35,36]:

$$\theta = f\frac{K_f x\{1-(1+\beta)x^{\beta}+\beta x^{\beta+1}\}}{(1-x)\left\{1+\left(K_f-1\right)x-K_f x^{\beta+1}\right\}}+(1-f)\frac{K_\mu x^{\alpha}}{1+K_\mu x^{\alpha}};\beta\geq\alpha. \quad (4)$$

The R^2 values for the two modeling curves were calculated as follows.

$$R^2 = 1-\frac{\sum_{i=1}^{m}(q_{meas,i}-q_{cal,i})^2}{\sum_{i=1}^{m}\left(q_{meas,i}-\bar{q}_{meas}\right)^2} \quad (5)$$

$$\bar{q}_{meas} = \frac{1}{m}\sum_{i=1}^{m} q_{meas,i} \quad (6)$$

where m is the number of measured isotherms.

The isosteric enthalpy of adsorption (ΔH_{ads}) was calculated using the following equation [37]:

$$\Delta H_{ads} = RT^2\left\{\frac{\partial lnP}{\partial T}\right\}_{q^a} = -R\left\{\frac{\partial lnP}{\partial \frac{1}{T}}\right\}_{q^a} \quad (7)$$

where q^a denotes the adsorption capacity at a given temperature, and the enthalpy of adsorption at that point is calculated.

Repeated water adsorption and desorption analyses were conducted using a DVS Intrinsic (Surface Measurement Systems, Wembley, UK) to test the water stability of the MDCs. Prior to the analysis, the samples were dried with N_2 gas (99.999%) at a flow rate of 200 cc/min for 300 min until the dm/dt value reached 0.001. The cutoff condition for each step was set at dm/dt = 0.005, and the relative humidity range was set at 10–90%. The water adsorption and desorption cycle tests were repeated 100 times.

3. Results and Discussion

Figure 1 depicts a schematic of the synthesis of MDC, which was derived from a precursor using MOF as a template via carbonization and activation. MDC maintains its morphology without significant changes even after heat treatment, which is essential for utilizing the porous properties of MOF.

Figure 2 shows the XRD patterns of CAU-10-H (which was used as a precursor) and C-MDC (which was synthesized via carbonization and activation). As shown in Figure 2, (a) sharp peaks were observed at 2 theta angles of 8.2°, 9.2°, 11.6°, 12.3°, 15°, 16.4°, 17°, 17.5°, 18.4°, 18.5°, 18.9°, and 22.3°, which correspond to the (200), (101), (220), (211), (301), (400), (231), (330), (240), and (202) planes of CAU-10-H [38]. As shown in Figure 2b, the sharp peaks of CAU-10-H were not observed after carbonization and the noise of the peaks increased, whereas broad peaks at approximately 22° and 43° were observed. The broad peak at 22° corresponds to an amorphous carbon structure, whereas the broad peak near 43° corresponds to a graphitic structure (JCPDS No. 41-1487). These results suggest that the bond between the aluminum clusters and organic ligands in the precursor was broken during carbonization, leading to the dispersion of pure aluminum metal inside the structure and the formation of an unstable amorphous carbon structure. The increase in the noise of the peak is attributed to the impurities generated during carbonization covering the surface. The intensity of the broad peak at 22°, corresponding to amorphous carbon, and the broad peak near 43°, corresponding to the graphitic structure, were higher, and the noise of the peaks was reduced after activation via wet etching. These results are attributed to the removal of impurities that remained on the surface during activation.

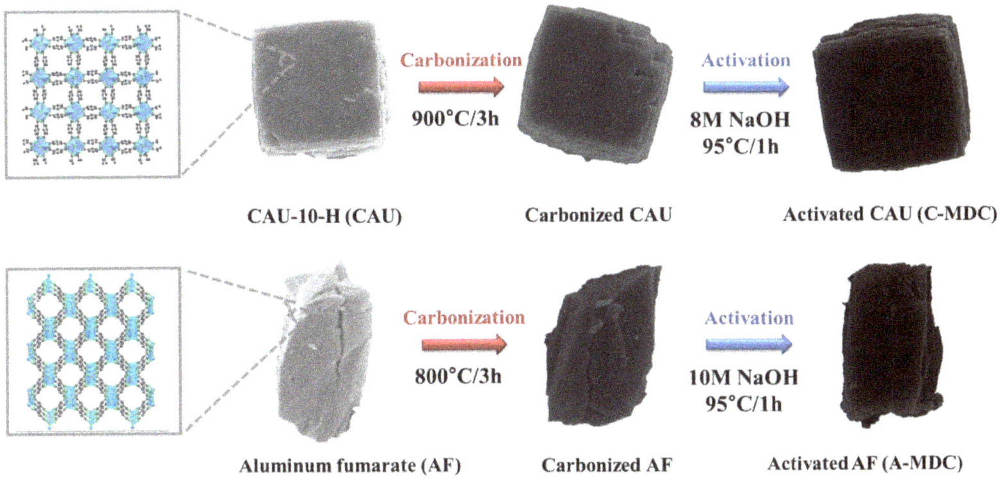

Figure 1. The schematics of fabrication of Al-based MOF-derived carbon.

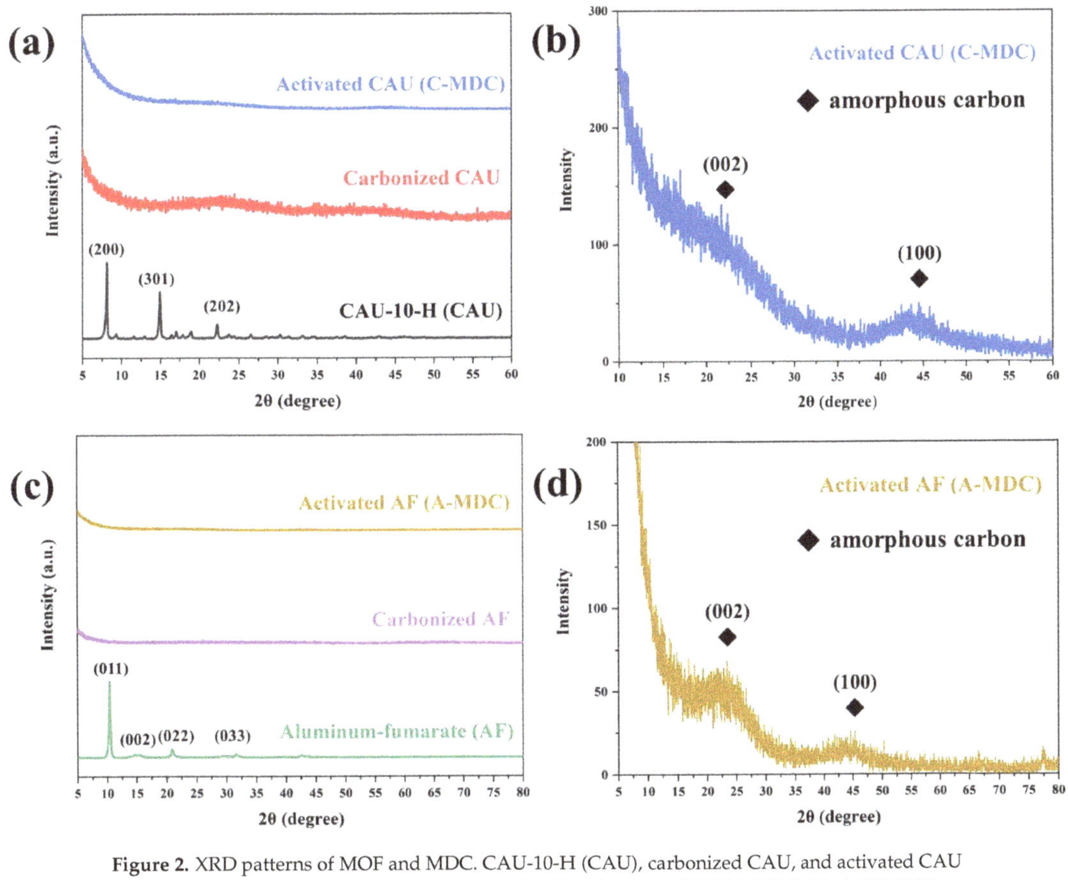

Figure 2. XRD patterns of MOF and MDC. CAU-10-H (CAU), carbonized CAU, and activated CAU (C-MDC) (**a**); C-MDC (**b**); Al-fu (AF), carbonized AF, and activated AF (A-MDC) (**c**); A-MDC (**d**).

As shown in Figure 2c, sharp peaks were observed at 2 theta angles of 10.4°, 15°, 21°, 32°, and 43°, which correspond to the (011), (002), (022), (033), and (212) planes of Al-fu [39]. As shown in Figure 2d, similar to that in the C-MDC XRD pattern, diffraction peaks corresponding to the original crystal structure of MOF were not observed due to the decomposition of the aluminum clusters and organic ligands of the precursor during heat treatment, and an unstable amorphous carbon structure was formed. After activation via wet etching, impurities were removed, and broad peaks were observed at 22° and 43°. Unlike CAU-10-H, which has a benzene ring bond in its ligands, Al-fu has a different ligand structure. Hence, the (002) plane corresponding to the amorphous carbon structure was more distinct than the (101) plane corresponding to the graphitic structure. These results indicate that the structure of the MDC produced through heat treatment and etching can vary depending on the type of ligand.

Table 1 and Figure 3 present the results of the XPS analysis of the surface elements and chemical bonding of CAU-10-H and its derivatives produced via carbonization and activation. Figure 3b shows the carbon bonding state, wherein the peaks at 284.88 and 286.08 eV, respectively, indicate the presence of sp^2 and sp^3 carbon, in the ligands [40]. The peak at 288.68 eV indicates the bond between carbon and oxygen in the framework. The carbonization led to a partial cleavage of the sp^2 carbon bonds and a relative increase in the proportion of sp^3 carbon. The proportion of sp^3 carbon was higher than that of the precursor even after the removal of impurities during activation. Figure 3c shows the oxygen bonding states, wherein the peak at 532.18 eV represents the carbon–oxygen bonding. The new peak at 530.38 eV corresponds to aluminum oxide formed during carbonization [41,42]. Aluminum oxide can block the surface and pores, potentially hindering the full utilization of pore characteristics. However, all impurities were removed after the activation process, as confirmed by the absence of an aluminum oxide peak in Figure 3c. As a result, it was confirmed that the formation of a carbonaceous structure on the surface of the C-MDC, wherein the proportion of carbon was 90.35 (Table 1, Figure 3a).

Table 1. Elemental composition of CAU-10-H and its derivatives analyzed via XPS.

Samples (at. %/XPS)	C1s	O1s	Al2p	Total
CAU-10-H (CAU)	72.75	23.49	3.77	100
Carbonized CAU	69.23	20.2	10.55	100
Activated CAU (C-MDC)	90.35	9.65	0	100

Table 2 and Figure 4 present the results of XPS analysis of the surface elements and chemical bonding of Al-fu and its derivatives produced via carbonization and activation. The peaks at 284.08 eV and 285.18 eV, shown in Figure 4b, correspond to sp^2 and sp^3 carbon, respectively, as the ligands consist of both single and double carbon bonds [43]. The peak at 288.68 eV indicates the bond between carbon and oxygen in the framework. However, compared to CAU-10-H, the carbonization of Al-fu resulted in a significant reduction in sp^2 carbon bonding, leading to a remarkable increase in the proportion of sp^3 carbon. Furthermore, even after the removal of impurities during activation, the proportion of sp^3 carbon remained higher than that in the precursor. These differences in recombination suggest that the structure of A-MDC may be more flexible than that of C-MDC, which may affect the pore characteristics. As shown in Figure 4c, the peak at 532.08 eV corresponds to the bond between carbon and oxygen. The new peak at 531.08 eV indicates the formation of aluminum oxide during carbonization [44,45]. All these impurities were removed during activation, as evidenced by the absence of the aluminum oxide peak in Figure 4d. As a result, it was confirmed that the formation of a carbonaceous structure on the surface of the A-MDC, wherein the proportion of carbon was 89.32% (Table 2, Figure 4a). Based on these results, it is anticipated that the activation process will secure maximum pore volume by removing impurities such as aluminum metals. Additionally, increasing carbon purity is expected to enhance water adsorption capability.

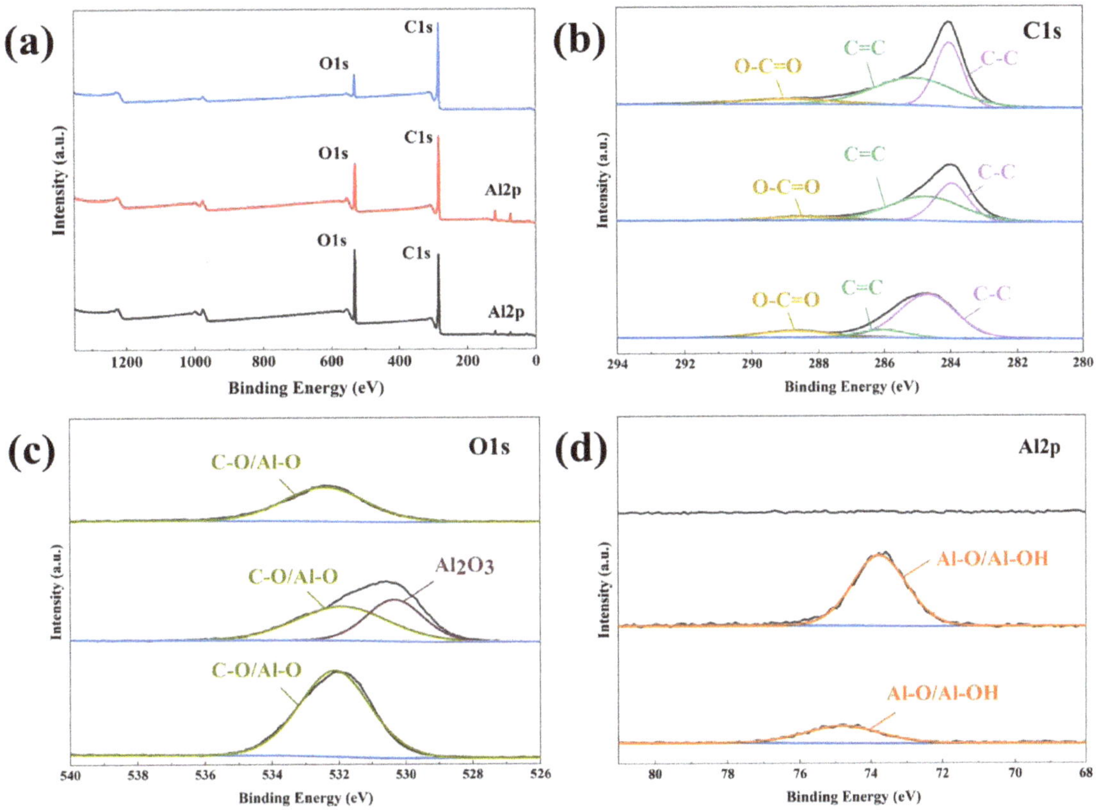

Figure 3. Results of XPS analyses of CAU-10-H and its derivatives. Survey scan spectrum (**a**) and high-resolution XPS scan spectrum over C1s (**b**); O1s (**c**); and Al2p spectra. The black line indicates the absence of any peaks (**d**).

Table 2. Elemental composition of Al-fu and its derivatives analyzed via XPS.

Samples (at. %/XPS)	C1s	O1s	Al2p	Total
Al-fu (AF)	58.98	33.59	7.44	100
Carbonized AF	46.88	33.45	19.68	100
Activated AF (A-MDC)	89.32	10.68	0	100

Figure 5 shows the FE-SEM images of CAU-10-H, Al-fu, and their derivatives. As shown in Figure 5a, CAU-10-H exhibits a cubic morphology with an average particle size of 3 μm. As shown in Figure 5b,c, no significant morphological changes were observed after carbonization and activation. The morphology of the MDC obtained from the precursor MOF remained largely intact, demonstrating that the inherent porosity of the MOF structure was preserved.

The Al-fu synthesis method used herein was different from previously reported methods. Hence, significant differences were observed between the morphologies of Al-fu synthesized in this study and those reported in the literature. As shown in Figure 5d, rectangular particles with a size of 1.5 × 2.5 μm aggregated to form a flower-like shape, which is attributed to the controlled nucleation and nucleation growth rate in our synthesis method. Resting for 24 h before the reaction at a temperature lower than the reaction temperature makes nucleation growth delay and allows sufficient nucleation. The temperature was subsequently increased to initiate nucleation growth after sufficient nucleation. Thus, Al-fu with a uniform particle size was synthesized. As shown in Figure 5e, the particles that constituted the flower-like shape separated into individual entities after carbonization; however, the change in morphology was not significant. In addition, the morphology of the precursor did not change after activation (Figure 5f). However, compared to C-MDC, cracks were observed in the morphology of A-MDC after carbonization and activation.

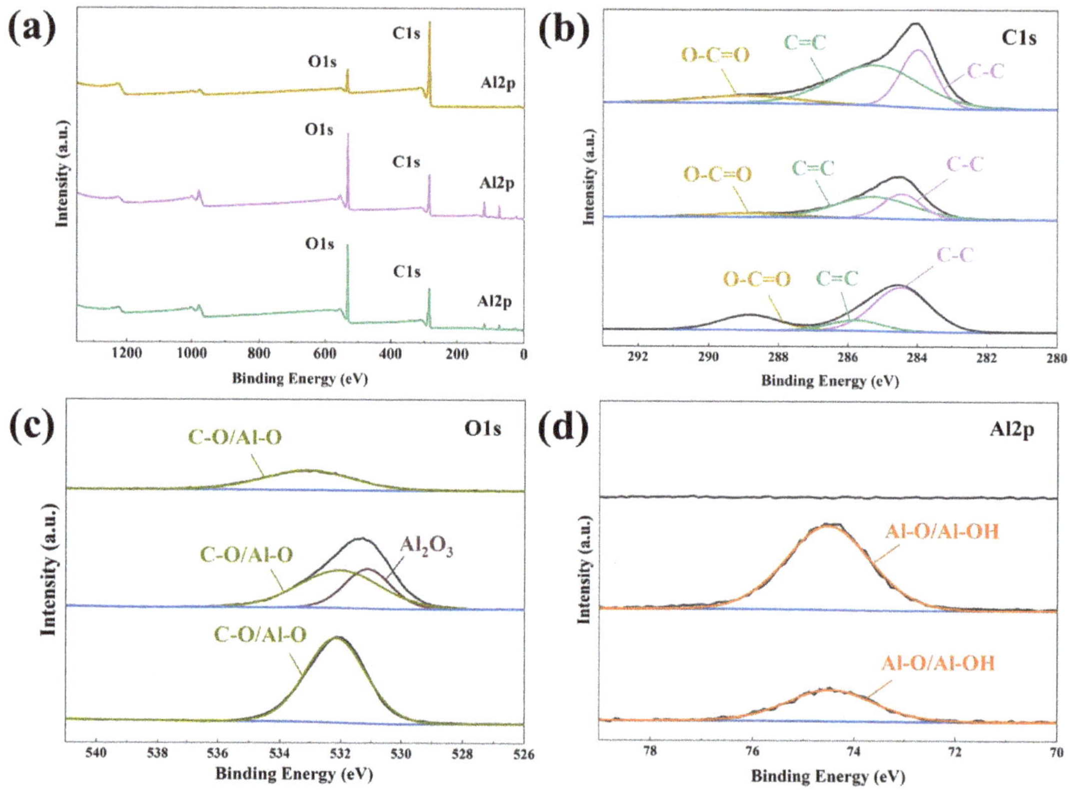

Figure 4. Results of XPS analysis of Al-fu and its derivatives. Survey scan spectrum (**a**) and high-resolution XPS scan spectrum over C1s (**b**); O1s (**c**); and Al2p spectra. The black line indicates the absence of any peaks (**d**).

The elemental compositions of CAU-10-H and its derivatives, obtained via EA and EDS, are presented in Table 3. Structural changes during MDC synthesis were predicted using the EA analysis results. The proportion of C, N, H, and O in carbonized CAU was 23.81 wt.% lower than that in CAU-10-H. This is expected because the cleavage of MOF bonds results in an increase in the proportion of aluminum. In addition, the proportion of carbon increases significantly owing to the removal of aluminum during activation. The results of EDS analysis similarly indicate that the proportions of aluminum and oxygen increase after carbonization. This suggests that the broken oxygen bonds combine with aluminum to form aluminum oxide, which is expected to be freely dispersed on the surface. Aluminum oxides are removed upon activation, and the EDS results indicate a higher proportion of carbon, which is consistent with the EA results. Both EA and EDS can be used for quantitative analysis of elements; however, EDS can only analyze the elements present on the surface, and hence, there may be some differences in the obtained proportions.

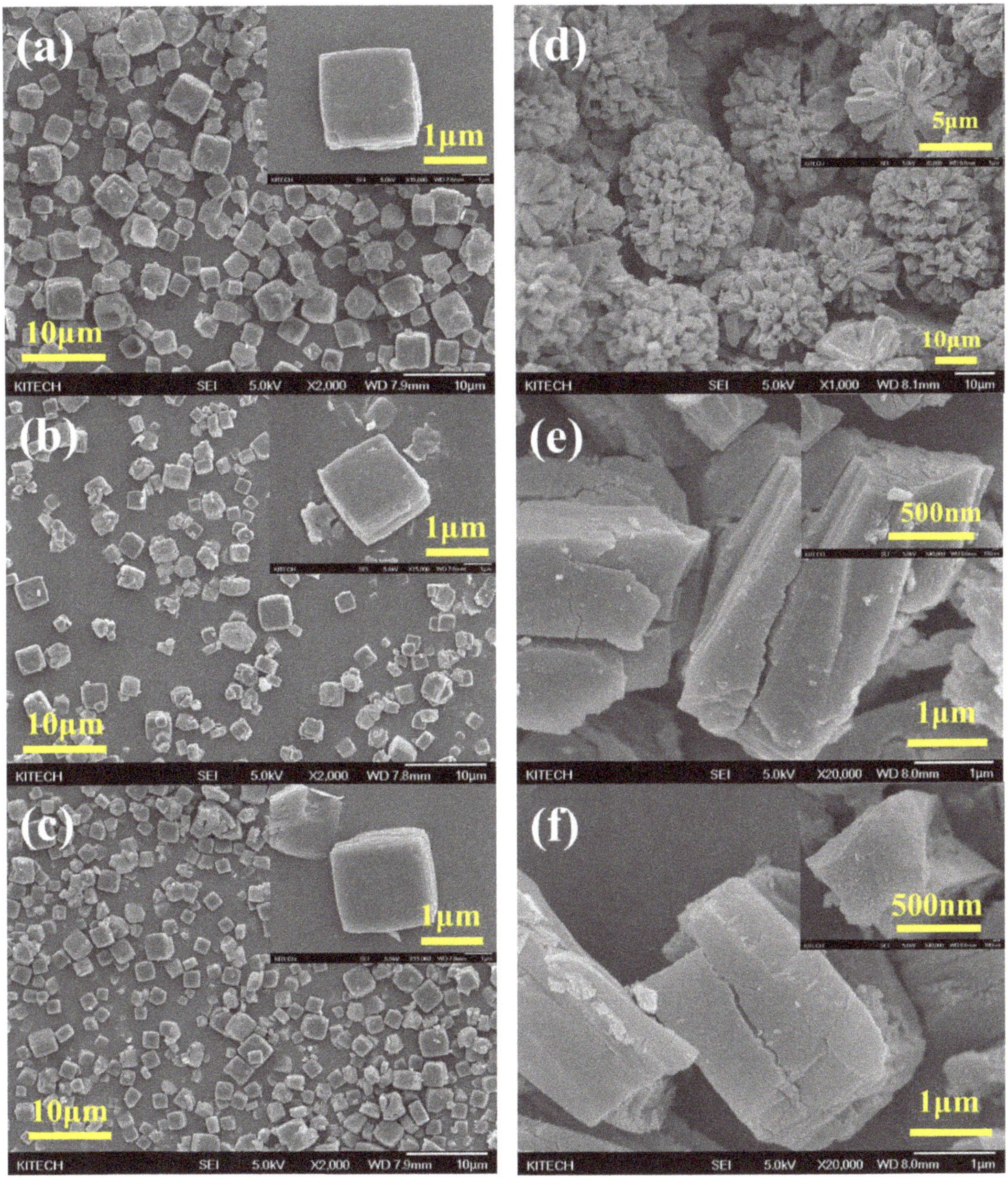

Figure 5. FE-SEM images of CAU-10-H, Al-fu, and their derivatives. CAU-10-H (CAU) (**a**); carbonized CAU (**b**); and activated CAU (C-MDC) (**c**); Al-fu (AF) (**d**); carbonized AF (**e**); and activated AF (A-MDC) (**f**).

The elemental compositions of Al-fu and its derivatives are shown in Table 4. The results are consistent with the results of the analyses of CAU-10-H and its derivatives. The EA results confirmed that the increased aluminum generated during carbonization was largely removed via activation, resulting in a higher carbon ratio in the final product. EDS analysis indicated that A-MDC was primarily composed of carbon, with aluminum and oxygen present on the surface. The aluminum and oxygen on the surface formed aluminum oxide during carbonization, which was removed during activation. These results suggest that carbon is the primary constituent element of A-MDC.

Table 3. Elemental composition of CAU-10-H and its derivatives analyzed via EA and EDS.

Samples (wt.%/EA)	C	O	N	H	Total
CAU-10-H (CAU)	45.44 (±0.01)	27.31 (±0.18)	0	2.66 (±0.03)	75.41
Carbonized CAU	41.66 (±0.09)	6.17 (±0.23)	0	0.83 (±0.06)	48.67
Activated CAU (C-MDC)	91.57 (±0.08)	3.88 (±0.08)	0	0.65 (±0)	96.10
Samples (wt.%/EDS)	**C**	**O**	**N**	**Al**	**Total**
CAU-10-H (CAU)	66.61	30.12	0	3.27	100
Carbonized CAU	56.89	32.79	0	10.32	100
Activated CAU (C-MDC)	86.48	13.52	0	0	100

Table 4. Elemental composition of Al-fu and its derivatives analyzed via EA and EDS.

Samples (wt.%/EA)	C	O	N	H	Total
Al-fu (AF)	29.90 (±0.05)	44.07 (±0.74)	0	1.96 (±0.09)	75.93
Carbonized AF	25.91 (±0.02)	7.74 (±0.30)	0	0.92 (±0.07)	34.34
Activated AF (A-MDC)	90.30 (±0.13)	5.17 (±0.29)	0.17 (±0.00)	1.07 (±0.00)	96.72
Samples (wt.%/EDS)	**C**	**O**	**N**	**Al**	**Total**
Al-fu (AF)	29.84	52.32	0	17.84	100
Carbonized AF	37.80	40.10	0	22.10	100
Activated AF (A-MDC)	88.27	11.73	0	0	100

Fourier-transform infrared (FT-IR) analysis was conducted to identify the organic ligands of the MOF and the functional groups on the surface of the materials. The results for CAU-10-H, C-MDC, Al-fu, and A-MDC are shown in Figure 6. As shown in Figure 6a, various peaks were observed for CAU-10-H. The peak at 3621 cm^{-1} corresponds to μ-OH vibrations of the aluminum oxide (AlO_6) clusters and water, whereas the peaks at 1558 cm^{-1} and 1403 cm^{-1} correspond to the asymmetric and symmetric vibrations of the coordination carboxylate groups, respectively [10,46]. In contrast, no peaks were observed for C-MDC, which indicates the formation of a hydrophobic carbon material without any functional groups on the surface.

Various peaks were also observed for Al-fu (Figure 6b). The peak at 3706 cm^{-1} corresponds to the μ-OH vibrations of the aluminum oxide (AlO_6) clusters and water, whereas the peaks at 1596 cm^{-1} and 1423 cm^{-1} correspond to the asymmetric and symmetric stretching of carboxylic groups, respectively. The peak at 987 cm^{-1} corresponds to O-H bending [47–49]. Similar to that in C-MDC, the absence of peaks in the FTIR results of A-MDC after chemical treatment suggests the presence of a hydrophobic carbon material.

The structural defects and crystallinity of the carbon materials were determined from Raman spectroscopy results. The D band (1320 cm^{-1}) corresponds to structural defects or amorphous carbon, whereas the G band (1590 cm^{-1}) corresponds to graphitic ordered carbon. The ratio I_D/I_G, which represents the intensity of the D band normalized to that of the G band, indicates the presence of structural defects in carbon. A higher I_D/I_G ratio suggests a higher level of structural imperfections [50–52]. The I_D/I_G ratio of A-MDC is higher than that of C-MDC. This implies that the structural changes in C-MDC are minimal, even after thermal and chemical treatments, whereas the structural changes in A-MDC are expected to be considerable. These structural defects may influence the pore structure of the carbon materials, suggesting that A-MDC may have larger pores than C-MDC.

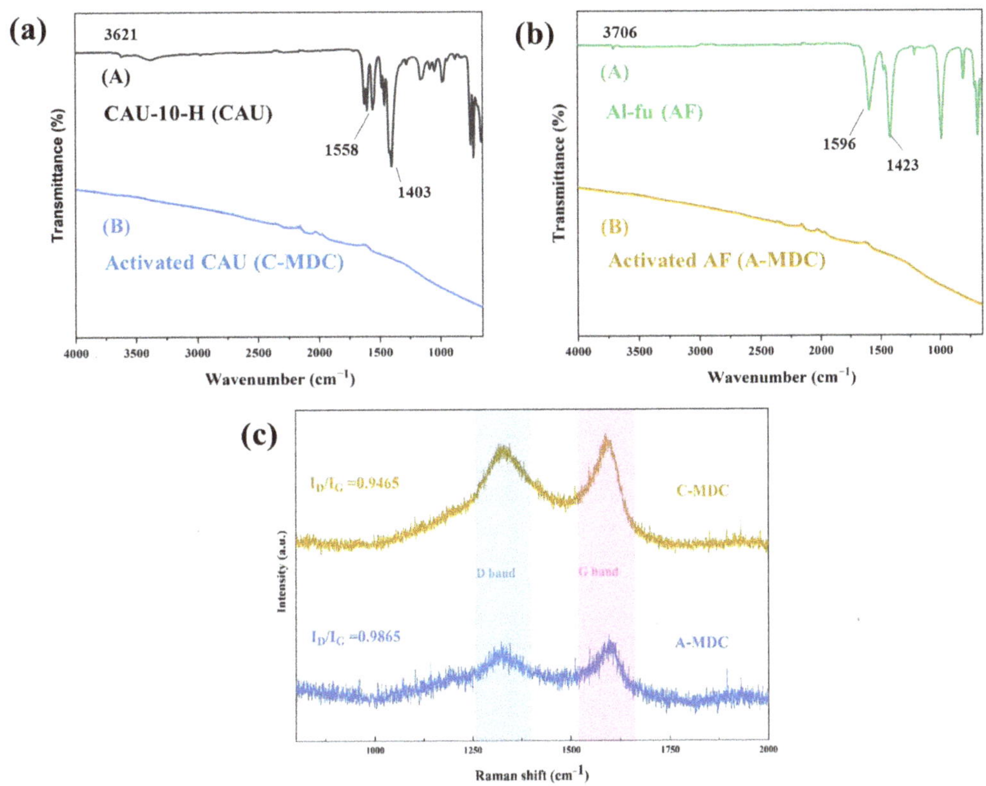

Figure 6. FT-IR results of CAU-10-H and C-MDC (**a**); FT-IR results of Al-fu and A-MDC; (**b**), Raman spectroscopy of C-MDC and A-MDC (**c**).

The TEM images and results of EDS mapping revealed that both synthesized C-MDC and A-MDC predominantly consisted of carbon, which is consistent with the particle size observed in SEM analysis (Figure 7a,c). The SAED patterns, which are in good agreement with the XRD results, further confirm the presence of amorphous carbon in both the MDC materials (Figure 7b,d).

Figure 8a–c show the results of the isothermal nitrogen adsorption and desorption analyses for CAU-10-H, Al-fu, and its derivatives. The specific surface areas, pore sizes, and pore volumes of CAU-10-H and its derivatives are summarized in Table 5. As shown in Figure 8a, both CAU-10-H and C-MDC exhibit adsorption characteristics at relative pressures below 0.05, indicating microporous characteristics. The nitrogen adsorption capacity decreased during carbonization because pure aluminum metal and other impurities produced by the decomposition of aluminum clusters and organic ligand linkers during carbonization occupy pores of the structure, thereby blocking the pores. In contrast, the nitrogen adsorption capacity increased significantly after activation because the metallic and impurity components that blocked the pores were removed by the alkaline NaOH solution. Compared to the precursor, C-MDC exhibited a considerably higher nitrogen adsorption capacity, which is primarily attributed to the additional surface area resulting from the removal of aluminum clusters. This hypothesis can be further supported by the Horvath–Kawazoe (HK) plot in Figure 8b. The pore size of CAU-10-H, as confirmed by the HK plot, was 1.708 nm, whereas that of C-MDC was 2.043 nm. Figure 8c shows the overall pore size distribution obtained from NLDFT calculations. The results further confirm the increase in pore size and volume after activation. Consequently, the specific surface area of C-MDC (1653.2 m^2g^{-1}) was approximately 2.5 times higher than that of the precursor (677.77 m^2g^{-1}).

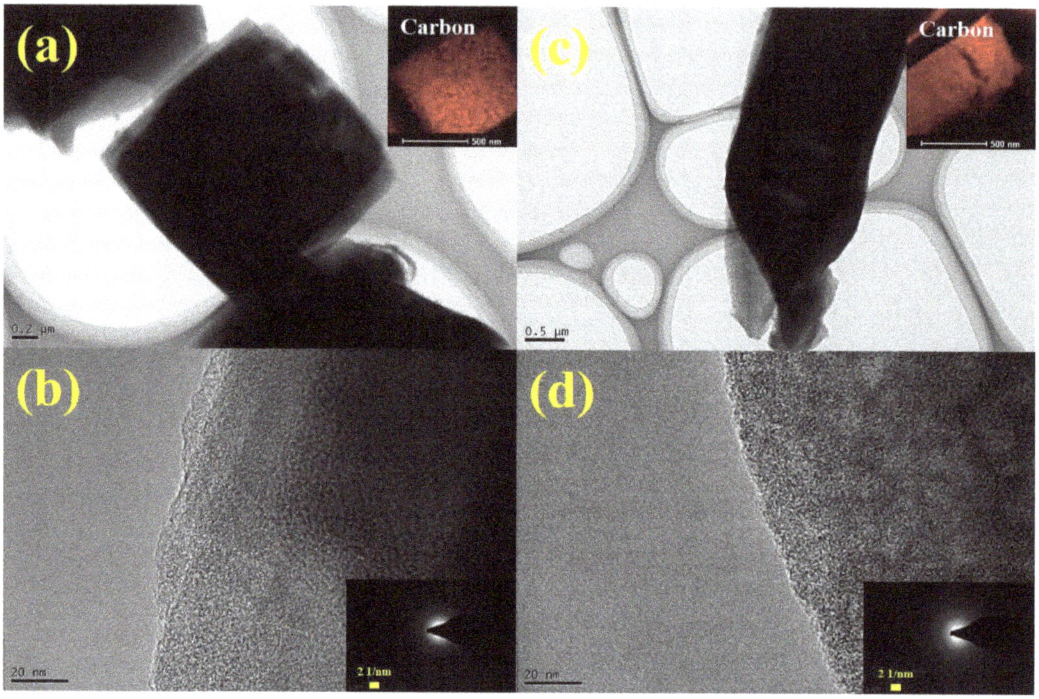

Figure 7. Results of TEM analysis. TEM images of C-MDC (**a**,**b**) and A-MDC (**c**,**d**). The insets show the EDS mapping and SAED pattern.

Table 5. BET surface area, pore size, and pore volume of CAU-10-H and its derivatives.

	S_{BET} [m^2g^{-1}]	V_{total} [cm^3g^{-1}]	D_{ave} [nm]
CAU-10-H (CAU)	677.77	0.2894	1.7078
Carbonized CAU	241.51	0.1234	2.0430
Activated CAU (C-MDC)	1653.2	0.7291	1.7640

The specific surface areas, pore sizes, and pore volumes of Al-fu, and its derivatives are summarized in Table 6. Similar to that observed for C-MDC, aluminum oxides produced during carbonization blocked the pores, leading to a decrease in overall pore characteristics (Figure 8d). However, in contrast to C-MDC, A-MDC exhibited a hysteresis loop and nitrogen adsorption characteristics at relative pressures of 0.5 or higher. The hysteresis loop is more evident once the impurities were removed via activation. This demonstrates mesoporous characteristics that were not observed in Al-fu, as indicated by the Barrett–Joyner–Halenda (BJH) plot shown in Figure 8e. The BJH plot shows that, compared to the precursor, the number of mesopores (4, 8 nm) increased considerably after carbonization and activation. Figure 8f shows the pore sizes and volumes obtained from NLDFT calculations. In A-MDC, the number of mesopores and micropores has significantly increased. Consequently, the BET (Brunauer, Emmett, and Teller) surface area (1736.6 m^2g^{-1}) of A-MDC produced through carbonization and activation was approximately 1.5 times higher than that of the precursor (1103.1 m^2g^{-1}).

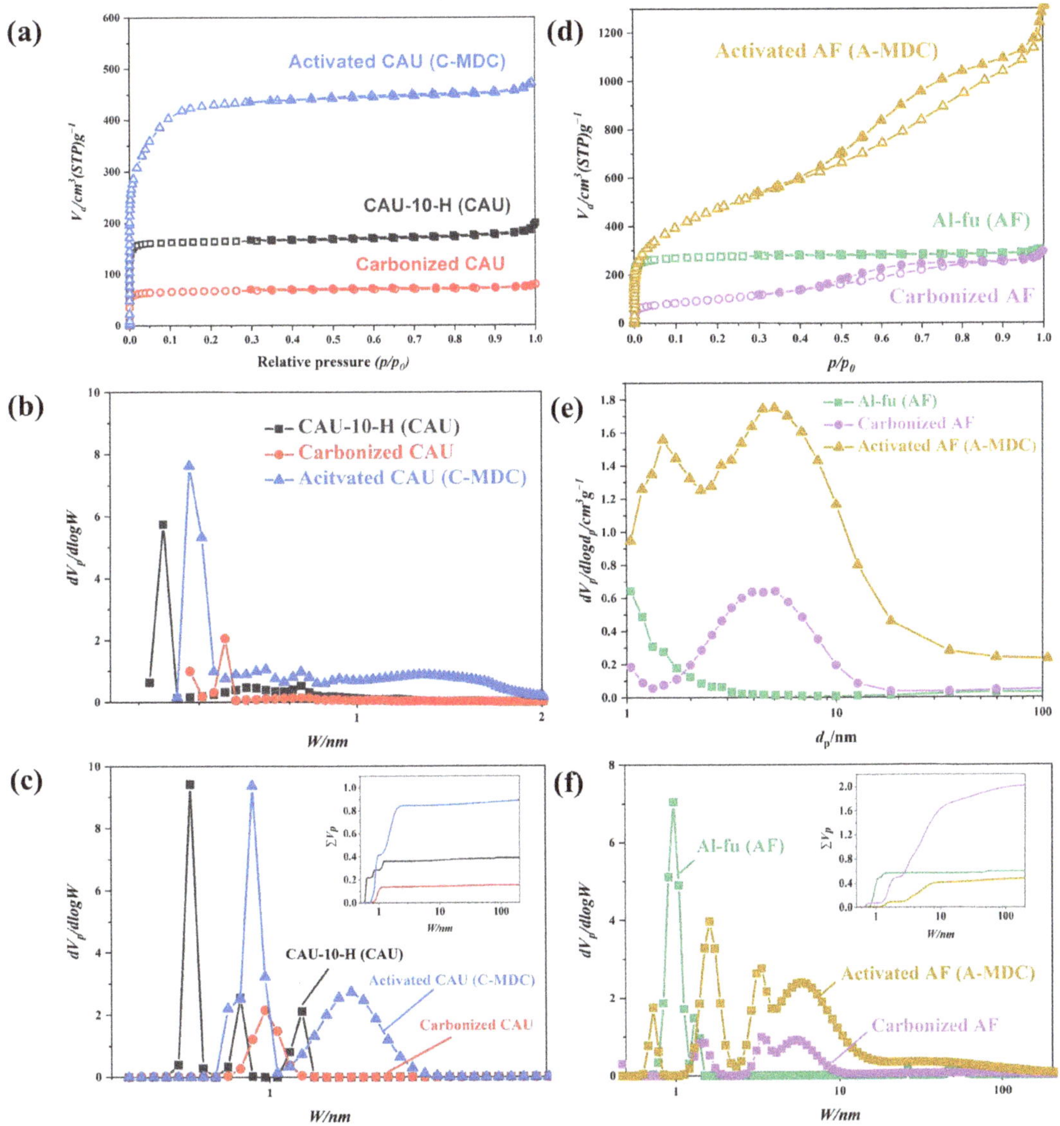

Figure 8. N_2 adsorption and desorption isotherm at 77 K. Isotherm adsorption and desorption curve (**a**); HK Plot (**b**); NLDFT calculation (**c**) of CAU-10-H and its derivatives, isotherm adsorption and desorption curve (**d**); BJH Plot (**e**); NLDFT calculation (**f**) of Al-fu and its derivatives.

Table 6. BET surface area, pore size, and pore volume of Al-fu and its derivatives.

	S_{BET} [m^2g^{-1}]	V_{total} [cm^3g^{-1}]	D_{ave} [nm]
Al-fu (AF)	1103.1	0.4664	1.6911
Carbonized AF	365.37	0.4328	4.7385
Activated AF (A-MDC)	1736.6	1.8394	4.2366

The pore structures of the synthesized MDCs were significantly different despite the identical aluminum metal clusters of the CAU-10-H and Al-fu precursors. This difference can be attributed to the organic ligands of MOFs. Isophthalic acid, employed in the synthesis of CAU-10-H, is an organic ligand with a benzene ring. In contrast, fumaric acid, utilized in the synthesis of Al-fu, is an organic ligand with carbon double bonds but without a benzene ring. Hence, CAU-10-H retains its structural integrity even after heat treatment because the benzene rings are resistant to thermal degradation. In contrast, Al-fu, which comprises ligands with carbon double bonds, is susceptible to structural changes after heat treatment because carbon double bonds are susceptible to thermal cleavage. This difference is closely related to the cracking observed on the surface of A-MDC but not on the surface of C-MDC in the FE-SEM images.

CAU-10-H and Al-fu MOF have been extensively studied as water adsorbents owing to their excellent stability and adsorption capacity for water. The water adsorption mechanism of carbonaceous materials is well known, and the water adsorption characteristics of MOFs and carbon materials are closely related to their pore structures and functional groups on the surface. CAU-10-H exhibited steeply increasing adsorption properties at relative pressures between 0.1 and 0.2 and no hysteresis loops in the adsorption and desorption curves. This suggests that the water adsorption mechanism of CAU-10-H is primarily influenced by its pore characteristics. However, the water adsorption properties of C-MDC were completely different. C-MDC exhibited continuous water adsorption at relative pressures below 0.3, with a remarkable increase in water adsorption around relative pressures of 0.4–0.6 (Figure 9a). The increase in water adsorption at relative pressures greater than 0.8 was not significant. This trend closely resembles the trends observed for type 5 carbon. Water adsorption by type 5 carbon occurs via (i) the adsorption of water molecules on functional groups on the carbon surface and the growth of clusters at low relative pressures ($P/P_0 > 0.3$). This is due to the fact that as more water is adsorbed, individual water molecules begin to interact with each other by hydrogen boing, leading to the formation and growth of clusters. And (ii) the adsorption of water in micropore pores around relative pressures $P/P_0 < 0.8$ [26]. In particular, there are differences in the adsorption range based on the size of micropores, and the synthesized C-MDC has pore sizes ranging from 1.1 nm to 2 nm. Therefore, it exhibits a rapid adsorption ability at relative pressures of 0.6 and above. Carbon materials with micropores smaller than 1.1 nm start adsorbing water at relative pressures below 0.6, but the quantity is not significant (Figure 9c). On the other hand, C-MDC with micropores larger than 1.1 nm demonstrates a higher adsorption capacity [27]. Hence, a limited number of functional groups might be present on the surface of C-MDC. These results can be predicted because most functional groups on the surface were removed by wet etching The hypothesis on the microporous characteristics was substantiated by analyzing the pore properties of C-MDC that were obtained from the nitrogen adsorption isotherm analyses.

Al-fu shows a steeply increasing water adsorption behavior in the relative pressure range of 0.2–0.3, similar to the behavior observed in CAU-10-H MOF, indicating that the water adsorption mechanism is governed primarily by pore characteristics. In contrast, the water adsorption of A-MDC is relatively low at relative pressures lower than 0.3, begins to increase at a relative pressure of 0.6, and increases significantly at relative pressures higher than 0.8 (Figure 9b). This behavior resembles the behavior observed for type 3 carbon. Type 3 carbon denotes hydrophobic carbon which exhibits (i) minimal water adsorption in the relative pressure range of $P/P_0 < 0.3$ owing to the scarcity of functional groups on the surface; (ii) water adsorption in the micropores of the material in the relative pressure range of 0.6–0.8; and (iii) water adsorption in the mesopores at relative pressures $P/P_0 > 0.8$. Moreover, the adsorption and desorption processes exhibit a hysteresis loop, which is a typical feature of mesoporous carbon materials and is attributed to capillary phenomena. Hence, it can be inferred that A-MDC possesses both micropores and mesopores, rendering it a hydrophobic carbon material, which agrees well with its previously described attributes.

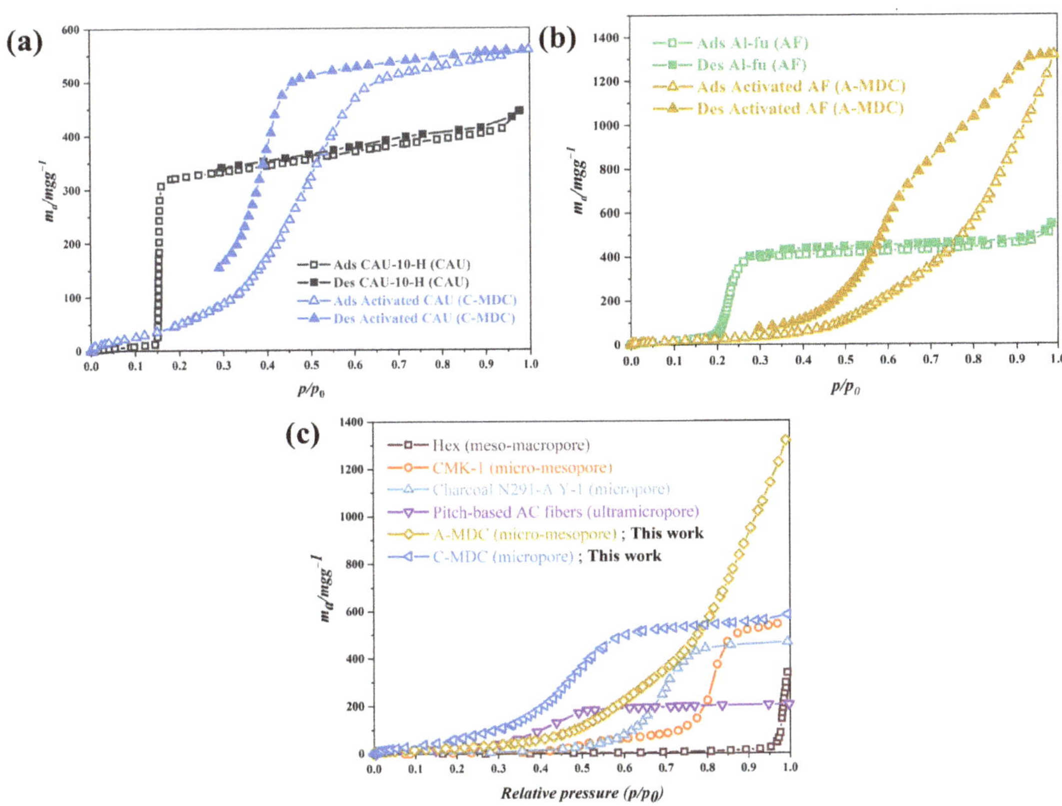

Figure 9. Water vapor adsorption and desorption isotherm curve at 298 K. CAU-10-H, C-MDC (**a**), Al-fu, A-MDC (**b**), compared with various carbon materials which have different sizes of pores (**c**) [27].

Although the total amount of water adsorption is much higher in A-MDC than in C-MDC, the relative pressure range of adsorption is crucial for commercially used adsorbents. This is because the energy required for adsorption at different relative pressures varies considerably, which is important from an economic perspective. Therefore, C-MDC may be a better adsorbent candidate than A-MDC. Furthermore, C-MDC exhibits high water adsorption characteristics in the relative pressure range of 0.45–0.65, which is the desired range for indoor dehumidification adsorbents. Therefore, in this study, C-MDC was selected as the candidate water adsorbent. Additional water adsorption modeling simulations were conducted, and the adsorption enthalpy was calculated.

Comparing experimental and analytical data with existing theories increases the reliability of the data and is crucial for broader applications. Moreover, relying solely on a single standardized model to compare and predict the properties of a newly synthesized material may lead to significant errors. Therefore, two models related to the characteristics of pores and functional groups, which are important features of the synthesized samples, have been applied in this study.

The first model, known as the Henry–Sips model, is a hybrid adsorption model that combines the theories of two water adsorption mechanisms. The model equation includes parameters representing Henry's law, which describes physical site adsorption from a thermodynamic perspective, and the Sips parameter, which represents adsorption by filling only the micropores. The parameters of the Henry–Sips model are presented in Table 7, and the analysis and modeling results at 298 K and 308 K are depicted in Figure 10a. Among the parameters, the value of K_s is larger than the values of β and K_H. This is because the values of β and K_H are determined by Henry's law, which represents adsorption by functional group sites. However, the value of K_s is determined by the Sips law, which represents adsorption due to the filling of micropores. The Henry–Sips model exhibits a good fit, with R^2 values of 0.9929 and 0.9921 at 298 K and 308 K, respectively.

Table 7. The Henry–Sips model fitting parameters.

Henry–Sips	β	K_H	q_m	K_s	A	B
	0.6594	0.4568	0.9141	2.0937	0.0099	0.2897

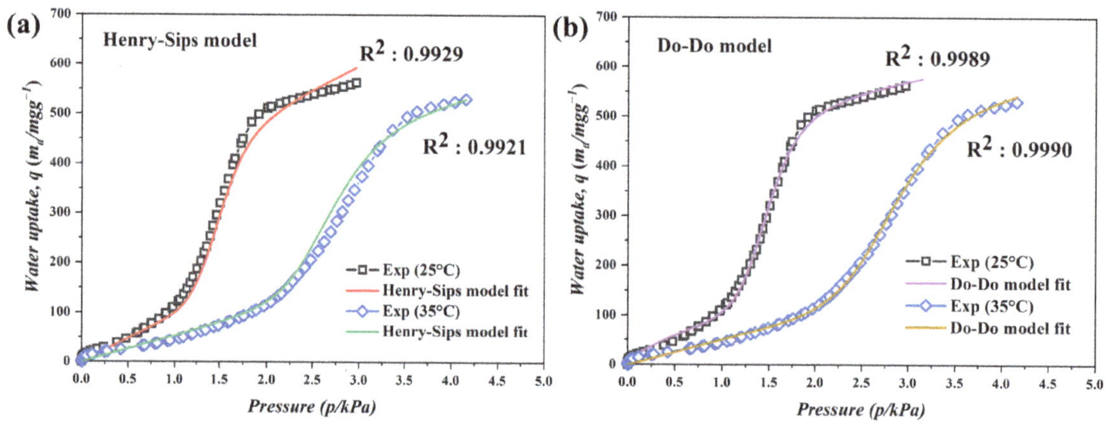

Figure 10. Equilibrium adsorption isotherm curves of water vapor on C-MDC. Symbols and lines represent experimental data and predicted isotherms data by Henry–Sips model (**a**) and Do–Do model (**b**), respectively.

The second model, known as the Do–Do model, demonstrates excellent compatibility of water adsorption for type 5 carbons, which are hydrophobic microporous carbon materials. Similar to the Henry–Sips model, the Do–Do model suggests that the initial site adsorption occurs on the functional groups, followed by a sharp increase in adsorption at high pressure through capillary condensation. The parameters of the Do–Do model are listed in Table 8, and the analysis and modeling results at 298 K and 308 K are compared in Figure 10b. In the Do–Do model, f and K_f represent the adsorption at the sites, while K_μ, α, and β represent the pore adsorption. Hence, the values of f and K_f are small enough to be ignored, whereas the values of K_μ, α, and β related to pore adsorption are relatively large. This observation adequately explains the water adsorption characteristics of hydrophobic C-MDC. The Do–Do model exhibits excellent compatibility, with R^2 values of 0.9989 and 0.9990 at 298 K and 308 K, respectively. Therefore, it can be concluded that the C-MDC synthesized in this study possesses properties that closely resemble those of hydrophobic microporous materials.

Table 8. The Do–Do model fitting parameters.

Do–Do	f	K_f	K_μ	α	β
298 K	0.0383	0.0063	0.1058	0.6914	10.769
308 K	0.0388	2.9530×10^{-5}	0.0556	1.0466	9.6265

The heat of adsorption is a crucial factor in process design. The isosteric adsorption enthalpy of C-MDC was calculated based on the water adsorption data at 298 K and 308 K using the Clausius–Clapeyron equation, and the results are shown in Figure 11a. The initial adsorption enthalpy was −43.81 kJ/mol and did not change considerably with an increase in adsorption. Typically, values in the range of −80 to −400 kJ/mol indicate chemical adsorption, whereas values greater than −80 kJ/mol suggest physical adsorption [53]. Hence, it can be inferred that the water adsorption by C-MDC is predominantly physical, which supports the previously described mechanism wherein water adsorption is primarily attributed to physical adsorption.

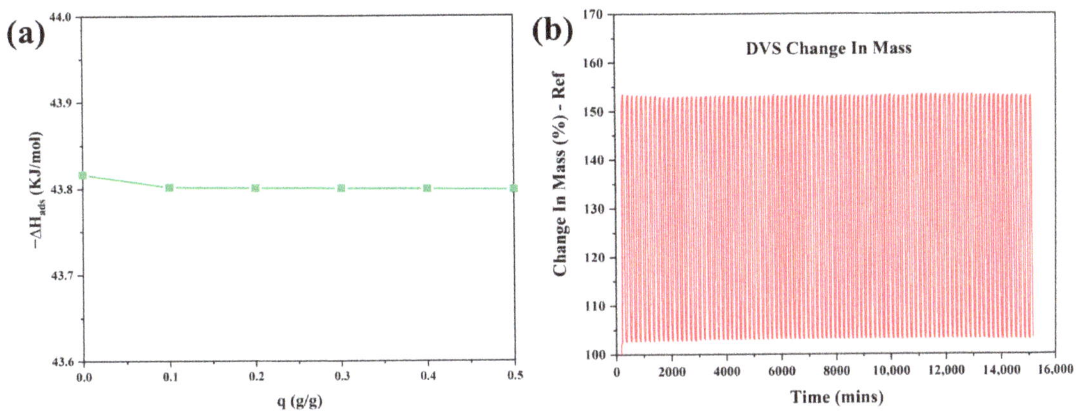

Figure 11. Isosteric adsorption enthalpy (**a**) and the adsorption and desorption cycle test to assess the stability of adsorption on C-MDC (**b**).

Water stability is an important factor for adsorbents. The water stability of C-MDC was analyzed via dynamic vapor sorption analysis by measuring the water sorption and desorption cycles in the P/P_0 range from 10–90 (Figure 11b). The changes in weight during the first and 100th cycles were 51.14% and 50.04%, respectively (Table 9). It was confirmed that 97.84% of the initial water sorption was maintained over 100 cycles, and the difference between the kinetics during the initial and subsequent cycles was not significant. This confirms the stability and consistency of the water sorption and desorption capacities of C-MDC. These results are considered valuable because water stability is a crucial factor for potential commercial applications of C-MDC.

Table 9. Changes in water adsorption and desorption after cycle test for C-MDC.

Change in Mass (%)		Target	C-MDC
Desorption	Sorption	P/P_0 (%)	
2.66	2.38	10.0	Cycle 1
53.52	53.52	90.0	
3.37	3.07	10.0	Cycle 100
53.11	53.11	90.0	

4. Conclusions

In this study, a comprehensive investigation was conducted to examine the pore characteristics of MDCs synthesized via carbonization and activation. The influence of various ligands used in the construction of aluminum-based organic/inorganic nanoporous materials (MOFs) was investigated. CAU-10-H, synthesized with isophthalic acid ligands containing benzene rings, maintains its structure without significant changes even after thermal treatment because of the preservation of its bonds. The removal of the aluminum clusters creates additional space, allowing for the expansion of the size and volume of the micropores. In contrast, Al-fu prepared using fumaric acid ligands containing double bonds exhibits structural changes during thermal treatment. The relatively weak carbon bonds break and reform, leading to the formation of mesopores which are absent in the MOF precursor. This results in a significant increase in pore volume and specific surface area. Based on these results, the microporous C-MDC was evaluated as a water adsorbent and isothermal water adsorption modeling was explored for extended usage. With its high water adsorption capacity in the relative pressure range of 45–65%, C-MDC demonstrates the potential for utilization as an indoor humidity control desiccant. In addition, the simulation results of the Do–Do model, specifically optimized for type 5 adsorption, showed excellent compatibility. The large-scale synthesis of MDC based on MOFs proposed in this study has the potential to guide the applications of MDC as water adsorbents.

Supplementary Materials: The following supporting information can be downloaded at: https://www.mdpi.com/article/10.3390/nano13162351/s1, S.1. Large-scale synthesis of aluminum fumarate MOF (Al-fu). S.2. Large-scale synthesis of A-MDCs; Figure S1: 100 L reflux reactor used for large-scale synthesis of Al-fu; Figure S2: HRTF-180/60 (HANTECH, Korea) rotary furnace used for large-scale carbonization of Al-fu; Figure S3: Characteristics of large-scale synthesized Al-fu MOF. XRD pattern (a), FE-SEM image (b), nitrogen adsorption isotherm curve at 77 K (c), and NLDFT calculation (d); Figure S4: Characteristics of carbonized AF and activated AF (A-MDC) (50 g of Al-fu MOF placed in the rotary furnace). XRD pattern (a) and FE-SEM image (b) of carbonized AF, XRD pattern (c) and FE-SEM image (d) of activated AF (A-MDC); Figure S5: Nitrogen adsorption isotherm of activated AF (A-MDC) (50 g of Al-fu MOF placed in the rotary furnace). Isotherm curve at 77 K (a), NLDFT calculation (b), and BJH plot (c); Figure S6: Characteristics of carbonized AF and activated AF (A-MDC) (300 g of Al-fu MOF placed in the rotary furnace). XRD pattern (a) and FE-SEM image (b) of carbonized AF, XRD pattern (c) and FE-SEM image (d) of activated AF (A-MDC); Figure S7: Nitrogen adsorption isotherm of activated AF (A-MDC) (300 g of Al-fu MOF placed in the rotary furnace). Isotherm curve at 77 K (a), NLDFT calculation (b), and BJH plot (c); Figure S8: Nitrogen adsorption isotherm of activated AF (A-MDC) (600 g of Al-fu MOF placed in the rotary furnace). Isotherm curve at 77 K (a), NLDFT calculation (b), and BJH plot (c); Table S1: BET surface area, total pore volume, and average pore size of large-scale synthesized Al-fu, and its derivatives; Figure S9: The structure of CAU-10-H; Figure S10: The structure of Al-fu MOF.

Author Contributions: Conceptualization, D.J., S.C.K. and J.P.; methodology, data curation, formal analysis, investigation, visualization, and writing—original draft preparation, D.J., S.C.K. and T.A.; writing—review and editing, D.J., S.C.K., H.H., S.Q.C. and J.P.; validation and supervision, D.L., H.H., S.Q.C. and J.P.; funding acquisition D.L. and J.P. All authors have read and agreed to the published version of the manuscript.

Funding: This work was supported by Korea Evaluation Institute of Industrial Technology(KEIT) grant funded by the Korea government(MOTIE) (No.20015430).

Data Availability Statement: Not applicable.

Conflicts of Interest: The authors declare no conflict of interest.

References

1. Mekonnen, M.M.; Hoekstra, A.Y. Four billion people facing severe water scarcity. *Sci. Adv.* **2016**, *2*, e1500323. [CrossRef]
2. AbdulHalim, R.G.; Bhatt, P.M.; Belmabkhout, Y.; Shkurenko, A.; Adil, K.; Barbour, L.J.; Eddaoudi, M. A Fine-Tuned Metal–Organic Framework for Autonomous Indoor Moisture Control. *J. Am. Chem. Soc.* **2017**, *139*, 10715–10722. [CrossRef] [PubMed]
3. Ashraf, S.; Sultan, M.; Bahrami, M.; McCague, C.; Shahzad, M.W.; Amani, M.; Shamshiri, R.R.; Ali, H.M. Recent progress on water vapor adsorption equilibrium by metal-organic frameworks for heat transformation applications. *Int. Commun. Heat Mass Transf.* **2021**, *124*, 105242. [CrossRef]
4. Furukawa, H.; Gándara, F.; Zhang, Y.-B.; Jiang, J.; Queen, W.L.; Hudson, M.R.; Yaghi, O.M. Water Adsorption in Porous Metal–Organic Frameworks and Related Materials. *J. Am. Chem. Soc.* **2014**, *136*, 4369–4381. [CrossRef]
5. Metrane, A.; Delhali, A.; Ouikhalfan, M.; Assen, A.H.; Belmabkhout, Y. Water Vapor Adsorption by Porous Materials: From Chemistry to Practical Applications. *J. Chem. Eng. Data* **2022**, *67*, 1617–1653. [CrossRef]
6. Bai, Y.; Liu, C.; Shan, Y.; Chen, T.; Zhao, Y.; Yu, C.; Pang, H. Metal-Organic Frameworks Nanocomposites with Different Dimensionalities for Energy Conversion and Storage. *Adv. Energy Mater.* **2021**, *12*, 2100346. [CrossRef]
7. Song, J.; Yang, W.; Han, X.; Jiang, S.; Zhang, C.; Pan, W.; Jian, S.; Hu, J. Performance of Rod-Shaped Ce Metal–Organic Frameworks for Defluoridation. *Molecules* **2023**, *28*, 3492. [CrossRef]

8. Wang, J.; Sun, Y.; Zhao, X.; Chen, L.; Peng, S.; Ma, C.; Duan, G.; Liu, Z.; Wang, H.; Yuan, Y.; et al. A poly(amidoxime)-modified MOF macroporous membrane for high-efficient uranium extraction from seawater. *e-Polymers* **2022**, *22*, 399–410. [CrossRef]
9. Reimer, N.; Bueken, B.; Leubner, S.; Seidler, C.; Wark, M.; De Vos, D.; Stock, N. Three Series of Sulfo-Functionalized Mixed-Linker CAU-10 Analogues: Sorption Properties, Proton Conductivity, and Catalytic Activity. *Chemistry* **2015**, *21*, 12517–12524. [CrossRef]
10. Solovyeva, M.V.; Shkatulov, A.I.; Gordeeva, L.G.; Fedorova, E.A.; Krieger, T.A.; Aristov, Y.I. Water Vapor Adsorption on CAU-10-X: Effect of Functional Groups on Adsorption Equilibrium and Mechanisms. *Langmuir* **2021**, *37*, 693–702. [CrossRef] [PubMed]
11. Canivet, J.; Fateeva, A.; Guo, Y.; Coasne, B.; Farrusseng, D. Water adsorption in MOFs: Fundamentals and applications. *Chem. Soc. Rev.* **2014**, *43*, 5594–5617. [CrossRef]
12. Zhang, S.; Yang, Q.; Xu, X.; Liu, X.; Li, Q.; Guo, J.; Torad, N.L.; Alshehri, S.M.; Ahamad, T.; Hossain, S.A.; et al. Assembling well-arranged covalent organic frameworks on MOF-derived graphitic carbon for remarkable formaldehyde sensing. *Nanoscale* **2020**, *12*, 15611–15619. [CrossRef] [PubMed]
13. He, Y.; Wang, Z.; Wang, H.; Wang, Z.; Zeng, G.; Xu, P.; Huang, D.; Chen, M.; Song, B.; Qin, H.; et al. Metal-organic framework-derived nanomaterials in environment related fields: Fundamentals, properties and applications. *Coord. Chem. Rev.* **2021**, *429*, 213618. [CrossRef]
14. Zheng, J.; He, W.; Hang, T.; Sun, Z.; Li, Z.; Jiang, S.; Li, X.; E, S.; Chen, Y. Flower-like bimetal-organic framework derived composites with tunable structures for high-efficiency electromagnetic wave absorption. *J. Colloid Interface Sci.* **2022**, *628 Pt B*, 261–270. [CrossRef]
15. Yang, S.J.; Kim, T.; Im, J.H.; Kim, Y.S.; Lee, K.; Jung, H.; Park, C.R. MOF-Derived Hierarchically Porous Carbon with Exceptional Porosity and Hydrogen Storage Capacity. *Chem. Mater.* **2012**, *24*, 464–470. [CrossRef]
16. Ahmed, I.; Panja, T.; Khan, N.A.; Sarker, M.; Yu, J.-S.; Jhung, S.H. Nitrogen-Doped Porous Carbons from Ionic Liquids@MOF: Remarkable Adsorbents for Both Aqueous and Nonaqueous Media. *ACS Appl. Mater. Interfaces* **2017**, *9*, 10276–10285. [CrossRef]
17. Liu, B.; Shioyama, H.; Akita, T.; Xu, Q. Metal-Organic Framework as a Template for Porous Carbon Synthesis. *J. Am. Chem. Soc.* **2008**, *130*, 5390–5391. [CrossRef] [PubMed]
18. Tang, J.; Salunkhe, R.R.; Zhang, H.; Malgras, V.; Ahamad, T.; Alshehri, S.M.; Kobayashi, N.; Tominaka, S.; Ide, Y.; Kim, J.H.; et al. Bimetallic Metal-Organic Frameworks for Controlled Catalytic Graphitization of Nanoporous Carbons. *Sci. Rep.* **2016**, *6*, 30295. [CrossRef]
19. Chu, X.; Meng, F.; Deng, T.; Zhang, W. Metal organic framework derived porous carbon materials excel as an excellent platform for high-performance packaged supercapacitors. *Nanoscale* **2021**, *13*, 5570–5593. [CrossRef]
20. Baby, R.; Saifullah, B.; Hussein, M.Z. Carbon Nanomaterials for the Treatment of Heavy Metal-Contaminated Water and Environmental Remediation. *Nanoscale Res. Lett.* **2019**, *14*, 341. [CrossRef]
21. Bläker, C.; Muthmann, J.; Pasel, C.; Bathen, D. Characterization of Activated Carbon Adsorbents—State of the Art and Novel Approaches. *ChemBioEng Rev.* **2019**, *6*, 119–138. [CrossRef]
22. Chairunnisa; Miksik, F.; Miyazaki, T.; Thu, K.; Miyawaki, J.; Nakabayashi, K.; Wijayanta, A.T.; Rahmawati, F. Development of biomass based-activated carbon for adsorption dehumidification. *Energy Rep.* **2021**, *7*, 5871–5884. [CrossRef]
23. Mohamed, A.R.; Mohammadi, M.; Darzi, G.N. Preparation of carbon molecular sieve from lignocellulosic biomass: A review. *Renew. Sustain. Energy Rev.* **2010**, *14*, 1591–1599. [CrossRef]
24. Thommes, M.; Kaneko, K.; Neimark, A.V.; Olivier, J.P.; Rodriguez-Reinoso, F.; Rouquerol, J.; Sing, K.S.W. Physisorption of gases, with special reference to the evaluation of surface area and pore size distribution (IUPAC Technical Report). *Pure Appl. Chem.* **2015**, *87*, 1051–1069. [CrossRef]
25. Ng, E.-P.; Mintova, S. Nanoporous materials with enhanced hydrophilicity and high water sorption capacity. *Microporous Mesoporous Mater.* **2008**, *114*, 1–26. [CrossRef]
26. Horikawa, T.; Tan, S.; Do, D.D.; Sotowa, K.-I.; Alcántara-Avila, J.R.; Nicholson, D. Temperature dependence of water adsorption on highly graphitized carbon black and highly ordered mesoporous carbon. *Carbon* **2017**, *124*, 271–280. [CrossRef]
27. Liu, L.; Tan, S.; Horikawa, T.; Do, D.D.; Nicholson, D.; Liu, J. Water adsorption on carbon—A review. *Adv. Colloid Interface Sci.* **2017**, *250*, 64–78. [CrossRef]
28. Fröhlich, D.; Henninger, S.K.; Janiak, C. Multicycle water vapour stability of microporous breathing MOF aluminium isophthalate CAU-10-H. *Dalton Trans.* **2014**, *43*, 15300–15304. [CrossRef] [PubMed]
29. Reinsch, H.; van der Veen, M.A.; Gil, B.; Marszalek, B.; Verbiest, T.; de Vos, D.; Stock, N. Structures, Sorption Characteristics, and Nonlinear Optical Properties of a New Series of Highly Stable Aluminum MOFs. *Chem. Mater.* **2012**, *25*, 17–26. [CrossRef]
30. Fröhlich, D.; Pantatosaki, E.; Kolokathis, P.D.; Markey, K.; Reinsch, H.; Baumgartner, M.; van der Veen, M.A.; De Vos, D.E.; Stock, N.; Papadopoulos, G.K.; et al. Water adsorption behaviour of CAU-10-H: A thorough investigation of its structure–property relationships. *J. Mater. Chem. A* **2016**, *4*, 11859–11869. [CrossRef]
31. Karmakar, S.; Dechnik, J.; Janiak, C.; De, S. Aluminium fumarate metal-organic framework: A super adsorbent for fluoride from water. *J. Hazard. Mater.* **2016**, *303*, 10–20. [CrossRef] [PubMed]
32. Alvarez, E.; Guillou, N.; Martineau, C.; Bueken, B.; Van de Voorde, B.; Le Guillouzer, C.; Fabry, P.; Nouar, F.; Taulelle, F.; de Vos, D.; et al. The Structure of the Aluminum Fumarate Metal-Organic Framework A520. *Angew. Chem. Int. Ed.* **2015**, *54*, 3664–3668. [CrossRef]

33. Lenzen, D.; Bendix, P.; Reinsch, H.; Fröhlich, D.; Kummer, H.; Möllers, M.; Hügenell, P.P.C.; Gläser, R.; Henninger, S.; Stock, N. Scalable Green Synthesis and Full-Scale Test of the Metal-Organic Framework CAU-10-H for Use in Adsorption-Driven Chillers. *Adv. Mater.* **2018**, *30*, 1705869. [CrossRef]
34. Kim, Y.-D.; Thu, K.; Ng, K.C. Adsorption characteristics of water vapor on ferroaluminophosphate for desalination cycle. *Desalination* **2014**, *344*, 350–356. [CrossRef]
35. Do, D.; Do, H. A model for water adsorption in activated carbon. *Carbon* **2000**, *38*, 767–773. [CrossRef]
36. Buttersack, C. Modeling of type IV and V sigmoidal adsorption isotherms. *Phys. Chem. Chem. Phys.* **2019**, *21*, 5614–5626. [CrossRef]
37. Chowdhury, S.; Mishra, R.; Saha, P.; Kushwaha, P. Adsorption thermodynamics, kinetics and isosteric heat of adsorption of malachite green onto chemically modified rice husk. *Desalination* **2011**, *265*, 159–168. [CrossRef]
38. Kianfar, E. Synthesis of characterization Nanoparticles isophthalic acid/aluminum nitrate (CAU-10-H) Using method hydrothermal. In *Advances in Chemistry Research*; Nova Science Publishers, Inc.: New York, NY, USA, 2020.
39. Tang, P.-H.; So, P.B.; Lee, K.-R.; Lai, Y.-L.; Lee, C.-S.; Lin, C.-H. Metal Organic Framework-Polyethersulfone Composite Membrane for Iodine Capture. *Polymers* **2020**, *12*, 2309. [CrossRef]
40. Estrade-Szwarckopf, H. XPS photoemission in carbonaceous materials: A "defect" peak beside the graphitic asymmetric peak. *Carbon* **2004**, *42*, 1713–1721. [CrossRef]
41. Alexander, M.R.; Thompson, G.E.; Beamson, G. Characterization of the oxide/hydroxide surface of aluminium using X-ray photoelectron spectroscopy: A procedure for curve fitting the O 1s core level. *Surf. Interface Anal.* **2000**, *29*, 468–477. [CrossRef]
42. Kumar, N.; Biswas, K. Cryomilling: An environment friendly approach of preparation large quantity ultra refined pure aluminium nanoparticles. *J. Mater. Res. Technol.* **2019**, *8*, 63–74. [CrossRef]
43. Wang, Y.; Qu, Q.; Liu, G.; Battaglia, V.S.; Zheng, H. Aluminum fumarate-based metal organic frameworks with tremella-like structure as ultrafast and stable anode for lithium-ion batteries. *Nano Energy* **2017**, *39*, 200–210. [CrossRef]
44. Manakhov, A.; Michlíček, M.; Nečas, D.; Polčák, J.; Makhneva, E.; Eliáš, M.; Zajíčková, L. Carboxyl-rich coatings deposited by atmospheric plasma co-polymerization of maleic anhydride and acetylene. *Surf. Coatings Technol.* **2016**, *295*, 37–45. [CrossRef]
45. Rojas-Luna, R.; Amaro-Gahete, J.; Gil-Gavilán, D.G.; Castillo-Rodríguez, M.; Jiménez-Sanchidrián, C.; Ruiz, J.R.; Esquivel, D.; Romero-Salguero, F.J. Visible-light-harvesting basolite-A520 metal organic framework for photocatalytic hydrogen evolution. *Microporous Mesoporous Mater.* **2023**, *355*, 112565. [CrossRef]
46. Reinsch, H.; Waitschat, S.; Stock, N. Mixed-linker MOFs with CAU-10 structure: Synthesis and gas sorption characteristics. *Dalton Trans.* **2013**, *42*, 4840–4847. [CrossRef]
47. Mohamadian-Kalhor, S.; Edjlali, L.; Basharnavaz, H.; Es'haghi, M. Aluminum Fumarate Metal–Organic Framework: Synthesis, Characterization, and Application as a Novel Inhibitor against Corrosion of AM60B Magnesium Alloy in Ethylene Glycol Solution. *J. Mater. Eng. Perform.* **2021**, *30*, 720–726. [CrossRef]
48. Moumen, E.; Bazzi, L.; El Hankari, S. Aluminum-fumarate based MOF: A promising environmentally friendly adsorbent for the removal of phosphate. *Process. Saf. Environ. Prot.* **2022**, *160*, 502–512. [CrossRef]
49. Zayan, S.; Elshazly, A.; Elkady, M. In Situ Polymerization of Polypyrrole @ Aluminum Fumarate Metal–Organic Framework Hybrid Nanocomposites for the Application of Wastewater Treatment. *Polymers* **2020**, *12*, 1764. [CrossRef]
50. Zhang, Y.; Nahil, M.A.; Wu, C.; Williams, P.T. Pyrolysis–catalysis of waste plastic using a nickel–stainless-steel mesh catalyst for high-value carbon products. *Environ. Technol.* **2017**, *38*, 2889–2897. [CrossRef]
51. Del Corro, E.; Kavan, L.; Kalbac, M.; Frank, O. Strain Assessment in Graphene Through the Raman 2D′ Mode. *J. Phys. Chem. C* **2015**, *119*, 25651–25656. [CrossRef]
52. Wang, Y.; Zhou, X.; Jin, Y.; Zhang, X.; Zhang, Z.; Wang, Y.; Liu, J.; Wang, M.; Xia, Y.; Zhao, P.; et al. Strain-dependent Raman analysis of the G* band in graphene. *Phys. Rev. B* **2019**, *100*, 241407(R). [CrossRef]
53. Doğan, M.; Alkan, M. Removal of methyl violet from aqueous solution by perlite. *J. Colloid Interface Sci.* **2003**, *267*, 32–41. [CrossRef] [PubMed]

nanomaterials

MDPI

Article

Zeolitic Imidazolate Framework-8 (ZIF-8) as a Drug Delivery Vehicle for the Transport and Release of Telomerase Inhibitor BIBR 1532

Shunyu Zhang [1,2,†], Jinxia Li [2,†], Liang Yan [2], Yue You [2], Feng Zhao [2,*], Jixing Cheng [1], Limin Yang [2], Yanqi Sun [3], Qingchao Chang [2], Ru Liu [2] and Yunhui Li [1,*]

1 Key Laboratory of Environmental Medicine Engineering of Ministry of Education, School of Public Health, Southeast University, Nanjing 210000, China; syzhang@ihep.ac.cn (S.Z.); chengjx0210@163.com (J.C.)

2 CAS Key Lab for Biomedical Effects of Nanomaterials and Nanosafety, Institute of High Energy Physics, Chinese Academy of Sciences, Beijing 100049, China; lijinxia@ihep.ac.cn (J.L.); yanliang@ihep.ac.cn (L.Y.); youyue@ihep.ac.cn (Y.Y.); yanglm@ihep.ac.cn (L.Y.); changqc@ihep.ac.cn (Q.C.); liuru@ihep.ac.cn (R.L.)

3 Department of Prevention and Health Care, Rizhao 276800, China; tangdou888@163.com

* Correspondence: zhaof@ihep.ac.cn (F.Z.); yhli@seu.edu.cn (Y.L.); Tel.: +86-10-88236456 (F.Z.); +86-135-125-38992 (Y.L.)

† These authors contributed equally to this work.

Abstract: Telomerase is constitutively overexpressed in the majority of human cancers and telomerase inhibition provides a promising broad-spectrum anticancer therapeutic strategy. BIBR 1532 is a well-known synthetic telomerase inhibitor that blocks the enzymatic activity of hTERT, the catalytic subunit of telomerase. However, water insolubility of BIBR 1532 leads to low cellular uptake and inadequate delivery and thus, limits its anti-tumor effects. Zeolitic imidazolate framework-8 (ZIF-8) is considered as an attractive drug delivery vehicle for improved transport, release and anti-tumor effects of BIBR 1532. Herein, ZIF-8 and BIBR 1532@ZIF-8 were synthesized, respectively, and the physicochemical characterizations confirmed the successful encapsulation of BIBR 1532 in ZIF-8 coupled with an improved stability of BIBR 1532. ZIF-8 could alter the permeability of lysosomal membrane probably by the imidazole ring-dependent protonation. Moreover, ZIF-8 encapsulation facilitated the cellular uptake and release of BIBR 1532 with more accumulation in the nucleus. BIBR 1532 encapsulation with ZIF-8 triggered a more obvious growth inhibition of cancer cells as compared with free BIBR 1532. A more potent inhibition on hTERT mRNA expression, aggravated G0/G1 arrest accompanied with an increased cellular senescence were detected in BIBR 1532@ZIF-8-treated cancer cells. Our work has provided preliminary information on improving the transport, release and efficacy of water-insoluble small molecule drugs by using ZIF-8 as a delivery vehicle.

Keywords: ZIF-8; drug loading; transport; protonation effect; telomerase inhibitor

1. Introduction

Telomerase, a reverse transcriptase ribonucleoprotein complex, plays an important role in maintaining the telomere length in cells with high replicative potentials. Since its discovery in 1985 [1], telomerase has become an attractive target for new and more effective anticancer therapies due to its high associations with tumor development [2,3]. Cancer cells could acquire the self-proliferative capability via telomerase activation-dependent telomere length maintenance mechanisms. Considering that telomerase is constitutively overexpressed in the vast majority of human cancers [4–7], targeting the telomere/telomerase machinery offers a promising broad-spectrum anticancer therapeutic strategy by promoting telomere shortening, senescence and apoptosis. Human telomerase reverse transcriptase (hTERT) [8], the catalytic subunit of telomerase, plays a major role in telomerase activation in human cancers. Therefore, hTERT is considered a key target for telomerase inhibition-based cancer treatment [9].

Much work has been done on targeting telomerase to cure cancer, but unfortunately, no sufficiently potent and highly selective telomerase inhibitor has been available in clinic yet [10]. First, it takes a long time for telomerase inhibitors to shorten telomeres. Especially in tumors with long telomeres or long doubling times, the efficacy is always delayed [11–13]. Second, telomerase-mediated telomere elongation is not the only mechanism involved in tumorigenesis. In a small number of human tumors, telomerase shortening caused by long-term and low-dose treatment with telomerase inhibitors can be compensated by activation of alternative lengthening of telomere (ALT) [14–16]. It is clear that time and dose are important factors influencing the efficacy of telomerase inhibitors. BIBR 1532 is a well-known non-nucleosidic, non-peptidic telomerase inhibitor. It can bind to a conserved hydrophobic pocket (FVYL motif) of hTERT and inhibits enzymatic activity [17]. Accumulating evidence has shown the inhibition of BIBR 1532 on tumor cells via progressive telomere shortening and the consequent inhibition on cell proliferation [18–20]. However, BIBR 1532 is poorly soluble in water, which greatly reduces its uptake and utilization by cells. Moreover, growth inhibition is usually insignificant by short-term treatment with low concentration of BIBR 1532 [21], a common drawback of telomerase inhibitors.

Several studies have attempted to improve the bioeffects of BIBR 1532 by modifying its structure or combining it with other drugs or therapies. In one study, structural optimization based on BIBR 1532 was performed for an improved efficacy of telomerase inhibitors. A series of new compounds with different structural characteristics were designed, synthesized and screened. The compounds which used cyanide to replace methyl and simultaneously retained amide moieties were reported to have an improved telomerase inhibitory activity and moderate cytotoxicity as compared with BIBR 1532 [22]. Besides, several studies have adopted a combination of BIBR 1532 and other chemotherapeutics or therapies for a synergistic anti-tumor effect. For instance, the combination of BIBR 1532 and paclitaxel was reported to increase telomere dysfunction and inhibit cancer cells, enabling the observation of therapeutic outcome in a shorten time interval [23]. A similar study on the combinational application of BIBR 1532 and emodin demonstrated that, based on emodin-induced temporary telomere dysfunction by stabilizing the G4 structure at the protruding segment of the telomere end, the combined use of BIBR 1532 induced a permanent telomere dysfunction [24]. Another study found that low concentrations of BIBR 1532 were not toxic to lung cancer cells but could enhance ionizing radiation-induced apoptosis and cellular senescence [18]. However, little information is available concerning the drawbacks of low cellular uptake, inadequate delivery and limited bioeffects arising from the water insolubility of BIBR 1532 itself. Therefore, a proper drug delivery system is considered for improved transport, release and anti-tumor effects of BIBR 1532.

Metal-organic frameworks (MOFs) with three-dimensional pore structures are generally composed of metal ions (such as transition metals and lanthanide metals) and organic ligands (such as carboxylates, phosphonates, imidazole and phenols) [25]. To date, many attractive physicochemical characteristics as drug delivery tools have been identified for MOFs, including the ultra-high surface area, tunable porosity, diversified chemical functionalization, high loading capacity, good biocompatibility and the controllable release of the cargos [26–30].

Both traditional and emerging MOFs showed potentials for drug delivery and bio-orientation. The formation of a MOF layer by biomimetic mineralization on the surface of biological macromolecules can improve the stability of biological macromolecules, which is an important application of MOF in the field of biotechnology. Liang et al. demonstrated the ability to biomimetically mineralize several MOF materials, including ZIF-8, HKUST-1, Eu/Tb-BDC and MIL-88A. The results also showed that ZIF-8 has a good protective effect against the enzymatic activity of HRP, (PQQ)GDH and urease [31]. An emerging MOF, chiral MOF [32–34], can also be used as a drug carrier, enabling better biocompatibility, bioavailability and biodegradability of chiral drugs [35]. A study focused on and highlighted the necessity for MOF-based nanoplatforms in tumor treatment, summarizing the effects of MOF-based radiotherapy, chemotherapy, chemo-dynamic therapy, photothermal therapy, photodynamic therapy, starvation therapy and immunotherapy, and combination therapy against tumors [36]. MOF also have shown great potentials in clinical translation. In a study based on MOF combined with photothermal therapy and H_2 therapy for the treatment of rheumatoid arthritis, Pt-MOF@Au@QDs/PDA had high biosafety and could maximize the treatment of rheumatoid arthritis, with great potential for clinical translation [37].

Zeolitic imidazolate framework-8 (ZIF-8), a member of the MOF family, consists of zinc metal properly coordinated with imidazole ligands. It exhibits as a cage-like three-dimensional network with a unique porosity similar to the zeolite. Thus, it possesses the dual ideal properties of zeolites and MOFs, such as tunable structure, monodisperse micropores, ultra-high surface area and structural flexibility [38]. More importantly, substantial data have demonstrated ZIF-8 as a promising drug delivery vehicle owning to its protonation effect induced by the contained imidazole ring. In the lysosomes, the imidazole ring is readily protonated in the acidic environment and causes the swelling of lysosomes, which increases the lysosomal membrane permeability and ultimately helps the lysosomal escape of the cargos. Actually, there have been many studies on the use of ZIF-8 as drug delivery vehicles. Scientists have encapsulated the genome editing machinery CRISPR/Cas9 with ZIF-8 and found that much more internalized CRISPR/Cas9 was released from lysosomes into cytoplasm and escaped from the lysosomal degradation when using ZIF-8 as a nanovehicle. This was found to be protonation-dependent lysosome escape [39]. In Hou's work, encapsulating glucose oxidase with ZIF-8 could greatly improve the sensitivity and catalytic activity of glucose oxidase towards glucose with an enhanced stability [40]. Ding's group encapsulated cytochrome C with ZIF-8 and showed that ZIF-8 encapsulation could protect cytochrome C from lysosomal proteolysis and ensure its safe delivery to cancer cells in an integral and active form [41]. Moreover, naked nucleic acids (DNA or RNA) are generally too large to penetrate into the cell membrane and are easily degraded by serum nucleases. In the study of Li's group, plasmid DNA was successfully encapsulated in ZIF-8. ZIF-8 encapsulation greatly improved cellular uptake of plasmid DNA, increased the transfection efficiency and thus, achieved a more successful gene expression [42]. A recent study synthesized ZIF-8 on ZnO surface and the obtained ZnO@ZIF-8 structure enabled a high drug-loading efficacy of ciprofloxacin within the pore network of ZIF-8 and thus, achieved a controlled release and high antimicrobial efficiency of ciprofloxacin [43]. Ju et al. used ZIF-8 as a drug vector to deliver miR-491-5p into cancer cells and the results demonstrated that the ZIF-8 nanovector could stably release miR-491-5p for a long period and greatly improved the efficiency of miR-491-5p in inhibiting tumor growth [44]. All these reports have pointed out that ZIF-8 has great potential as a drug delivery vehicle.

Considering the water insolubility of the telomerase inhibitor BIBR 1532, the functional advantages of ZIF-8 as well as the lack of studies on addressing insignificant bioeffects of BIBR 1532, our study was the first to encapsulate BIBR 1532 with ZIF-8 and explore the effects of ZIF-8 as a drug delivery vehicle (Scheme 1). Our present work demonstrated that the synthesized ZIF-8 facilitated the transport and release of BIBR 1532 within cells by altering the permeability of the lysosomal membrane. With ZIF-8 encapsulation, more BIBR 1532 accumulated in the nucleus and triggered a much more obvious growth inhibition of tumor cells as compared with free BIBR 1532.

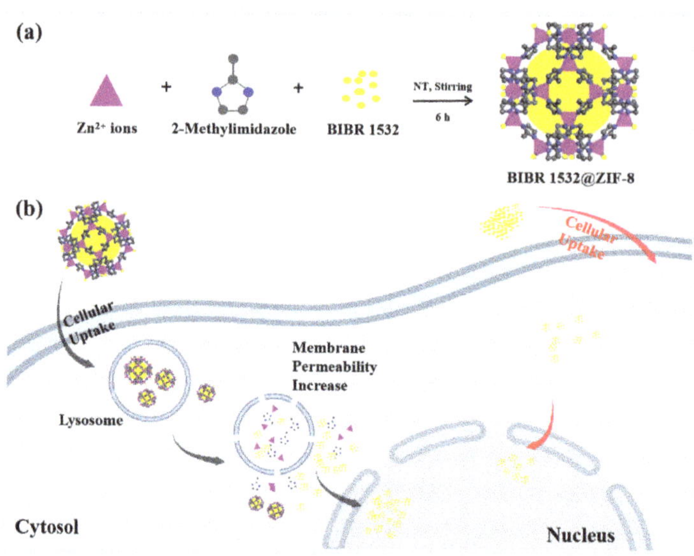

Scheme 1. Preparation, transfer and bioeffects of BIBR 1532@ZIF-8. (**a**) Encapsulating BIBR 1532 with ZIF-8 to form BIBR 1532@ZIF-8. RT means room temperature. (**b**) Transport, release and bioeffects of BIBR 1532@ZIF-8.

2. Materials and Methods

2.1. Chemicals and Materials

Zinc nitrate hexahydrate ($Zn(NO_3)_2$ $6H_2O$) was purchased from Sigma-Aldrich. 2-methylimidazole was purchased from Macklin. Trypsin-EDTA was obtained from MACGENE. PBS buffer and DMEM (high glucose) were purchased from Hyclone. BIBR 1532 was obtained from Aladdin and the structure of BIBR 1532 was shown in Figure S1. Fluorescein, free acid (green fluorescence signal) was purchased from Coolaber. 6-aminofluorescein (6-AF, green fluorescence signal), N-hydroxysuccinimide (NHS) and 1-ethyl-3-(3-dimethylaminopropyl) carbodiimide (EDC) were obtained from Yuanye Biological. Hoechst 33,342 and Lyso-Tracker Red were purchased from Beyotime Biotechnology. Recombinant anti-lysosomal associated membrane protein 1 (LAMP1) antibody, recombinant anti-cathepsin B antibody and recombinant anti-GAPDH antibody were purchased from Abcam. Anti-rabbit IgG, HRP-conjugated antibody was purchased from Cell Signaling Technology. A549 cell line was obtained from American Type Culture Collection (ATCC).

2.2. Synthesis of Nanoparticles

ZIF-8 was synthesized according to a reported procedure with some modifications [45]. A measure of 0.2925 g of $Zn(NO_3)_2 \cdot 6H_2O$ was dissolved in 3 mL of deionized water, and then 5.675 g of 2-methylimidazole was dissolved in another 20 mL of deionized water. The zinc nitrate solution was added to 2-methylimidazole solution with vigorous stirring. After stirring for 6 h, the product was collected and centrifuged at 15,000$\times$ g for 15 min. Finally, the as-synthesized nanoparticles were washed alternately with different ratios of anhydrous methanol/deionized water mixture (1:3 and 3:1) for 5 times. The product was then dried in a lyophilizer overnight to obtain ZIF-8 powder. All operations were performed at room temperature.

The loading of BIBR 1532 was achieved by synthesis of BIBR 1532@ZIF-8. Briefly, 0.011 g of $Zn(NO_3)_2 \cdot 6H_2O$ was dissolved in 0.113 mL of deionized water, then 0.86 mg of BIBR 1532 was dissolved in zinc solution; the mixed solution of zinc and BIBR 1532 was added to 0.752 mL of deionized water containing 0.213 g of 2-methylimidazole with vigorous stirring. After stirring for 6 h, the product was collected and washed. The product was then dried in a lyophilizer overnight to obtain the powder and stored at −80 °C. All operations were performed at room temperature.

To explore the transfer and localization of ZIF-8 in A549 cells, we labeled ZIF-8 with fluorescein (fluorescein@ZIF-8) according to a reported procedure with some modifications [46]. Briefly, 3.3 mg of fluorescein was dissolved in 0.3 mL of methanol and then mixed with 3 mL of deionized water containing 0.2925 g of $Zn(NO_3)_2 \cdot 6H_2O$. 5.675 g of 2-methylimidazole was dissolved in a further 20 mL of deionized water. The mixed solution of zinc nitrate and fluorescein was added to 2-methylimidazole solution with vigorous stirring. After stirring for 6 h, the product was collected and washed for several times. The product was then dried in a lyophilizer overnight to obtain fluorescein@ZIF-8 powder and stored at −20 °C. All operations were performed at room temperature.

The localization and accumulation of BIBR 1532 was observed by labeling BIBR 1532 with 6-AF (6-AF@BIBR 1532). BIBR 1532 was labeled with 6-AF according to a reported procedure with some modifications [47]. A 3.31 mg measure of BIBR 1532 was dissolved in 3 mL of absolute ethanol and 12 mL of deionized water, and then 5 mg of 6-AF was added while stirring. After stirring for 10 min, 10 mg EDS and 5.75 mg NHS were added to activate the carboxyl group of BIBR 1532. Stirring was continued for 6 h at room temperature. The reaction solution was then collected and washed several times. The product was then dried in a lyophilizer overnight to obtain the 6-AF@BIBR 1532 powder and stored at −80 °C. The whole experiment was performed in the dark.

The localization and accumulation of BIBR 1532@ZIF-8 was observed by synthesizing 6-AF@BIBR 1532@ZIF-8. A 0.011 g measure of $Zn(NO_3)_2 \cdot 6H_2O$ was dissolved in 0.113 mL of deionized water, then 250 µL of 6-AF@BIBR 1532 obtained in the previous step was dissolved in the zinc solution. While stirring vigorously, the above mixed solution was added to 0.752 mL of deionized water containing 0.213 g of 2-methylimidazole. After stirring for 6 h, the product was collected, centrifuged at 15,000× *g* for 15 min and washed several times. The product was then dried in a lyophilizer overnight to obtain the powder and stored at −80 °C. The whole experiment was performed in the dark.

2.3. Characterization

The size and morphology of nanoparticles were observed by scanning electron microscope (SEM, s-4800, HITACHI, Tokyo, Japan) and transmission electron microscope (TEM, JEM2100Plus, JEOL, Tokyo, Japan). The size distribution and zeta potential of the nanoparticles were analyzed using dynamic light scattering (DLS, NanoBrook Omni, Holtsville, NY, USA). Powder X-ray diffraction (XRD) of the samples was recorded using a D8 Advance X-ray diffractometer (Bruker, Karlsruhe, Germany) at a voltage of 40 kV in the range of 2θ = 15–90° and a current of 40 mA. The formation of nanoparticles in the powder form was determined using a Fourier transform infrared (FT-IR) spectrometer (Thermo Scientific, Waltham, MA, USA) coupled with a Nicolet iN10 MX spectrograph. Thermal Gravimetric Analysis (TGA) was performed on TGA-Q50 (TA Instruments, New Castle, DE, USA). In total, 10 mg of samples were placed in an alumina crucible and heated in a continuous flow of nitrogen gas at a ramp rate of 10 °C/min from 0 °C up to 800 °C. The intracellular Zn content was determined using an inductively coupled plasma optical emission spectrometer (ICP-OES, Thermo Fisher Scientific, Pittsburg, PA, USA) to reflect the cellular uptake of ZIF-8.

2.4. BIBR 1532 Loading

6-AF@BIBR 1532 was dispersed in deionized water to get a stock concentration of 800 µg/mL and then diluted in a gradient to 400, 200, 100, 50, 25 and 12.5 µg/mL. The fluorescence intensity value of each well was measured using a microplate reader (Thermo Scientific, Waltham, MA, USA) with water as a blank control. The data were analyzed to obtain the standard curve equation: Y = 42066X + 17429, R^2 = 0.9997 (Figure S2).

The fluorescence intensity value of the supernatant after centrifugation of 6-AF@BIBR 1532@ZIF-8 dispersion was measured and then substituted into the equation, giving a concentration of 0.269 mg/mL.

Drug loading was calculated using the following formula: Drug Loading = $\frac{0.6675 - 0.269 \times (0.1128 + 0.752)}{7.333}$ = 6%. It means that each 100 mg of BIBR 1532@ZIF-8 contains 6 mg of BIBR 1532.

In this formula, 0.6675 mg was the weight of 6-AF@BIBR 1532 added for 6-AF@BIBR 1532@ZIF-8 synthesis. The concentration of 6-AF@BIBR 1532 in the supernatant after centrifugation of 6-AF@BIBR 1532@ZIF-8 dispersion was 0.269 mg/mL. The total volume of the synthetic 6-AF@BIBR 1532@ZIF-8 solvent was 0.8648 mL. The weight of the 6-AF@BIBR 1532@ZIF-8 product was 7.333 mg.

In the following study, we designed the incubation concentrations of BIBR 1532 and BIBR 1532@ZIF-8 at an equivalent concentration of BIBR 1532. We selected the concentrations of 10, 50, 100 μg/mL for BIBR 1532@ZIF-8, and the equivalent concentrations for BIBR 1532 were 0.6, 3, 6 μg/mL, respectively, with the corresponding ZIF-8 concentrations of 9.4, 47, 94 μg/mL.

2.5. Cell Viability Assay

A549 cells were maintained in Dulbecco's modified Eagle's Medium (DMEM) containing 10% fetal bovine serum, 100 U/mL penicillin and 100 μg/mL streptomycin at 37 °C in a humidified atmosphere of 5% CO_2 and 95% air. The effects of ZIF-8, BIBR 1532 and BIBR 1532@ZIF-8 on cell viability were evaluated using CCK-8 assay (Dojindo Laboratories, Tokyo, Japan). Cells were seeded in 96-well plates (5×10^3 cells/well) and cultured for 24 h. Cells were then treated with different concentrations of ZIF-8 (0, 50, 75, 100, 150 μg/mL) or different concentrations of ZIF-8 (9.4, 47, 94 μg/mL), BIBR 1532 (0.6, 3, 6 μg/mL) and BIBR 1532@ZIF-8 (10, 50, 100 μg/mL) for 24 h. Cell viability was assessed using standard procedures. Absorbance was measured at 450 nm using a microplate reader (Thermo Scientific, Waltham, MA, USA).

2.6. Inductively Coupled Plasma-Optical Emission Spectrometer (ICP-OES)

Cellular uptake of ZIF-8 was determined by measuring intracellular Zn content using ICP-OES. A549 cells were cultured on 35 mm petri dishes at a density of 2×10^5 cells/dish. After adhesion, cells were incubated with 50 μg/mL ZIF-8 for different time intervals (1, 3, 6, 9, 12 and 24 h). After incubation, A549 cells were harvested and counted and then transferred to quartz beakers. A 3 mL measure of HNO_3 was added to each quartz beaker. The quartz beaker was placed on a hot plate and kept at 150 °C for 2 h. Then, 1 mL of H_2O_2 was added for further heating digestion until digestion was complete. The solution volume was diluted to 3 mL with 2% HNO_3. A series of Zn standard solutions (0.1, 0.5, 1, 5, 10, 50 and 100 ng/mL (ppb)) were prepared. Both standards and test samples were measured by ICP-OES. Cellular uptake of ZIF-8 was calculated from cell counts (μg/10^5 cells).

2.7. Confocal Laser Scanning Microscopy

To explore the transfer and localization of ZIF-8, BIBR 1532 and BIBR 1532@ZIF-8, A549 cells were cultured on 35 mm dishes at a density of 4×10^5 cells/dish. After attachment, the cells were treated with 100 μg/mL fluorescein@ZIF-8 or 6 μg/mL 6-AF@BIBR 1532 and 100 μg/mL 6-AF@BIBR 1532@ZIF-8 for 1 h in the dark. After incubation, the cells were incubated with fresh medium for a further 0, 2, 5 h. At each time point, the cells were fixed with 1mL of 2.5% glutaraldehyde for 15 min at room temperature. Afterwards, cells were stained with Lyso-Tracker Red and Hoechst 33342 at 37 °C for 20 min. The cells were then washed with PBS and loaded onto a laser scanning confocal microscope (Nikon A1, Tokyo, Japan) for observation.

To explore the nucleus accumulation of BIBR 1532 and BIBR 1532@ZIF-8, A549 cells were cultured on 35 mm dishes at a density of 4×10^5 cells/dish and treated with 6 μg/mL 6-AF@BIBR 1532 or 100 μg/mL 6-AF@BIBR 1532@ZIF-8, respectively, for different time intervals. At 6, 12 and 24 h, cells were fixed with 4% paraformaldehyde for 15 min at room temperature followed by Hoechst 33342 staining at 37 °C for 10 min. Then cells were washed with PBS and loaded onto a laser scanning confocal microscope for observation.

2.8. Acridine Orange (AO) Staining

AO staining was used to reflect the change of lysosomal membrane permeability induced by ZIF-8. Cells were seeded onto 35 mm petri dishes at a density of 4×10^5 cells/well. After attachment, cells were stained with AO at a final concentration of 10 μg/mL at 37 °C in the dark for 15 min. Afterwards, the cells were treated with or without 100 μg/mL ZIF-8 for different time periods (1, 3, 6, 12 and 24 h). At each time point, cells were fixed with 1 mL of 4% paraformaldehyde for 15 min at room temperature. The samples were then washed and loaded onto a fluorescence microscope for observation.

2.9. Western Blot Analysis

A549 cells were seeded on 6-well plates at a density of 9×10^5 cells/well and cultured for 24 h. Then, cells were treated with different concentrations of ZIF-8 (0, 10, 25, 50, 75 and 100 μg/mL) for 24 h. At the end of the incubation period, the cells were lysed in RIPA lysis buffer (Beyotime Biotechnology, China) on ice for 15 min and protein was extracted. The lysate was centrifuged at 18,000× *g* for 5 min at 4 °C and the protein was collected. The protein concentrations of the samples were measured using a bicinchoninic acid assay (BCA) kit (Beyotime Biotechnology, China). Equal amounts (20 μg) of protein from the samples were separated by 12% sodium dodecyl sulfate-polyacrylamide gel electrophoresis. Proteins were transferred to 0.22 μm PVDF membranes (Millipore, Boston, MA, USA). Proteins were then blocked for 1 h in 5% BSA and incubated overnight at 4 °C with primary antibodies of lysosomal-associated membrane protein 1 (LAMP1) and cathepsin B, followed by 1 h of incubation with HRP-conjugated secondary antibody. Target protein bands were visualized using ECL-chemiluminescent kit (ECL-plus, Thermo Scientific, Waltham, MA, USA). Proteins levels were semi-quantitatively analyzed using ImageJ software. Glyceraldehyde-3-phosphate dehydrogenase (GAPDH) was used as a loading protein. The relative expression levels of LAMP1 and cathepsin B were normalized to the level of GAPDH.

2.10. Real-Time Quantitative Polymerase Chain Reaction (RT-qPCR) Analysis

hTERT mRNA level was detected using RT-qPCR analysis. Cells were seeded on 35 mm dishes at a density of 9×10^5 cells/well and cultured for 24 h. Then, cells were treated with different concentrations of ZIF-8 (9.4, 47, 94 μg/mL), BIBR 1532 (0.6, 3, 6 μg/mL) and BIBR 1532@ZIF-8 (10, 50, 100 μg/mL) for 24 h (drug-loading efficiency is approximately 6%). After incubation, RNA was extracted from cells, followed by reverse transcription (FastFire Rapid Fluorescence PCR Premix (SYBR Green), TIANGEN, Beijing, China) and PCR amplification (Telomerase Activity Fluorescence Real-time Quantitative PCR Detection Kit (Human), KeyGEN BioTECH, Nanjing, China). Measurements were performed according to the manufacturer's instructions. GAPDH was used as an internal control.

2.11. Cytokinesis-Block Micronucleus (CBMN) Assay

The micronucleus assay (MN) is one of the most popular methods to assess genotoxicity. The formation of micronucleus is easily recognizable in cytokinesis-blocked cells with their binucleate appearance [48]. Cell division block and telomere dysfunction induced by BIBR 1532 or BIBR 1532@ZIF-8 was explored using the cytokinesis-block micronucleus assay. In detail, cells were seeded in 35 mm dishes at a density of 4×10^5 cells/well, then treated with 94 μg/mL ZIF-8, 6 μg/mL BIBR 1532 or 100 μg/mL BIBR 1532@ZIF-8. After incubation for 2 h, cytochalasin B was added with a final concentration of 5 μg/mL for one cell cycle (22 h for A549 cells) to ensure inhibition of cytoplasmic division to form binucleated cells, making it easier to observe micronucleus formation. Cytochalasin B was then removed and the cells were incubated with ZIF-8, BIBR 1532 or BIBR 1532@ZIF-8, respectively, for a further 48 h. At the end of the incubation period, the cells were fixed and then stained with Hoechst 33342 (100×) at 37 °C for 10 min. Then cells were washed and loaded onto a laser scanning confocal microscope for observation. Representative images were captured. The number of cells containing micronucleus and the total amount of micronucleus among the 300 cells in different groups were compared.

2.12. Cell Cycle Analysis

Cells were seeded in 6-well plates at a density of 4 $\times 10^5$ cells/well and cultured for 24 h. Then, the cells were treated with different concentrations of ZIF-8 (9.4, 47, 94 μg/mL), BIBR 1532 (0.6, 3, 6 μg/mL) or BIBR 1532@ZIF-8 (10, 50, 100 μg/mL) for 24 h. At the end of the treatment time, all the cells were collected and centrifuged at 1000× g for 5 min. Afterwards, cells were resuspended with 300 μL of pre-cooled PBS and then fixed with 700 μL of absolute ethanol at −20 °C overnight. Then cells were centrifuged, washed and stained with propidium iodide (PI) at 37 °C in the dark for 30 min. Cell cycle analysis was performed by flow cytometry (BD LSR-Fortessa, San Jose, CA, USA).

2.13. Senescence-Associated β-Galactosidase Detection

The effects of BIBR 1532 and BIBR 1532@ZIF-8 on cellular senescence were evaluated based on senescence-associated β-galactosidase (SA-β-Gal) staining using the Senescence Cells Staining Kit (Beyotime Biotechnology, Shanghai, China). Cells were seeded in 24-well plates at a density of 8 × 10^4 cells/well and cultured for 24 h. Then, cells were treated with different concentrations of ZIF-8 (9.4, 47, 94 μg/mL), BIBR 1532 (0.6, 3, 6 μg/mL) or BIBR 1532@ZIF-8 (10, 50, 100 μg/mL) for 7 days. Cells treated with 100 nM DOX for 7 days were used as the positive controls. At the end of the treatment intervals, cells were subjected to SA-β-Gal staining according to the manufacturer's instructions. For quantitative analysis of the β-galactosidase positive rate, 200 cells were counted in each group.

2.14. Statistical Analysis

Statistical analysis was performed using two-way ANOVA (GraphPad Prism 7.00) followed by Tukey's multiple comparison test. All statistical parameters were presented as mean ± standard deviation (SD). $p < 0.05$ was considered significant.

3. Results

3.1. Physicochemical Characterizations of ZIF-8 and BIBR 1532@ZIF-8

The observation of ZIF-8 using scanning electron microscopy showed that ZIF-8 had a regular hexagonal or regular octagonal structure with a particle size of approximately 100–200 nm, which was further confirmed by the results from transmission electron microscopy (Figure 1A,B). The mean hydrated particle size of ZIF-8 was 129.98 ± 0.25 nm with a polydispersity index (PDI) of about 0.12 ± 0.01, indicating a relatively good dispersion. The zeta potential of ZIF-8 was tested as being 17.8 ± 1.3 mV (Figure 1E). The FT-IR results showed that most absorption peaks of the ZIF-8 sample came from the vibration of 2-methylimidazole: the band observed at 1589 cm^{-1} was assigned to the stretching vibration of the C=N group of 2-methyl imidazole. The bands in the range 1350–1500 cm^{-1} were assigned to the stretching vibration of the imidazole ring. Other bands in the 900–1350 cm^{-1} range were assigned to the stretching and bending modes of the imidazole ring (Figure 1H). The infrared spectroscopic results clearly demonstrated the successful synthesis of ZIF-8. The XRD patterns of ZIF-8 together with the corresponding simulated XRD patterns are shown in Figure 1I. The observed sharp and strong characteristic peaks were consistent with the simulated XRD patterns, suggesting that ZIF-8 had a rhombohedral dodecahedral structure. Few miscellaneous peaks were detected, indicating a relatively pure ZIF-8 crystal (Figure 1I). The mass change curve of ZIF-8 from TG analysis showed a long stable platform in the temperature range 25–400 °C, and the mass was decreased sharply as the temperature increased further, suggesting that the crystal skeleton of ZIF-8 began to decompose at higher temperatures (Figure 1J).

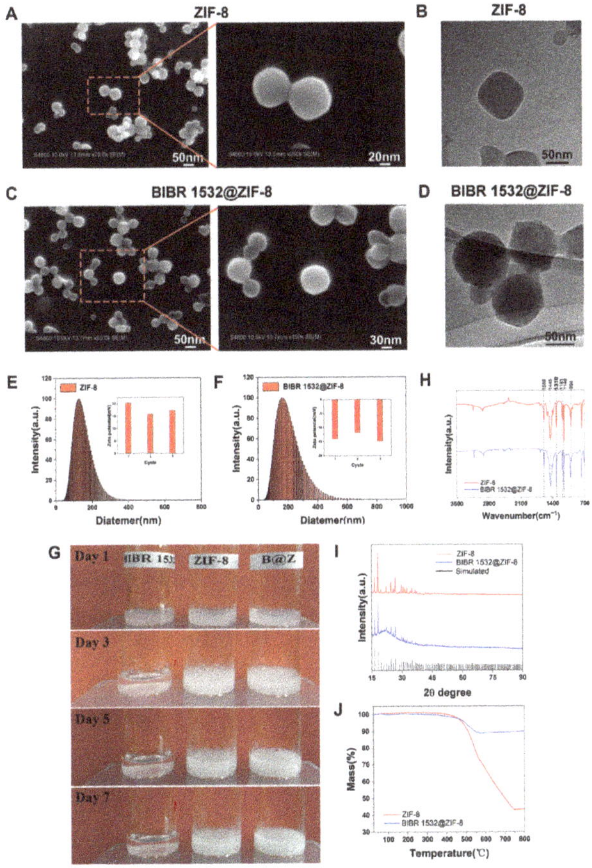

Figure 1. Physicochemical characterizations of ZIF-8 and BIBR 1532@ZIF-8. (**A**) SEM images of ZIF-8. (**B**) TEM images of ZIF-8. (**C**) SEM image of BIBR 1532@ZIF-8. (**D**) TEM image of BIBR 1532@ZIF-8. (**E**) Particle size distribution and zeta potential of ZIF-8. (**F**) Particle size distribution and zeta potential of BIBR 1532@ZIF-8. (**G**) Stability of ZIF-8, BIBR 1532 and BIBR 1532@ZIF-8 in aqueous solutions. (**H**) FT-IR analysis of ZIF-8 and BIBR 1532@ZIF-8. (**I**) XRD analysis of ZIF-8 and BIBR 1532@ZIF-8. (**J**) TG analysis of ZIF-8 and BIBR 1532@ZIF-8.

Then, ZIF-8 was loaded with BIBR 1532. The obtained BIBR 1532@ZIF-8 also maintained a regular hexagonal or octagonal shape as observed using SEM and TEM (Figure 1C,D). The mean hydrated particle size of BIBR 1532@ZIF-8 was 165.37 $\pm$ 0.48 nm, slightly larger than that of ZIF-8, with a polydispersity index (PDI) about 0.21 $\pm$ 0.01. The potential value of BIBR 1532 @ZIF-8 was -13.65 $\pm$ 0.93 (Figure 1F). BIBR 1532 was insoluble and poorly dispersed in water. It would precipitate rapidly within 1 day with a transparent supernatant appearing. In contrast, the dispersion of BIBR 1532@ZIF-8 in water remained relatively stable without precipitation for at least 1 week, suggesting an improved dispersion and stability of BIBR 1532 in water with ZIF-8 as a vector (Figure 1G). The FT-IR results of BIBR 1532@ZIF-8 showed that the bands observed at 1589 cm^{-1}, 1350–1500 cm^{-1} and 900–1350 cm^{-1} were assigned to the vibration of 2-methylimidazole. In addition, the specific band of BIBR 1532@ZIF-8 was observed at 1300 cm^{-1}, which corresponded to the stretching absorption of -OH in BIBR 1532. The results of FT-IR results suggested the successful synthesis of BIBR 1532@ZIF-8 (Figure 1H). The XRD results indicated that BIBR 1532@ZIF-8 had a rhombohedral dodecahedral structure similar to that of ZIF-8 (Figure 1I). The TGA curve of BIBR 1532@ZIF-8 showed that the weight decreased only by 10% even at temperature up to 750 °C (Figure 1J). Strikingly, the successful encapsulation of BIBR 1532 with ZIF-8 was again demonstrated by the changes in the curve.

3.2. *ZIF-8 Localized in Lysosomes*

First, the results from CCK-8 assay demonstrated that ZIF-8 did not induce obvious cytotoxicity in A549 cells in the concentration range of 0–100 μg/mL within the observed 24 h. However, when the concentration of ZIF-8 reached 150 μg/mL, a little cytotoxicity was exhibited (Figure 2A).

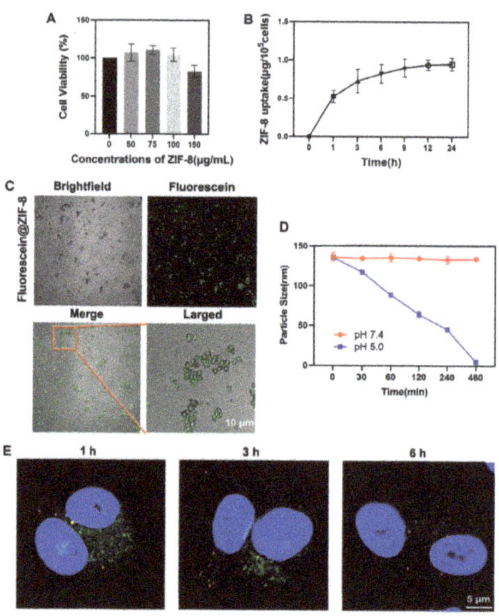

Figure 2. Effects of ZIF-8 on A549 cells. (**A**) The effect of ZIF-8 on cell viability. Cells were treated with 0, 50, 75, 100 and 150 μg/mL ZIF-8 for 24 h. Cell viability was measured using CCK-8 assay. (**B**) Cellular uptake of ZIF-8 within 24 h. A549 cells were treated with 50 μg/mL for 24 h. (**C**) CLSM images of fluorescein@ZIF-8. fluorescein@ZIF-8 (100 μg/mL) was dispersed in water. (**D**) pH sensitivity of ZIF-8. Evaluation of the average particle sizes of ZIF-8 at pH 7.4 and 5.0 over 480 min. Results were shown as mean ± SD (n = 3). (**E**) The localization of ZIF-8 in cells. Cells were treated with 100 μg/mL fluorescein@ZIF-8 (green) for 1 h, and then incubated with fresh medium for further 0, 2, 5 h. Then, cells were stained with Lysotracker (red) and Hoechst 33342 (blue) for CLSM observation.

Then, the cellular uptake of ZIF-8 was examined within 24 h. A549 cells were exposed to 50 μg/mL ZIF-8 for 24 h and the cellular uptake of ZIF-8 was determined indirectly by the analysis of the intracellular zinc content using ICP-OES. The intracellular Zn concentration increased invariably with increasing exposure time of cells to ZIF-8. The result demonstrated that during the incubation period, cellular uptake of ZIF-8 was accelerated within 1 h, then gradually decreased between 1 and 12 h and reached saturation after 12 h (Figure 2B).

To study the transfer and localization of ZIF-8 in A549 cells, we labeled ZIF-8 with fluorescein (green). By comparing bright-field and fluorescence phase images, it confirmed that ZIF-8 was successfully labeled with fluorescein and moreover, the fluorescence signal could remain stable for at least 7 days (Figure 2C and Figure S3A,B). Our experiment also explored the pH-responsiveness of ZIF-8 in vitro. ZIF-8 was dispersed in neutral (pH 7.4) and acidic (pH 5.0) solution, respectively, and its average particle size was monitored at 0, 30, 60, 120, 240 and 480 min. As shown in Figure 2D, the average particle size of ZIF-8 remained basically stable at neutral pH whereas it gradually decreased till undetectable at acidic pH, indicating that ZIF-8 was degraded in acidic solution.

A549 cells were exposed to 100 μg/mL fluorescein@ZIF-8 for 1 h. Fluorescein@ZIF-8 was then discarded and the cells were incubated with fresh medium. The transfer and localization of fluorescein@ZIF-8 was observed after addition of fluorescein@ZIF-8 for 1, 3 and 6 h. At 1 h, fluorescein@ZIF-8 was localized inside lysosomes, as demonstrated by the obvious yellow dots of lysosomes (Figure 2E). At 3 h, the lysosomes still showed a yellow fluorescence signal, but the green fluorescence signal in the cytoplasm decreased. At 6 h, the green fluorescence signal of ZIF-8 almost disappeared in both lysosomes and cytoplasm.

We further investigated the effect of ZIF-8 internalization on lysosomes. First, lysosomal membrane permeability was examined using AO staining. AO, a fluorescent dye that can enter lysosomes, is used to label lysosomes. It is present in intact lysosomes as an oligomer with red fluorescence and as a monomer in the cytoplasm with green fluorescence. As the lysosomal membrane stability decreases, AO is released from lysosomes into the cytoplasm, and an increase in the green fluorescence signal in the cytoplasm can be observed using laser scanning confocal microscopy. Therefore, the alteration on the ratio of green/red fluorescence intensity can reflect the change in lysosomal membrane permeability. In the present study, no obvious change on the ratio of green/red fluorescence intensity was observed in the control cells within 24 h. However, after 24 h of co-incubation with ZIF-8, the ratio of green/red fluorescence intensity in the cells increased, that is, the red fluorescence signal of AO in intact lysosomes gradually decreased and the green fluorescence signal of AO in the cytoplasm gradually increased, indicating that more AO dye was released from the lysosomes into the cytoplasm. The results suggested that ZIF-8 exposure induced a change in lysosomal membrane permeability (Figure 3A,B).

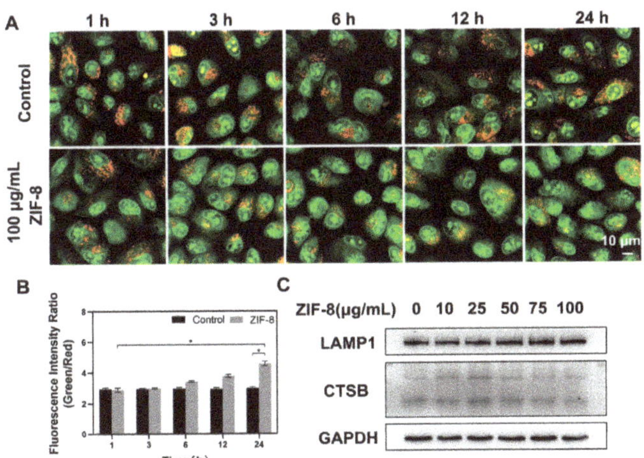

Figure 3. Effects of ZIF-8 on lysosomes. (**A**) The effect of ZIF-8 on lysosomal membrane integrity in A549 cells. Cells were treated with 10 μg/mL AO for 15 min followed by 100 μg/mL ZIF-8 treatment for 1, 3, 6, 12 and 24 h. Representative results of AO staining were shown. (**B**) The ratio of green/red fluorescence in AO-stained cells at 1, 3, 6, 12 and 24 h as described in (**A**). Image J was used to quantitatively analyze the intensity changes of red and green fluorescence within 24 h. The date was representative of three independent experiments. * $p < 0.05$. (**C**) The effect of ZIF-8 on the protein expression levels of LAMP1 and CTSB. Cells were treated with different concentrations of ZIF-8 for 24 h. GAPDH was used as a loading protein. Representative results were shown here.

The expressions of cathepsin B (CTSB) and lysosomal associated membrane protein 1 (LAMP1) were detected to further explore the effect of ZIF-8 on lysosomal function. Cathepsin B (CTSB), an active hydrolase inside lysosomes, can mediate the degradation of extracellular particles caused by endocytosis. The expression of CTSB could reflect the degradative function of lysosomes [49]. Lysosomal-associated membrane protein 1 (LAMP1), the most abundant protein on the lysosomal membrane, plays an important role in lysosomes trafficking, maintenance of lysosomal acidification and homeostasis [50–52]. A549 cells were harvested after treatment with different concentrations of ZIF-8 (0, 10, 25, 50, 75 and 100 μg/mL) for 24 h. The results from western blot analysis showed that ZIF-8 treatment had no apparent effect on the expression of CTSB and LAMP1 in cells (Figure 3C).

3.3. BIBR 1532 Encapsulation by ZIF-8 Increased Delivery Efficacy

First, we labeled BIBR 1532 with 6-AF (green). The FT-IR results confirmed the successful labeling of BIBR 1532 by 6-AF (Figure 4A). Due to the coupling of the amino group of 6-AF with the carboxyl group of BIBR 1532, 6-AF@BIBR 1532 exhibited a weaker band at 3500–3350 cm^{-1} and 1700–1500 cm^{-1} associated with the stretching vibration of the amino group of 6-AF. Compared with BIBR 1532, 6-AF@BIBR 1532 exhibited a weaker band corresponding to the stretching of -COOH at 1300 cm^{-1}, which fully demonstrated the successful labeling of BIBR 1532 with 6-AF.

Then, we encapsulated 6-AF@BIBR 1532 with ZIF-8 and confirmed the successful synthesis of 6-AF@BIBR 1532@ZIF-8 by FT-IR results (Figure 4B). The infrared spectrum of 6-AF@BIBR 1532@ZIF-8 showed that the band observed at 1589 cm^{-1} was assigned to the stretching vibration of the C=N group of 2-methylimidazole; the bands in the range 1350–1500 cm^{-1} were assigned to the stretching vibration of the imidazole ring; the bands in the range 900–1350 cm^{-1} were assigned to the stretching and bending modes of the imidazole ring; the band at 1300 cm^{-1} corresponded to the stretching of -COOH. Due to the carboxyl-amino coupling, 6-AF@BIBR 1532@ZIF-8 exhibited weaker stretching bands of -COOH at 1300 cm^{-1}, indicating the successful synthesis of 6-AF@BIBR 1532@ZIF-8. Furthermore, by comparing the images of the bright-field and fluorescence images, we demonstrated that the encapsulation of ZIF-8 did not affect the detection of the fluorescence signals (Figure 4C) and the fluorescence signal remained stable for at least one week (Figure S4A,B).

Figure.4D illustrated the different transfer pathways and localization of 6-AF@BIBR 1532 and 6-AF@BIBR 1532@ZIF-8 in A549 cells. At 1 h after the addition of 6-AF@BIBR 1532 or 6-AF@BIBR 1532@ZIF-8, the green fluorescein was localized inside lysosomes, as shown by the obvious yellow dots in lysosomes (Figure 2D). After 3 h of incubation, green fluorescence of 6-AF@BIBR 1532 and 6-AF@BIBR 1532@ZIF-8 in lysosomes decreased. In 6-AF@BIBR 1532@ZIF-8-incubated cells, the fluorescence was distributed in the cytoplasm and even a very small amount of fluorescence signal appeared in the nuclear region. In 6-AF@BIBR 1532-incubated cells, the decreased fluorescence in lysosomes was not accompanied by an increase in the cytoplasm or nucleus. After 6 h of incubation, only a very small amount of fluorescence was detected in the nuclear region of 6-AF@BIBR 1532-treated cells. Comparatively, more fluorescence was concentrated in the nucleus in 6-AF@BIBR 1532@ZIF-8-treated cells. The results showed that encapsulation of BIBR 1532 in ZIF-8 led to an increased release of BIBR 1532 into the cytoplasm and the sequent transfer into the nucleus.

We also evaluated the accumulation of 6-AF@BIBR 1532 and 6-AF@BIBR 1532@ZIF-8 in the nucleus within 24 h. Free BIBR 1532 accumulated slowly in the nucleus. The fluorescence intensity in the nucleus increased slowly from 6 to 24 h, suggesting a relatively limited cumulative amount of BIBR 1532 within 24 h. With ZIF-8 encapsulation, more BIBR 1532 accumulated in the nucleus. The fluorescence intensity in the nucleus increased sharply from 6 h to 24 h, suggesting a steady accumulation of BIBR 1532 in the nuclear region (Figure 4E,F). Herein, the encapsulation of BIBR 1532 in ZIF-8 successfully increased the nuclear accumulation of BIBR 1532.

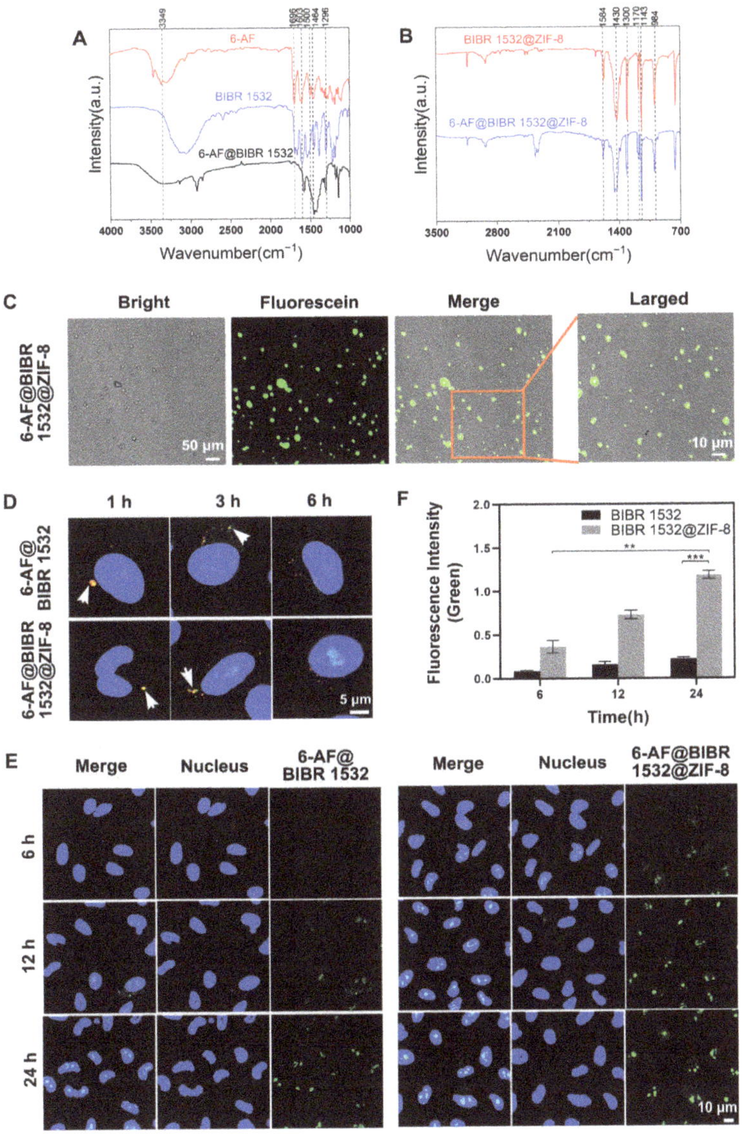

Figure 4. The difference in delivery efficacy between BIBR 1532 and BIBR 1532@ZIF-8. (**A**) The successful labeling of BIBR 1532 with 6-AF. FT-IR analysis results of 6-AF, BIBR 1532 and 6-AF@BIBR 1532. (**B**) The successful encapsulation of 6-AF@BIBR 1532 with ZIF-8. FT-IR analysis of BIBR 1532@ZIF-8 and 6-AF@BIBR 1532@ZIF-8. (**C**) CLSM images of 6-AF@BIBR 1532@ZIF-8. 6-AF@BIBR 1532@ZIF-8 was dispersed in water at the concentration of 100 μg/mL. (**D**) The localization of 6-AF@BIBR 1532 and 6-AF@BIBR 1532@ZIF-8 in cells. Cells were treated with 6 μg/mL 6-AF@BIBR 1532 (green) or 100 μg/mL 6-AF@BIBR 1532@ZIF-8 (green) for 1 h, then incubated with fresh medium for a further 0, 2, 5 h. Then, cells were stained with Lysotracker (red) and Hoechst 33,342 (blue) for CLSM observation. (**E**) The accumulation of 6-AF@BIBR and 6-AF@BIBR 1532@ZIF-8 in the nucleus. Representative CLSM images of A549 cells after incubation with 6 μg/mL 6-AF@BIBR 1532 (green) and 100 μg/mL 6-AF@BIBR 1532@ZIF-8 (green) for 6, 12 and 24 h were shown. ** $p < 0.01$. *** $p < 0.001$. (**F**) Image J was used to quantitatively analyze the changes of green fluorescence at 6, 12 and 24 h as described in (**E**).

3.4. ZIF-8 Encapsulation Enhanced the Treatment Efficacy of BIBR 1532

The advantage of encapsulating BIBR 1532 with ZIF-8 was evaluated from several aspects. As demonstrated in Figure 5A, the treatment with 0.6 μg/mL BIBR 1532 did not cause a decrease in cell viability, whereas the treatment with the equivalent BIBR 1532 encapsulated by ZIF-8 (100 μg/mL BIBR 1532@ZIF-8) caused an 18% decrease in cell viability. This result suggested a remarkable increase in the cytotoxicity of BIBR 1532 with ZIF-8 as a delivery vector. This kind of enhanced cytotoxicity was more evident in the higher concentration group. Only a 20% decrease of cell viability was induced by the treatment of 6 μg/mL BIBR 1532 within 24 h whereas the equivalent BIBR 1532 contained in 100 μg/mL BIBR 1532@ZIF-8 caused an 80% decrease of cell viability. Therefore, it demonstrated that the cytotoxicity of BIBR 1532 could be enhanced significantly by ZIF-8 encapsulation.

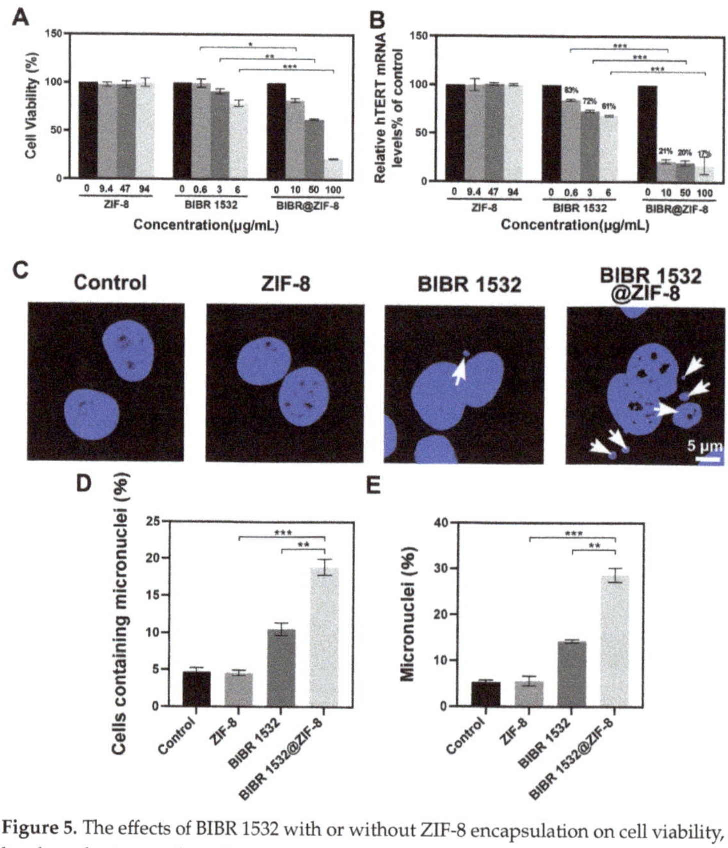

Figure 5. The effects of BIBR 1532 with or without ZIF-8 encapsulation on cell viability, hTERT mRNA levels and micronucleus formation. (**A**) The effect of ZIF-8 encapsulation on BIBR 1532-induced cytotoxicity. A549 cells were treated with different concentrations of ZIF-8 (9.4, 47, 94 μg/mL), BIBR 1532 (0.6, 3, 6 μg/mL) or BIBR 1532@ZIF-8 (10, 50, 100 μg/mL) for 24 h. (**B**) The hTERT mRNA levels in A549 cells with different treatments. Cells were treated with different concentrations of ZIF-8 (9.4, 47, 94 μg/mL), BIBR 1532 (0.6, 3, 6 μg/mL) or BIBR 1532@ZIF-8 (10, 50, 100 μg/mL) for 24 h. Data are presented as mean ± SD of three independent experiments. * $p < 0.05$. ** $p < 0.01$. *** $p < 0.001$. (**C**) The formation of micronucleus in cells with different treatments observed using CLSM. Cells were treated with 94 μg/mL ZIF-8, 6 μg/mL BIBR 1532 or 100 μg/mL BIBR 1532@ZIF-8 for 72 h. The percentage of cells with micronucleus in 300 cells (**D**) and the percentage of micronucleus (**E**) in 300 cells were statistically analyzed. ** $p < 0.01$. *** $p < 0.001$.

To explore the inhibitory effects of BIBR 1532 and BIBR 1532@ZIF-8 on telomerase activity, hTERT mRNA levels were detected after cells were incubated with different concentrations of BIBR 1532 or BIBR 1532@ZIF-8. The results showed that the mRNA levels of hTERT in A549 cells treated with 0.6, 3 and 6 μg/mL BIBR 1532 were decreased to ~83%, ~72% and ~61%, respectively. A dose-dependent inhibition of hTERT mRNA levels was induced by BIBR 1532 alone. An equivalent amount of BIBR 1532 encapsulated by ZIF-8 induced a much more pronounced decrease in hTERT mRNA levels. It showed that when A549 cells were treated with 10, 50 and 100 μg/mL BIBR 1532@ZIF-8 for 24 h, the hTERT mRNA levels were decreased to ~21%, ~20% and ~17%, respectively. Moreover, no decreases in hTERT mRNA levels were caused by ZIF-8 alone (Figure 5B). Thus, ZIF-8 encapsulation could significantly enhance the inhibitory effect of BIBR 1532 on hTERT mRNA levels.

Micronucleus (MN) is the small circular or elliptical nucleus located in the cytoplasm, completely separated from the nucleus. When cells were exposed to genotoxic insults, the chromosomes in the cells are usually lost or broken and fail to enter the daughter cell with mitosis. The end-to-end fusion of the broken chromosomes occurs, forming one or more small nucleus in the cytoplasm [53,54]. The appearance of micronucleus may reflect telomere dysfunction [55,56]. In the present study, cells were treated with 94 μg/mL ZIF-8, 6 μg/mL BIBR 1532 and 100 μg/mL BIBR 1532@ZIF-8 for 72 h, and then the formation of the micronucleus was observed using CLSM. As a result, ZIF-8 treatment alone did not promote the formation of the micronucleus in cells. A micronucleus was observed in BIBR 1532-treated cells and a much more obvious micronucleus formation was exhibited in BIBR 1532@ZIF-8-treated cells (Figure 5C). Then, a statistic was made on the micronucleus formation of 300 cells. As a result, neither the number of cells containing a micronucleus nor the total amount of micronuclei among the 300 cells showed any difference between the control group and the ZIF-8-treated group, suggesting that ZIF-8 treatment alone had no impact on micronucleus formation. As expected, treatment with free BIBR 1532 induced an obvious micronucleus formation, but most binuclear cells contained only one micronucleus, whereas BIBR 1532@ZIF-8 led to the formation of a micronucleus in more cells and multiple micronucleus in one cell. According to the statistic, treatment with 100 μg/mL BIBR 1532@ZIF-8 induced an 8.3% increase in the percentage of micronucleus-containing cells and a 14.3% increase in the percentage of the micronucleus as compared with the group treated with the equivalent free BIBR 1532 (6 μg/mL). Taken together, the encapsulation of BIBR 1532 in ZIF-8 promoted the formation of the micronucleus, indicating an increased telomere dysfunction.

Next, the effects of BIBR 1532 and BIBR 1532@ZIF-8 on cell cycle distribution were explored. Compared with that of the control group, no alteration on cell cycle distribution was detected in ZIF-8-treated cells. A slight G0/G1 arrest was induced by the treatment of free BIBR 1532 relative to that of the control group. The percentages of cells at the G0/G1 phase upon the treatment of 0.6, 3 and 6 μg/mL BIBR 1532 were 45.82%, 45.36% and 54.50%, respectively. Notably, the percentages of cells at the G0/G1 phase after treatment with 10, 50 and 100 μg/mL BIBR 1532@ZIF-8 were increased to 47.00%, 80.54% and 82.88%, respectively, indicating that the encapsulation of BIBR 1532 with ZIF-8 aggravated BIBR 1532-induced G0/G1 phase arrest (Figure 6A,B).

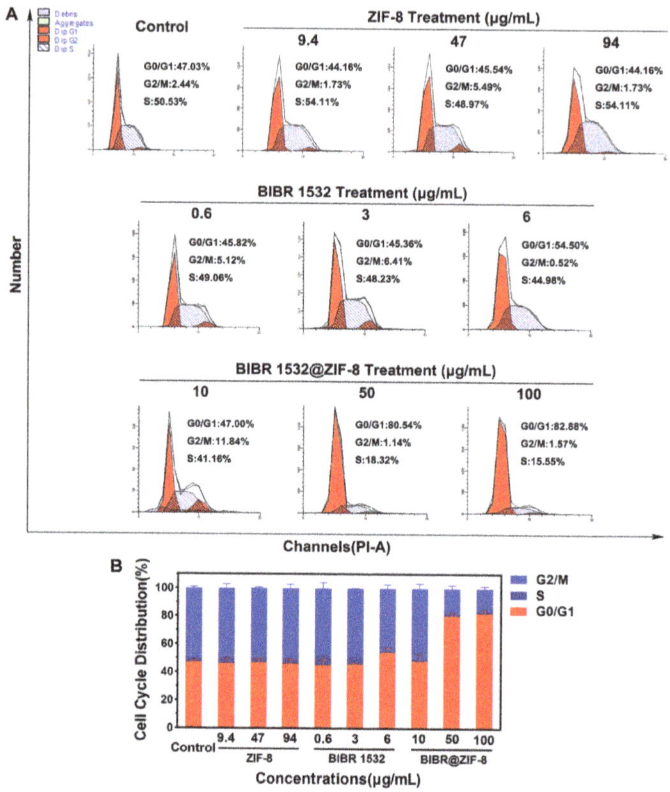

Figure 6. Effect of BIBR 1532 and BIBR 1532@ZIF-8 on cell cycle distribution. A549 cells were treated with different concentrations of ZIF-8 (9.4, 47, 94 μg/mL), BIBR 1532 (0.6, 3, 6 μg/mL) and BIBR 1532@ZIF-8 (10, 50, 100 μg/mL) for 24 h. (**A**) Representative results of cell cycle analysis were shown. (**B**) Data analysis of cell cycle distribution in each group.

The effect of BIBR 1532 with or without ZIF-8 encapsulation on cellular senescence was assessed using senescence-associated β-galactosidase (SA-β-Gal) staining. As demonstrated in Figure 7A, an obvious cellular senescence was exhibited in doxorubicin (DOX)-treated cells, which was used a positive control. Meanwhile, no obvious cellular senescence occurred in either the control group or ZIF-8-treated group. Free BIBR 1532 treatment induced a dose-dependent cellular senescence as evidenced by the increased number of green-stained cells with the increasing BIBR 1532 concentrations. The percentages of SA-β-Gal positive cells upon the treatment of 0.6, 3 and 6 μg/mL BIBR 1532 were 1.8%, 2.8% and 4.5%, respectively (Figure 7B), whereas in the BIBR 1532@ZIF-8-treated group, cell density obviously decreased with increasing exposure concentrations. Notably, the percentages of SA-β-Gal positive cells after treatment with 10, 50 and 100 μg/mL BIBR 1532@ZIF-8 were increased to 3.2%, 30.7% and 89.0%, respectively. The inhibition of cell proliferation induced by BIBR 1532@ZIF-8 was again confirmed here. Therefore, the encapsulation of BIBR 1532 in ZIF-8 aggravated BIBR 1532-induced cellular senescence.

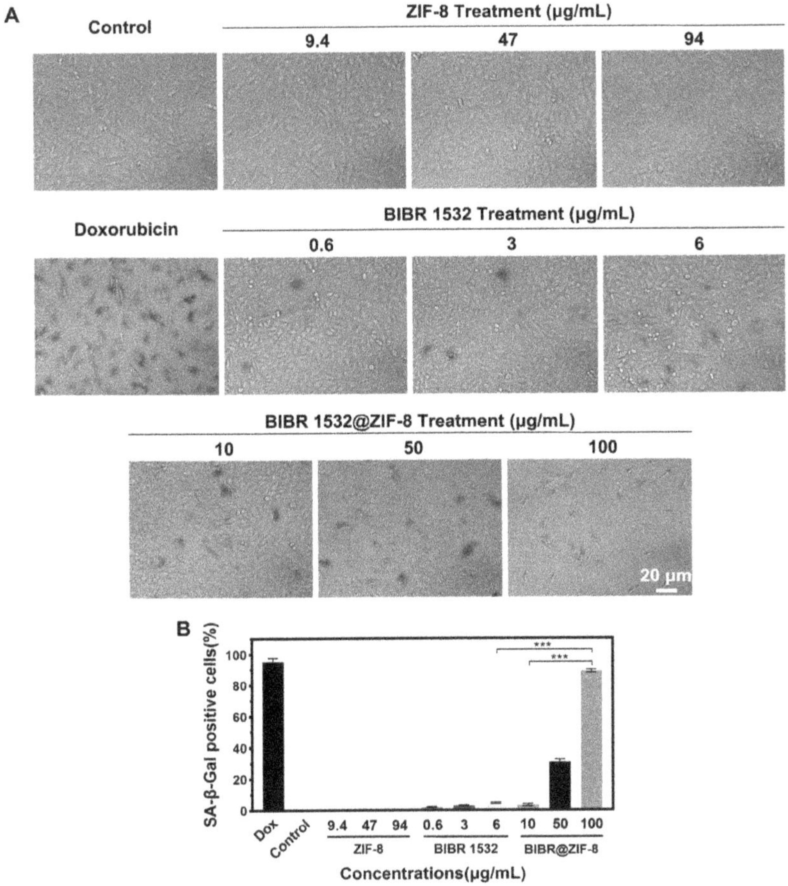

Figure 7. Cellular senescence induced by ZIF-8, BIBR 1532 and BIBR 1532@ZIF-8. A549 cells were treated with different concentrations of BIBR 1532 (0.6, 3, 6 μg/mL), ZIF-8 (9.4, 47, 94 μg/mL) and BIBR 1532@ZIF-8 (10, 50, 100 μg/mL) for 7 days. Cells treated with 100 nM doxorubicin (DOX) for 7 days were used as positive. Green-stained cells were identified as senescent cells. (**A**) Representative results of cellular senescence detection were shown. (**B**) Data analysis of cellular senescence in each group. Two hundred cells were counted in each group. *** $p < 0.001$.

4. Discussion

It has been widely accepted that poor solubility poses great limitations for the bioavailability and efficiency of many drugs. Fortunately, the rapid nanotechnology has provided powerful solutions to this problem owing to the great advantages of nanocarriers. For instance, polymer nanocomposites were used to improve the solubility of curcumin for an enhanced antioxidant and anti-inflammatory effect in vitro [57]. The loading of water-insoluble tripterin with bilayer structured liposomes has achieved a better solubility, bioavailability and lower toxicity [58]. Other studies have also achieved more significant therapeutic effects by improving drug solubility [59,60]. In our present study, encapsulating BIBR 1532 with ZIF-8 was found to increase the stability of BIBR 1532 in water as well as its cellular uptake.

ZIF-8 is pH sensitive because the cleavage of the ligand coordination bond between the zinc ion and 2-methylimidazole in ZIF-8 readily occurs in acidic environments. Thus, ZIF-8 is considered to a promising drug vehicle for pH-sensitive release of the loaded drugs [61].

Our experiment also confirmed its pH-responsiveness. As shown in Figure 2D, the average particle size of ZIF-8 remained unaltered when dispersed in the neutral environment with pH 7.4, whereas in the acidic environment with pH 5.0, its average particle size decreased significantly till undetectable within the tested 480 min. Combined with our experimental results in Figure 2E and the pH-responsiveness capability of ZIF-8, it figured out that ZIF-8 was captured by the lysosome after entering the cell as shown by the colocalization of red fluorescence (Lyso-Tracker) and green fluorescence (fluorescein-labeled ZIF-8) at 1 h. As the incubation time increased, the green fluorescence intensity in the cytoplasm gradually decreased, indicating that ZIF-8 in the cytoplasm was continuously captured by lysosomes. After entering the acidic environment of the lysosomes, the coordination bond between zinc and imidazole was broken. After 6 h, no green fluorescence was observed in the cytoplasm or lysosomes, suggesting that most of the ZIF-8 and fluorescein had been cleared by the cells.

The loading efficiency of ZIF-8 largely depends on the loaded content. In a study where curcumin was encapsulated by ZIF-8, the drug loading capacity was up to 10.89% [62]. Another report showed that ZIF-8 encapsulated the chemotherapy drug rapamycin with a high drug loading rate of 9.39% [63]. However, the loading rate of ZIF-8 for the gene editing platform CRISPR/Cas9 was only 1.2% [39], and the loading rate of plasmid DNA was only 2.5% [42]. The drug loading rate of ZIF-8 for BIBR 1532 in our present study was 6%. The loading capability of ZIF-8 towards BIBR 1532 was not very striking from the above data. However, after being encapsulated into ZIF-8, BIBR 1532 actually showed much more significant inhibitory effects towards cancer cells. This suggested the feasibility of using ZIF-8 as a drug delivery vehicle for an improved BIBR 1532 efficacy in spite of a relatively general loading rate.

The proton sponge effect is acknowledged as one of the important mechanisms of lysosomal escape. Histidine-rich molecules exhibit a buffering effect upon protonation of the imidazole ring. The protonation induces the flux of ions and water into the lysosomal environment, increasing the lysosomal membrane permeability. The lysosomal membrane then ruptures and releases the trapped components [64,65]. In addition to the proton sponge effect, other mechanisms of lysosomal escape have also been reported including the formation of pores in the lysosomal membrane [66], lysosomal membrane fusion [67] and the photochemical disruption of lysosomal membrane [68]. Our work has suggested that ZIF-8-facilitated lysosomal escape predominately depended on the proton sponge effect of 2-methylimidazole. Actually, the proton sponge effect has also been exploited by other scientists and many novel materials have been developed with the aim of promoting lysosomal escape of functional molecules and reducing the unnecessary lysosomal degradation. For instance, Chen's group reported the use of metal (Fe^{III} or Al^{III})-phenolic (polyphenol tannic acid) networks as a coating for nanoparticles, which could achieve lysosomal escape of nanoparticles through the proton sponge effect [69]. Xi et al. also designed the functional nanoparticles (PG-FAPEP) consisting of insulin with dual surface decorations of folate and charge-convertible tripeptide. Such functional nanoparticles could also trigger the proton sponge effect in acidic lysosomes and help the lysosomal escape of insulin [70]. Another report demonstrated that by immobilizing the highly branched and rigid cationic polymers on silica nanoparticles, the proton sponge effect with osmotic swelling and lysosome rupture could be achieved [71].

Nanoparticles typically enter cells via an energy-dependent process known as endocytosis. During endocytosis, nanomaterials are usually initially confined to vesicular structures such as endosomes, lysosomes, phagosomes or macropinosomes [72]. Different endocytic pathways will lead to different fates after nanomaterials enter the cell [73]. To date, endocytic pathways can be classified and summarized into the following 7 types: macro pinocytosis, phagocytosis, clathrin independent/dynamin-dependent endocytosis, clathrin-mediated endocytosis, caveolae-mediated endocytosis, clathrin independent/dynamin-independent endocytosis and fusion [74,75]. Different internalization pathways are mediated and regulated by different lipids and transporters. Our experiments showed that ZIF-8 and BIBR 1532@ZIF-8 were first captured by lysosomes after entering the cell, suggesting that ZIF-8 and BIBR 1532@ZIF-8 entered the cell by endocytosis. However, the actual endocytosis pathway needs to be further investigated in the following work.

In our work, the concentrations of BIBR 1532 used were 0.6, 3 and 6 μg/mL, corresponding to the molar concentrations of 1.81, 9.05 and 18.11 μM, respectively. No obvious G0/G1 phase arrest was induced in cells treated with 1.81 or 9.05 μM BIBR 1532 within 24 h, while cells treated with 18.11 μM BIBR 1532 showed a slight cell cycle arrest. Similarly, the number of senescent cells increased with the increasing concentrations of BIBR 1532. With ZIF-8 as a drug carrier, more BIBR 1532 accumulated in the nucleus and a remarkable cell cycle arrest was observed at 24 h especially in 50, 100 μg/mL BIBR 1532@ZIF-8-treated cells. No cell apoptosis was detectable (data not shown). With the incubation of BIBR 1532@ZIF-8 for 7 days, a remarkable decrease in cell number accompanied with increased cellular senescence was observed. Cellular senescence is described as a state of permanent cell cycle arrest in response to different damaging stimuli [76]. In the present study, it was reasonable that BIBR 1532@ZIF-8 induced cell cycle arrest followed by cellular senescence. Besides destroying tumor cells directly, pro-cellular senescence is an alternative to limit tumor cells. The relatively high concentrations of BIBR 1532 have been reported to induce apoptosis or destroy tumor cells directly. A study showed that LN18 cells treated with 25 μM BIBR 1532 showed flat and enlarged cell morphology. When the concentration of BIBR 1532 was increased to 100 μM, cell morphology was narrowed and the nuclear membrane was lost [77]. It figured out that the encapsulation of BIBR 1532 in ZIF-8 not only decreased the dosage of BIBR 1532 needed for a comparable inhibition effect towards tumor cells but also acts in a milder manner of promoting cell cycle arrest and senescence of tumor cells rather than destroying tumor cells directly.

Our work was the first to encapsulate hydrophobic BIBR 1532 with ZIF-8 for an improved BIBR 1532 efficacy. Besides of our study, only one study was available with the involvement of BIBR 1532 loading. In this study, peptide dendrimeric prodrug nanoassembly (PDPN) acted as a nanocarrier to load BIBR 1532 and DOX, thus constructing a telomerase-terminated nanoplatform to reverse DOX resistance [78]. As illustrated in Table S1, the similarities and differences were analyzed between this study and our work. The similarities between the two papers reflected the common theoretical basis which clarified the practical problems in the application of BIBR 1532 and the necessity of loading BIBR 1532 with nanocarriers. The differences between the two papers were highlighted in the aspect of key observations, drug release, the bio-effects of BIBR 1532 and the significance of the work. In our present study, after BIBR 1532-bearing ZIF-8 was endocytosed into cells and localized in lysosome, ZIF-8 was degraded in the acidic environment and BIBR 1532 was released, causing the proliferation inhibition, cell cycle arrest and cellular senescence of tumor cells; whereas in this reported study [78], BIBR 1532 was released from the PEG layer and then weakened mitochondria protection, enhanced ROS production and meanwhile, promoted the apoptosis of DOX-resistant tumor cells. In any case, both studies addressed the essentials of proper drug delivery vehicles for an improved efficacy of hydrophobic BIBR 1532.

5. Conclusions

In the present study, ZIF-8 and BIBR 1532@ZIF-8 were synthesized, respectively, and the physicochemical characterizations confirmed the successful encapsulation of BIBR 1532 by ZIF-8 coupled with an improved stability of BIBR 1532. ZIF-8 facilitated the transport and release of BIBR 1532 within cells via the increased permeability of lysosomal membrane. As compared with free BIBR 1532, ZIF-8 encapsulation facilitated more of an accumulation of BIBR 1532 in the nucleus, leading to a more obvious growth inhibition of cancer cells. A more potent inhibition on hTERT mRNA expression, an aggravated G0/G1 arrest accompanied with an increased cellular senescence was detected in BIBR 1532@ZIF-8-treated cancer cells. Our work has provided preliminary information on improving the transport, release and efficacy of water-insoluble small molecule drugs by using ZIF-8 as a drug delivery vehicle.

Supplementary Materials: The following supporting information can be downloaded at: https://www.mdpi.com/article/10.3390/nano13111779/s1, Figure S1: The structure of BIBR 1532; Figure S2: Plot of the standard curve; Figure S3: Fluorescence stability of fluorescein@ZIF-8; Figure S4: Fluorescence stability of 6-AF@BIBR 1532@ZIF-8; Table S1: A summary of the similarities and differences between our present study and others' study ([78]).

Author Contributions: Conceptualization, F.Z., S.Z. and J.L.; Formal analysis, F.Z., S.Z. and J.L.; Funding acquisition, F.Z.; Investigation, S.Z., J.L., F.Z., L.Y. (Limin Yang), Y.Y. and J.C.; Methodology, S.Z., F.Z., J.L., Y.Y., Q.C., L.Y. (Limin Yang), J.C. and Y.S.; Project administration, F.Z.; Software, S.Z., J.C., L.Y. (Liang Yan) and Y.Y.; Supervision, F.Z., J.L. and Y.L.; Visualization, J.L., Y.L., L.Y. (Liang Yan), R.L. and S.Z.; Writing—original draft, S.Z. and J.L.; Writing—review and editing, S.Z., J.L., F.Z., Y.Y. and L.Y (Liang Yan). All authors have read and agreed to the published version of the manuscript.

Funding: This work was funded by the National Key Research and Development Program of China (2021YFA1200904), the National Key Research and Development Program of China (2020YFA0710700), and the Directional Institutionalized Scientific Research Platform relies on the Beijing Synchrotron Radiation Facility of the Chinese Academy of Sciences.

Institutional Review Board Statement: Not applicable.

Informed Consent Statement: Not applicable.

Data Availability Statement: Not applicable.

Conflicts of Interest: The authors declare no conflict of interest.

References

1. Greider, C.W.; Blackburn, E.H. Identification of a specific telomere terminal transferase activity in Tetrahymena extracts. *Cell* **1985**, *43*, 405–413. [CrossRef] [PubMed]
2. Hahn, W.C.; Meyerson, M. Telomerase activation, cellular immortalization and cancer. *Ann. Med.* **2001**, *33*, 123–129. [CrossRef] [PubMed]
3. Gomez, D.L.; Armando, R.G.; Cerrudo, C.S.; Ghiringhelli, P.D.; Gomez, D.E. Telomerase as a cancer target. Development of new molecules. *Curr. Top. Med. Chem.* **2016**, *16*, 2432–2440. [CrossRef] [PubMed]
4. Villa, R.; Folini, M.; Perego, P.; Supino, R.; Setti, E.; Daidone, M.G.; Zunino, F.; Zaffaroni, N. Telomerase activity and telomere length in human ovarian cancer and melanoma cell lines: Correlation with sensitivity to DNA damaging agents. *Int. J. Oncol.* **2000**, *16*, 995–1002. [CrossRef] [PubMed]
5. Nakajima, A.; Tauchi, T.; Sashida, G.; Sumi, M.; Abe, K.; Yamamoto, K.; Ohyashiki, J.H.; Ohyashiki, K. Telomerase inhibition enhances apoptosis in human acute leukemia cells: Possibility of antitelomerase therapy. *Leukemia* **2003**, *17*, 560–567. [CrossRef]
6. Nowak, J.; Januszkiewicz, D.; Lewandowski, K.; Nowicka-Kujawska, K.; Pernak, M.; Rembowska, J.; Nowak, T.; Wysocki, J. Activity and expression of human telomerase in normal and malignant cells in gastric and colon cancer patients. *Eur. J. Gastroenterol. Hepatol.* **2003**, *15*, 75–80. [CrossRef]
7. Ohmura, Y.; Aoe, M.; Andou, A.; Shimizu, N. Telomerase activity and Bcl-2 expression in non-small cell lung cancer. *Clin. Cancer Res. Off. J. Am. Assoc. Cancer Res.* **2000**, *6*, 2980–2987.
8. Rosen, J.; Jakobs, P.; Ale-Agha, N.; Altschmied, J.; Haendeler, J. Non-canonical functions of telomerase reverse transcriptase—impact on redox homeostasis. *Redox Biol.* **2020**, *34*, 101543. [CrossRef]
9. Leão, R.; Apolónio, J.D.; Lee, D.; Figueiredo, A.; Tabori, U.; Castelo-Branco, P. Mechanisms of human telomerase reverse transcriptase (hTERT) regulation: Clinical impacts in cancer. *J. Biomed. Sci.* **2018**, *25*, 22. [CrossRef]
10. Guterres, A.N.; Villanueva, J. Targeting telomerase for cancer therapy. *Oncogene* **2020**, *39*, 5811–5824. [CrossRef]
11. Krupp, G.; Bonatz, G.; Parwaresch, R. Telomerase, immortality and cancer. *Biotechnol. Annu. Rev.* **2000**, *6*, 103–140. [PubMed]
12. Gurung, R.L.; Balakrishnan, L.; Bhattacharjee, R.N.; Manikandan, J.; Swaminathan, S.; Hande, M.P. Inhibition of poly (ADP-Ribose) polymerase-1 in telomerase deficient mouse embryonic fibroblasts increases arsenite-induced genome instability. *Genome Integr.* **2010**, *1*, 5. [CrossRef] [PubMed]
13. Fakhoury, J.; Nimmo, G.A.; Autexier, C. Harnessing telomerase in cancer therapeutics. *Anti-Cancer Agents Med. Chem.* **2007**, *7*, 475–483. [CrossRef] [PubMed]
14. Bryan, T.M.; Englezou, A.; Gupta, J.; Bacchetti, S.; Reddel, R.R. Telomere elongation in immortal human cells without detectable telomerase activity. *EMBO J.* **1995**, *14*, 4240–4248. [CrossRef]
15. Cesare, A.J.; Reddel, R.R. Alternative lengthening of telomeres: Models, mechanisms and implications. *Nat. Rev. Genet.* **2010**, *11*, 319–330. [CrossRef]

16. Grach, A.A. Alternative telomere-lengthening mechanisms. *Cytol. Genet.* **2011**, *45*, 121–130. [CrossRef]
17. Bryan, C.; Rice, C.; Hoffman, H.; Harkisheimer, M.; Sweeney, M.; Skordalakes, E. Structural basis of telomerase inhibition by the highly specific BIBR1532. *Structure* **2015**, *23*, 1934–1942. [CrossRef]
18. Ding, X.; Cheng, J.; Pang, Q.; Wei, X.; Zhang, X.; Wang, P.; Yuan, Z.; Qian, D. BIBR1532, a selective telomerase inhibitor, enhances radiosensitivity of non-small cell lung cancer through increasing telomere dysfunction and ATM/CHK1 inhibition. *Int. J. Radiat. Oncol. Biol. Phys.* **2019**, *105*, 861–874. [CrossRef]
19. Pascolo, E.; Wenz, C.; Lingner, J.; Hauel, N.; Priepke, H.; Kauffmann, I.; Garin-Chesa, P.; Rettig, W.J.; Damm, K.; Schnapp, A. mechanism of human telomerase inhibition by BIBR1532, a synthetic, non-nucleosidic drug candidate. *J. Biol. Chem.* **2002**, *277*, 15566–15572. [CrossRef]
20. Damm, K.; Hemmann, U.; Garin-Chesa, P.; Hauel, N.; Kauffmann, I.; Priepke, H.; Niestroj, C.; Daiber, C.; Enenkel, B.; Guilliard, B.; et al. A highly selective telomerase inhibitor limiting human cancer cell proliferation. *EMBO J.* **2001**, *20*, 6958–6968. [CrossRef]
21. El-Daly, H.; Kull, M.; Zimmermann, S.; Pantic, M.; Waller, C.F.; Martens, U.M. Selective cytotoxicity and telomere damage in leukemia cells using the telomerase inhibitor BIBR1532. *Blood* **2005**, *105*, 1742–1749. [CrossRef] [PubMed]
22. Liu, C.; Zhou, H.; Sheng, X.B.; Liu, X.H.; Chen, F.H. Design, synthesis and SARs of novel telomerase inhibitors based on BIBR1532. *Bioorg. Chem.* **2020**, *102*, 104077.
23. Shi, Y.; Sun, L.; Chen, G.; Zheng, D.; Li, L.; Wei, W. A combination of the telomerase inhibitor, BIBR1532, and paclitaxel synergistically inhibit cell proliferation in breast cancer cell lines. *Target. Oncol.* **2015**, *10*, 565–573. [PubMed]
24. Liu, R.; Liu, J.; Wang, S.; Wang, Y.; Zhang, T.; Liu, Y.; Geng, X.; Wang, F. Combined treatment with emodin and a telomerase inhibitor induces significant telomere damage/dysfunction and cell death. *Cell Death Dis.* **2019**, *10*, 527. [PubMed]
25. Giliopoulos, D.; Zamboulis, A.; Giannakoudakis, D.; Bikiaris, D.; Triantafyllidis, K. Polymer/metal organic framework (MOF) nanocomposites for biomedical applications. *Molecules* **2020**, *25*, 185.
26. Ma, Y.; Su, Z.; Zhou, L.; He, L.; Hou, Z.; Zou, J.; Cai, Y.; Chang, D.; Xie, J.; Zhu, C.; et al. Biodegradable metal-organic-framework-gated organosilica for tumor-microenvironment-unlocked glutathione-depletion-enhanced synergistic therapy. *Adv. Mater.* **2022**, *34*, e2107560. [CrossRef]
27. Yang, J.; Ma, S.; Xu, R.; Wei, Y.; Zhang, J.; Zuo, T.; Wang, Z.; Deng, H.; Yang, N.; Shen, Q. Smart biomimetic metal organic frameworks based on ROS-ferroptosis-glycolysis regulation for enhanced tumor chemo-immunotherapy. *J. Control. Release Off. J. Control. Release Soc.* **2021**, *334*, 21–33. [CrossRef]
28. Pu, Y.; Yin, H.; Dong, C.; Xiang, H.; Wu, W.; Zhou, B.; Du, D.; Chen, Y.; Xu, H. Sono-controllable and ROS-Sensitive CRISPR-Cas9 genome editing for augmented/synergistic ultrasound tumor nanotherapy. *Adv. Mater.* **2021**, *33*, e2104641.
29. Huang, R.; Cai, G.Q.; Li, J.; Li, X.S.; Liu, H.T.; Shang, X.L.; Zhou, J.D.; Nie, X.M.; Gui, R. Platelet membrane-camouflaged silver metal-organic framework drug system against infections caused by methicillin-resistant Staphylococcus aureus. *J. Nanobiotechnol.* **2021**, *19*, 229.
30. Yang, H.; Yu, Z.; Ji, S.; Yan, J.; Han, L.; Liu, Y.; Wang, Y.; Niu, Y.; Huo, Q.; Xu, M. Construction and evaluation of detachable bone-targeting MOF carriers for the delivery of proteasome inhibitors. *RSC Adv.* **2022**, *12*, 14707–14715. [CrossRef]
31. Liang, K.; Ricco, R.; Doherty, C.M.; Styles, M.J.; Bell, S.; Kirby, N.; Mudie, S.; Haylock, D.; Hill, A.J.; Doonan, C.J.; et al. Biomimetic mineralization of metal-organic frameworks as protective coatings for biomacromolecules. *Nat. Commun.* **2015**, *6*, 7240. [CrossRef] [PubMed]
32. Nasi, H.; di Gregorio, M.C.; Wen, Q.; Shimon, L.J.W.; Kaplan-Ashiri, I.; Bendikov, T.; Leitus, G.; Kazes, M.; Oron, D.; Lahav, M.; et al. Directing the morphology, packing, and properties of chiral metal-organic frameworks by cation exchange. *Angew. Chem.* **2022**, *61*, e202205238.
33. Ma, M.; Chen, J.; Liu, H.; Huang, Z.; Huang, F.; Li, Q.; Xu, Y. A review on chiral metal-organic frameworks: Synthesis and asymmetric applications. *Nanoscale* **2022**, *14*, 13405–13427. [CrossRef]
34. Di Gregorio, M.C.; Singh, V.; Shimon, L.J.W.; Lahav, M.; van der Boom, M.E. Crystallographic-morphological connections in star shaped metal-organic Frameworks. *J. Am. Chem. Soc.* **2022**, *144*, 22838–22843. [CrossRef] [PubMed]
35. Gong, W.; Chen, Z.; Dong, J.; Liu, Y.; Cui, Y. Chiral metal-organic frameworks. *Chem. Rev.* **2022**, *122*, 9078–9144. [CrossRef]
36. Yang, J.; Dai, D.; Zhang, X.; Teng, L.; Ma, L.; Yang, Y.W. Multifunctional metal-organic framework (MOF)-based nanoplatforms for cancer therapy: From single to combination therapy. *Theranostics* **2023**, *13*, 295–323. [CrossRef]
37. Pan, W.; Li, Z.; Qiu, S.; Dai, C.; Wu, S.; Zheng, X.; Guan, M.; Gao, F. Octahedral Pt-MOF with Au deposition for plasmonic effect and Schottky junction enhanced hydrogenothermal therapy of rheumatoid arthritis. *Mater. Today Bio.* **2022**, *13*, 100214. [CrossRef]
38. Tanaka, S.; Kida, K.; Okita, M.; Ito, Y.; Miyake, Y. Size-controlled synthesis of zeolitic imidazolate framework-8 (ZIF-8) crystals in an aqueous system at room temperature. *Chem. Lett.* **2012**, *41*, 1337–1339. [CrossRef]
39. Alsaiari, S.K.; Patil, S.; Alyami, M.; Alamoudi, K.O.; Aleisa, F.A.; Merzaban, J.S.; Li, M.; Khashab, N.M. Endosomal escape and delivery of CRISPR/Cas9 genome editing machinery enabled by nanoscale zeolitic imidazolate framework. *J. Am. Chem. Soc.* **2018**, *140*, 143–146. [CrossRef]
40. Hou, C.; Wang, Y.; Ding, Q.; Jiang, L.; Li, M.; Zhu, W.; Pan, D.; Zhu, H.; Liu, M. Facile synthesis of enzyme-embedded magnetic metal-organic frameworks as a reusable mimic multi-enzyme system: Mimetic peroxidase properties and colorimetric sensor. *Nanoscale* **2015**, *7*, 18770–18779. [CrossRef]

41. Ding, L.; Lin, X.; Lin, Z.; Wu, Y.; Liu, X.; Liu, J.; Wu, M.; Zhang, X.; Zeng, Y. Cancer cell-targeted photosensitizer and therapeutic protein co-delivery nanoplatform based on a metal-organic framework for enhanced synergistic photodynamic and protein therapy. *ACS Appl. Mater. Interfaces* **2020**, *12*, 36906–36916. [CrossRef]
42. Li, Y.; Zhang, K.; Liu, P.; Chen, M.; Zhong, Y.; Ye, Q.; Wei, M.Q.; Zhao, H.; Tang, Z. Encapsulation of plasmid DNA by nanoscale metal-organic frameworks for efficient gene transportation and expression. *Adv. Mater.* **2019**, *31*, e1901570. [CrossRef] [PubMed]
43. Costa, B.A.; Abuçafy, M.P.; Barbosa, T.W.L.; da Silva, B.L.; Fulindi, R.B.; Isquibola, G.; da Costa, P.I.; Chiavacci, L.A. ZnO@ZIF-8 nanoparticles as nanocarrier of ciprofloxacin for antimicrobial activity. *Pharmaceutics* **2023**, *15*, 259. [CrossRef]
44. Ju, G.; Liu, B.; Ji, M.; Jin, R.; Xu, X.; Xiao, Y.; Li, J.; Xu, D.; Huang, Y.; Hou, J. Folic acid-modified miR-491-5p-loaded ZIF-8 nanoparticles inhibit castration-resistant prostate cancer by regulating the expression of EPHX1. *Front. Bioeng. Biotechnol.* **2021**, *9*, 706536. [CrossRef] [PubMed]
45. Pan, Y.; Liu, Y.; Zeng, G.; Zhao, L.; Lai, Z. Rapid synthesis of zeolitic imidazolate framework-8 (ZIF-8) nanocrystals in an aqueous system. *Chem. Commun.* **2011**, *47*, 2071–2073. [CrossRef]
46. Zhuang, J.; Kuo, C.H.; Chou, L.Y.; Liu, D.Y.; Weerapana, E.; Tsung, C.K. Optimized metal-organic-framework nanospheres for drug delivery: Evaluation of small-molecule encapsulation. *ACS Nano* **2014**, *8*, 2812–2819. [CrossRef] [PubMed]
47. Novio, F.; Lorenzo, J.; Nador, F.; Wnuk, K.; Ruiz-Molina, D. Carboxyl group functionalized coordination polymer nanoparticles as efficient platforms for drug delivery. *Chemistry* **2014**, *20*, 15443–15450. [CrossRef]
48. Sommer, S.; Buraczewska, I.; Kruszewski, M. Micronucleus assay: The state of art, and future directions. *Int. J. Mol. Sci.* **2020**, *21*, 1534. [CrossRef]
49. Yoon, M.C.; Hook, V.; O'Donoghue, A.J. Cathepsin B dipeptidyl carboxypeptidase and endopeptidase activities demonstrated across a broad pH range. *Biochemistry* **2022**, *61*, 1904–1914. [CrossRef]
50. Eskelinen, E.L. Roles of LAMP-1 and LAMP-2 in lysosome biogenesis and autophagy. *Mol. Asp. Med.* **2006**, *27*, 495–502. [CrossRef]
51. Li, C.; Wang, X.; Li, X.; Qiu, K.; Jiao, F.; Liu, Y.; Kong, Q.; Liu, Y.; Wu, Y. Proteasome inhibition activates autophagy-lysosome pathway associated with TFEB dephosphorylation and nuclear translocation. *Front. Cell Dev. Biol.* **2019**, *7*, 170. [CrossRef] [PubMed]
52. Ranjbar, M.A.; Jamshidi, M. Overexpression of lysosome-associated membrane protein 1 in oral squamous cell carcinoma and its correlation with tumor differentiation and metastasis. *Iran. J. Otorhinolaryngol.* **2022**, *34*, 3–8. [PubMed]
53. Surrallés, J.; Catalán, J.; Creus, A.; Norppa, H.; Xamena, N.; Marcos, R. Micronuclei induced by alachlor, mitomycin C and vinblastine in human lymphocytes: Presence of centromeres and kinetochores and influence of staining technique. *Mutagenesis* **1995**, *10*, 417–423. [CrossRef] [PubMed]
54. Garaj-Vrhovac, V.; Kopjar, N.; Besendorfer, V.; Papes, D. Induction of micronuclei in human lymphocytes after occupational exposure to ultrasound. *Chemosphere* **1999**, *38*, 3541–3553. [CrossRef] [PubMed]
55. Vohhodina, J.; Goehring, L.J.; Liu, B.; Kong, Q.; Botchkarev, V.V., Jr.; Huynh, M.; Liu, Z.; Abderazzaq, F.O.; Clark, A.P.; Ficarro, S.B.; et al. BRCA1 binds TERRA RNA and suppresses R-loop-based telomeric DNA damage. *Nat. Commun.* **2021**, *12*, 3542. [CrossRef] [PubMed]
56. Lee, Y.W.; Arora, R.; Wischnewski, H.; Azzalin, C.M. TRF1 participates in chromosome end protection by averting TRF2-dependent telomeric R loops. *Nat. Struct. Mol. Biol.* **2018**, *25*, 147–153. [CrossRef]
57. Lee, J.Y.; Lee, S.; Choi, J.H.; Na, K. ι-Carrageenan nanocomposites for enhanced stability and oral bioavailability of curcumin. *Biomater. Res.* **2021**, *25*, 32. [CrossRef]
58. Que, H.; Hong, W.; Lan, T.; Zeng, H.; Chen, L.; Wan, D.; Bi, Z.; Ren, W.; Luo, M.; Yang, J.; et al. Tripterin liposome relieves severe acute respiratory syndrome as a potent COVID-19 treatment. *Signal Transduct. Target. Ther.* **2022**, *7*, 399. [CrossRef]
59. Tian, C.; Guo, J.; Wang, G.; Sun, B.; Na, K.; Zhang, X.; Xu, Z.; Cheng, M.; He, Z.; Sun, J. Efficient intestinal digestion and on site tumor-bioactivation are the two important determinants for chylomicron-mediated lymph-targeting triglyceride-mimetic docetaxel oral prodrugs. *Adv. Sci.* **2019**, *6*, 1901810. [CrossRef]
60. Mundaca-Uribe, R.; Karshalev, E.; de Ávila, B.E.-F.; Wei, X.; Nguyen, B.; Litvan, I.; Fang, R.H.; Zhang, L.; Wang, J. A Microstirring pill enhances bioavailability of orally administered drugs. *Adv. Sci.* **2021**, *8*, 2100389. [CrossRef]
61. Yan, L.; Chen, X.; Wang, Z.; Zhang, X.; Zhu, X.; Zhou, M.; Chen, W.; Huang, L.; Roy, V.A.L.; Yu, P.K.N.; et al. Size controllable and surface tunable zeolitic imidazolate framework-8-poly(acrylic acid sodium salt) nanocomposites for pH responsive drug release and enhanced in vivo cancer treatment. *ACS Appl. Mater. Interfaces* **2017**, *9*, 32990–33000. [CrossRef] [PubMed]
62. Duan, S.; Zhao, X.; Su, Z.; Wang, C.; Lin, Y. Layer-by-layer decorated nanoscale ZIF-8 with high curcumin loading effectively inactivates gram-negative and gram-positive bacteria. *ACS Appl. Bio Mater.* **2020**, *3*, 3673–3680. [CrossRef] [PubMed]
63. Xu, M.; Hu, Y.; Ding, W.; Li, F.; Lin, J.; Wu, M.; Wu, J.; Wen, L.P.; Qiu, B.; Wei, P.F.; et al. Rationally designed rapamycin-encapsulated ZIF-8 nanosystem for overcoming chemotherapy resistance. *Biomaterials* **2020**, *258*, 120308. [CrossRef]
64. Moreira, C.; Oliveira, H.; Pires, L.R.; Simões, S.; Barbosa, M.A.; Pêgo, A.P. Improving chitosan-mediated gene transfer by the introduction of intracellular buffering moieties into the chitosan backbone. *Acta Biomater.* **2009**, *5*, 2995–3006. [CrossRef] [PubMed]
65. Pack, D.W.; Putnam, D.; Langer, R. Design of imidazole-containing endosomolytic biopolymers for gene delivery. *Biotechnol. Bioeng.* **2000**, *67*, 217–223. [CrossRef]
66. Jenssen, H.; Hamill, P.; Hancock, R.E. Peptide antimicrobial agents. *Clin. Microbiol. Rev.* **2006**, *19*, 491–511. [CrossRef]

67. Wiley, D.C.; Skehel, J.J. The structure and function of the hemagglutinin membrane glycoprotein of influenza virus. *Annu. Rev. Biochem.* **1987**, *56*, 365–394. [CrossRef]
68. Berg, K.; Selbo, P.K.; Prasmickaite, L.; Tjelle, T.E.; Sandvig, K.; Moan, J.; Gaudernack, G.; Fodstad, O.; Kjølsrud, S.; Anholt, H.; et al. Photochemical internalization: A novel technology for delivery of macromolecules into cytosol. *Cancer Res.* **1999**, *59*, 1180–1183.
69. Chen, J.; Li, J.; Zhou, J.; Lin, Z.; Cavalieri, F.; Czuba-Wojnilowicz, E.; Hu, Y.; Glab, A.; Ju, Y.; Richardson, J.J.; et al. Metal-phenolic coatings as a platform to trigger endosomal escape of nanoparticles. *ACS Nano* **2019**, *13*, 11653–11664. [CrossRef]
70. Xi, Z.; Ahmad, E.; Zhang, W.; Li, J.; Wang, A.; Wang, N.; Zhu, C.; Huang, W.; Xu, L.; Yu, M.; et al. Dual-modified nanoparticles overcome sequential absorption barriers for oral insulin delivery. *J. Control. Release Off. J. Control. Release Soc.* **2022**, *342*, 1–13. [CrossRef]
71. Wojnilowicz, M.; Glab, A.; Bertucci, A.; Caruso, F.; Cavalieri, F. Super-resolution imaging of proton sponge-triggered rupture of endosomes and cytosolic release of small interfering RNA. *ACS Nano* **2019**, *13*, 187–202. [CrossRef]
72. Donahue, N.D.; Acar, H.; Wilhelm, S. Concepts of nanoparticle cellular uptake, intracellular trafficking, and kinetics in nanomedicine. *Adv. Drug Deliv. Rev.* **2019**, *143*, 68–96. [CrossRef] [PubMed]
73. Linnane, E.; Haddad, S.; Melle, F.; Mei, Z.; Fairen-Jimenez, D. The uptake of metal-organic frameworks: A journey into the cell. *Chem. Soc. Rev.* **2022**, *51*, 6065–6086. [CrossRef] [PubMed]
74. Sousa de Almeida, M.; Susnik, E.; Drasler, B.; Taladriz-Blanco, P.; Petri-Fink, A.; Rothen-Rutishauser, B. Understanding nanoparticle endocytosis to improve targeting strategies in nanomedicine. *Chem. Soc. Rev.* **2021**, *50*, 5397–5434. [CrossRef]
75. Rennick, J.J.; Johnston, A.P.R.; Parton, R.G. Key principles and methods for studying the endocytosis of biological and nanoparticle therapeutics. *Nat. Nanotechnol.* **2021**, *16*, 266–276. [CrossRef] [PubMed]
76. Muñoz-Espín, D.; Serrano, M. Cellular senescence: From physiology to pathology. *Nat. Rev. Mol. Cell Biol.* **2014**, *15*, 482–496. [CrossRef]
77. Lavanya, C.; Venkataswamy, M.M.; Sibin, M.K.; Srinivas Bharath, M.M.; Chetan, G.K. Down regulation of human telomerase reverse transcriptase (hTERT) expression by BIBR1532 in human glioblastoma LN18 cells. *Cytotechnology* **2018**, *70*, 1143–1154. [CrossRef]
78. Wu, Y.; Zhong, D.; Li, Y.; Wu, H.; Xu, X.; Yang, J.; Gu, Z. Tumor-oriented telomerase-terminated nanoplatform as versatile strategy for multidrug resistance reversal in cancer treatment. *Adv. Healthc. Mater.* **2020**, *9*, e1901739. [CrossRef]

MDPI AG
Grosspeteranlage 5
4052 Basel
Switzerland
Tel.: +41 61 683 77 34

Nanomaterials Editorial Office
E-mail: nanomaterials@mdpi.com
www.mdpi.com/journal/nanomaterials

www.ingramcontent.com/pod-product-compliance
Lightning Source LLC
Chambersburg PA
CBHW040312270926
51850CB00007BC/1103
9783725875160